# First International Geographical Conference on China-Mongolia-Russia Economic Corridor: Geographical and Environmental Factors and Territorial Development Opportunities 2018

IOP Conference Series: Earth and Environmental Science
Volume 190



Irkutsk, Russia
20 – 26 August 2018

ISBN: 978-1-5108-7377-3
ISSN: 1755-1307

**Printed from e-media with permission by:**

Curran Associates, Inc.
57 Morehouse Lane
Red Hook, NY 12571

**Some format issues inherent in the e-media version may also appear in this print version.**



Printed by Curran Associates, Inc. (2018)

For permission requests, please contact the Institute of Physics
at the address below.

Institute of Physics
Dirac House, Temple Back
Bristol BS1 6BE UK

Phone: 44 1 17 929 7481
Fax: 44 1 17 920 0979

techtracking@iop.org

**Additional copies of this publication are available from:**

Curran Associates, Inc.
57 Morehouse Lane
Red Hook, NY 12571 USA
Phone: 845-758-0400
Fax: 845-758-2633
Email: curran@proceedings.com
Web: www.proceedings.com

# Table of contents

## Volume 190

**First International Geographical Conference of North Asian Countries "China-Mongolia-Russia Economic Corridor: Geographical and Environmental Factors and Territorial Development Opportunities"**

**20–26 August 2018, Irkutsk, Russia**

**Accepted papers received: 8 October 2018**
**Published online: 30 October 2018**

### Preface



### Papers

This volume contains papers submitted to the First International Geographical Conference of North Asian countries on «China–Mongolia–Russia» Economic Corridor: Geographical and Environmental Factors and Territorial Development Opportunities" devoted to the analysis of fundamental scientific problems in the field of optimal ecologically oriented resource providing of the livelihoods and the quality of life of the population, the development of modern urbanized and cultural landscapes and the preservation of the natural environment in transboundary and outlying territories. The ultimate goal of the Conference is to find common ground and make constructive proposals for the solution of the geoecological problems facing the border regions, the development of methods for component and integral assessment of natural resources and the anthropogenic influence of the transboundary territories on the environment, the modernization of the economy of the regions and the transport and social infrastructure, the improvement of the population distribution systems in outlying territories as well as to establish new scientific relations and strengthen the existing international collaboration in the sphere of territorial development, and the environmental and ecological protection within the China–Mongolia–Russia Economic Corridor.

The volume is intended for a wide range of researchers in the field of geography and economics, university teachers, students, and practitioners with an interest in problems of ecology, effective environmental management, and territorial development.

**Foreword**

Dear Readers:

You will find in this volume the contributions at the First International Geographical Conference on "China–Mongolia–Russia Economic Corridor: Geographical and Environmental Factors and Territorial Development Opportunities".

An agreement to create the China–Mongolia–Russia Economic Corridor was reached by the leadership of three countries on the margins of the Shanghai Cooperation Organization Summit as early as June 2016 held in Tashkent (Uzbekistan).

Northeast Asia is well endowed with natural resources and has diverse ecosystems and enormous potential for development. The countries of Northeast Asia reached consensus on the need to enhance regional cooperation. Such a joint initiative opens up brand new vistas and advantages for sustainable development as well as presenting various economic, social and ecological challenges to be met. Specifically, there is an ongoing transformation of the ecosystems and biota in Siberia, the Far East, Northeast China, Mongolia and Central Asia. Environmental pollution continues to escalate, especially in industrial centers. Conflicts (including inter-State conflicts) occur because of the uneven distribution and unsustainable use of water resources. There is an increasing lack of good quality drinking water.

The Conference provides a platform for a joint search for solutions of the scientific problems in the field of optimal ecologically oriented resource provisioning of the livelihoods and the quality of life of the population, the development of modern urbanized and cultural landscapes and the preservation of the natural environment in transboundary and outlying territories.

The themes of the Conference include the geoecological and socioeconomic problems of the border regions, the issues concerning the component and integral assessment of natural resources and the anthropogenic influence on the environment of the transboundary territories, and the characteristic features of the inter-regional impact of the geographical, socioeconomic and ecological factors. An explicit focus is placed on spatiotemporal models of the distribution and use of natural resources and comprehensive geographical analysis of the system of population distribution within the China–Mongolia–Russia Transport and Economic Corridor, including its influence on the potential for development of neighboring regions.

The organizers hope that not only will this Conference become an important event to promote the expansion of international cooperation in the field of geographical and ecological research, but it also make it possible to obtain new knowledge and reach agreed decisions during the consideration of the reports and working-level discussions.

Director, V.B. Sochava Institute of Geography SB RAS, Cand. Sci. (Geogr.)
*Igor Nikolaevich Vladimirov*

IOP Conf. Series: Earth and Environmental Science **190** (2018) 011002 doi:10.1088/1755-1315/190/1/011002

# Peer review statement

All papers published in this volume of *IOP Conference Series: Earth and Environmental Science* have been peer reviewed through processes administered by the proceedings Editors. Reviews were conducted by expert referees to the professional and scientific standards expected of a proceedings journal published by IOP Publishing.

# Socio-economic potential of municipalities in the context of natural risk (case study – Southern Siberian regions)

**S V Badina**

Institute of Economic Forecasting, Russian Academy of Sciences; Lomonosov Moscow State University, Faculty of Geography, Moscow, Russia

E-mail: bad412@yandex.ru

**Abstract**. A method for natural risk assessment is proposed. It based on data of socio-economic potential and natural hazards spatial distribution on the study area. This methodology for assessing the economic component of natural risk consists by two integral indices. The first is the index of social and economic density on the territory. It shows the probable maximum direct and indirect damage from natural hazards per unit area only the economically exploited part of total municipal territory. Index parameters are: population, fixed assets value and gross production volume, which divided into the built-up area of municipality. The second is the index of socio-economic spatial vulnerability of the territory. It demonstrates the probability of crossing the concentrated social and economic potential areas and disaster zone. Index calculated as multiplication of population, fixed assets value and gross production volume to the spatial vulnerability coefficient. This coefficient is equal to the share of built-up area in the total area of municipal land. We have identified potentially the most vulnerable municipalities in South Siberia with indices values above the macro-regional average. In addition, the maximum debris flow risk zones were singled out.

## 1. Introduction

In recent years, there has been a steady trend to increase the number of natural disasters and their catastrophic consequences in Russia. The primary reason for this is not only a change in natural conditions, but a factor of the socio-economic development intensification in the most dangerous areas. The lack of land resources in cities, investment projects related to mineral resources extraction, tourism and recreation, new industrial plants and infrastructure building, etc., in places with a high natural danger level is one of the reasons for natural risk increasing. At the beginning of 21 century natural risk is recognized as one of the main strategic risks in Russian Federation [1]. Thus, scientifically based natural risk management is a necessary condition for country sustainable development. Economic approach to natural hazards assessment is currently poorly elaborated and it is basically reduced to make predictions attempt based on damage estimates from already occurred natural disasters, which is methodologically incorrectly [2]. Natural hazards studies are always associated with a high degree of uncertainty. The risk and dangerous natural phenomena stochastic nature determines preferred use a probabilistic (but not a widely used method of direct calculations) methods for assessing the society and economy vulnerability.

Natural risk is a probability function of natural hazard occurrence and its negative consequences for economy and society as a whole in the specific territory. The most general risk indicator is the mathematical expectation (average value) of the damage from a dangerous event for the year, [3].

According to this, we offer a hypothesis: there is close to a direct functional relationship between the economic development level of the territory and the risk level. It allows the expression of natural risks through the socio-economic parameters.

In this study the "socio-economic potential of the territory" was chosen as a complex parameter in order to assess the social component of natural risk. It is defined as the territorial socio-economic system ability to support long-term balanced development through rational use of all resources diversity, which concentrated in it [4]. The socio-economic potential of the territorial system can be destroyed in whole or partly by the natural disaster impact. Thus, based on the adopted approach, it is important to identify the internal regional territorial differentiation according to the degree of catastrophic consequences probability. For these purposes, it is necessary not only to quantify the potential in absolute value, but also to understand features of its spatial distribution in a territory with its unique physiographical characteristics. First of all, we are talking about the finding differences in level of the potential concentration at each point on considered space.

The territory of South Siberian regions, located within two large mountainous countries (Altai-Sayan and Baikal), is characterized by a high natural risk level, taking into account the most destructive natural hazards for Russia: earthquakes, floods, hurricanes, avalanches, landslides and debris flows [5]. In this study, in order to test our approach we chose the debris flows. Southern Siberia characterize as a region with very high level of debris flow activity in Russia [6]. Due to relief specificity debris flows are widespread here, although a most part of the territory is only potentially dangerous (where the debris flows may be activate by vegetation cover destruction). The main factor of dangerous slope processes activation in this territory is not a climatic change, but first of all it is a human activity in areas of industrial, agricultural, and silvicultural development [7].

## 2. Models and Methods

Based on the conceptual approach described above, a methodology for estimating the maximal likely socio-economic damage from natural hazards was developed. Municipalities' level was chosen as an optimal territorial level. Socio-economic municipal statistic data do not have a link with relevant economic use areas (especially in eastern Russian regions, where the municipality's size is extremely large and the territory has only local development). This factor complicates the ability to compare socio-economic data with nature data. According to this, we created a special methodology that eliminated this drawback.

The socio-economic potential density of territory was assessed by the ratio to only most intensive economic development areas (not to the full administrative-territorial unit area, in the borders of which statistical information is provided). Thanks to this it became possible to identify the real concentration of potential and to link population and economy with real areas of their concentration. The spatial vulnerability of socio-economic potential is defined as probability of its crossing with natural hazard zone.

Our methodology for assessing the economic component of natural risk based on two integral indexes: the index of socio-economic density and the index of socio-economic spatial vulnerability of the territory. Indices consist three blocks: population, fixed assets and gross production of the municipality. Parameters choice is conditioned by the fact that the social and economic consequences of catastrophes characterized by the direct and indirect damages. For population and fixed assets the damage from natural disasters is direct in general and manifests itself immediately (people death, production capacities failure, buildings and structures destruction, etc.). Destructive effect of natural hazard to gross product is closely related to the first two parameters, but it can be also manifested directly, independently of them. It means the reduction of goods and services production by reducing the workforce, destruction of infrastructure, plants, industrial chains, agricultural lands etc., additional expenses in the social sphere and some economy sectors, reducing the tax payments in the emergencies territory and so on.

A summary algorithm for calculating the index of socio-economic spatial vulnerability on municipal level:

Assessment of population spatial vulnerability (*Ip*):

$$I_p = kP \tag{1}$$

where *P* – population of municipality (people), *k* – spatial vulnerability coefficient (share of economic developed territory in total municipal lands area ($k = S_{e1} / S_t$, where $S_{e1}$ – built-up area, $S_t$ – total area of municipality)).

Assessment of fixed assets spatial vulnerability ($I_f$):

$$I_f = kF \tag{2}$$

where *F* – estimated value of municipal fixed assets (million rubles), *k* – coefficient of spatial vulnerability.

Assessment of gross production spatial vulnerability ($I_{gp}$):

$$I_{gp} = kGP \tag{3}$$

where *GP* – estimated volume of municipal gross production (million rubles), *k* – coefficient of spatial vulnerability ($k = (S_{e1} + S_{e2}) / S_t$, where $S_{e1}$ – built-up area, $S_{e2}$ – area of agricultural land (used if the share of agricultural sector in regional economy structure and potential damage to agriculture are high (for example, the share is higher than Russian average), $S_t$ – total area of municipality).

Then all of indicators are normalized. The sum of three resulting normalized indicators (population – $NI_p$, fixed assets – $NI_f$, gross production – $NI_{gp}$) gives the required integrated index of socio-economic spatial vulnerability value ($SEP_v$):

$$SEP_v = NI_p + NI_f + NI_{gp} \tag{4}$$

Further, based on the index values a rating of municipalities is compiled (distribution from maximum to minimum values); boundaries of the index intervals (which demarcate municipality's types) are delineated by graphoanalytical method.

The socio-economic density index is calculated by a similar algorithm.

Assessment of population density ($P_d$):

$$P_d = P / S_{e1} \tag{5}$$

Assessment of gross production density ($GP_d$):

$$GP_d = GP / (S_{e1} + S_{e2}) \tag{6}$$

Assessment of fixed assets density ($F_d$):

$$F_d = F / S_{e1} \tag{7}$$

The socio-economic density index of municipalities ($SEP_d$) is calculated by aggregating of three normalized indicators (population density – $NP_d$, fixed assets density – $NF_d$, gross production density – $NGP_d$):

$$SEP_d = NP_d + NF_d + NGP_d \tag{8}$$

Comparison of indices calculation results with natural hazards data at the consider territory makes it possible to identify municipalities with the maximum risk level.

## 3. Results and Discussion

The socio-economic space of South Siberia is characterized by its compressed. According to our calculations, the first 10 municipalities (Krasnoyarsk, Novokuznetsk, Irkutsk, Kemerovo, Barnaul, etc.) concentrate 38% of macroregion population, 49% of gross production and 46% of fixed assets (figure 1, figure 2, figure 3). Besides the study area is characterized by a pronounced internal contrast of three socio-economic potential elements density values. So, in terms of gross production density the value in the first and last municipality in the ranking has a 14.5 thousand times difference (by

IOP Conf. Series: Earth and Environmental Science **190** (2018) 012001 doi:10.1088/1755-1315/190/1/012001

population density – 3.3 thousand times, by fixed assets density – 53 thousand times). For comparison, in another Russian more evenly developed mountain macroregion – in the North Caucasus – the differences in gross production density and population density are only 2.6 and 1 thousand times, respectively. Strong internal differences in density characteristics show the heterogeneity of South Siberian spatial development, and, therefore, the heterogeneity of its spatial vulnerability. So, lands of poorly developed peripheral districts are poorly involved to the economic turnover. The most developed zones are characterized by this feature: high vulnerability with rather low density (agrarian and densely populated Altai Krai, plain part of Kemerovo Oblast, etc.). For least developed zones: low vulnerability with high density (peripheral mountainous areas of Irkutsk Oblast, Buryatia Republic, Zabaykalsky Krai with its local resource type of development).

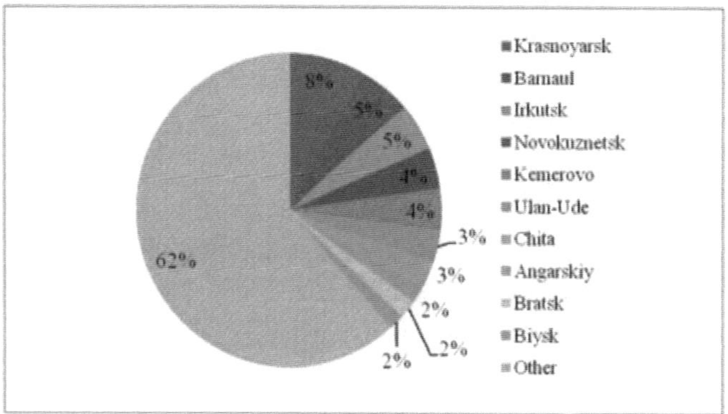


**Figure 1.** Population structure of South Siberian municipalities.

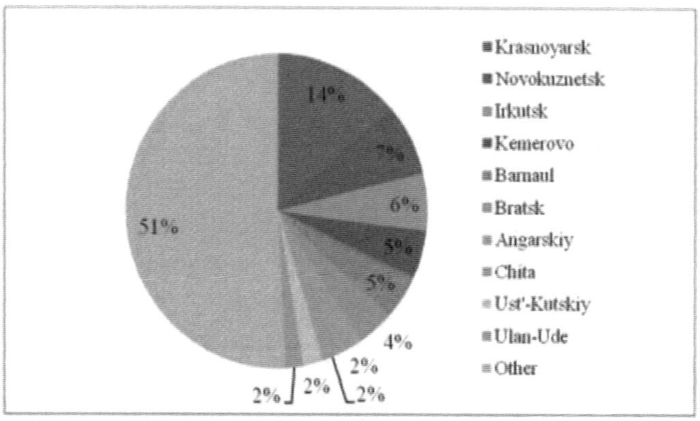


**Figure 2.** Gross production structure of South Siberian municipalities.

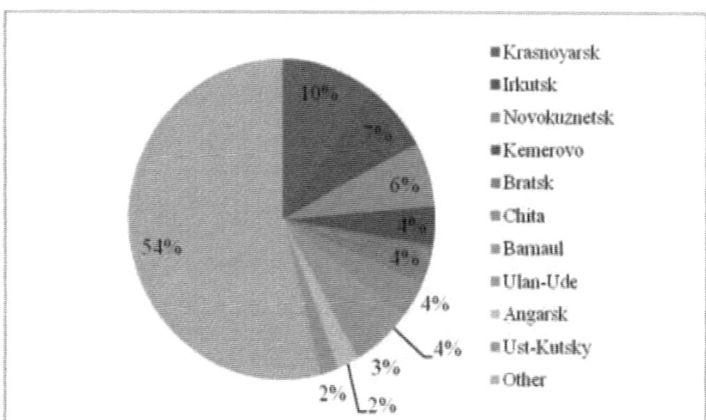


**Figure 3.** Fixed assets structure of South Siberian municipalities.

The map (figure 4) illustrates the results of two indices comparison, it shows an integrated spatial vulnerability and most vulnerable to natural hazards municipalities in South Siberia. Among potentially the most vulnerable municipalities with values of density and spatial vulnerability socio-economic potential indexes the higher than the average macro-region one we can single out those: all regional capitals (Krasnoyarsk, Kemerovo, Irkutsk, Abakan, Barnaul, Chita, Gorno-Altaisk, Kyzyl, Ulan-Ude) and other large in population and in scale of economy cities (Novokuznetsk, Bratsk, Achinsk, Sayanogorsk, etc.).

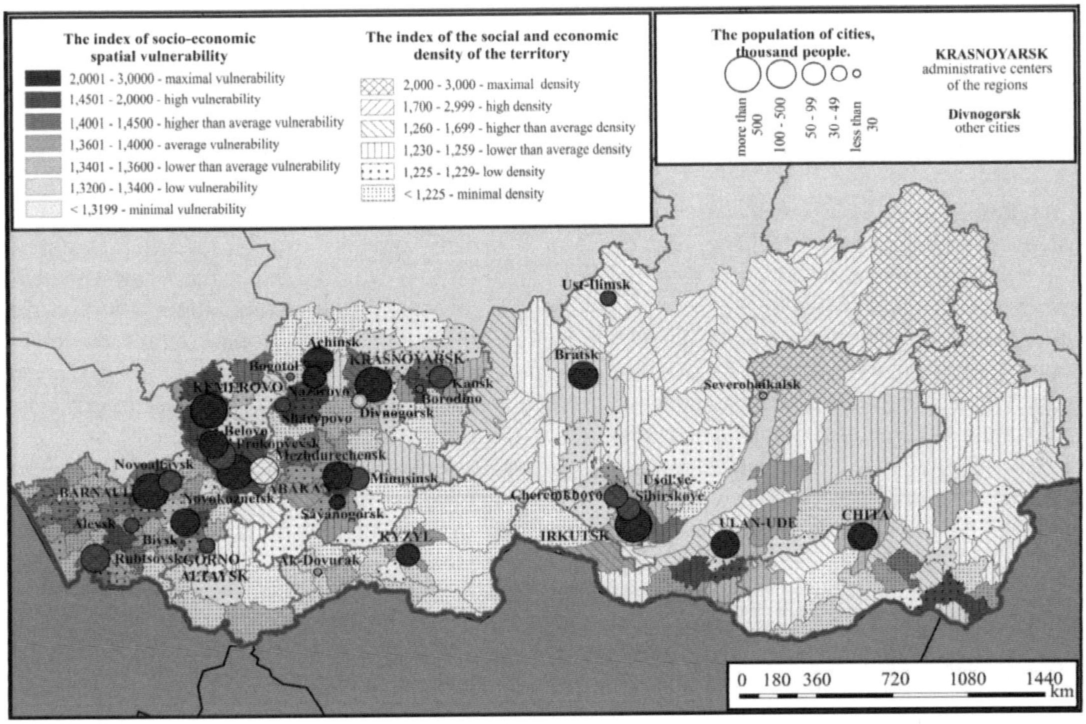


**Figure 4.** Integrated spatial vulnerability of socio-economic potential in South Siberian municipalities, 2017.

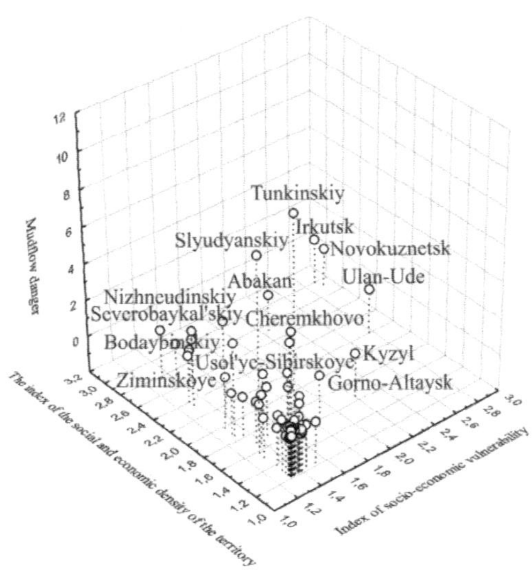


**Figure 5.** South Siberian municipalities groups by the ratio of socio-economic potential density and spatial vulnerability indices with debris flows hazard.

In the municipalities group with high values of potential density with its low vulnerability disposed primarily mountain territories (Sorsk, Sayansk, Severobaikalsk, Bodaibo District, etc.). These are mainly peripheral areas with local development, where the presence of mineral resources historically determines the economic activity areas. In the first approximation, the level of vulnerability to natural hazards looks low there due to small developed lands share in total municipal area or due to low absolute values of the potential parameters. But the calculations shows that it is namely high potential concentration, if other things being equal, makes the territory more vulnerable.

Analysis of the South Siberian municipalities' combinations results by the ratio of socio-economic potential density and spatial vulnerability indices with debris flows hazard (figure 5), helps to identify municipalities with the highest risk level. These are: Irkutsk, Novokuznetsk, Abakan, Ulan-Ude, Tunkinsky and North Baikal districts of Buryatia Republic.

## 4. Conclusion

Created integral indices of socio-economic potential density and its spatial vulnerability help to identify the probability of maximal potential concentration and hypothetical natural hazards areas crossing at each point of the space under consideration. In general, the method approbation on the South Siberian territory allowed to detect an important pattern: high absolute values of socio-economic potential components with its high spatial concentration and, finally, a high spatial vulnerability degree determine the maximum integral potential vulnerability in space and, accordingly, the maximum risks in case of hypothetical disaster realization (under other equal conditions). It is very important to note, that municipalities with very different socio-economic and spatial development levels may have the same vulnerability level. In all South Siberian regions there is a debris flows hazard. But the risk level is lower here than, for example, in the North Caucasus, because there is no lack of land resources that's why the most dangerous areas are rather poorly developed.

## Acknowledgments

The work was funded by the subsidy of the Russian Science Foundation («Snow avalanches and debris flows risk at the territory of Russia: estimation, forecast and mitigation measures» (RSF project № 16-17-00104, Moscow, 2016–2018) and «Environmental changes in the Arctic and their impacts on the human wellbeing and infrastructure» (RSF project № 14-37-00038-П, Moscow, 2017–2018)).

## References

[1] Vorobiev Y L et al 2005 *Russia's Strategic Risks: Estimation and Forecast* (Moscow: Delovoy ekspress) p 392 (in Russian)

[2] Vladimirov V A et al 2002 *Natural Hazards of Russia. Natural Dangers and Society* (Moscow: KRUK) p 245 (in Russian)

[3] Myagkov S M 1995 *Geography of Natural Risk* (Moscow: Moscow State University) p 222 (in Russian)

Porfiryev B N 2011 *Nature and Economy: Risks of Interaction. Ecological and Economic Essays* (Moscow: Ankil) p 352 (in Russian)

[4] Baburin V L and Badina S V 2015 Evaluation of the social-economic potential of natural hazard-subjected territories *Vestnik Moskovskogo Universiteta. Seriya 5: Geografiya* **5** 9–16 (in Russian)

Badina S V 2017 Quantification of Russian Arctics socio-economic potential vulnerability in the zone of permafrost degradation *Regional'nyye Issledovaniya* **3** 107–16 (in Russian)

[5] Osipov V I 2016 Technogenesis and modern tasks of earth sciences *Ecology and Industry of Russia* **20**(3) 4–12 (in Russian)

Borodko A V et al 2004 *National Atlas of Russia. Volume 2. Nature and Ecology* (Moscow: GOSGISTSENTR) p 495 (in Russian)

[6] Perov V et al 2017 Debris flow hazards for mountain regions of Russia: regional features and key events *Natural Hazards* **88**(1) 199–235

[7] Socratov S A, Seliverstov Y G, Shnyparkov A L and Koltermann K P 2013 Anthropogenic impact on avalanche and debris flows activity *Ice and Snow* **122**(53) 121–8 (in Russian)

# Socio-economic mobility of the population of the Baikal region in current conditions

**A V Bilgaev**

Institute of Geographic Sciences and Natural Resources Research CAS, Beijing, China

E-mail: bilgaev.alexey@gmail.com

**Abstract.** Human factor plays an important role in the effective implementation of the socio-economic policy of the state and it is a key component of sustainable development of the regions. This study considers the current living standards of the population of the Baikal region. The analysis allowed identifying trends in the increase in the unemployment rate of the population living on its territory. It revealed that in recent years there has been a tendency to increase the poverty level of the population of Baikal region, which remains high compared to the average Russian level. Also, this paper analyses the quality of life of the population; determines the key factors of influence on the socio-economic mobility of the population of the region. It was determined that the fastest and most accessible channel of social mobility is education, which provides an opportunity to obtain a high-paid profession and rapid progress on the social ladder. In order to change the negative trends in the human development of the Baikal region, this article considers the role and prospects of implementation of the "green" development mode of the Baikal region as a driving mechanism on the way to sustainable development and strengthening of socio-economic mobility of the population.

## 1. Introduction

The human factor plays an important role in the effective implementation of the socio-economic policy of the state. It is also a key component of sustainable development of the regional economic system.

Within the framework of the Eastern orientation of the Russian policy, the study of the modern socio-economic mobility of the population of the Baikal region is of the greatest interest. At the same time, it is necessary to give a new qualified and systematic scientific assessment of social mobility of the population, the state of labor potential, the quality of life of the population.

The study determined that the fastest and most accessible channel of social mobility is education, which makes it possible to obtain a high-paid profession and rapid progress on the social ladder (other favorable conditions). The future of the country depends on it. All this actualizes the research aimed at analyzing the socio-economic mobility of the population of the Baikal region as a factor for improving the quality and standard of living of the population.

The aim of the study was to assess the current trends in the development of socio-economic mobility of the population of the Baikal region. The achievement of this goal determined the solution of the following tasks: description of the characteristic geographical and economic features of the Baikal region, comparative assessment of socio-economic mobility of the population of the Baikal region, analysis of the quality of life of the population of the Baikal region.

Systematic monitoring of the level and dynamics of social development of the region allows realistically assessing the current state and timely taking effective management decisions aimed at improving the quality and standard of living of the population of the region.

## 2. Models and methods

This paper proposes a model of socio-economic mobility of the Baikal region in the "green" economy. The main methodological approach of the study is based on the system analysis of processes, interrelations, and results of the studied problems. In this work was used the method of comparative analysis of indicators of subjects of the Baikal region characterizing social and economic mobility and quality of life of the population. During the study author analysed dynamics of the considered indicators by means of construction of time series and their chronological comparison, assessed the growth rates of indicators using the index method and the dynamics of changes in structural proportions by structural analysis.

## 3. Results and Discussion

### *3.1. Results*

In order for a society to truly tackle the issue of opportunity to advance socioeconomically for all young people, there must be open communication, investment, and innovative partnerships between government, business, community centers, schools, and crucially the youth themselves. We must carefully consider each barrier faced in the life of a young person and implement and scale up successful initiatives for each step of the way. We must carefully consider each barrier faced in the life of a young person and implement and scale up successful initiatives for each step of the way.

According to the rating of the regions of the Russian Federation on the quality of life in 2017, Buryatia and the Irkutsk oblast retained their positions compare to last year and took 76th and 69th positions, Zabaykalskii Krai fell one line lower, being in the rating by 81 positions out of 85 (Rating of regions of Russian Federation on quality of life 2017, 2018).

*3.1.1. Baikal region and its characteristics.* Baikal region includes three subjects of the Russian Federation, united by belonging to the Lake Baikal basin to the phenomenon of the world importance. The total area of the region is 1558.1 thousand km$^2$, population – 4464.5 thousand. By subjects of the Russian Federation, the territory and population are distributed as follows: Republic of Buryatia – 22.5%, Zabaykalskii Krai – 27.7%, Irkutsk Oblast – 49.7% of the Baikal region territory; the population of the Republic of Buryatia accounts for 22.1%, Zabaykalskii Krai - 24.0%, Irkutsk Oblast - 53.9%. Lake Baikal has a special status, fixed not only at the federal but also at the world level as a part of the UNESCO world natural heritage (1996). Baikal natural area is located on the territory of the Republic of Buryatia, Irkutsk oblast, Zabaykalskii Krai. On the protected Baikal natural area lives 93% of the population of Republic of Buryatia, 55% of Irkutsk oblast, 9% of Zabaykalskii Krai. The 40-year environmental regulation regime has had a restrictive impact on the socio-economic development of the region. The region has all the features of landscape-ecological and economic-cultural integrity. The geographical and geopolitical position of the Baikal region compares favorably with neighboring regions. Located between Siberia and the Far East, this territory, which has good access to the countries of Central and East Asia, has always been used by Russia to establish contacts and trade with these countries. At the present time, when Russia became open to the world community, the strategic role of the geographical location of the Baikal region has further strengthened. It was promoted, first of all, by its transport and geographical position. The region is a link of Intercontinental transit Railways – Trans-Siberian and Baikal-Amur has direct rail links to China and Mongolia. On its territory are all the main transport and communication lines, international and domestic airlines, including cross-polar routes between the countries of South and South-East Asia, as well as the American continent (Baikal natural area. http://irkipedia.ru/content/baykalskaya_ prirodnaya_territoriya). Lake Baikal region is a world natural heritage, which is not only beautiful, unique and complex for its landscape ecosystem and culture, rich in freshwater, flora, and fauna, but also one of the most attractive international tourism destination well recognized by international tourists [9].

*3.1.2. Socio-economic mobility.* Social-economic mobility is the movement of individuals, families, or groups through a system of social hierarchy or stratification. If such mobility involves a change in position, especially in occupation, but no change in social class, it is called "horizontal mobility". If, however, the move involves a change in social class, it is called "vertical mobility" and involves either "upward mobility" or "downward mobility" (Social mobility. https://www.britannica.com/topic/social-mobility).

*3.1.3. Socio-economic mobility of the population of Baikal region.* In order for a society to truly tackle the issue of opportunity to advance socioeconomically for all young people, there must be open communication, investment, and innovative partnerships between government, business, community centers, schools, and crucially the youth themselves. We must carefully consider each barrier faced in the life of a young person and implement and scale up successful initiatives for each step of the way [1].

A key component of the functioning and development of the region is the human factor, acting as the leading subject and object of the changes. In this regard, the study and assessment of labor potential, measuring the quality of life of the population, etc., will allow better assessing and determining further actions for the implementation of regional socio-economic policy. Figure 1 shows a graph of occupation level of the Russian Federation as a whole and of the subjects of Baikal region

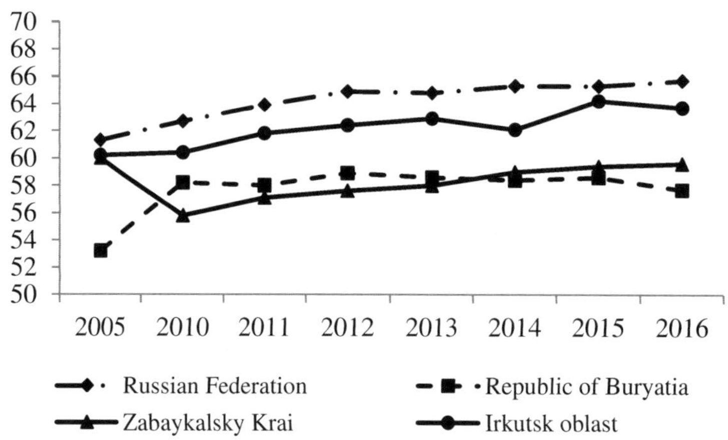


**Figure 1.** Occupational level, %.

The graph shows that the employment rate in the Russian Federation as a whole gradually increased until 2012, in 2013 there was a slight decline (by 0.1%), and by 2016 increased by 0.8% compared to 2012. In the subjects of the Baikal region, the picture is rather uneven, for example, in the Irkutsk Oblast, occupation level increased from 2010 to 2013 with slight declines in 2014 (by 0.8%) and by 2016 increased by 1.3% compared to 2012. In the Republic of Buryatia we can see increasing of occupation level until 2012, and in subsequent years – a gradual decline in employment and by 2016 is 57.7%, which is 0.8% less than in 2012. In Zabaykalskii Krai, the decline in employment was in 2010 compared to 2005, and since 2010 there has been a gradual increase and compare to 2012 in 2016 the employment rate increased by 2%. The data for the period 2014-2016 show that the dynamics of occupation level in the Baikal region is uneven. So, if in Buryatia there is a decrease in employment, then in Zabaykalskii Krai and Irkutsk oblast it is growing. Territorial differences in economic activity and employment largely depend on the characteristics of the region-demographic, socio-cultural, settlement, environmental, the current structure of the economy, as well as the heterogeneity of territories determined by natural and geographical conditions [2].

Thus, the material wealth and social status of the population depend on the availability of jobs. Table 1 shows the unemployment rates of the population for 2005-2016 in Russia as a whole and subjects of the Baikal region.

**Table 1.** Unemployment rate, %.

| | 2005 | 2010 | 2011 | 2012 | 2013 | 2014 | 2015 | 2016 |
|---|---|---|---|---|---|---|---|---|
| **Russian Federation** | 7.1 | 7.3 | 6.5 | 5.5 | 5.5 | 5.2 | 5.6 | 5.5 |
| **Republic of Buryatia** | 12.2 | 10.4 | 9.0 | 7.9 | 8.0 | 8.4 | 9.2 | 9.6 |
| **Zabaykalsky Krai** | 8.6 | 11.1 | 10.6 | 10.6 | 10.5 | 10.0 | 10.4 | 10.8 |
| **Irkutsk oblast** | 9.8 | 10.1 | 9.1 | 7.8 | 8.3 | 8.8 | 8.2 | 8.8 |

In recent years, the sharpest changes in the unemployment rate are observed in the Republic of Buryatia. During the period 2015-2016, there has been a tendency to increase the unemployment rate of the population living in the Baikal region. The dynamics of general unemployment in the regions coincides with the all-Russian trend, but significantly exceeds it, and the periods of growth and fall were sharper. However, in all three regions, there is a steady increase in the unemployment rate, while in Russia as a whole there is a slight decline.

Solving the issues of increasing and reducing unemployment requires: an active state policy in the field of employment and regulation of demand for labor; assistance to employers in creating effective jobs that meet modern requirements, as well as increasing the cost of retraining of personnel of organizations in order to improve its quality and competitiveness and further employment; social protection of workers [3].

The fastest and most accessible channel of social mobility is education, which makes it possible to obtain a highly paid profession and to move quickly up the social ladder (under other favorable conditions). And passing it will be sure to promote for any scale of social hierarchy [4][4]. Education is a very important factor in moving up the social ladder. Table 2 shows the structure of the employed population by level of education in 2016.

**Table 2.** Structure of the employed population by level of education in 2016.

| Region | Employed, in total | Including education | | | | |
|---|---|---|---|---|---|---|
| | | Higher | Vocational secondary, in total | High school | Middle school | No middle school |
| **Russian Federation** | 100 | 33.5 | 45.1 | 18.1 | 3.2 | 0.2 |
| **Republic of Buryatia** | 100 | 32.6 | 44.7 | 16.5 | 5.9 | 0.4 |
| **Zabaykalskii Krai** | 100 | 24.5 | 40.5 | 27.5 | 6.9 | 0.4 |
| **Irkutsk oblast** | 100 | 29.1 | 43.6 | 20.9 | 6.2 | 0.3 |

The table shows that among the subjects of the Baikal region in Buryatia, about a third (32.6% ) of the employed population have higher education, which is 0.9% less than the national level. For comparison, in Irkutsk Oblast the share of employed with higher education is 29.1%, and in the Zabaykalskii Krai 24.5%.

*3.1.4. Quality of life of the population of Baikal region.* We have analyzed the main indicators of the standard of living of the population, such as the share of the urban population, real incomes of the population and the population with monetary income below the subsistence minimum. Table 3 shows the share of urban population in the total population.

**Table 3.** Share of urban population in total population (year-end estimate, in %).

| | 2005 | 2010 | 2012 | 2013 | 2014 | 2015 | 2016 | 2016 to 2005, % |
|---|---|---|---|---|---|---|---|---|
| **Russian Federation** | 73.2 | 73.8 | 74 | 74.2 | 74 | 74.1 | 74.3 | 101.5 |
| **Republic of Buryatia** | 56.4 | 58.5 | 59 | 58.8 | 58.9 | 59 | 58.9 | 104.4 |
| **Zabaykalskii Krai** | 64.2 | 66 | 66.8 | 67.1 | 67.4 | 67.7 | 68 | 105,9 |
| **Irkutsk oblast** | 79.3 | 79.6 | 79.5 | 79.4 | 78.9 | 79 | 78.9 | 99.5 |

An analysis of the data in table 3 showed that the rural population was declining during the period under review. The current trend of outflow of rural residents to the city, despite the measures taken in the social sphere, is typical for the Baikal region. One of the main reasons is the irrational location of industrial production, the ratio of the arrived and decreased population and natural population growth. As it is known, Siberian regions are characterized by high differentiation in the standard of living of the population. One of the main indicators of the characteristics of the standard of living of the population is the value of per capita income. Analysis of the dynamics of real incomes (table 4) showed. From 2011 to 2013, all subjects of the Baikal region had stable income growth, but in 2014 there was a decline, which repeats the all-Russian trend. In the period 2005-2014 in Russia, there was an increase in real income of the population, since 2014 there is a decline. Similar dynamics can be seen in the Baikal region.

**Table 4.** Real incomes of the population (% to the previous year).

| | 2005 | 2010 | 2011 | 2012 | 2013 | 2014 | 2015 | 2016 |
|---|---|---|---|---|---|---|---|---|
| **Russian Federation** | 111.7. | 105.4 | 101.2 | 105.8 | 104.8 | 99.5 | 95.9 | 94.4 |
| **Republic of Buryatia** | 111.6 | 102.5 | 100.9 | 103.3 | 109.5 | 99.8 | 100.2 | 93.3 |
| **Zabaykalskii Krai** | 109.2 | 103.2 | 102.1 | 103.1 | 104.8 | 95.3 | 95.3 | 92.4 |
| **Irkutsk oblast** | 117.9 | 100.8 | 96.8 | 104.8 | 102.2 | 97.5 | 97.2 | 91.1 |

In recent years, there has been a tendency to increase the level of poverty of the population, both in the Baikal region and in Russia as a whole. Table 5 shows the indicators of population with monetary income below the subsistence minimum in % of the total the population of the subject (Regions of Russia. Socio-economic indicators, 2017)**Ошибка! Источник ссылки не найден.**.

**Table 5.** Population with monetary income below the subsistence minimum (% of the total population of the subject).

| | 2005 | 2010 | 2012 | 2013 | 2014 | 2015 | 2016 |
|---|---|---|---|---|---|---|---|
| **Russian Federation** | 17.8 | 12.5 | 10.7 | 10.8 | 11.2 | 13.3 | 13.4 |
| **Republic of Buryatia** | 32.6 | 19.2 | 17.7 | 15.9 | 16.9 | 17.7 | 18.3 |
| **Zabaykalskii Krai** | 26.1 | 19.0 | 17.6 | 16.2 | 18.0 | 20.4 | 21.4 |
| **Irkutsk oblast** | 21.3 | 18.1 | 16.8 | 17.0 | 18.0 | 20.1 | 20.6 |

As the table shows, the poverty level in Russia decreased until 2014 and began to increase in 2015. In the Baikal region, the poverty level in such regions as the Republic of Buryatia and Zabaykalskii

Krai decreased until 2013, and from 2014 to 2016 increased by 2.4% and 5.2%, respectively. In Irkutsk oblast, level of the population with monetary incomes below the subsistence minimum began to increase earlier, since 2013. If we consider the same period with other subjects, the poverty level in the Irkutsk region increased by 3.6%, and in Russia as a whole – by 2.6%. It should be noted that the level of poverty of the population of the subjects of the Baikal region in comparison with the average Russian indicators remains high.

Today, human development is one of the main factors of the successful functioning of individual territories and countries of the world. Low incomes and a high cost of housing determine the quality of life of the population. Income inequality causes inequality in the level, structure, and quality of consumption among different groups and has a negative impact on human development. That is, there is a significant differentiation of opportunities and social inequality in human development. [5]. In fact, social inequality is a constraint on human capital development.

*3.2. Discussion*

In modern conditions, the role of science, education, innovation has increased, and this also applies to the Baikal region. Therefore, one of the main tasks in the near future should be the investment in human and the creation of favorable conditions for life and activities, taking into account local natural and climatic characteristics [6].

Thus, in the context of improving the quality of life in the Baikal region, it is very important to use the principles of the "green" economy and to include them in new development projects. It should be noted that:

- "Green" economy is a means to achieve sustainable development;
- "Green" economy should promote the creation of green jobs and decent work;
- "Green" economy - equality, justice between countries and between generations;
- "Green" economy protects biodiversity and ecosystems;
- "Green" economy reduces poverty, ensures the well-being of the population, social protection and access to basic;
- "Green" economy improves governance and the rule of law. It is democratic, has principles of equal decision-making, responsible for the result, transparent and stable;
- The green development path should preserve, enhance and, where necessary, restore natural capital as an essential economic asset and source of public goods, especially for the poor [7].

The world, and in particular the Baikal region, continues to move along an unstable path of development. This is accompanied by the degradation of all natural life-support systems, increased inequality between countries and regions, and the aggravation of multilateral crises [8]. Therefore it is necessary to reduce the economic losses caused by environmental degradation by incorporating the value of natural resources into strategic national priorities, sound policies and institutional reforms; create quality jobs and reduce poverty - improve social equity by greening investment in sectoral policies such as renewable energy and climate-resilient infrastructure.

Obviously, the transition to a green economy varies from country to country and region to region, as it depends on the specific nature of the natural and human capital of each country or region and the relative level of development. In this transition, all countries have a number of new opportunities. Some countries have achieved high levels of human development, but this is often due to their natural resources, environmental quality and increased greenhouse gas emissions. These countries face the challenge of reducing their per capita environmental impact without compromising the quality of life. Other countries still have a relatively small per capita impact on the environment, but need to improve the quality of services and the material well-being of their citizens. They face the challenge of doing so without significantly increasing their environmental impact.

One of the most important conditions for poverty reduction is the transition to a green economy. Chronic poverty is the most visible manifestation of social injustice, resulting from unequal access to education and healthcare, unequal access to credit and income, and non-equitable protection of property rights. A key feature of the green economy is that it creates different opportunities for economic development and poverty reduction without destroying or destroying natural assets (UNEP Towards a Green Economy: Pathways to Sustainable Development and Poverty Eradication – A Synthesis for Policy Makers, 2011).

## 4. Conclusion

The study of socio-economic mobility of the population of the Baikal region allows us to draw the following conclusions. It is obvious that for the transition to the sustainable development we need investment in human and physical capital. In practical terms, the state and the region, in particular, need to sharply increase investment in science, education, health, innovative development [10]. Today, human development is a determinative factor in the successful functioning of individual territories and countries of the world. The social and economic mobility of the population is one of the factors of improving the quality and standard of living of the population. Income inequality causes inequality in the level, structure, and quality of consumption among different groups and has a negative impact on human development. To improve the quality of life of the population, it is possible to apply the principles of "green" economy. The principles of the "green" economy should be taken into account developing the socio-economic development strategy of the region.

## References

[1] Hanna A L 2015 Socio-economic mobility of youth: factors, obstacles, and potential solutions *Journal of Youth Development* **10**(1) 72-84
[2] Saktoev V E and Sadykova E T 2011 *Sustainable Development of Regional Economic System with Environmental Regulation* (Moscow: Ekonomika) p 295
[3] Efimova L A 2015 Employment and unemployment in Russia regional economics theory and practice *Socioeconomic policy* **395**(20) 14-22
[4] Melanina M V 2013 Institutionalization factors of social and economic mobility of Russian population *Economics: Yesterday, Today and Tomorrow* **1-2** 70-85
[5] Kalugina Z I 2018 Siberia from the perspective of human development *Region: Economics & Sociology* **98**(2) 110-32
[6] Kryukov V A 2018 Studying Siberia's economy: continuity and complexity *Region: Economics & Sociology* **98**(2) 3-32
[7] Bobylev S N, Goryacheva A A and Nemova V I 2017 Project Management in Green Economy *Public Administration (E-journal)* **64** 34-44
[8] Sdasyuk G V and Tishkov A A 2012 Rio +20: The concept of green economy sustainable development and problems of its implementation in Russia *Sustainable Development: Problems and Prospects* **5** 12-46
[9] Dong S, Li Y, Li Z, Li F, Bilgaev A V, Yang Y and Xia B 2017 Green development mode of Baikal region *Baikal as a World Natural Herritage Site:20 Years Later: Proc. of the International Scientific Conf. July 26-30, 2017* (Ulan-Ude: BSU Press) pp 31-38
[10] Bobylev S N 2014 The ecological challenges and the "green" economics *Vestnik MITHT* **1**(1) 5-14

# Geoinformation mapping of transport in the Baikal region

**T B Dashpilov**

Sochava Institute of Geography SB RAS, Russia, Irkutsk

E-mail: tsirik81@mail.ru

**Abstract.** Transportation of the Baikal region is one of the components of a united transport complex of our country and Mongolia. We developed a semantic classification of cartographic symbols of transport systems with a view to streamlining the structure of thematic layers and the system of cartographic sign mapping of transport. Using this classification, geoinformation mapping of the transport of the Baikal region was carried out. To represent the transport of this territory, several semantic parameters were chosen that characterize both the transport complex and its possible impact on the environment (main transport types, infrastructure facilities, density of highways). The existing and projected constructions of railway tracks, highways, as well as pipelines are shown as linear objects. Point features show major airports, ports and piers. The background is the density of motor roads in the administrative regions of the Baikal region and aimaks of Mongolia. Therefore, the transport of the Baikal region was analyzed with an integrative approach on the basis of the created map, and also using other literary sources and statistical data. The main types of transport in the region, their spatial location, as well as infrastructure of transport facilities are considered. Prospects for the development of regional transport are presented.

## 1. Introduction

The development of the Baikal region has some specificity due to the fact that the region has a limited nature management regime. The economic and geographical position of the Baikal region determined its strategic role in the development of the country, since the main highways in the region are part of the intercontinental transport corridors. The regional development strategy is to maximize the beneficial effects of territorial factors, including transport. The study of the transport development system can be carried out effectively using a modern information system based on digital thematic maps and atlases.

To develop the content of maps, we built a semantic database. The structure of the database is revealed with the help of our transport-defined semantic classification of cartographic signs [1]. This classification is based on the research of scientists in the field of mapping semiotics [2]. The structure of our semantic classification is largely determined by the sequence of the logic of spatial analysis of transport networks and structures.

## 2. Objects, data and methods

The territory of mapping covers a macroregional transboundary level - the subjects of the Baikal region and aimaks of Mongolia located in the basin of Lake Baikal. The initial layers characterizing the transport of the Baikal region are the different types of transport routes and their features. Railway lines are divided into operating, under construction and contemplated. The routs are shown with line thickness as the main and others, and highlighted with a color as roads and tracks. Also, the dotted line shows the roads to be designed for construction. Pipeline transportation is represented by operating oil pipelines and gas pipelines designed for construction. The next thematic layer is the characterization of the quality parameters of transport knotes - airports, river ports and piers. The background layer shows the density of paved roads in the administrative regions of the Baikal region and the aimaks of Mongolia.

The transport map of the Baikal region was prepared for a cartographic information system based on the scientific concept of system ecological mapping, developed in the V.B. Sochava Institute of Geography SB RAS [3].

## 3. Results and discussion

Cartographic analysis showed that in the Baikal region almost all types of transport, except for sea transport, are represented. Railway transport is represented by the section of the Transsiberian Railway, the western section of the Baikal-Amur MainlineRailway, the Khrebtovaya–Ust'-Ilimsk, Ulan-Ude-Naushki, Karymskaya–Zabaikal'sk, Borzya-Solovyovsk, Kharanor-Priargunsk, Kuenga-Sretensk, Chernyshevsk- Bukachach, the Circum-Baikal railway, the section of the branch line Rechoty-Karabul, the eastern section Abakan-Taishet. In Mongolia, the railway base is represented by a section from the Russian border through Ulaanbaatar to the border with China, leading to Beijing. The main railway within the Lake Baikal basin is connected to Erdenet, Sharyn Gol and Baganuurby the branch lines.

The railways pass through the most developed regions of the Baikal region with majority of population, main industrial centers and agricultural lands located here. In 2015, most of all cargo and passengers transported by rail were in the Irkutsk region (table 1).

**Table 1.** Rail transport statistics in 2015

| | Freight, MT | Passengers, thou.ppl |
|---|---|---|
| Republic of Buryatia | 14.8 | 905 |
| Irkutsk region | 50.6 | 12967 |
| Transbaikal region | 9.5 | 1856 |
| Mongolia | 19.99 | 2600 |

The road network is represented by the federal roads Novosibirsk – Taishet – Irkutsk, Irkutsk - Ulan-Ude - Chita, Kultuk - Mondy, Ulan-Ude – Kyakhta - the border with Mongolia, Chita - Zakamensk, and regional highways. On the territory of Mongolia, paved roads are represented by the road from the border with Russia to Ulaanbaatar and further to the border with China, as well as its branch: Darkhan -Erdenet - Bulgan. From Ulaanbaatar two roads leadto Arwaikheer and Underkhaan. The remaining roads on the territory of Mongolia are unpaved or just rural tracks. Up-to-date the technical condition of roads in Russia and Mongolia is being improved.

Irkutsk region is also the leader in the transportation of goods by road, and Mongolia in passenger transportation (table 2).

**Table 2.** Road transport statistics in 2015

| | Freight, MT | Passengers, thou. ppl |
|---|---|---|
| Republic of Buryatia | 12.1 | 78.8 |
| Irkutsk region | 55.0 | 147.7 |
| Transbaikal region | 33.3 | 50.9 |
| Mongolia | 20.4 | 260.7 |

IOP Conf. Series: Earth and Environmental Science **190** (2018) 012003 doi:10.1088/1755-1315/190/1/012003

In recent years, there has been a steady increase in the motorization of population. However, the development tempo of roads lags behind the motorization tempo, which leads to the overload of roads (table 3).

**Table 3.** Private cars in the subjects of the Baikal region and Mongolia, per 1000 ppl

| | 2012 | 2013 | 2014 | 2015 |
|---|---|---|---|---|
| Republic of Buryatia | 208.5 | 231.1 | 234.1 | 238.9 |
| Irkutsk region | 229.5 | 237.6 | 247.5 | 253.6 |
| Transbaikal region | 251.5 | 271.8 | 270.5 | 271.3 |
| Mongolia | 120.5 | 131.3 | 146.1 | 157.6 |

Inland waterway transport provides mainly the delivery of goods to the northern hard-to-reach areas of both the Irkutsk region and the Republic of Sakha (Yakutia). Within the Baikal region, the river network is represented by basins of the rivers Yenisei, Angara, Lena and Amur. Regular navigation is carried out along the rivers Angara and its tributaries, Lena, Lower Tunguska, Nepa, Vitim, as well as Lake Baikal and Lake Khubsugul in Mongolia.

The Angara river is divided into three sections by two dams of the Irkutsk and Bratsk hydroelectric power stations. The largest ports are located in the cities: Irkutsk, Usol'e-Sibirskoe, Svirsk, Bratsk, and Ust-Ilimsk. On the Lenariver regular navigation with guaranteed depths is carried out to Ust'-Kut (upstream to the village of Zhigalovo navigation is possible only with small-sized vessels). On the Kirenga river from the mouth (of the Lena river) to the Karam village shipping is not regular. On the Vitim river regular navigation is available from the mouth (of the Lena river) to the city of Bodaibo. Here, large ports are located in the town of Ust'-Kut (the Osetrovo port) and in the town of Kirensk. On the Lower Tunguska river goods are delivered from Podvololoshino up to Nakanno. Also, the regular delivery of goods takes place along the river Nepa from the mouth (the Lower Tunguska river) to the Nepavillage. The passenger traffic dominates mainly on Lake Baikal to Nizhneangarsk, Ust'-Barguzin and other settlements.

The tourism-oriented administration of the Irkutsk region and the Republic promotes the growth of passenger transportation. In the water area of Lake Baikal the largest port is the port of Baikal, as well as piers are in the villages Kultuk, Ust'-Barguzin and Nizhneangarsk.

The fleet involved in Lake Baikal transportations and in 2015 comprised on the classification record of the East Siberian Branch of the Russian River Register is represented by: 1) traveling, walking, search and rescue vessels; 2) dry cargo, passenger, expedition, scientific research vessels; 3) cargo and cargo-passenger ferries; 4) self-propelled tugs; 5) dynamically positioned vessel; 6) buoy tenders [4]. The largest ship owner, engaged in economic activities in the water area of Lake Baikal and the Angara river, is the "East Siberian River Shipping Company" (in 2015, 20 units of the fleet were operated, including passenger ships - 9, cargo-passenger vessels - 2, tugs - 1, dry cargo barges - 2, auxiliary vessels -2, 4 units of the racket fleet).

On the Lena river cargo and passengers are transported by OJSC "Lena United River Shipping Company" (328 units). Passenger transportation is effected from Ust'-Kutto Peledui by a subsidiary LLC "Lenaturflot".

Many navigable waterways ceased to be used in the post-Soviet period. Potentially suitable waterways for navigation, which were used earlier, are the following areas: in the Irkutsk region - Kirenga, Lena from Zhigalovo to Kachug, Vitim above Bodaibo, lower currents of the Chara, Zhuya and Mama; in the Republic of Buryatia-Selenga, the lower reaches of the Chikoi, Khilok, Barguzin and the Upper Angara, the upper and middle course of the Vitim at the border with Transbaikal region; in the Transbaikal Territory the Ingoda, Shilka and Argun'.

For population and economy of the Baikal region aviation transport plays an important role. This is due to the large area of the region, poor conditions for the construction of land communications (mountain relief, permafrost, high seismicity of the territory), and low population density. Air transport at long and ultra-long distances carries out passenger communications of the region with the cities of the European part of the country, Siberia and the Far East, and at short distances within the region with isolated areas and settlements that do not have other types of regular communication. In the last decades of the last century, due to uneven growth in tariffs, there have been changes in the structure of passenger transportation. A significant increase in the cost of air tickets led to the switching of passenger traffic to more affordable rail transport.

There are 14 airports operating in the Baikal region, including 9 airports in the Irkutsk region (Bodaibo, Bratsk, Erbogachen, Irkutsk, Irkutsk-2 (Vostochnyi), Kirensk, Mama, Nizhneudinsk and Ust'-Kut), 3 in the Republic of Buryatia (Nizhneangarsk, Taksimo and Ulan-Ude (Mukhino)), 2 in the Transbaikal Territory (Chita (Kadala) and Chara). The main share of air transportation is carried out by five international airports - Irkutsk, Ulan-Ude, Chita, Ulaanbaatar and Bratsk. There is direct communication of both domestic inter-regional air traffic (Moscow and other cities of Russia) and international (China, Japan, South Korea and Germany)through these ports. Prior to 1995, 72 secondary airports operated, 46 of which were transferred to the landing sites, and 22 were closed.

On the territory of the region pipeline transport is represented by the underground main oil pipeline Omsk-Angarsk (two lines), providing the West Siberian oil to the Angarsk petrochemical company; Petrochemical pipelines Angarsk-Irkutsk and Angarsk-Sayansk, intended for supplying respectively the aviation kerosene of the Irkutsk airport and ethylene for the enterprise Sayanskkhimplast, as well as the East Siberia-Pacific Ocean (ESPO) oil pipeline, through which oil is exported to the APAC countries.

The perspective development of transport in the Baikal region is determined by several documents approved by the government of the Russian Federation. The further development of transport and transport network of the Baikal region is connected both with an intensive use of its transit position between Europe and Asia on the basis of the existing railways, and with the construction of new railways and highways, oil and gas pipelines. According to the Transport Strategy of the Russian Federation for the period up to 2030, and the Strategy for the Development of Railway Transport in the Russian Federation until 2030, the following railways are planned to be built in the region: the eastern section of the North Siberian Railway Ust'-Ilimsk – Lesosibirsk–Nizhnevartovsk, the lines of Ust'-Kut – Kirensk – Nepa – Lena, New Uoyan – Mogzon, Naryn – Lugokan. The construction of the Priargunsk-Berezovskoye railway was started. Also one of the significant projects is the project of modernization and expansion of the capacity of the Trans-Siberian and Baikal-Amur Mainlines. At the Baikal-Amur Main Line, the construction of the second railway line has already begun on some sections. The construction of the Vilyui highway in the direction of Us't-Kut – Nepa–Mirnyi will give the Republic of Sakha (Yakutia) another year-round access to the main road network of Russia. A draft of the pipeline "Power of Siberia" from the Kovykta gas condensate field was prepared, as well as gas pipelines from this field to Sayansk and Angarsk and oil pipelines from the northern oil fields (Verkhnechonskoe and others) to the main oil pipeline ESPO.

## 4. Conclusion

The compiled original of the transport map of the Baikal region was included in the "Ecological Atlas of the Baikal Region" in the form of an electronic resource on the geoportal "Atlas"[10]. The transport map is presented in both raster and vector form. The vector map layers on the geoportal contain the necessary database of transport indicators. Secondary transport-related sector maps of the region can be compiled using the basic digital maps.

The mapping of the transport network of the Baikal region has shown that the southern most developed and inhabited regions of the Baikal region have a developed transport network. In the northern regions inland waterways and air transport are of great importance. There are a number of intra-regional problems: weak development of the northern territories of the Baikal region; low density

IOP Conf. Series: Earth and Environmental Science **190** (2018) 012003 doi:10.1088/1755-1315/190/1/012003

of highways in comparison with the all-Russian; reduction of navigable sections of rivers; closing of airports or their transfer to temporarily operated landing sites. Solving these problems will improve the efficiency of the transport infrastructure.

**Acknowledgements**
The work was done within the program of research works of the V.B. Sochava Institute of Geography, SB RAS with partial support of the Russian Foundation for Basic Research (projects No. 0347-2016-0004 and No. 17-29-0504\17).

**References**

[1] Dashpilov T B 2014 Mapping of the transport systems in the region: semantic classification of signs *Vestnik Irkut. State Technical University* **12** 153–60
[2] Batuev D A 2012 Creation of a complex cartographic model of nature management in the region *Izvestiya Irkutsk State University Series of Earth Sciences* **5**(2) 33 – 55
Volodchenko A 2009 *Mapping Semiotics* (Drezden: e-Lexicon) p 61
Lyutyi A A 1988 *Map Language: Entity, System, Functions* (Moscow: IG AS USSR) p 292
Prokhorova E A 2010 *Socio-Economic Maps: a Tutorial* (Moscow: KDU) p 424
[3] Batuev A R, Beshentsev A N, Bogdanov V N, Dorzhgotov D, Korytny L M and Plyusnin V M 2015 Ecological atlas of the Lake Baikal basin: cartographic innovation *Geography and Natural Resources* **1** 5–16
Koshkarev A V, Antipov A N, Batuyev A R, Yermoshin V V and Karakin V P 2008 Geoportals as part of spatial data infrastructures: Russian Academy-supported resources and geoservices *Geography and Natural Resources* **1** 21–31
Vladimirov I N, Korytny L M, Plyusnin V M and Sorokovoi A A 2016 Institute of Geography SB RAS into the Baikal natural territory *Geography and Natural Resources* **5** 6–14 DOI: 10.21782/GiPR0206-1619-2016-5(6-14).
[4] *State Report "On the State of Lake Baikal and Measures to Protect it in 2015* 2016 (Moscow: Ministry of Natural Resources of Russia) 232–8
[5] *Ecological Atlas of the Baikal Region* ed Plyusnin V M et al (Irkutsk: V B Sochava IG SB RAS, V M Matrosov ISDCT SB RAS) Income accessed online on 15$^{th}$ June 2018 http://atlas.isc.irk.ru

# Development and marketing of ecotourism in Baikal region

**D V Dayneko[1,2], V V Dayneko[2]**

[1]Irkutsk Science Center, Siberian Branch of Russian Academy of Science
[2]Institute of Architecture, Construction, and Design, Irkutsk National Research Technical University, Irkutsk, Russia

E-mail: [1]dayneko@oresp.irk.ru, [2]ddayneko@gmail.com

**Abstract.** The necessity of development and marketing of ecological tourism in Baikal region is substantiated today by requirements for sustainability of entrenched economic activities in regions of Russia, as well as, by problems of environment preservation. Therefore, the modern ecological tourism concept and its principles suggested for Baikal region are described. The ecological tourism with active involvement of local residents in particular should become an important element for development of the Baikal region, promoting both economic development and preservation of natural and cultural heritage in the region. The work presents the model and approaches to ecological tourism, as well as marketing concept for ecological tourism products and services, the implementation of which is required for sustainable development of the given entrenched kind of activity in Baikal region territory. The issues of marketing planning and marketing strategy in the sphere of ecological tourism are discussed. The importance of nature preservation and of ecological tourism development for Baikal region is presented, as the most prospective and profitable way of tourism and entrenched for local people activity, which will result in new jobs and sustainable incomes to support local nature and communities.

## 1. Introduction

Today, when Russia undergoes EC and USA sanctions, the support of innovative activities is required in various industries of economy in Baikal region. After a profound study of the foreign and domestic tourism experience [1], we concluded that it is the ecological tourism, which is to be an important component to facilitate economic development of Baikal region and to preserve its natural and cultural values.

Ecological tourism is a relatively new concept in Russia and in Baikal region in particular. The distinguishing feature of ecological tourism is that the common tourism in general implies just travelling a distance from home, while ecological tourism is defined specifically as responsible travel to natural areas that conserves the environment and improves the welfare of local people. Other definitions of the ecological tourism presented in modern dictionaries, including American Heritage Dictionary of the English Language, Collins English Dictionary, Random House Kernerman Webster's College Dictionary, are the following:

- "tourism involving travel to areas of natural or ecological interest, typically under the guidance of a naturalist, for the purpose of observing wildlife and learning about the environment";

- "tourism that is designed to contribute to the protection of the environment or at least minimize damage to it, often involving travel to areas of natural interest in developing countries or participation in environmental projects";
- tourism to places having unspoiled natural resources;
- tourism to exotic or threatened ecosystems to observe wildlife or to help preserve nature.

## 2. Development of Ecological Tourism in Baikal Region

What we know about the Ecological tourism in Baikal region:

- Tourism in Baikal region is on the rise but there is a lack of ecological tours despite the many ecological sights and nature attractions;
- The entire Baikal region is one of the Russia's biggest regions. About 1.5 million overnight stays per year only, of which over 70% are during the summer season. Less than 10% of the tourists visit the region in the winter. The number of foreign tourists fluctuates from 40 to 50 thousand visitors per year.
- Very few residents of the Baikal region live on tourism. Despite the fact that many local residents are aware of the prospective incomes to be brought by tourists only few are in the tourism industry and even fewer are aware of the ecological tourism potential.
- The local products, including souvenirs, are still of low interest for tourists. Ecological and regional products can be marketed better by ecological tourism.

The understanding of the tourism and of ecological tourism in particular in Baikal region by local residents is usually strongly correlated with backpacking, recreation activities and leisure time. Recreation is considered as a sort of amusement or an activity a person takes part in for pleasure or relaxation rather than as work. Hence local tourist areas are mostly areas for recreational stay, sports, and games, often with children. Considering the climatic conditions of the region, with average annual temperatures below zero degrees Celsius, the outdoor recreation is more closely related to the mass tourism and usually abroad to the ocean. For example, sunbathing, ashore and in-water attractions visited are usually both recreational activities and tourist activities, with no regard to the duration and location of the trip. Whereas, a traveler who is going out or backpacking, for example, to the nearby forests, for a day or for a weekend, or a longer trip, if visiting further destinations, can be considered as a tourist participating in an ecological tour.

Every year ecological tours only to the Lake Baikal itself attract many thousands tourist from Russia and abroad. Lake Baikal is quite interesting for the visitors not only during the summer but during other seasons too. Though, most of the visits are definitely during the summer months. Some like to stay in comfortable cottages and bungalows, others prefer to spent nights in tents or aboard the boats. However, there is still no such a developed ecological tourism infrastructure in Baikal region as we can see in other eco-attractive destinations.

Today we can name at least seven defining points to describe ecotourism in international practice, which are to be adopted and implemented in Russia and in Baikal region in particular too. The profound studies of ecotourism including University programs use this as the following defining points for the working definition [2]:

a. *It assumes travels to sites of real nature:* These places are usually quite remote areas, whether inhabited or uninhabited, and are officially under the environmental protection at the international, national, regional, municipal or private level.

b. *Minimum impact:* Ecological tourism always strives to minimize the adverse anthropological impact, which can be achieved via, for example, well established and organized tours and use of recycled materials or local construction materials for hotels and other buildings required to provide the tourists infrastructure and environmentally and culturally sensitive architectural design, as well as, alternative energy sources, recycling and/or non-harmful disposal of wastes. Minimization of impact also requires that the behavior of tourists and number of visits are regulated to ensure no damage to the local ecosystem.

c. *It educates environmental awareness:* Ecological tourism always means education, for both tourists and residents of the local communities. It is very important to provide travelers with the relevant information about the country, environment and people, as well as about the formal and informal institutions involved well before the departure to a destination. According to Ecotourism Societies guidelines it is required "to learn about the places and peoples visited" and "to minimize their negative impacts while visiting sensitive environments and cultures". Therefore, good eco-touristic programs should be provided by well-trained, multilingual guides with the knowledge of natural and cultural history, environmental interpretation, ethical principles and effective ways of communication. All ecotourism projects are supposed to educate tourists and local community members, including adults and kids in the host country.

d. *It provides direct financial benefits for nature protection and conservation:* Ecotourism not only promotes environmental awareness, but is directed to raise some funds required for the environmental protection, research and education. This can be achieved via a variety of financial mechanisms, including park entrance fees, eco-tour and transportation companies' contributions, special offers in hotels, eco-taxes and voluntary contributions.

e. *It should provide financial benefits and job advantages to local people:* It is a common believe that National Parks and other conservation areas will only survive if there are "happy people" around the area. It is getting obvious that local communities must be involved with and receive income and other tangible benefits related to state of life improvement and local infrastructure as a result of the conservation areas promotion and construction of it's tourist facilities. Actually, people from the local communities surrounding an ecotourism attraction should run campsites, lodges, guide services, restaurants and other concessions. Such a partnership is suggested for all sorts of investors planning their activities in parks or other tourist destination. More importantly, since Ecotourism is often viewed as a tool for rural development, it should also allocate most of economic and political control to the local communities. This is the most common mistake of the operators and the foreign partners to neglect the role of the local officials and business people, including those living in villages and nearby settlements.

f. *It respects local culture:* Local culture is another concern of the ecotourism. Ecological tourism is not only "greener" but also less culturally intrusive and exploitative than any other forms of tourism. Ecotourism always strives to be culturally respectful and have a minimal effect on both the natural environment and the human population of a host country. Sometimes it is not that easy, since ecotourism often involves travel to distant and secluded places where small and isolate communities have had little experience interacting with foreign visitors. Sometimes, even ecotourism involves an unequal relationship of the visitors and the hosts and a commoditization of the relationship through exchange of money. Actually, responsible eco-tourists are learning beforehand about the local customs, dress codes and other social norms and do not intrude into the community unless either invited or as part of a well organized tour.

g. *It should support human rights and local democracy:* Traditionally, it is tourism that is loudly hailed as a tool for building international understanding and world peace promotion. However, it is not always the fact or it does not necessarily happen immediately upon a visit. Most often tourism just bolsters the economies of various countries including even those that tend to be or are undemocratic and repressive. While regular tourists pay low attention to the political system of a country they visit, unless a civil unrest brakes out in there and which is harming the tourists too. In fact, rural population living around national parks and close to other ecotourism attractions are in contradictions with the national and regional laws and regulations, and multinational corporations for control of the assets and possible benefits.

## 3. Marketing in the Sphere of Ecological Tourism

Marketing in the sphere of ecological tourism is the same activity of planning, development, promotion, and selling of touristic products, including goods and services, stimulating their demand and price setting, as for mass tourism. Marketing in ecological tourism also can be defined as a complex of measures to increase the market share of eco-touristic products and services. Touristic marketing is a system of management and organization of touristic organization for development and provision of new ecological and eco-effective kinds of touristic services and excursions for their further sale to make a better profit resulting from eco-oriented attitude to services and products provided and considering the processes happening in tourism markets. The major task of Ecological tourism marketing is to stimulate eco-oriented demand for recreation activities, to set the prices right, to provide market segmentation and develop a favorable assortment of eco-touristic services and excursions and attractive advertisement.

The specifics of the marketing in the sphere of ecological tourism are that the major product is actually intangible, the source is inseparable from the service offered itself, the great correlation of the service quality with the local culture and level of training for the personnel, and that none of the eco-product can be stored or kept in a storage. The touristic services offered in the ecotourism market are intangible, you cannot taste or try them, really smell or hear before an actual purchase of such a product.

The marketing of any eco-tour should start with marketing analysis of the environment. The overall and detailed perception of the markets and professional feedback is the responsibility of marketing specialists, whose job is not only to stimulate the demand and increase sales but to develop a firm considering employees' potential, financing, management, price formation, customers' feedback. It implies the creation of channels of distributions, marketing programs development and implementation and budgeting.

Marketing is a good tool to stimulate the demand for ecological tourism, the knowledge of which is a responsibility for everyone involved in tourism industry. The major marketing task in ecological tourism is creation of an ecological tour/product/service to motivate and satisfy clients eco-oriented demands. The major task of a professional marketing specialist in the sphere of ecological tourism is to attract and keep aware customers about the advantages of ecological tours and services, to satisfy the existing and create new demand to increase the profitability. Therefore, marketing in the sphere of ecological tourism is to solve such tasks, as, for example, determination and/or formation of needs and demands for ecological services and touristic products in the market, development and offer of high quality products and services at a market price, stimulation of demand and distribution of supply and its availability. Hence, we can name major tasks of the marketing management in the sphere of ecological tourism. These are demand regulation, analysis, planning, ecological tourism products sales and control, and touristic markets monitoring.

The major feature of the concept of ecological tourism marketing is the nature protection compound. Marketing in the sphere of ecological tourism always assumes a high level of responsibility of marketing specialists for the environment and nature. The nature oriented concept of marketing implies environmental friendliness and careful management of nature resources. Touristic companies and ecological tour operators have to consider their potential negative influence on the environment and strive, if not fully eliminate it, than, rather to minimize it via related environmental activities. This is what is called the environmental friendliness. The care for the clients health in places of their stay is the task of managers in eco-oriented touristic companies.

The modern marketing strategy always requires the inclusion of the ecological component. The issue of social responsibility and of ethical behavior are especially important in tourism business today. The social concept of marketing means employees and specialists awareness about the lawful and acceptable business practices and have to know the business dealing principles, which are based on ethics, corporate culture and interests of clients.

The customer oriented approach is the key to success for any touristic company, since, touristic business consists not only in the development of a final touristic product, but, rather, in satisfaction of customers' demands. Herewith, customer demands can be changed and corrected, and specifically due to professional work of marketing specialists. Marketing can manage the demand and promote ecological tourism products at the target markets, can help to evaluate dynamics of sales in every strategic direction and determine prospective incomes. Once the ecological tour aims of a company and possible financing are determined, the marketing can solve the tasks, which have been set, effectively to maximize the profitability.

Marketing planning in ecological tourism is required to forecast levels of ecological tours sales and incomes. Therefore, current and planned sales and their time are analyzed and related strategies are suggested. First of all, marketing specialists have to consider the existing markets shares and further market penetration with ecological tour products and only than the possibilities of new markets development. The following steps are development of new products for existing markets, and offer of new products in new target markets, i.e. diversification.

Once a marketing strategy is developed, the related programs for its implementation are defined. Therefore, the strategy includes a frame, resources and personnel to achieve the strategy aims. The qualified and professional employees are involved, training and educational programs for the personnel are introduced, products and services quality control and improvement programs are implemented, the services and touristic products are promoted, informing about the positive and innovative features.

It is worth to mention the particular importance of the personnel awareness and involvement in touristic companies into the development and implementation of corporate strategies. The firm employees in touristic business are the most important element of successful product and service. Formation of terms and conditions, which encourage good services for clients, is often called *internal marketing* [3]. Today, the most successful tourism enterprises have to offer in the markets not only ecological tourism products, but to develop corporate ethics and mentality to satisfy the customers' specific ecological demands. The employees of travel agencies should understand prestige value of ecological tours and advantages of their promotion. The key aspect of the successful internal marketing is employees' motivation. Therefore, along with various financial advantages, the positive influence plays credibility and prospects of their workplace. Other reinforcements and motivators are, for example, a flexible work schedule, child and kids care programs, and other non-financial benefits. The internal marketing will help tourist companies attract and keep good employees too.

We can specify at least three major current tasks of ecological tourism marketing in Baikal region: 1) maintaining and developing existing markets of ecological tourism; 2) search and development of new markets (new clients and destinations); 3) reduction of seasonality influence on demand. In order for a touristic company to achieve these tasks there is a need to perform the following at least annually:

- to correct the marketing plan;
- to assign group of professionals to develop a program of the company's specific, collaborative, research, and other related activities to improve the marketing plan and attract more customers;
- to cooperate with other touristic companies to allocate, provide, and serve customers in the market of ecological tourism;
- to test new ecological tourism products and services;
- to develop mutual measures to promote ecological tourism products in the target markets;
- to develop target segments of the market with seasonal character of services;
- to offer ecological tour packages to multiple destinations according to customers preferences and inquiries.

The major task of the marketing of ecological tourism is to understand the specific ecological interest of customers and offer them the best suiting ecological destinations and services. As a matter of fact, the most successful touristic companies offering ecological tourism products and services, first of all, study from the customers and their demands. It is called "getting closer to the customer". Later they have just to monitor their client and their demands to improve their capabilities and services offered.

Hence, the basic functions of ecological tourism marketing suggested for Baikal region should be the following:

- creation of ecological tourism products and services to be offered to clients in Baikal region;

- promotion of the ecological touristic products and services in the market, including advertisement, and consequent sale;
- arrangement of direct and in person sales;
- setting prices and achieving an allowable level of sales and incomes.

In order to comply with international norms and standards, marketing specialist of ecological tours in Baikal region should follow three main tourism marketing functions proposed by International Tourism Organization:

- Customer relations and feedback establishment. The major task is to convince clients that ecological destinations offered fully satisfy their expectations and demands;
- Development, including design of innovations;
- Control, which is analysis of the results of activities to promote ecological tourism services in the market and verification of results real compatibility to existing touristic capabilities.

Therefore, the especial importance in the marketing of ecological tourism agency plays the relations with the customers. The long lasting relations with clients can be more lucrative and cost less than the marketing expenses, incurred to attract the interest of prospective consumers of ecological tourism services.

If the long lasting relations with ecological tourist are established, it provides a synergy effect, expressed even in adherence and devotion of clients to the touristic agency and their reluctance to change the tour company providing ecological products and services when repeating purchases. It is an interesting fact, that the Pareto effect is clearly observed in the sphere of ecological tourism: 20% of consumers make 80% of a company's income.

Tourism and, especially, ecological tourism is totally different from other consumer industries. Hence, we can indicate its specific characters in Baikal region to be considered by marketing specialists:

- While planning ecological touristic activities, the central issue to be considered are the preferences and demands of the final consumers;
- Non-primary essence of ecological tourism services and lack of knowledge about Baikal region. The ecological tourism is not an essential product, though, the importance of the tourism and of ecological tourism, in particular, as the means of recreation and healthy life style, is on the rise;
- Marketing plays a crucial role in ecological tourism industry. Since, there is no a reference samples of touristic tourism products and the ecological tours sellers have to provide the proof and sound arguments in advantage of their products and services. It is possible only with a well planned marketing program that uses tailored for the target markets instruments of advertising and promotion;
- There is a need to constantly monitor the quality of ecological tourism products and services in Baikal region due to their unstable provision and seasonality;
- Ecological tourism services are unique. The repetitions of all its aspects and provisions are very uncommon. The schedules, routes, terms and conditions of services and prices usually always vary.

Therefore, marketing of ecological tourism for a touristic agency operating in Baikal region should become a system of continuous harmonization of services and products provided with services and products demanded, which an ecology oriented company can offer with a profit and more effectively than the competitors in the market. Today, only a firm with creative and ecology oriented marketing concept can succeed in competitive environment, constantly searching new ways to promote their services in changing markets, actively influencing and interacting with the consumers.

## 4. Ecotourism Model to be implemented in Baikal Region

Since ecological tourism is always described as an industry developed specifically to promote the sustainable use of resources, the question is how this actually can be achieved in Baikal region. What should be done if the local people often disregard the ecological importance of the flora and fauna in favor of the mass tourists' visits and advantages, who often ignore the harm and damage possible to the environment, which is done to maximize the profits per visit to a site. The ecological approach to the tourism suggested in Baikal region needs to meet a certain set of conditions, and it is required to comply with rules and an overall ecotourism concept. If the conditions and rules of the ecological tourism are followed, then the sustainability of the tourism and use of the resources involved is achieved.

Actually, there are two traditional approaches to ecotourism in developed world. The first view puts a premium on the conservation of local wildlife, culture and other resources and uses ecotourism as a method to achieve that preservation. Another approach centers on eco-efficient tourism, which is based on the use of resources and depends on their continued existence in order to be marketable and therefore seeks ways of preserving the attractive or pristine state of the destination. Therefore, the proper management of ecotourism programs requires a somewhat different framework from the mass tourism. Hence, participation of scientists and practice practitioners, government officials, and local community is always important for ecotourism. So, in our model, there are at least five sectors involved in ecological tourism development in the order of their importance assumed for the Baikal region [4]:

1. *Local citizens:* These are the local people with their everyday use of the natural, cultural and historic resources who should play the crucial role in local ecological tourism development in Baikal region. Since, they do not only monitor the use of these resources, but are also the hosts to visitors.

2. *Scientists:* The scientists' role is to look at the local resources and their evaluation from economic, social, and natural points of view. The importance of the scientists' involvement is their advisory position, based on their thorough scientific, cultural and historical knowledge, and their ability to disseminate the relevant information. They can also work on planning and elaborating of various tour programs with minimum harm to the nature in mind.

3. *Travel industry:* The travel industry should develop products based on the scientists researches and advices. The travel industry helps stimulate local economies through running its businesses, such as, for example, guided tours.

4. *Tourists:* The tourists, thanks to the knowledge imparted to them by local guides, become more informed on the environment and, of course, they bring both direct and indirect profits to the local businesses and community. They are essential in helping local people realize that sustaining and conserving their natural resources is crucial for the stabilization of their economy

5. *Government:* Government bears the responsibility of supporting the establishment of ecotourism: for example, contributing to the drafting of guidelines, providing education for staff, and supporting non-profit organizations.

## 5. Results and Discussion

Baikal region is the most prospective centre of ecological tourism in Russia. These days we can offer a number of purely ecological tours. The visitors wishing to see the cleanest and most beautiful lake can easily reach it and stay at the modern ecological facilities few days or weeks and enjoy the comfort of local small resorts and campings.

Ecotourism to natural attractions, which is usually accompanied with recreational and educational activities, actually, is a promising means of advancing social, economic, and environmental objectives in Baikal region. It offers Baikal region new opportunities for small-enterprise investment and employment and increases of the national stake in protecting biological resources. However, making ecotourism a positive economic and environmental tool requires policies that foster responsible nature tourism development, broad-based and active local participation in its benefits, and conservation of Baikal region heritage. It also helps educate travelers and locals and make them environmentally responsible; it even supply funds if not for the conservation but for the minimization of the negative anthropogenic influence; and can directly benefit the economic development and social empowerment of local communities; and it also fosters respect for different cultures and for human rights.

## 6. Conclusion

Wildlife and its habitats in Baikal region are becoming increasingly popular attractions for international tourism. Baikal region can provide many of the richest, in terms of unspoiled nature, places with bio diverse inhabitants conserved [5]. The numbers of eco-tourists to our region are growing, despite the crisis and sanctions. The new wave of nature tourists is expected to bring with them the finances to spend. This money will support the local travel industry, which already exists and will create new jobs and incomes for households and local communities in and around places of attractions, such as lake Baikal and adjoining forests and other protected areas. There is a need to establish ecotourism enterprises, special tour agencies and guide services, lodges and private reserves, as well as, such activities *as arts and crafts industries*, already present in Baikal region. There is also a need to develop transportation and food services at the local levels to improve ecological tourism. Local administrations can use the incomes generated in this way for more efficient operating and protection of natural attractions. There is a need to implement modern effective and efficient ecotourism practices in Baikal region.

## References

[1] Zorin I and Kvartalnov V 2000 *Enciclopedia Turizma: Spravochnik* (Moscow: Financy i statistica) p 368

Zorin I and Kaverina T 2005 *Menedzhment Turizma: Turizm kak vid Deyatelnosti* (Moscow: Finansy i statistika) p 288

Kvartalnov V, Zorin I and Irisova T 2001 *Menedzhment Turizma. Turizm i Otraslevie Systemy* (Moscow: Finansy i statistika) p 272

Oppermann M 1993 Tourist space in developing countries *Annals of Tourism Research* **20**(3)

Harrison D, King V and Eades J 2018 *Tourism in East and Southeast Asia* (London: Routledge) 1404

Price M and Martin F 1995 Mountain Research in Europe: An Overview of MAB Reserch from the Pyrenees to Siberia *Man and the Biosphere Series* vol 00214 UNESCO (New York: Parthenon Publishing Group) p 230

[2] Honey M 2008 *Ecotourism and Sustainable Development* (Washington: Island Press) p 549

[3] Prourzin L 2003 *Turizm kak Economichesky Prioritet* (Moscow: Novy Vek) p 120

[4] Dayneko D 2017 Podhody k ecologicheskomy turizmu v Baikalskom regione *Geosistemy i ih Komponenty v Severo-Vostochnoi Azii: Evolitcia i Dinamica Prirodnyh, Prirodno-Resursnyh i Socialno-Economicheskih Otnoshenyi* (Vladivostok: Dalnauka) pp. 430-434

[5] Chernyh A, Kuznetcova E and Kozlov V 2010 *Turisticheskie Resursy Baikalskogo Regiona. Kulturno-Istoricheskie Resursy: Uchebnoe Posobie* (Irkutsk: Isdatelstvo INRTU) p 248

IOP Conf. Series: Earth and Environmental Science **190** (2018) 012005 doi:10.1088/1755-1315/190/1/012005

# Formation and development factors of Altai ethno-cultural landscapes

**D A Dirin**[1], **C Madry**[2]

[1]Department of economic geography and cartography, Altai State University, Barnaul, Russia
[2]Institute of Socio-Economic Geography and Spatial Management, Adam Mickiewicz University, Poznan, Poland

E-mail: denis_dirin@mail.ru, cezary_madry@o2.pl

**Abstract.** In the article major factors of culture genesis in the Altai Mountains are analyzed. Natural and social laws of formation and evolution of the Altai's ethno-cultural landscapes are revealed. Ten basic factors in the formation and development of the ethno-cultural landscapes of the Altai have been identified through the joint application of natural-scientific and humanitarian research methods. Such are the High landscape diversity, the Naturally caused low population of Altai, the Barrier and contact role of the Altai Mountains, the External expansion, the Natural isolation of the Altai geo-cultural space, the Natural environment dynamics in Altai and adjacent contiguous territories, the State regulation of economic activity, the Change in the geoeconomical and geopolitical situation of the region, the Change in market conditions, the Change in the traditional worldview of the local population. According to the results of the work, it is determined that the culture genesis of Altai is largely determined by natural factors. Now in the Altai Mountains there is a great variety of ethno-cultural landscapes, the laws of development of which are related both to the features of the environment and to social processes. The isolation and inaccessibility of the Altai Mountains territory, its remoteness from civilizational centers, significant internal differentiation determine the low intensity of processes cultural diffusion, and significant sustainability and the preservation of traditional ethno-cultural landscapes. Globalization processes lead to simplification and unification structure of the Altai's geo-cultural space and loss of cultural diversity.

## 1. Introduction

Most of the planet mountain systems attract the researchers' attention not only by the diversity and uniqueness of their natural landscapes, but also by the social systems specificity. In the era of globalization in all spheres of human life, the issues of studying ethnocultural diversity and the specifics of the space development by different cultures and ethnoses are becoming increasingly important.

These issues are becoming more relevant as the erosion of local cultural traditions, the unification of perception and behavior stereotypes among different peoples. In addition to the cultural heritage loss, these processes also lead to the loss of ethnic adaptation mechanisms to the natural environment developed by centuries and are consequently the causes of environmental and social-economic crises.

IOP Conf. Series: Earth and Environmental Science **190** (2018) 012005 doi:10.1088/1755-1315/190/1/012005

In this aspect, special attention is paid to the regions, where landscapes are noted along with polyethnicity. One of these regions is the Republic of Altai (administrative region of Russia). This article is devoted to the analysis of the formation and development factors of the ethno-cultural landscapes of Altai Mountains (within the Altai Republic boundaries), as well as to identify trends and causes of their modern transformation.

## 2. Models and Methods

The studying of the human communities' interaction processes (particulaly ethnoses) with the geographical environment of their habitat has always been one of the most important problems of Geography. At the same time, the issue of developing ethnoses adaptation to the transforming environment is of a particular interest. To study this issue and some others close to it, many concepts and entire science directions arose.

### *2.1. The concept of ethno-cultural landscape*

At the beginning of the 20th century, at the intersection of Geography, Cultural studies and Ethnography, the concept of a cultural landscape began to develop almost simultaneously in Russia [1], Germany [2] and the USA [3]. This concept has undergone considerable evolution during its existence, but its basis remained unshakable. The objects of research in it are "cultural landscapes" as natural-social systems where all the components (the human community, its economy, the natural environment, the elements of material and spiritual culture that are simultaneously a form of adaptation to the environment and its result, etc.) are inextricably linked and interdependent and the cultural landscapes themselves represent the outcome of the societies and nature interaction. Thus, the main properties of cultural landscapes are their systematic and spatial. In addition to purely utilitarian material connections of the community with the natural environment, spiritual interrelations in cultural landscapes are taken into account.

The understanding of the ethno-historical essence of the cultural landscape appeared very quickly, in the framework of this concept [4], which brought Cultural geography closer to Ethnography and led to a new "ethno-cultural landscape" concept. The term "ethno-cultural landscape" (ECL) is referred in this article to the natural and cultural territorial complex formed as a result of the evolutionary interaction of nature and the local ethnic people community, utilitarian, semantic and symbolically mastering and transforming the geographical space according to their spiritual and material needs. An important feature of the ethno-cultural landscape functioning is the culture interpenetration in space and space into culture.

Most studies of ECL are devoted to their classification, delimitation of ECL boundaries, characterization of their current state and trends in their development. At the same time, there are not many publications that reveal the factors and patterns of the formation of such spatial systems.

This article is devoted to the determination of the factors of the culture-genesis in the Altai Mountains, as well as the specific features of the formation and development of ECL of this territory.

### *2.2. Basic research methods and dates*

To achieve this goal, the following methods were used:

*2.2.1. Statistical.* The analysis of regional statistical information reflecting the peculiarities of spatial organization of the population and economy, as well as their dynamics was carried out. For this purpose, official statistical data of the regional statistical agency (http://akstat.gks.ru/) were used.

*2.2.2. The landscape-cartographic method.* The study of landscape structure of the territory, settlement systems and nature management was carried out with the definition of the degree of their adaptability and efficiency of functioning. For the analysis, a landscape map was used for the entire study area [5]. At the same time, tools for processing spatial geodata in the ArcGIS 10.1 and MapInfo 12.0 environment were used.

*2.2.3. Unincorporated surveillance.* During the expeditionary studies in the period from 2012 to 2018, modern methods of adapting different ethno-territorial groups of the Altai's population to different types of landscapes were observed; study of the regime of daily and periodic household activity of people related to the functioning of ECL.

*2.2.4. Questioning.* Special questionnaires were developed, which were completed by representatives of local communities, made it possible to identify the main features of the spatial and functional organization of modern ECL, as well as identify key trends in their dynamics. A total of 1,138 people took part in the survey, living in 51 settlements of the Altai Republic. Among the respondents – representatives of 12 ethnic groups.

*2.2.5. Narrative interview.* A sociological survey of representatives of different ethnic groups of the Altai's population was conducted (43 interviews in the period from 2012 onwards), which made it possible to identify significant factors in the spatial organization and development of the Altai's ECL. The interviews allowed to detail the results of the questionnaire and to reveal some factors of ECL development, which were not obvious by the results of the questionnaire.

## 3. Results and Discussion

The formation and evolution features of the Altai's ethno-cultural landscapes are determined by its geographical conditions specificity and socio-historical processes direction. The analysis made it possible to identify the main factors of the Altai's culture, which determined the current state of the ECL s of the given territory, as well as the trends of their further development.

### *3.1. The High landscape diversity*

The Altai Mountains are the most important climatic and watershed of Eurasia, situated at the junction of several physiographic countries and climatic sectors. This determines the natural conditions diversity within one mountainous country, and, consequently, the need for a various ways of economic adaptation. Thus, in the Altai, with the combination of different types of semi-desert, steppe, forest, meadow, tundra and glacial-nival landscapes, the corresponding traditional cultural and economic types are represented: hunting and fishing, semi-nomadic, pasture and stables, farming, etc. In Altai, depending on the height and exposure of the slopes, the territories are zoned for winter, spring-autumn and summer pastures, hunting and agricultural lands. However, with a certain economic specialization, all the Altai's ethno-cultural landscapes are characterized by the diversified nature management.

### *3.2. The Naturally caused low population of Altai*

The territory of Altai is characterized by low population density, which is historically associated with the relatively low natural productivity of mountain landscapes, their vulnerability to anthropogenic impact, and also inaccessibility. At present, just over 218 thousand people live on the territory of the Altai Republic (92.9 thousand km$^2$) (according to: Approved estimation of the population of the Republic of Altai as of 01.01.2018 Territorial body of state statistics for the Altai Republic: official site Access mode: http://akstat.gks.ru/wps/wcm/connect/rosstat_ts/akstat/ru/statistics/altayRepubStat/population/). At the same time, the population is very differentiated within the territory. A complex relief for transport and communal construction determined the settlement of only a small part of the territory (large river valleys and intermountain basins). For a long time, the main limiting factor in the settlement of Altai was the high snow cover (the cause also in the height of the terrain and mountainous terrain), which limits the livestock breeding possibilities and, correspondingly, the population growth with the predominant livestock type of the farm.

### *3.3. The Barrier and contact role of the Altai Mountains*

The mountain structures always represent natural obstacles to the cultural interaction of the areas they share, but they themselves often become zones of interpenetration of different cultures. Altai historically was the territory of interethnic contacts. It was the natural border of the distribution areas of the nomadic Turkic-Mongolian peoples of Inner Asia and the Finno-Ugric peoples of Siberia, who were forest hunters and fishers. Ethnic mixing and interpenetration of cultures took place in Altai. The peoples introduced their own traditions when interacting with other peoples. Thus, Altai is located at the junction of the peripheral zones of several cultural worlds (civilizations): a) Turkic-Mongolian Buddhist-Islamic nomadic (Inner Asia), b) Slavic Christian agricultural (Russia), c) Turkic-Finno-Ugrian Christian-animistic forestry (Siberia). In its geocultural space, elements of all the above-mentioned civilizational types were intertwined.

### *3.4. The external expansion*

Throughout history Altai was repeatedly subordinated to various empires and other state formations. Of course, involving the territory in the sphere of someone's geopolitical influence can lead to significant changes in existing ethno-cultural landscapes and even to the emergence of new ones. This process is accompanied by the foreign people influx into the given territory, the exchange of knowledge between the aboriginal and the newcomers, the emergence of new types of wildlife management, previously untypical for the area, and often the targeted modification of traditional ethno-cultural landscapes. So the Altai accession into the Russian state (the Russian Empire – the USSR – the Russian Federation) led to profound changes in its geocultural space. Firstly, with the Russian population influx, the ethnic structure of the Altai population has dramatically changed (the share of Russians here is now 60%). Secondly, nature management types atypically earlier for the territory – plow farming, maral breeding, forestry and mining industry, etc. appeared there. Thirdly, the traditional way of life of the indigenous population has changed. Particularly significant transformations were associated with the actual forcible transfer of Altaians to a settled way of life and a change in the type of economic relations from the communal-clan to the collective-planned (during the Soviet period of history) and then to the market (in the post-Soviet period).

### *3.5. The Natural isolation of the Altai's geo-cultural space*

The geocultural space of the Mountainous Altai is differentiated by a system of ridges into individual areas within the river valleys and intermountain basins. This factor has several consequences: a) cultural diffusion in the mountains is slow, but its results are more stable (for Altai, the identity and safety of local cultures is characteristic); b) the relict traditional ECL are preserved within the Altai; c) isolation also determines the social and economic gap from the flat territories.

### *3.6. The Natural environment dynamics in Altai and adjacent contiguous territories*

The natural environment dynamics (for example, primarily climate change) in a pre-industrial society provokes mass migration. According to the studies conducted in Altai [6], the change of material cultures is synchronous with climatic cycles. For example, when the drought occurred in the steppes of the Inner Asia, Altai became an attractive place for nomads, and in connection with it, in the past a change of cultures took place (due to assimilation, mixing, or displacement of some peoples by others). Consequently, the waves of cultural genesis can be linked to the climate dynamics.

However, during the last several decades, global climate changes have been clearly manifested in Altai. At present, the main trend of changes in the Altai environment is the climate warming, which is reflected in the transformation of the territory landscape structure. There is an aridization of the landscapes of the South-Eastern, Eastern and partly Central Altai [7]. This leads to a decrease in the landscapes productivity and a decrease in their resistance to anthropogenic impact. At the same time, the thickness reducing of the snow cover determines the possibility of year-round pasture use. Accordingly, in a number of Altai regions, the hayfields are gradually transformed into pasturelands. The case of wild animals decreases in winter, which improves hunting conditions. Other consequences of climate warming are an increase in the upper and lower boundaries of the forest and intensive melting of glaciers. All this leads to territorial and structural shifts in the land use by the local population. The instability of natural geosystems functioning in the context of global climate change leads to an increase in the number of natural disasters – fires, floods, mudflows, avalanches, etc. [8], which affect not only the conditions of economic activity, but also the resettlement, migration situation etc.

*3.7. The State regulation of economic activity*

State management of the territory development is the most powerful factor in the transformation of ethnocultural landscapes. Government bodies regulate land relations. The state determines even the ideology of the space development and the population life.

The most significant transformations in the nature management of Altai, during the state administration of the territory, were associated with the abolition of private and communal and clan land ownership in the post-revolutionary period (after 1917) and the population transfer to a settled way of life, as well as the economy collectivization in the 1930s. After the collapse of the USSR, most of the collective farms disintegrated. Today, there is a return to traditional forms of management on the basis of private property and the formation of public (often family-clan) brigades in order to jointly conduct economic activities [9]. At the same time, state socio-economic projects are being implemented in the area under investigation, which modify landscapes and change the structure of nature management, introducing new branches of the economy. One of such innovative industries in the nature use of Altai is tourism [10].

*3.8. The Change in geoeconomical and geopolitical situation of the region*

The change in the geoeconomical and geopolitical situation of Altai is primarily due to the improvement of its transport accessibility, which allows it to sell its products in rather remote markets, and also to use its territory and resources to provide services (for example, tourist and recreational ones) to the population of other regions and countries. The geopolitical significance of Altai also changes as a result of Russia's international interaction with neighboring countries – China, Mongolia, Kazakhstan – x ose vectors determine its role as an important region of cross-border cooperation or as a frontier outpost on the borders of the country. These processes, in particular, contribute to the introduction of innovations that also transform the ethno-cultural landscapes of the Altai. Nowadays major projects for the transport network development in Altai are being developed. In particular the construction of a gas pipeline and road to China through the Ukok Plateau and the Kanas Pass are actively discussed.

*3.9. The Change in market conditions*

In market economy a change in demand for certain products often leads to a change in the nature management systems that are reoriented to a new more profitable specialization. Maral breeding, fur trade, goat breeding, beekeeping and other branches of economic activity of the Altai population greatly depend on market demand. The change in the structure of the Altai regional economy is strongly influenced by the rapid tourism development (especially in the most developed northern lowland part and along the main transport arteries). The greater part of the local population is involved in tourist service, and also produces goods that are popular among the tourists (agricultural goods, souvenirs) [9].

*3.10. The Change in the traditional worldview of the local population*

The change in the traditional outlook of the local population is caused by the processes of socio-ethnic integration and globalization along with the spread of the so-called "mass" technogenic and consumer culture, against the backdrop of degradation and sometimes the disappearance of the spiritual and ecological distinctive ethnos culture. In this case, the following consequences are seen: the loss of "roots" deforms the consciousness and self-consciousness of an individual, deprives him of the centuries-old cultural and historical experience of his ancestors. A person no longer accepts moral and ethical norms and principles (including nature use) of his ethnos, although he can still identify himself with it. An important consequence of the technogenic consumer culture spreading is a constantly rising psychologically conditioned level of population material needs. However, the growing needs of the population, leading the traditional nature management, leads to the most serious environmental problems. So, in this case with the preservation of management forms (for example, transhumance) its intensification (the number of livestock) is multiplied, that leads to a sharp increase in loads on landscapes and rapid degradation of natural complexes [11].

## 4. Conclusions

To sum it up we can draw a number of conclusions:

Modern cultural landscapes of Altai are embodied in space as a result of a constantly ongoing process of cultural genesis, caused by a multitude of interrelated natural and social nature factors, the combination of which is unique for each specific ethnic area.

Altai is characterized by a significant diversity of cultural landscapes, determined by the nature diversity and the boundary position of this mountainous country, as well as the stability of traditional culture forms associated with the natural isolation and inaccessibility of the territory, its remoteness from civilizational centers.

The formation and development specifics of Altai cultural landscapes is still largely determined by ethnic factors. It is ethnic identity that is the foundation that unites people of this territory into original cultural communities creating specific ethno-cultural landscapes.

The differences between the ethno-cultural landscapes of Altai are well expressed, in particular, in the existing systems of traditional nature management, adapted to local natural conditions.

The globalization processes lead to simplification and unification structure of the Altai's geo-cultural space and loss of cultural diversity.

## References

[1] Berg L I 1915 The subject and tasks of geography *Izvestiya IRGO* **51**(9) 463-75

[2] Shlüter O 1920 Die Erdkunde in ihrem Verhältnis zu den Natur – und Geisteswissenschaften *Geographische Anzeiger* B 21

[3] Sauer C 1925 The Morphology of Landscape *University of California Publications in Geography* **22** 19–53

[4] Lowenthal D and Prince H C 1965 English landscape tastes *Geographical Review* **55** 186–222
Haeyrynen M 1994 National landscapes and their making in Finland *Topos* **3** 6-15
Yamskov A N 2003 Ethno-ecological research of culture and the concept of cultural landscape *Cultural Landscape: Theoretical and Regional Studies* ed V N Kalutskov, T M Krasovskaya (Moscow: Moscow State University Press) pp 62-77
Kalutskov V N 2006 Ethno-cultural landscape science *Bulletin of Moscow University. Series 5: Geography* **2** 6-12

[5] Chernykh D V and Samoylova G S 2011 *Landscapes of Altai The Map M 1: 500000* (Novosibirsk: Novosibirsk Cartographic Factory)

[6] Bykov N I and Bykova V A 2007 *Altai-Sayan Mountainous Country and the History of its Development by Nomads* (Barnaul: Altai State University Press) pp 6-8

[7] Kharlamova N F 2013 *The Assessment and Forecast of Modern Climate Changes in the Altai Region* (Barnaul: Altai State University Press) p 156

[8] Gerasimchuk I V 2011 *Climate Change and its Impact on Ecosystems, Population and Economy of the Russian Part of the Altai-Sayan Ecoregion: an Evaluation Report* Section 6 ed A O Kokorin (Moscow: World Wildlife Fund – Russia) pp 105-48

[9] Dirin D and Madry C 2017 Transformations processes in traditional nature management systems in the Altai mountain *region Proc. Int. Scientific GeoConf. Surveying Geology and Mining Ecology Management, SGEM vol 17, Ecology, Economics, Education and Legislation* (Albena) pp 1033-42

[10] Lubenets L F and Manysheva T V 2006 Polyethnic communities of mountain territories under conditions of social and economic transformation (on the example of Ust-Koksinsky and

Kosh-Agach districts of the Altai Republic) *Geography and Nature Management of Siberia* **8** 124-9
[11] Dirin D A 2008 Ethnoecological Problems of the Mountainous Altai *Geography and Nature Management of Siberia* **10** 43-71

# Ecological barriers to tourism development (a case study of the Baikal region)

**O V Evstropyeva, T I Zabortseva, L M Korytny**

Sochava Institute of Geography, Siberian Branch, Russian Academy of Sciences, Irkutsk, Russia

E-mail: ledotop@irigs.irk.ru, zabti@irigs.irk.ru, kor@irigs.irk.ru

**Abstract.** We examine results of recreation-geographical investigations in the Baikal region, where in Irkutsk oblast and Republic of Buryatia different measures are implemented, aimed at tourist flow increase. An analysis is made of the leading natural-ecological and socio-economic factors for the territorial organization of tourist and recreational activity. This study is based on theoretical-methodological approaches of recreation geography using the landscape planning, zoning and mapping approaches. It is confirmed that the development of the tourist system limits its capacity, which is determined by the natural properties of the natural environment and the state of the basic infrastructure of tourism and the associated infrastructure complexes - transport, social, household, environmental protection. Three types of ecological barriers to tourism development are identified: the existing legislatively established standards regulating recreational pressure on them; the resilience of ecosystems to tourism-related anthropogenic impacts, and the present status of natural ecosystems under recreational pressure. Territorial features of the system are determined by the tourist-recreational zoning, it is proposed to allocate eleven tourist-recreational zones. Based on the results of the research, it is planned to develop the Rules for tourism and recreation adjustment in the Baikal region.

## 1. Introduction

Tourism as a multicomponent and hierarchical ecological-economic system and, at the same time, a large-scale socio-cultural phenomenon has an active influence on nature, the economy, politics and culture. The development of the world's tourism system reflects the exponential growth of the tourist flows, the annual volume of which barely reached 100 thousand tourists in the mid-20$^{th}$ century, and in 2012 it exceeded 1 billion for the first time. Furthermore, the existing entities of tourist activity, with their continued efforts to increase the tourist flows, must also preserve natural ecosystems, which is not always possible.

An illustrative example at the regional level is provided by the Baikal region with a full-fledged Baikal-centered model of tourism. On the shores of this World Natural Heritage site, over the course of many years two federal subjects of Russia (Irkutsk oblast and Republic of Buryatia) have implementing measures aimed to increase tourist flows amounting to almost 2.5 million tourists per year to date. The web of legislative-legal restrictions introduced within the Baikal Natural Territory (BNT) with a special modality of nature management, coupled with the explosive growth of the tourist flow and recreational pressures on Baikal's shores as well as the accumulated problems with land use and infrastructure deficits have led to an exacerbation of ecological-economic contradictions.

IOP Conf. Series: Earth and Environmental Science **190** (2018) 012006 doi:10.1088/1755-1315/190/1/012006

## 2. Models and Methods

The investigations are based on theoretical-methodological approaches of recreation geography including the scientific views of territorial recreational systems, recreational properties of the natural environment, tourist-recreational functions of territories, and a combination of ecological, economic and social priorities of their development [1]. A number of approaches of landscape planning (ecologically oriented land use planning) were used, which open up possibilities for decision-making concerning the development, improvement or non-use of particular areas of Baikal's shores [2]. Data processing and interpretation used key methods of geographical research: mapping and geoinformation modeling.

The development of the tourism system limits its capacity which, on the one hand, is determined by the natural properties of the natural environment and, on the other, by the state of the basic infrastructure of tourism and accompanying infrastructure complexes: transport, social amenities and environmental protection [3].

Territorial characteristics of the system are determined by tourist-recreational regionalization. Consideration is also given to the priority of the possible kinds of tourism and recreation, and to the current and future development level of the recreational, transport and ecological infrastructure. In the Central Ecological Zone (CEZ) of Irkutsk oblast, it is suggested that eleven tourist-recreational zones be identified (table 1).

**Table 1.** Tourist-recreational zones of CEZ BNT of Irkutsk oblast

| Municipal entity | Tourist-recreational zones | Area, thou ha | Number of places in accommodation facilities, thou, as of 2017 | Population size, thou, as of 2017 |
|---|---|---|---|---|
| | | Irkutskii district | | |
| **Listvyanskoe** | Listvyanskaya | 1.8 | 2.6 | 2.1 |
| | Bol'shekotskaya | 0.067 | 0.2 | 0.05 |
| **Goloustnenskoe** | Goloustnenskaya | 2.6 | 1.8 | 0.6 |
| | | Ol'khonskii district | | |
| **Bugul'deiskoe** | Bugul'deiskaya | 1.7 | 0.1 | 0.9 |
| **Shara-Togotskoe** | Malomorskaya | 9.7 | 6.5 | 0.9 |
| **Khuzhirskoe** | Ol'khonskaya | 1.7 | 3.6 | 1.5 |
| | | Slyudyanskii district | | |
| **Kultukskoe** **Slyudyanskoe** | Kultuksko-Slyudyanskaya | 6.1 | 0.9 | 18.4 |
| **Utulikskoe** **Baikal'skoe** | Utuliksko-Baikal'skaya | 1.8 | 3.2 | 14.1 |
| **Portbaikal'skoe** | Portbaikal'skaya | 0.1 | 0.1 | 0.4 |
| **Snezhninskoe** | Murinskaya | 1.1 | 0.1 | 0.2 |
| | Snezhninskaya | 2.5 | 0.3 | 0.4 |

## 3. Results and Discussion

Our investigations determined three types of ecological barriers to tourism development on BNT: 1) the existing legislatively approved standards meeting the natural properties of natural ecosystems and regulating recreational pressure on them; 2) the naturally conditioned resilience of ecosystems to anthropogenic tourism-associated impacts, and 3) the present state of natural ecosystems undergoing real recreational pressures. The parameters of these barriers are closely linked and, taken together, determine the restrictions to tourism development.

A complex of legislative-legal restrictions aimed to preserve the unique ecosystem of Lake Baikal defines the frame conditions for the tourist-recreational development of CEZ BNT. They may be characterize as complicated and strict. The key provisions are defined by the Federal law "On the protection of Lake Baikal" (of 5/1/1999 no. 94-FZ) and associated legislative legal acts [4]. However, some regional documents show substantial differences in approaches to governance of recreational activity between Irkutsk oblast and the Republic of Buryatia. To reconcile the legal framework requires developing the unified Rules for tourist-recreational activity on BNT aimed to form and improve agreed actions for the preservation of the unique ecosystem of Lake Baikal.

For assessing the stability of landscapes and separate components of the natural environment to recreational impacts as well as the tourist-recreational potential of using them, M.V. Tsygankova and E.A. Istomina developed the 1:500 000 landscape-typological map of CEZ BNT, using the landscape cartographic basis at a scale of 1:1 500 000 [5]. The fact that CEZ BNT is located in the Baikal rift zone represented by the Baikal basin and the mountains surrounded it is responsible for a high landscape diversity. In particular, for the Irkutsk part of CEZ: goletz landscapes – 3%, subgoletz – 11%, mountain-taiga – 62%, submontane and subtaiga – 9%, and steppe landscapes over 10%).

The water bodies, soil, vegetation cover and fauna were assessed according to the criteria of recreational resilience reflecting their sensitivity to recreational impacts and capacity for self-recovery; after that, an integral assessment was made. Low-mountain landscapes of pine, larch-pine and Siberian stone pine-larch forests, pine steppizated forests and light- and dark-coniferous valleys were identified as the most stable landscapes. The least stability is characteristic for goletz and subgoletz landscapes as well as dark- and light-coniferous and steppe landscapes occurring on steep mountain slopes, including those facing Lake Baikal.

The study of existing recreational pressures and their consequences is essential. Landscape-ecological investigations of the locations of tourist activity provided a preliminary averaged assessment of the capacity and allowable pressure on terrestrial natural ecosystems in different parts of the shores of Lake Baikal [6]. For instance, 5–6 pers/ha for the southern (within the Slyudyanskii district) and western (Irkutskii district) shores of Lake Baikal, 6–7 pers/ha for the shores of Maloe More (figure 1), and 2 pers/ha for Olkhon Island.

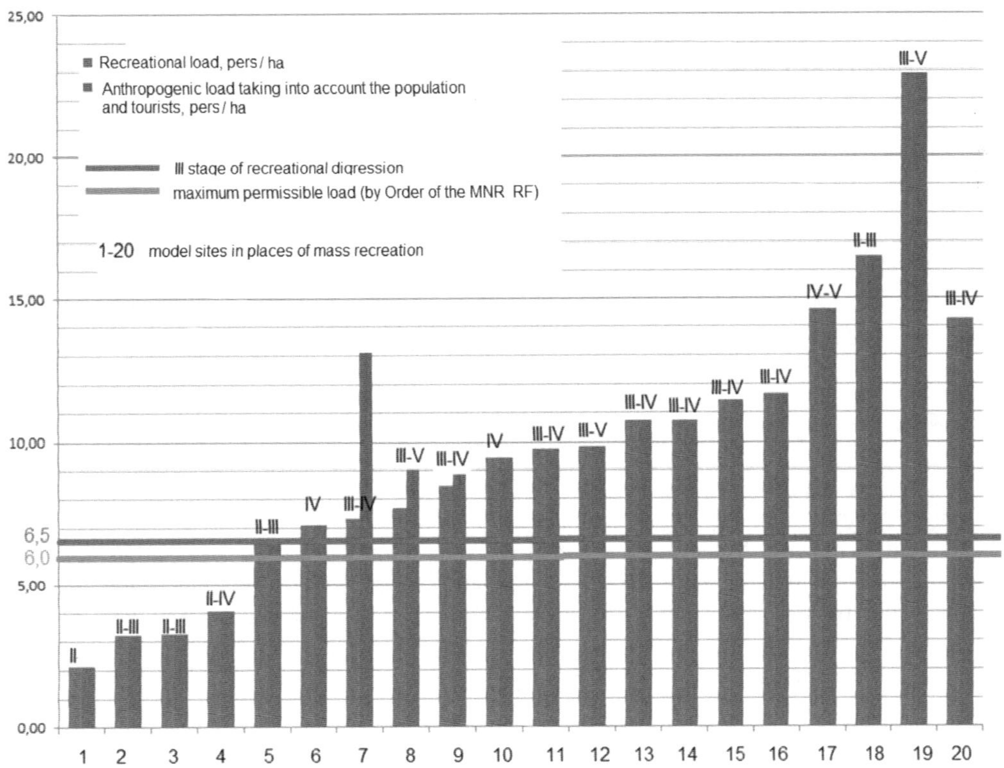


**Figure 1.** Recreational pressure on the shores of Maloe More of Lake Baikal

The analysis of the current ecological-economic problems and conflicts in environmental management aimed at the preservation of landscapes and sustainable development of the ecological-social-economic space in CEZ BNT identified today's overwhelming challenge of waste handling. Tourism development, coupled with an appropriate establishment of the "industry of cleanliness", is a necessary condition for the preservation of the basic tourism product, i.e. attractive landscapes and favorable residential territories. The modern approach to waste handling on the scientific-methodological basis of social geography builds upon sectoral principles– through the environmental protection infrastructure (EPI). The environmental protection infrastructure is the territorial system of works, production facilities and enterprises for waste utilization (recycling, regeneration and recuperation), disposal and neutralization (including the institutional as well as normative-legislative accompaniment for monitoring and management of waste flows) [7]. By coping with the identified upgrading deficits (of the environmental protection infrastructure), it will be possible to preserve a favorable human habitat and a rational use of spatial combinations of resources and natural environment in CEZ BNT (table 2).

**Table 2.** Infrastructural upgrading deficits (exemplified by the environmental protection infrastructure)

| | Deficits | Suggested solutions |
|---|---|---|
| **1.** | 1. Environmental costs imposed on tourist business and local authorities that are declared but not reimbursable. | 1. Accounting of expenses connected with nature conservation and environmental protection in the financial balance for the purposes of their subsequent reimbursement (development of legislative and normative-legal mechanisms for reimbursement of environmental costs for CEZ BNT). |
| **2.** | 1. Continuing problems with the handling of municipal solid waste (MSW) and liquid domestic waste (LDW). | 1. Implementation of investment projects of the creation of environmental protection infrastructure facilities and the introduction of environmentally friendly technologies. |
| | 2. Remoteness of existing and planned MSW disposal areas (up to 100 km) and unreasonably high cost of environmental protection measures, etc. | 2. Formation of modern logistic territorial schemes for the handling of MSW approved, including with municipal entities of district and settlement status, etc. |
| | 3. Mainstreaming of the uniqueness of the natural site, Lake Baikal, and of the equally significant necessary infrastructure facilities on its shores, the central ecological zone. | 3. Development of the environmental protection infrastructure on the basis of medium- and long-term planning of territorial schemes of ecological-social-economic development of CEZ BNT. |

## 4. Conclusions

By analyzing the natural and socio-economic conditions on the territory of the Baikal region, and primarily on Baikal's shores, in the context of tourism issues, it was possible to identify the driving factors behind the territorial organization of tourist-recreational activity in the presence of different types of ecological barriers. Some of these issues are long overdue and are to be immediately dealt with, namely the problems of improving the legislative-legal framework, optimizing recreational pressures with due regard for the resilience of landscapes, and determining the territories for priority tourism development and the measures for reconstruction of the environmental protection infrastructure. It is only within the suggested and approved (in territorial general plans) tourist-recreational zones that tourist-recreational activity becomes legitimate and regulated. On the basis of the findings of this investigation, it is planned to develop and implement the Tourism and Recreation Rules in the Baikal region.

IOP Conf. Series: Earth and Environmental Science **190** (2018) 012006 doi:10.1088/1755-1315/190/1/012006

**Acknowledgements**
The research was funded by the project the project №17-29-05043 of the Russian Foundation for Basic Research офи_м 2017

**References**
[1] Preobrazhensky V S 1975 *Theoretical Foundations of the Recreational Geography* (Moscow: Science) (in Russian)
Veal A J 1992 *Research Methods for Leisure and Tourism. A Practical Guide.* (London: Longman) p 320
Theobald W F 2005 The meaning, scope, and measurement of travel and tourism *Global Tourism* (Amsterdam: Butterworth–Heinemann) Third edition p 561
[2] *Landscape Planning: Tools and Experience in Implementation* 2005 ed A N Antipov (Irkutsk: V.B. Sochava Institute of Geography SB RAS Publisher) p 150
Ryashchenko S V, Bogdanov V N and Romanova O I 2008 *Regional Analysis of Recreational Activity* (Irkutsk: V.B. Sochava Institute of Geography SB RAS Publisher) p 143 (in Russian)
[3] Evstropyeva O V 2016 Development of the tourist system on the Baikal natural territory *Geogr. Prir. Resur.* **5** 184–95 DOI: 10.21782/GIPR0206-1619-2016-5 (in Russian)
Lentell B 1999 An excursion in sport, leisure and tourism research: the changing face of research in sport, leisure and tourism *Sport, Leisure and Tourism Information Sources: A Guide for Researchers* (Boston: Butterworth Heinemann) pp 1–26
[4] Bychkov I V, Gagarinova O V, Orlova I I, Korytny L M, Plyusnin V M and Bogdanov V.N. 2017 Water protection zoning of the Baikal natural territory: Conflicts of laws, landscape-hydrological approach *Geogr. Nat. Resour.* **38**(4) 364–70 DOI: 10.1134/S187537281704008
[5] Mikheev V S and Ryashin V A 1977 *Landscapes of the South of Eastern Siberia: The 1:1 500 000 Map* (Moscow: GUGK) (in Russian)
[6] Ponomarenko E A and Solodyankina S V 2013 Transformation of coastal geosystems of Lake Baikal under the influence of recreational activity. *Izv. Irkutsk. Gos. Univ. Ser. Nauki o Zemle* **6**(1) 147–60 (in Russian)
[7] Zabortseva T I 2014 Problems of the environmental protection infrastructure of Siberia: Economic-geographical Approach *Izv. RAN. Ser. Geogr.* **5** 47–55 (in Russian)

# Landscape-hydrological basis for water resources protection and development of the Baikal territory

**O V Gagarinova**

Sochava Institute of Geography, Siberian Branch of the Russian Academy of Sciences, Irkutsk, Russia

E-mail: whydro@irigs.irk.ru

**Abstract.** Landscape-hydrological researchin the Lake Baikal catchment areahas identified individual hydrological properties of landscapes. We analysed physicogeographical conditions, ecological and socioeconomic features of the territory and identified the areas responsible for the formation of qualitative and quantitative characteristics of natural waters, determined areas with significant loads on the aquatic environment, revealed areas with different resistance to anthropogenic influences. Functional zoning of the territory provides the basis for the schemes of water resources use and protection as well as recommendations for the development of the Baikal Natural Territory in relation to the water factor. Our experience in the practical implementation of the landscape-hydrological approach for the water resources management within the Lake Baikal catchment area is represented in the projects of ecological justification of economic activity, water protection zoning of natural and residential areas.Assessing opportunities for the development of the territory based on the water factor rooted in the analysis of water and ecological natural potential, in the stability of hydrological functions of landscapes to natural and anthropogenic changes we examined using the example of zoning the southern shore of Lake Baikal for the recreational purposes.The conception of moisture transformation mechanisms in landscapes as it moves to the draining water body is taken as a basis of water protection zoning.

## 1. Introduction

Water resources management based on water utilization measures does not fully ensure the reproduction and preservation of natural waters of high quality, since they are aimed only at preventing and eliminating negative water-ecological situations. The objectives of a preventive focus are important as they take into account the consequences of possible natural and climatic changes and anthropogenic impacts on water bodies and their catchment areas. The research experience here shows that basin management should be based on knowledge of the formation and transformation patterns of natural waters in terms of possible changes. Such studies are expedient on the basis of the landscape-hydrological approach, concerning landscape as a spatially localized territory, with certain water-balance relationships. Within the Lake Baikal catchment area we, being guided by landscape-hydrological principles, carried out various projects aimed at optimizing the use and protection of water resources [1].

IOP Conf. Series: Earth and Environmental Science **190** (2018) 012007 doi:10.1088/1755-1315/190/1/012007

## 2. Theory and methods

The landscape-hydrological analysis of the territory is a research of the process water yield from a unit of area with an individual physical and geographical structure that determines the formation of hydrological characteristics within the natural complex.

The key-notes of the landscape-hydrological approach are as follows.

1. Landscape is a single object with specific internal connections which determine the transformation of precipitation into surface and underground runoff, i.e. possessing individual hydrological functions.

2. The process of watershed yield from landscape is caused by its physical and geographical parameters, especially by steepness of slopes, type of soils and vegetation, composition of soils, geomorphological features, etc.

3. The quantitative measure of the hydrological function of a landscape is the overland runoff, that shows the rate of water yield from a local area for a certain period of time and can be calculated on the basis of the model equation: $\sum qifij = Qj$, where Qj is the river runoff at the main stream station from the m-th catchment; qi is the overland runoff of i-th landscape; fnj is the area of the i-th landscape within the boundaries of the m-th catchment [2].

Differences in the hydrological functions of landscapes and spatial differentiation of hydrological characteristics in terms of magnitude and character in the intra-annual section determine the landscape-hydrological organization of the territory, which is the basis for solving practical issues related to water and environmental issues, as well as to water protection and managementof the territory. The specific nature of natural water formation and transformation in a specific landscape represents a landscape-hydrological basis for assessing the water-ecological status of the territory and determining the functional hydrological changes resulting from the transformation of nature [3].

Spatial differentiation of runoff values from the landscapes of the Lake Baikal catchment area evaluates the water-resource and water-ecological potential of the catchment area and enables zoning the territory in accordance with these factors [4]. The individual hydrological functions of each landscape determine the quantitative and qualitative characteristics of water supply from the catchment to the draining water body, determine the floods and high water areas formation, as well the areas of low moisture and water-ecological tension, and other hydrological characteristics of the catchment area [5].

## 3. Results and discussion

Combination of physicogeographical settings of the catchment area is determined by the individual hydrological regime and the water-ecological potential of the landscapes. Water-ecological potential due to the flow control mechanisms within the natural complex characterizes the sustainability of landscapes to external impact, the ability to stabilize the water regime, regulate the quantitative and qualitative characteristics of natural waters [6]. The level of water-ecological potential determines different degree of landscapes stability to anthropogenic impact and the possibility of economic development of the territory without compromising the quality of water resources.

Zoning the Lake Baikal catchment area according to the level of water-ecological potential shows areas with various parameters of inland landscape runoff regulation and natural complex resiliency to natural and anthropogenic changes [7]. Low level of water and ecological potential is characteristic for high-mountainous and steppe landscapes with a low-power soil and vegetation cover, where precipitation forms an increased surface run-off, causing erosion processes and possible rapid arrival of pollutants into water bodies.

Dark coniferous landscapes on watersheds, gentle slopes with moss-grassy soil cover, meadow-marsh and floodplain-valley complexes on foothill plains and lowlands a high water-ecological capacity. Significant filtration and accumulating properties of these communities cause slow water loss, providing a high level of self-purification, water protection and environmental protection capabilities of the territory, including significant resistance of landscapes to changes (figure 1).

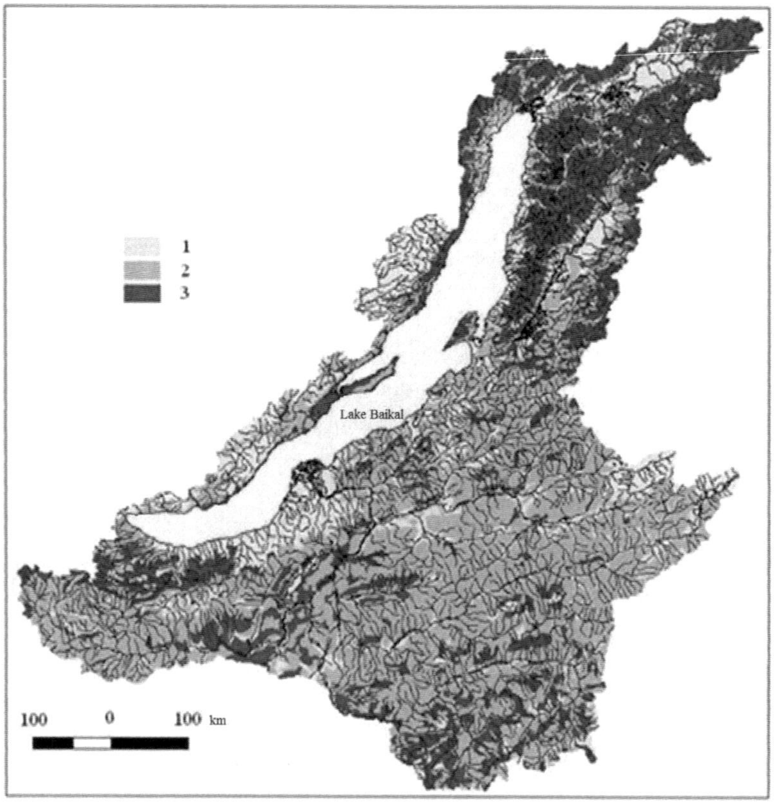


**Figure 1.** Water-ecological capacity of landscapes of the Lake Baikal catchment area. (within the borders of the Russian Federation): 1 - high, 2 - medium, 3 - low.

Based on this analysis, it is possible to assess the development of the territory with respect to the water factor, to develop schemes for utilization and protection of water resources of the Baikal Natural Territory [7]. In accordance with these principles, the rationale for the recreational development of the Lake Baikal shore within the boundaries of the Irkutsk region and a project of the water protection zone of Lake Baikal have been carried out.

Complex analysis of structural landscape elements, geomorphological and hydrological features of the shores, mechanisms of landscape planning of the territory [8] identify areas preferable for recreational activities and areas prohibited for economic use with regard to the environmental protection. The zone of current state preservation is an area with landscapes which determine the formation of quantitative and qualitative parameters of water resources. In these areas it is necessary to preserve the existing landscape structure and it is advisable to prohibit all types of economic activities, including recreation areas, placement of permanent camps and tourist bases, economic infrastructure that contribute to the violation of vegetation and soil cover.

The zone of possible adjusted development includes natural complexes that do not have the environment forming and water protection role, which have an average sensitivity and resistance to external influences. The low level of water and ecological capacity of these sites assumes possible recreational development of the territory subject to the implementation of environmental measures aimed at preventing surface disturbance of soil and vegetation cover and the entry of pollutants into surface waters and subterranean horizons.

Intensive development of recreation is possible in areas with low hydrological functions, but optionally high or medium level of resistance to external impact. Anthropogenically transformed residential and agricultural territories are also included in this zone. The main water-ecological patterns of nature management are the self-restoration of natural hydrological properties, the observance of water protection measures and the prevention of the receipt of pollutants into water bodies and on the terrain (figure 2).

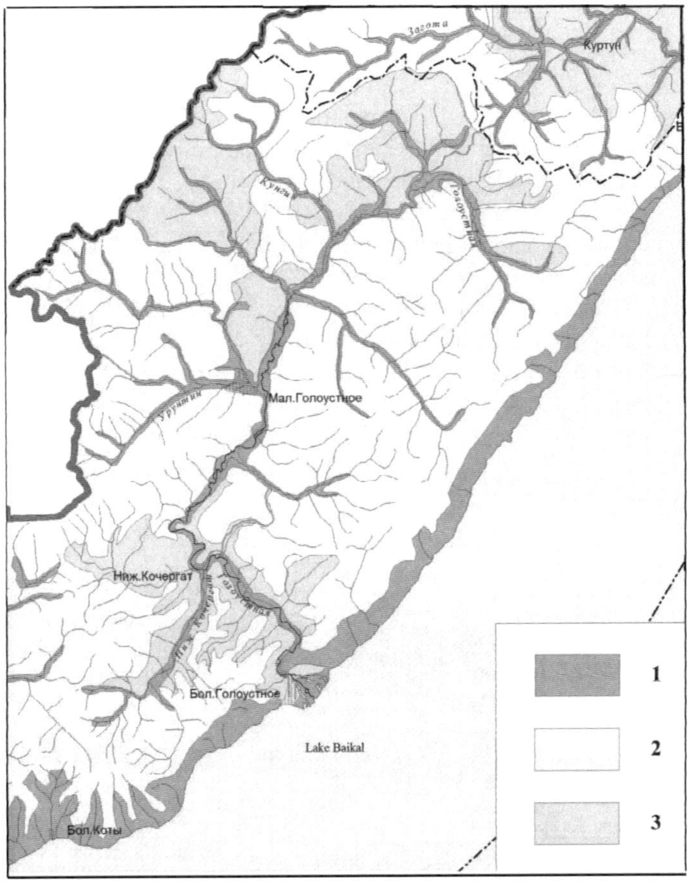


**Figure 2.** Landscape-hydrological zoning for the purposes of recreational development of the Central ecological zone of the Baikal Natural Territory. Zones: 1 - preservation of the current state; 2 - regulated use; 3 - development.

Recreation development on the Lake Baikal shore should not contribute to reducing the quantitative and qualitative characteristics of surface and groundwater entering the lake reservoir. The task is especially urgent up-to-date, under a significant deterioration of the environment within the lake and coastal area.

One of the conservation tools for Lake Baikal and its shore is the water protection zone, the project of which is scientifically based and developed in accordance with landscape and hydrological principles and assumes elements of additional internal functional differentiation for the economically developed areas at the shore [9]. Namely the hydrological functions of landscapes are the main factors that determine the quantitative and qualitative characteristics of surface and groundwater entering the lake and, hence, are the parameters that determine the size and configuration of the water protection zone of the water body.

The individual project of the water protection zone of Lake Baikal presupposes prevention of water pollution and water quality and coastal ecosystem preservation. The dimensions of the water protection zone should ensure the protection of the lake from the receipt of pollutants through the realization of the natural properties of self-cleaning of surface and ground runoff in landscape elements, and also minimize the consequences of dangerous hydrological and geomorphological processes and anthropogenic impact on the water body.

In accordance with the landscape-hydrological principles, the water protection zone limits the territory which drains directly into the lake. The water protection zone includes coastal areas occupied by landscapes with high water and ecological capacity, as well as areas adjacent to the reservoir and characterized by low resistance to natural and anthropogenic changes. The frame of the water protection zone of Lake Baikal is composed of dark coniferous landscapes on gentle slopes, humidified estuary areas of tributaries and wetlands on the shores, which act as a natural filter on the way of pollutants entering the lake. To limitthe anthropogenic transformation of certain coastal areas, the landscapes subject to exogenous processes and the potential threat of technogenic emergency situations are included into the water protection zone.

On the natural territory of the Baikal mountain fringe the boundary of the water protection zone is drawn along the points of inflection of the relief, the mountain ledges and the peaks of the first watershed, including all areas of direct flow of slope flow into the lake. The expansion of the water protection zone beyond the first watershed is carried out in areas of high exogenous activity (floods, mudflows, collapses, debris flows, etc.). The water protection zone includes basins of small elementary watercourses that directly flow into the lake, wetlands of lowland and meadow-marsh complexes of the coastal, mouth areas of tributaries. In coastal areas with slopes of 3-7 °, the boundary is drawn along the contour of landscapes, the structural components of which have a high water conservation capacity (figure 3).

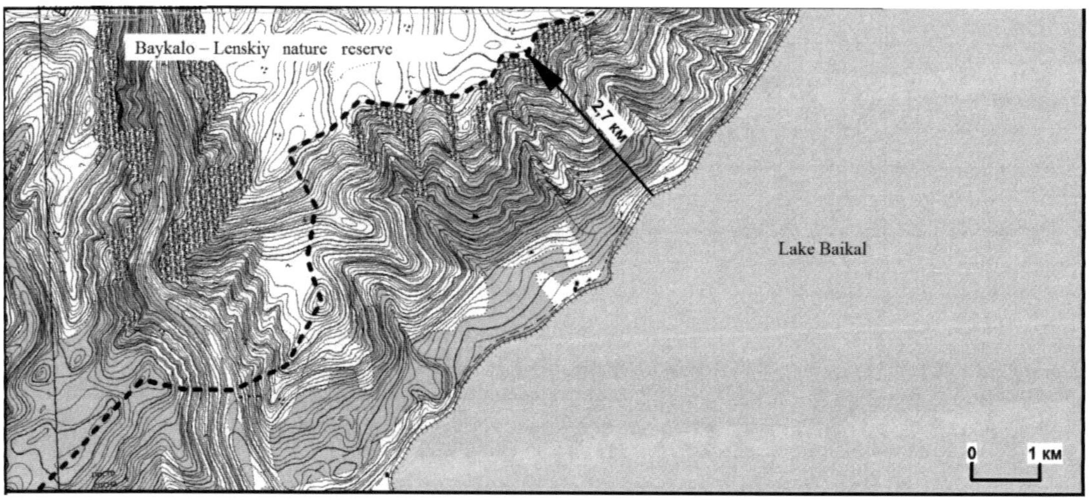


**Figure 3.** The boundary of the water protection zone of Lake Baikal

Using landscape-hydrological principles in combination with landscape planning tools, we developed the methods for water protection zoning of settlements [10]. In urbanized areas, as well as on natural ones, the priority criterion for justifying the size of the water protection zone is the possibility of natural self-purification of natural waters. Settlement areas requires an assessment of current state of natural complex and analysis of anthropogenic transformations of its components, which makes it possible to carry out functional zoning of the urbanized area according to the water factor. This zoning implies the establishment of different restriction levelswithin the zones in order to prevent the entry of pollutants into Lake Baikal and to minimize the negative anthropogenic impact on the ecosystem of the reservoir, taking into account the economic development opportunities for the areas [11].

## 4. Conclusion

The landscape-hydrological approach enables to consider the landscape as a single object with a set of cause-effect relationships and a specific water-balance structure that determine the mechanism of precipitation transformation into surface and underground runoff with specific quantitative and qualitative characteristics.

Changes in the structure of landscapes as a result of fires, logging, and economic use of the territory result in transformation of properties and all hydrological characteristics of the territory. Analysis of disturbances in natural conditions allows predicting changes in the quantity and quality of the basin's water resources.

Landscape-hydrological zoning is the basis for developing targets for different territories from the standpoint of preserving the hydrological functions of landscapes and developing schemes for the utilization and protection of natural waters.

**Acknowledgements**
The work was done within the Research Project 0347-2016-003 and the RFBR grant No. 16-05-00286

**References**

[1] Antipov A N and Fedorov V N 2000 *Landscape-Hydrological Organization of the Territory* (Novosibirsk: Publishing house of the SB RAS) p 254
[2] Fedorov V N 1992 *Evaluation of the Hydrological Functions of Landscapes on the Basis of Indicative Multiparametric Models of the Catchment Area* (Landscape-Hydrological Analysis of the Territory) ed A N Antipov and G V Bachurin (Novosibirsk: Nauka) pp 145-57
[3] Fedorov V N 2007 *Landscape Flow Indication* (Irkutsk – Moscow: Izd IG SB RAS) p 175
[4] Gagarinova O V 2012 Landscape-hydrological patterns of formation of runoff in the Lake Baikal catchment area *Geography and Natural Resources* **3** 55-60
[5] Preobrazhensky V S, Alexandrova T D and Kupriyanova T P 1988 *Fundamentals of Landscape Analysis* (Moscow: Nauka) p 190
[6] Antipov A N and Gagarinova O V 2002 Water protection zoning of territories of settlements *Geography and Natural Resources* **4** 30-7
[7] Antipov A N and Korytny L M 1981 *Geographical Aspects of Hydrological Studies (on the Example of the South Minusinsk Basin)* (Novosibirsk: Nauka) p 177
[8] *Landscape Planning: Tools and Experience of Application 2005* (Irkutsk: Publishing house of the Institute of Geography of the SB RAS) p 165
[9] *Landscape Planning: Principles, Methods, European and Russian Experience* 2002 (Bonn - Moscow - Irkutsk: Publishing House of the Institute of Geography of the SB RAS) p 141
[10] *Ecologically Oriented Land Use Planning in the Baikal Region. Territorial Development of the City of Baikalsk and Suburban Areas* 2003 (Irkutsk: Publishing house of the Institute of Geography of the SB RAS) p 192
[11] Bychkov I V, Gagarinova O V, Orlova I I, Korytny L M, Plyusnin V M and Bogdanov V N 2017 Water protection zoning of the Baikal natural territory: legal collisions, landscape-hydrological approach *Geography and Natural Resources* **4** 76-82 doi: 10.21782/GiPR0206-1619-2017-4(76-82)

# Transaction costs and green economy potential in Russian regions

**I P Glazyrina, L M Faleychik, I A Zabelina**

Institute of Natural Resources, Ecology and Cryology, Russian Academy of Sciences, Siberian Branch, Chita, Russia

E-mail: lfaleychik@bk.ru

**Abstract**. The concept of a "green" economy and "green" growth has a clearly expressed social context. It declares the improvement of well-being, the reduction of inequality, and the eradication of poverty in the development processes. This article aims to explore the problem of transaction costs in forest management in the different Russian regions in the context of the concept of "green economy". We can argue, on the base of results of this study, that Russian regions have very different "starting conditions" on the way to sustainability in the context of carbon balance. The calculations revealed a high degree of heterogeneity of the Russian regions in terms of the socio-environmental indicators. The analysis of spatial distribution of the transaction costs in the Russian regions also showed their high heterogeneity. Therefore, the establishment of a "carbon tax", which is now widely discussed by the scientific community, should not be based on uniform approaches and procedures for the whole country.

## 1. Introduction

Russian forests are not only a source of valuable resources. They perform environmental functions that are essential to the quality of life of the citizens, including the absorption of greenhouse gases and climate control. Forest conservation is an important state task that requires the functioning of special state institutions. In accordance with the already established consensus in the scientific literature [1], we relate the costs of maintaining these institutions to the transaction costs (TC) in the management of natural resources. This article aims to explore the issue of TC in forest management in the different Russian regions in the context of the concept of "green economy".

## 2. Methodology

The idea of our quantitative tools is based on the P. Victor's model proposed in [2] for gross domestic product (GDP) and carbon emission for time analysis. We have modified this model and adopted it for spatial analysis. The first results of the comparative analysis of regional heterogeneity in Russian forestry the readers can find in [3]. The detailed model used in this article is presented in forthcoming paper [4].

In this article we considered two quantitative factors that characterize the welfare of the population: the total wage fund and own revenues of the regional budgets per capita (*W/N*). The key ecological and economic indicators used in this work are indicators of eco-intensity (*EI*) [5]:

- *CO/tax* – the volume of carbon monoxide emissions per 1000 rubles of taxes, fees and other mandatory payments to the consolidated budgets of the Russian sub-federal regions (kg / 1000 rubles);

- *CO/wage* – the volume of carbon monoxide emissions per 1000 rubles of the wages fund (kg / 1000 rubles).

We also introduce indicator *CO/N* – the volume carbon monoxide in per capita terms. It is an important socio-ecological indicator for the regional economic systems. It shows the "environmental cost" of providing human habitation in the region.

On the basis of these indicators, ecological-economic zones have been identified; the characteristics of zones are given in table 1.

**Table 1.** Ecological-economic zones in the concept of "green growth".

| Zone[a] | Characteristics |
|---|---|
| "Green" zone Gr+: all three indicators are better than the Russian average | Higher per capita welfare (*W/N*) than the Russian average, with lower rates of both *EI* and anthropogenic pressure per capita (*CO/N*) |
| "Brown" zone Br+ | *EI* is lower than the Russian average with higher welfare per capita (*W/N*), but higher negative impact per capita (*CO/N*) |
| "Black" zone Bl+ | *EI* is higher than the Russian average with higher welfare per capita (*W/N*), and higher negative impact per capita (*CO/N*) |
| "Black" zone Bl –: all three indicators are worse than the average | Both environmental indicators –*EI* and negative impact per capita (*CO/N*) – are higher, i.e. worse, while the per capita welfare indicator (*W/N*) is lower. |
| "Green" zone Gr – | The indicators of welfare (*W/N*) and the per capita anthropogenic load (*CO/N*) are characterized by lower than the Russian average, but higher *EI*. The decline in the per capita anthropogenic load takes place on the background of increase of the *EI* indicators, i.e. by reducing indicators of well-being. |
| "Absolutely green" zone AGr – | Both environmental indicators *EI* and negative impact per capita (*CO/N*) – are lower, i.e. better than the national average, but per capita welfare (*W/N*) is also lower. |

[a] The symbols "+" and "–" mark the zones where per capita indicators of welfare are higher and lower than the Russian average, respectively.

The study used official data of the Federal state statistics service, the Federal tax service of Russia (monetary indicators were given to a comparable form) characterizing: the volume of carbon monoxide emissions from stationary sources; average annual number of employees and average monthly wages of employees, taxes, fees and other mandatory payments to the consolidated budget of the Russian regions; population.

The analysis of spatial distribution of the considered indicators in the study was performed using GIS tools.

## 3. Results and discussion

### *3.1. Transaction costs in forestry*

One of the assessments of the transaction costs in the forest sector of the region is the amount of expenses for its management, aimed at the conservation and reproduction of resources, and the performance of their environmental functions. It consists of the volumes of subventions of the Federal and regional budgets for the management of forest ecosystems and revenues from other sources.

The spatial distribution of the amount of TC per unit area of forest land (FL) of the regions is presented in figure 1. The analysis of this distribution shows that for Krasnoyarsk region and Irkutsk region which are the main exporters of forest products, the value of TC per 1 ha of FL is one of the lowest – less than 20 rubles per hectare (15.1 and 18.1 rubles, respectively). In Tyumen region and Khanty-Mansi Autonomous Area – Yugra, these indicators are significantly higher – 90 and 39 rubles, respectively. The highest value of the index is in Lipetsk region – more than 1 800 rubles. Here and below the region names are presented in accordance with the official translation of the Russian Constitution.

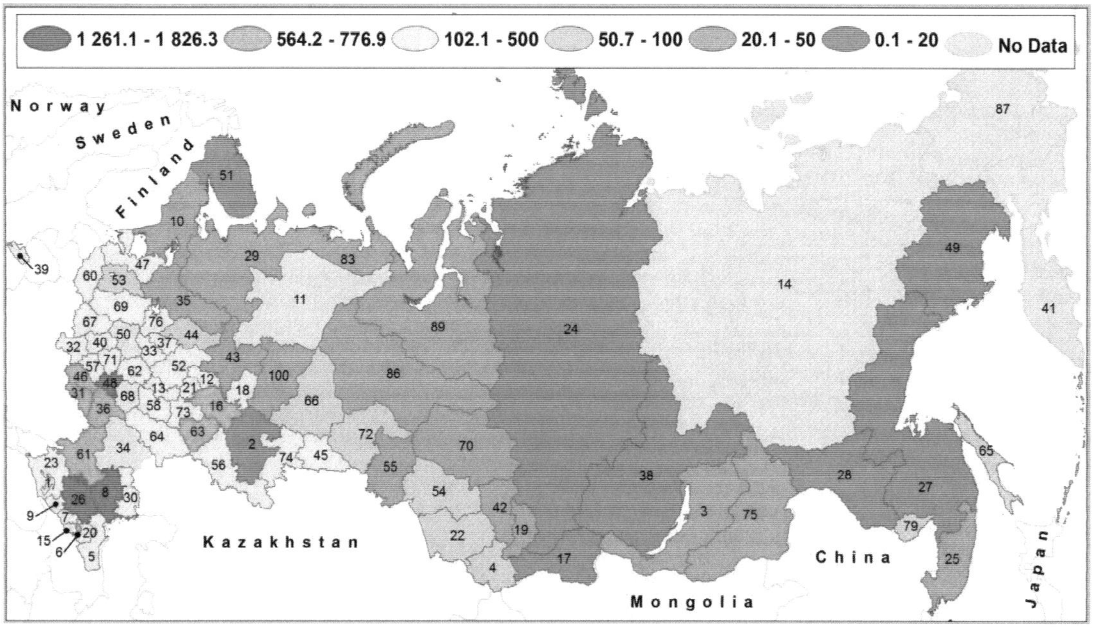


**Figure 1.** Total transaction costs per 1 hectare of forest land in the sub-federal Russian regions.

The next step in our study was an analysis of the spatial differentiation of the Russian regions by forest charges per 1 hectare for all types of forest use (figure 2).

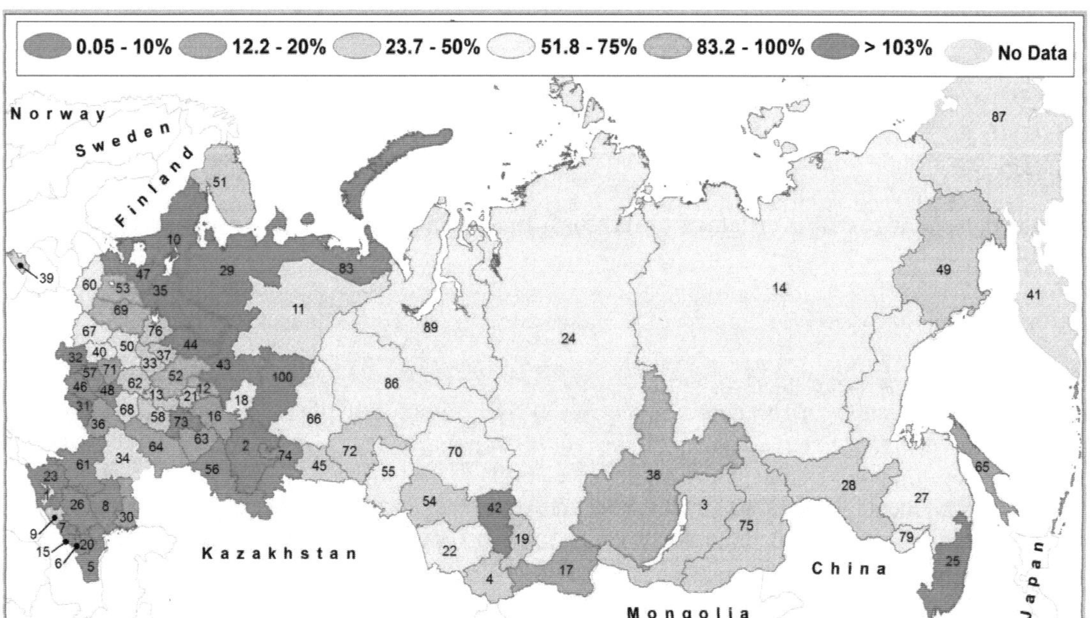


**Figure 2.** The contribution of private business to the costs of forest management in the regions.

In 22 regions, mainly the regions of Western and Eastern Siberia, charges for the use of FL are less than 20 rubles per hectare. In 12 regions of the European part of Russia, these charges exceed 100 rubles. Maximum value of this indicator is 881.4 rubles per 1 hectare of FL is in Belgorod region. The amounts of forest use fees for all types of forest management exceed the total TC in 14 forest regions (figure 2).

The different shares of business in the costs of forest management in the regions can be partly explained by the different shares of the leased forest land. More than 90% of all forest lands are the leased in Leningrad (99.8%), Tula (93.5%) and Kurgan (91.5%) regions, in Primorye Territory (93.4%).

Some indicators of the forest sector in these regions are given in table 2.

**Table 2.** Comparative characteristics of the forest sector in some Russian regions[a].

| Regions | FL area in the region, thousand hectares | Total TC in the region, thousand rubles | TC per 1 ha of FL, rubles | Forest use charge on all types of forest, per 1 ha of FL, rubles | Share of leased FL, % |
|---|---|---|---|---|---|
| Krasnoyarsk Territory (24) | 158 727.9 | 2 404 595.6 | 15.1 | 8.81 | 16.7 |
| Khabarovsk Territory (27) | 73 732.4 | 963 692.8 | 13.1 | 7.98 | 21.1 |
| Irkutsk Region (38) | 69 419.2 | 1 254 455.8 | 18.1 | 15.61 | 28.1 |
| Khanty-Mansi Autonomous Area – Yugra (86) | 49 350.3 | 1 922 963.4 | 39.0 | 25.75 | 8.2 |
| Trans-Baikal Territory (75) | 32 614.8 | 907 997.4 | 27.8 | 10.32 | 24.4 |
| Yamal-Nenets Autonomous Area (89) | 31 655.4 | 638 818.8 | 20.2 | 13.44 | 15.7 |
| Amur Region (28) | 30 515.2 | 609 670.7 | 20.0 | 9.04 | 13.8 |
| Tyumen Region (72) | 11 389.2 | 1 025 580.7 | 90.0 | 31.65 | 20.9 |
| Belgorod region (31) | 228.5 | 168 129.8 | 735.8 | 881.41 | 29.4 |
| Lipetsk Region (48) | 180.5 | 329 654.9 | 1 826.3 | 91.07 | 3.3 |

[a] The region numbers (indicated in brackets) are given in accordance with the map on figures 1 and 2.

In the Eastern Russian regions very little square of the leased lands are in Republic of Tuva – only 1.5%, less than 10% – in Omsk and Kemerovo regions, in the rest – from 10.3 – Republic of Buryatia – to 29.8 – on Sakhalin.

Thus, it can be concluded that the Eastern regions of Russia have lower TC and the maintenance of their costs for resource-saving and environmental functions are covered from public financial funds. In the forest regions of the North of the European part of the country, by contrast, most of the transaction costs are covered by business.

The results presented at figures 1-2 and table 2 show both a high spatial differentiation of per hectare TC share and of TC covered private business.

### *3.2. Development of the Russian regions in the context of "green" economy*

In numerous works devoted to the development of the "green" economy, great attention is paid to greenhouse gas emissions [6], the reduction of which is considered as a necessary condition for "green growth". Our analysis revealed not only a high degree of spatial heterogeneity in terms of *CO/tax* and *CO/wage*. The distribution of the Russian regions by ecological and economic zones is presented on figure 3 and in table 3. They show that only a very small number of regions belong to the "green zone", i.e. all three types of indicators are better than the average for the Russian Federation. This is true for both quantitative factors that characterize the welfare of the population. In the "brown" zone (Br+), which includes only a few Eastern regions, relative socio-environmental well-being (i.e. higher per capita incomes) occurs at higher eco-intensity. Thus, we cannot say that these regions are more successful in terms of social and environmental modernization.

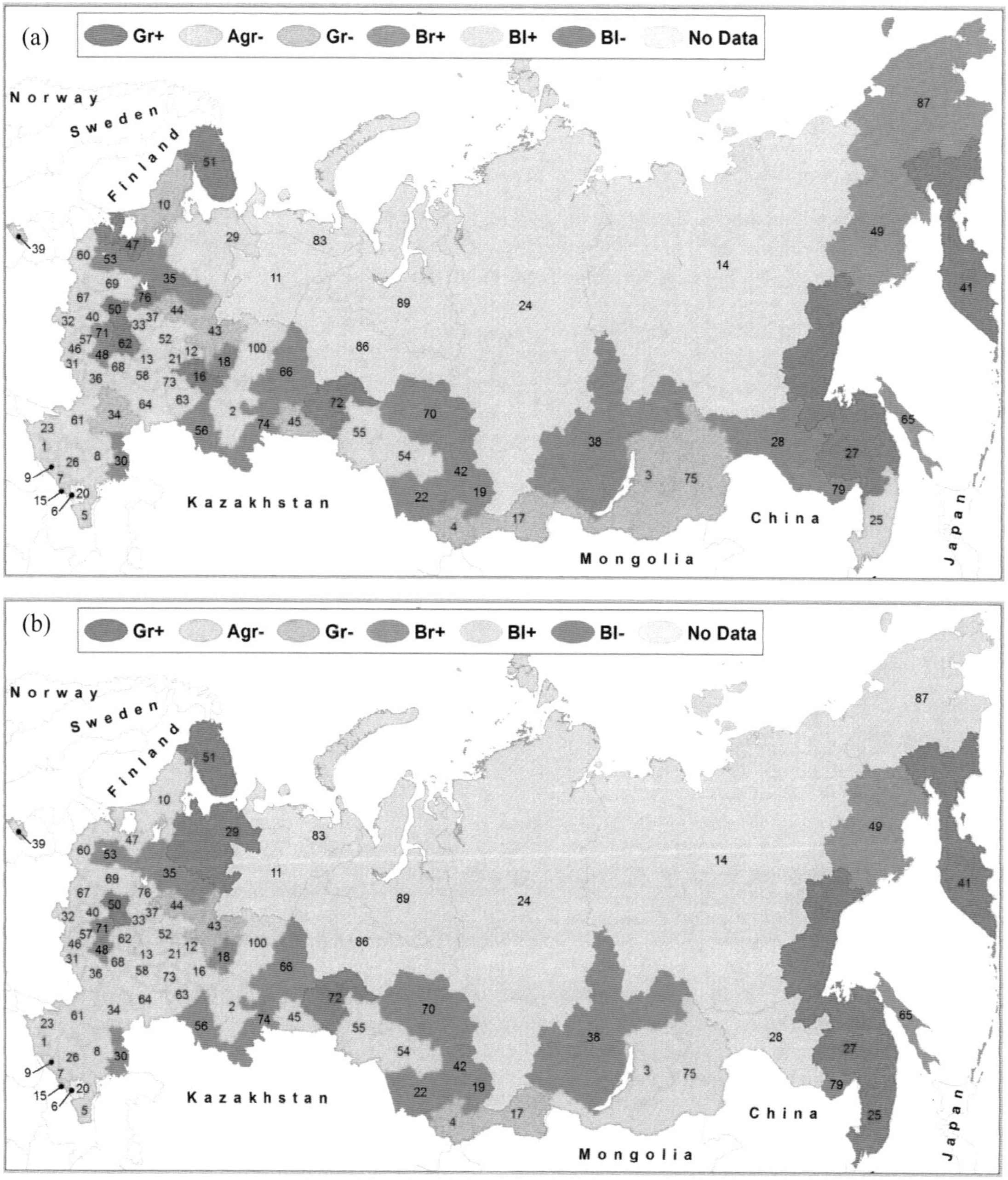


**Figure 3.** Allocation of ecological-economic zones in Russia: (a) by indicator *CO/tax*; (b) by indicator *CO/wage.*

Most of the oil and gas extracting areas, as well as some regions of Siberia and the Far East, rich with natural resource, fall into the Bl+ zone, i.e. per capita welfare factors are higher than the average Russian ones, but they have worse environmental indicators. However, for a number of industrialized regions (Irkutsk, Kemerovo, Sverdlovsk, Chelyabinsk, Vologda regions, etc.), these welfare indicators are lower than the national average, despite the fact that the socio-environmental indicators of eco-intensity are worse. This means that people living there have both environmental discomfort and the lower welfare conditions at the same time.

**Table 3.** Allocation of the Russian sub-federal regions by ecological-economic zones[a].

| Zone | Russian sub-federal regions |
|---|---|
| | (a)*CO/tax* |
| | "Black" zones – 22 sub-federal regions |
| Bl + | Khanty-Mansi Autonomous Area – Yugra (86); Komi Republic (11); Krasnoyarsk Territory (24); Nenets Autonomous Area (83); Republic of Sakha (Yakutia) (14); Yamal-Nenets Autonomous Area (89) |
| Bl – | Altai Territory (22), Amur Region (28); Astrakhan Region (30); Chelyabinsk Region (74); Irkutsk Region (38); Jewish Autonomous Region (79); Kemerovo Region (42); Lipetsk Region (48); Novgorod Region (53); Orenburg Region (56); Republic of Khakassia (19); Sverdlovsk Region (66); Tomsk Region (70); Tula Region (71); Udmurtian Republic (18); Vologda Region (35) |
| | «Brown» zone – 3 sub-federal regions |
| Br + | Chukotka Autonomous Area (87); Magadan Region (49); Sakhalin Region (65) |
| | «Green» zones– 56 sub-federal regions |
| Gr + | Kamchatka Territory (41);Khabarovsk Territory (27);Leningrad Region (47);Moscow; Moscow Region (50); Murmansk Region (51);Republic of Tatarstan (16);Ryazan Region (62);St. Petersburg; Tyumen Region (72);Yaroslavl Region (76) |
| Gr –, Agr – | other sub-federal regions |
| | (b)*CO/wage* |
| | "Black" zones– 23 sub-federal regions |
| Bl + | Amur Region (28); Chukotka Autonomous Area (87); Khanty-Mansi Autonomous Area – Yugra (86); Komi Republic (11); Krasnoyarsk Territory (24); Nenets Autonomous Area (83); Republic of Sakha (Yakutia) (14); Yamal-Nenets Autonomous Area (89) |
| Bl – | Altai Territory (22); Astrakhan Region (30); Chelyabinsk Region (74); Irkutsk Region (38); Jewish Autonomous Region (79); Kemerovo Region (42); Lipetsk Region (48); Novgorod Region (53); Orenburg Region (56); Republic of Khakassia (19); Sverdlovsk Region (66); Tomsk Region (70); Tula Region (71); Udmurtian Republic (18); Vologda Region (35) |
| | «Brown» zone – 2 sub-federal regions |
| Br + | Magadan Region (49); Sakhalin Region (65) |
| | «Green» zones– 56 sub-federal regions |
| Gr + | Arkhangelsk Region (29); Kamchatka Territory (41); Khabarovsk Territory (27); Moscow; Moscow Region (50); Murmansk Region (51); Primorye Territory (25); St. Petersburg; Tyumen Region (72) |
| Gr –, Agr – | other sub-federal regions |

[a] The region numbers (indicated in brackets after region names) are given in accordance with the map on figures 1-3.

## 4. Conclusions and Recommendations

A joint comparative analysis of the distribution of the Russian regions in the ecological and economic zones and the data presented in figures 1 and 2 shows that all regions (except for Lipetsk region) that are in the "brown" (Br+) and "black" (Bl+ and Bl–) zones are characterized by lower transaction costs in forestry. At the same time, the high level of atmospheric pollution by carbon monoxide emissions necessitates the maintenance of the sustainability of forest ecosystems, which play a key role in ensuring carbon balance. The increase in financing of forest management can be achieved through payments of users of natural resources for the forest cuts, whose contribution to forest management in certain regions is very high. Thus, in the Vologda and Kemerovo regions, which are part of the disadvantaged "black" zone Bl–, the contribution of forest users exceeds 100% but the total amount of TC is low.

Our analysis showed a high differentiation in per hectare transaction costs between regions of Russia. The differentiation of transaction costs for management, protection and restoration of forest ecosystems should also be the subject of special attention in the elaboration of state policy to reduce carbon emissions.

We can conclude that the carbon dioxide emissions per 1000 rubles of well-being factors are above the national average in most sub-federal Russian regions, i.e. the regional socio-ecological-economic systems are currently set up so that for the formation of units of socially important goods they have to exceed the national rates of carbon emissions in the production processes.

We can argue, based on the results of this study, that Russian regions have very different "starting conditions" on the way to sustainability in the context of carbon balance. Therefore, the establishment of a "carbon tax", which is now widely discussed by the scientific community, should not be based on common approaches and procedures for the whole country.

**Acknowledgments**
The methodology and tools for quantitative indicators of "green" economy was carried out within the frame work of the Program XI. 174 of Siberian Branch of Russian Academy of Sciences.Partial support from Russian Scientific Foundation (Subsection 3.1, Project 16-18-00073) is also gratefully acknowledged.

**References**

[1] Marshall G R 2013 Transaction costs, collective action and adaptation in managing socio-economic system *Ecol. Econ.* **88** 185-94
Daly H and Farley J 2003 *Ecological Economics: Principles and Applications* (Washington: Island Press) p 454
Daly H 1996 *Beyond Growth: The Economics of Sustainable Development* (Boston: Beacon Press) p 276
McCann L, Colby B, Easter K W, Kasterine A and Kuperan K V 2005 Transaction cost measurement for evaluating environmental policies *Ecol. Econ.* **52** 527-42
Challen R 2000 *Institutions, Transaction Costs, and Environmental Policy: Institutional Reform for Water Resource* (Aldershot: Edward Elgar) p 233
Ketokivi M and Mahoney J T 2016 Transaction cost economics as a constructive stakeholder theory *Academy of Management Learning ad Education* **15** 123–38

[2] Victor P A 2015 The Kenneth E. Boulding Memorial Award 2014: Ecological economics: A personal journey *Ecol. Econ.* **109** 93–100

[3] Glazyrina I P, Faleichik L M and Yakovleva K A 2015 Socioeconomic effectiveness and "green" growth of regional forest use *Geogr. Nat. Resour.* **36** 327-34

[4] Glazyrina I P and Zabelina I A 2018 Spatial heterogeneity of Russia in the light of the concept of green economy: the social context *Geogr. Nat. Resour.* **2** 14-22

[5] de Haan M 2004 *Accounting for Goods and for Bads: Measuring Environmental Pressure in a National Accounts Framework* (Voorburg: M. de Haan) p 216
Mazneva M 2005 Eco-intensity *Quality of Growth Indicators for Regional Economies* ed I Glazyrinaand and I Potravny (Moscow: NIA-Priroda) chapter 5 pp 126-47

[6] Shang Y, Si Y and Zeng G 2015 Black or green? Economic growth patterns in China under low carbon economy targets *J. Resour. Ecol.* **6** 310–7
Bobylev S N, Kudryavtseva O V and Yakovleva Y Y 2015 Regional priorities of green economy *Econ. Reg.* **2** 148–60
Dong S, Li Z, Li Y, Shi G, Yu H, Wang J, Li J, Mao Q and Huang Y 2015 Resources, environment and economic patterns and sustainable development modes of Silk Road Economic Belt *J. Resour. Ecol.* **6** 65-72
Fan Y, Liu L-C, Wu G, Tsai H-T and Wei Y-M 2007 Changesin carbon intensity in China: Empirical findings from 1980–2003 *Ecol. Econ.* **62** 683-91

IOP Conf. Series: Earth and Environmental Science **190** (2018) 012008 doi:10.1088/1755-1315/190/1/012008

Wang H, Ang B W and Su B 2017 A multi-region structural decomposition analysis of global $CO_2$ emission intensity *Ecol. Econ.* **142** 163-76

# Industrial impact on the environment (as exemplified by the Baikal region and Mongolia)

**N A Ippolitova**

Sochava Institute of Geography, Siberian Branch, Russian Academy of Sciences, Irkutsk, Russia
Irkutsk State University, Irkutsk, Russia

E-mail: nins-ip@list.ru

**Abstract**. The paper presents the results of ecological and geographical studies of the Baikal region and Mongolia to identify the negative industrial impact on the environment. The industrial impact on the environment depends on the nature of its territorial location, extraction and consumption of raw material volume, materials and energy, waste recycling and completion degree of energy production cycles. We analyzed location of the leading industries and its impact on the environment. The identified industrial centers have their own influence on different natural environments. Each industry and individual industrial facilities "invade" the environment and have a different features and impact, including on human health. The problem of the industrial impact on the environment is global, and makes it important to study this topic. During the research we took into account the specifics of industrial development of the territories, as well as the existing ecological situation around Lake Baikal, and based on theoretical and methodological approaches to socioeconomic geography. Zoning and mapping approaches we used in this research. Based on the analysis of the location of industrial facilities and their impact on various natural environments, the areas of maximum and local impact on the environment within the boundaries of the considered territories were identified. Ecological-geographical studies of the areas where industrial objects were located enabled an overall assessment of anthropogenic impact both on the environment and on certain natural environments, taking into account the demographic pressure. The territorial features of the heterogeneous impact on the natural environment are determined by the unevenness of economic development of the territory.

## 1. Introduction

Currently, human economic activity has a significant impact on the environment, where the main pollutant is industry. The higher the concentration of industrial facilities, the wider is the zone of environmental change.

The current borders of industrial impact areas on the natural environment are very relative, since the expansion of technogenesis from certain industrial sources, especially in the aquatic environment and atmosphere does not have clear borders and often reaches tens and hundreds of kilometers. The nature of the environmental impact is integral, since any contamination of one of the spheres of natural environment is reflected in others (violation of lithosphere indirectly affects the regime of surface and groundwater, predetermines dust and gas pollution of the atmosphere, etc.).

The industry has a leading role among the branches of the national economy; it determines the specificity of regions and its place in the territorial and sectoral division of labor. Industry remains the basic sector of the economy (which can also be interpreted as an element of sustainability), setting a vector for regional development and major cities, although in the latter case the industrial function is complemented by service, management, and scientific and educational specializations [1].

The industrial impact on natural environments is considered by the example of the subjects of the Baikal region (Irkutsk region, the Republic of Buryatia and the Transbaikal region) and Mongolia, which have a similar industry structure, but different natural conditions for its location and functioning.

## 2. Data and Methods

The research is based on theoretical and methodological approaches to social and economic geography, including scientific concepts on territorial systems for placing industrial objects, combining environmental and socioeconomic priorities for their location and further functioning.

Since the 80s of the last century, the increasing anthropogenic impact on the habitat has led to an active study of environmental issues by economic geographers [2]. The assessment of environmental factors for the location of industry formed the basis for ecological and economic regionalization [3]. The research used the approach of developing the ecological component of sustainable human development at the regional and national levels [4].

The most complete picture, with the allocation of zones of industrial objects location and its impact on the environment, can be represented by mapping where the industrial center serves as the object. It is one of the most widespread forms of territorial organization of industrial production and represents a local (within the boundaries of a populated area) group of enterprises [5]. Several lands were separately allocated, disturbed by the mining industry, and areas of maximum impact on the environment.

Based on the analysis of the influence of industry on various natural environments, the areas of maximum and local impact on the environment within the borders of the Baikal region and Mongolia have been identified. Using the cartographic method, the research results are presented in the form of a cartographic work [6, p. 74-75]. As the object of mapping serves the industrial center, since it is the most widespread form of territorial organization of industrial production and represents a local (within the boundaries of a populated area) group of enterprises.

## 3. Results and Discussion

Our research has determined that in all territorial units under consideration, the territories are mostly susceptible to anthropogenic impact from the industry of the fuel and energy complex, the pulp and paper industry and transport. Objects of the mining industry form local centers of anthropogenic impact.

On the territory of the Republic of Buryatia, the maximum industrial impact on the environment is observed in almost all large industrial settlements. The most significant ecological problems are in the areas: Zakamenskii, Kyakhtinskii, Gusinoozerskii, Nizhne-Selenginsky (Kamenskii), Severobaikalskii and Ulan-Udenskii. Industrial production of these territories is not only the main participants in the pollution of atmospheric air, but also suppliers of precursors of acid precipitation.

The main industries that have a negative impact include the enterprises of the fuel and energy complex, mining, pulp and paper and food industries. The main pollutants are Ulan-Udenskii CHPP-1, aircraft plant, locomotive and car repair plant, glass plant, Selenginskii pulp-and-cardboard mill, as well as light and food industries. Significant damage to the environment is caused by large and medium dumps of domestic and industrial waste [7].

The main share of emissions from stationary sources falls on the enterprises for the production and distribution of electricity, gas and water; processing industries and mining. In the structure of emissions, the share of solids is 24.8%, of gaseous and liquid is 76.2%.

The complexity of the ecological situation in the Transbaikal Territory is due to several reasons, primarily the active development and exploitation of uranium deposits, the extraction of brown and coal and precious metals. In large cities there are enterprises of food and light industry, building materials are produced. On the territory of the region, there are a number of features that exacerbate the negative impact of industry. First, in the region the largest, most powerful enterprises of the extractive and manufacturing industries were created for many years. Some of them are particularly dangerous for nature and people (non-ferrous metallurgy, oil refining, pulp and paper and chemical industry). At many of these enterprises, treatment facilities do not meet modern requirements. The second reason is that local meteorological conditions are unfavorable for purifying the atmosphere. Winter months are characterized by a large number of windless days and deep temperature inversions. The lowest air temperatures at this time are due to the maximum consumption of fuel and cause the highest pollution of the atmosphere by the products of their combustion. The largest area of negative impact on the environment is the Chita industrial complex, where the main pollutants are the enterprises of the fuel and energy complex (CHPP-1 and CHPP-2), machine building and metalworking, and transport.

On the territory of the Irkutsk region, the main sources of pollution are the mining, chemical and pulp and paper industries, as well as transport and energy companies. Irkutsk-Usolye-Sibirskoe, Ziminskii and Bratskii belong to the high-impact areas. Special areas with a high and very high impact on the lithosphere and hydrosphere were formed as a result of the mining industry in Bodaibo, Cheremkhovo, Tulun and Nizhneilimsk. In the Bodaibo district gold deposits have been exploited for almost a half a century (mainly by drainage method). In the Tulun and Cheremkhovo districts open-cast mining of coal is conducted, and in Nizhneyilimskii the iron ore mining.

The most unfavorable cities on the status of atmospheric air are: Bratsk, Angarsk, Shelekhov and Irkutsk. The air in these cities is oversaturated with such harmful substances as: benzapyrene, nitrogen dioxide, formaldehyde, hydrogen fluoride, carbon disulfide, suspended substances, carbon monoxide and sulfur dioxide [8]. The total amount of all pollutants in the atmosphere of the region by stationary sources of pollution amounted to 3126.687 tons in 2015. The highest rate in the cities: Angarsk is 1032.257 tons; Bratsk - 366.353; Sayansk – 557.841; Irkutsk – 390.830; Ust'-Ilimsk – 263.809; Shelekhov – 141.307 thou tons and in Usolye - Siberiskoe 130.619 tons.

The administrative center of the Irkutsk region - Irkutsk is included in the list of cities with the highest emission level. The main air pollutants of the city are car exhausts (52% of emissions), heat power sources that are not equipped with filters (46% of emissions) and production plants account for about 2% of all emissions.

The main wastewater discharges to the Angara are made by the Irkutsk Aviation Plant and the city's water utilities. The number of unauthorized landfills near the city is increasing.

A separate place is occupied by the Olkhon region, where the main source of environmental pollution is recreational activity. The problem of utilization of solid domestic waste, as well as pollution of the water area of Lake Baikal, is associated with tourist activities.

On the territory of Mongolia, the main areas of industrial impact on the environment are also represented by industrial centers where a large part of the population is concentrated and industrial enterprises are located (Ulan Bator, Darkhan, Erdenet, etc.) and local mining areas and light industry enterprises (wool and leather processing). In northern Mongolia, apart from Ulaanbaatar, Sukhbaatar and Altan-Bulak are distinguished, where the anthropogenic impact is less strong.

On the territory of Mongolia, the issue of the impact of industrial facilities on water resources is particularly acute. In the last 20 years, 852 rivers and 1131 lakes from five thousand rivers and lakes have dried up due to mining [9]. In addition, intensive pollution of water bodies is noted in all the allocated areas and large industrial centers (water quality belongs to 3-4 water pollution class), and the main elements-pollutants of water are petroleum products, phenols, and there is also increased oxidation [10].

The combination of extreme natural conditions and various types of anthropogenic impact establishes the most intense ecological situation in the central part of Mongolia. All natural areas in the area Naushki-Ulan-Bator and Kyakhta-Altan-Bulak are especially impacted. The main reasons of the environmental problems are the increased urbanization and industrialization of the country. The population in the country grew by 330 thousand people from 2012 to 2017 (in 0.8 times). In the metropolitan district Ulan Bator there is more than 1/3 of the total population of the country (1.405 thousand people for 2017).

## 4. Conclusion

The analysis of the location of industrial facilities and environmental conditions of the Baikal region and Mongolia enables identification of areas with highest industrial impact on all natural environments (large industrial centers), and with local impact on the environment of individual industries. Such spatial location is associated with the unevenness of economic development and population location. As a result, large industrial centers (Irkutsk, Bratsk, Ulan-Ude, Ulan-Bator, Darkhan, Erdenet, etc.) are mostly affected by a high concentration of industrial enterprises, which are characterized by significant emissions and sewage waters discharges in large volumes, as well as mining industries. It is the latter that pose the greatest danger of contamination of land, surface and groundwater by toxic substances from tailing dumps.

Ecological-geographical studies of industrial areas provide an assessment of anthropogenic impact both on the environment and on certain natural environments, taking into account the demographic pressure.

## References

[1] Gontar N M 2013 *Factors and Modern Features of Industrial Complex Location of Russia* (Moscow: Publishing House of the Plekhanov Russian Academy of Economics)

[2] Privalovskaya G A and Runova G G 1994 Regional approach to solving environmental problems *Izv. RAS. Ser. Geogr.* **5** 79-87

[3] Sharygin M D 1995 *Ecological and Economic Areas (Theoretical and Methodological Aspects of Development* (Perm: Perm-Perm University)

[4] Karakin V P and Baklanov P Y 2012 *The Far East: Formation of Sustainable Nature Management* (Moscow: Tov-vo Nauch. Izd., KMK) pp 400-24

[5] Ippolitova N A 2016 Mapping of industry as the basis of territorial organization of the economy *Geodesy and Cartography* **2** 38-45

[6] Ippolitova N A 2015 *Industrial Impact on the Environment* (Irkutsk: Izd-vo of the VB Sochava Institute of Geography SB RAS)

[7] Zabortseva T I 2014 Problems of the environmental protection infrastructure in Siberia: the economic-geographical approach *Izd. RAS. Ser. Geogr.* **5** 47-55

[8] Akhtimankina A V 2013 Atmospheric emission by industrial enterprises of Irkutsk *Izvestia of Irkutsk Gos. Universiteta* **6**(1) 3-19

[9] Basayev C 2018 *The gold rush in Mongolia struck Baikal* Accessed (in Russian) Income accessed online on July 30, 2018 via http://buryatia.asia/zolotodobyvayushhaya-promyshlennost-mongolii-udarila-po-ekologii-ozera-bajkal/

[10] National Atlas of Mongolia 2009 (Ulan-Bator: Mongolian Academy of Sciences)

# Problems of natural environment preservation of transboundary territories in uranium mining

**Z M Khasheva**[1], **V I Golik**[1,2]

[1]Southern Institute of Management, Krasnodar, Russia
[2]Vladikavkaz Science Center RAS, Vladikavkaz, Russia

E-mail: zarema_muratovna@mail.ru[1], v.i.golik@mail.ru[2]

**Abstract**. Within the transboundary territories of Far Eastern Russia, China and Mongolia, various non-ferrous and expensive rare earth metals are actively mined, with uranium being most actively developed among all of them. The extraction of radioactive ores creates an increased threat to population of transboundary territories and forms the problem that requires coordinated solution to life safety issues. Establishing unified extraction and mineral processing complex within these territories could potentially reduce danger to the environment and increase economic efficiency of metal mining. Decontaminated ore utilisation can become an effective factor in strengthening raw material resource base of metallurgy in both, mining and processing within consolidated regions of the Far Eastern regions. Leaching technology options are a real step towards creation of low and waste-free mining based on combination of physical and technical & chemical processes of metal mining. Peculiarities of natural resource localization and extraction within territories of Russian Far East, China and Mongolia create favorable conditions in establishing environmentally safe and highly profitable metal mining cluster development of deposits of exposed metal ores. Collective use of modern equipment, based on leaching processes can significantly improve environmental, economic and social indicators of regional metal deposits development.

## 1. Introduction

Mineral deposits of Russia, China and Mongolia contain significant reserves of non-ferrous metals and ores, and especially ores and minerals that are used in fastest growing industry – energy industry. Russia, China and Mongolia together produce 16% of the world uranium mining [1]. Uranium resides within a significant majority of complex ores of non-ferrous metals. In addition to the environmental threats during an extraction of non-ferrous metals, mining of radioactive ore has specific features that should be taken into consideration for environmentally oriented resource supply planning for transboundary territories [2].

The problem lies in the increase of the lost reserves amount due to the lack of cost - effective enrichment technologies. The rate of extraction of base metals from ores does not exceed 60%, and associated elements – 30%. Selective mining of rich ore in a market economy impoverishes remaining reserves and leads to an increased number of radioactive tailings on the surface. Tailings of radioactive ore and ore from underground mining contain valuable and dangerous metals, and constitute a significant portion in chemical pollution of the biosphere by natural leaching.

When in the process of production and economic activity arise contradictions between economic interests of industrial enterprises and ecosystems state, technogenic load reduction on the environment becomes an objective necessity. Key role in that – new technical policy where environmental interest of production and disposal of waste is of utmost importance.

Traditional development systems exclude off-balance reserves from production sphere. The emission for processing of all rock mass on the Earth's surface, separated from the array, meets modern conditions and leaving stocks, not meeting the requirements, in the depths creates conditions for natural leaching with migration of natural leaching solutions into the environment.

Cumulative impact of mining on the environment is carried out at all stages of mining, especially in stowage of substandard mineral raw materials, which are not subject to immediate processing, as although they contain valuable minerals they are too toxic to use.

Substandard mineral raw materials, located on the surface, are subject to dusting, and quite often concentration of dust in the air can repeatedly exceed sanitary standards at a distance of up to 10 km. Settling on the soil and water, dust generates excess concentration of heavy metals and toxic elements with negative consequences for living matter.

Current means to neutralize harmful effects of metals on living matter are ineffective, so the only real way to minimize danger of chemical and radiation pollution is waste-free metals extraction by means of "leaching" extracted minerals in underground blocks and disintegrators [3].

The purpose of present research is to substantiate creation of a "single cycle" mining and processing of mineral ore, that operates in a closed circuit.

## 2. Models and Methods

The object of research are methods of extraction and processing of substandard uranium ores to produce metals and components of hardening mixtures.

Present research aims to create technological basis for rational usage of technogenic waste in order to improve environmental situation of the Far Eastern enterprises.

To prove the possibility of achieving this goal, a retrospective analysis of leaching technologies practical application in the uranium mining Russian enterprises is conducted. Experimental leaching of metals in disintegrator is implemented. Based on the analysis of the modern concepts of radioactive ore mining, organizational and technological measures for nature and resource-saving technologies are proposed. A complex method of research is applied, including: system analysis and scientific generalization of literature sources, previous research results and production experience, data array processing, obtained as a result of researches using statistical methods and probability theory, mathematical modeling.

Comparative evaluation of possible solution options of the problem under study, taking into account results of predictive studies, conducted on similar topics, is carried out; research studies and selection of methods & tools, research directions and ways of solving problems; theoretical research of creation ways of highly effective innovative mining and ore processing technologies of ore recycling.

## 3. Results and Discussion

Traditional enrichment processes do not provide full minerals disclosure, therefore they cannot be claimed in the elimination of tailings enrichment by extracting metals from them. Combination of magnetic, gravitational and electrochemical separation and enrichment methods allows to emit metals from tailings into commercial products, but differs in complexity and increased costs. Therefore, for disposal of substandard reserves for today's technology, fundamentally new technologies are developed.

Underground leaching of uranium dates back to the second half of the last century. In the Charcase, Kiik-Tal, Taboshar, etc. mines uranium was leached in the underground blocks of ore containing 0,015-

IOP Conf. Series: Earth and Environmental Science **190** (2018) 012010 doi:10.1088/1755-1315/190/1/012010

0,017% [4]. From 1963 Bykogorsky deposit (Northern Caucasus) was developed by underground and heap leaching of off-balance ores. Technologies from leaching when developing Streltsovsky uranium deposits have been used since the 80s of the last century with the increase of their share to 30%. Leaching is carried out with sulfuric acid solution with 65% extraction. Off-balance ores heap leaching is conducted since 1974 to year-round operation extracting up to 50%. In heaps balance ores are leached with extraction of 85% with uranium content up to 0.15 %. In Tselinny mining and chemical plant leaching of uranium ore was carried out in a complex way (figure 1).

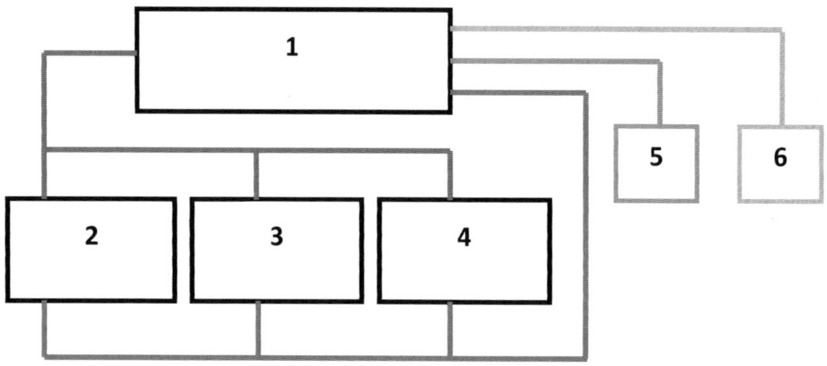


**Figure 1.** Scheme of complex metals leaching: 1-plant of solute preparation and processing; 2-pile; 3-leaching unit; 4-disintegrator; 5-pond-tailing; 6-concentrate.

The block of balance ores leaching of Vostok deposit in Kazakhstan is prepared according to the scheme (figure 2). Uranium extraction was 72% [5].

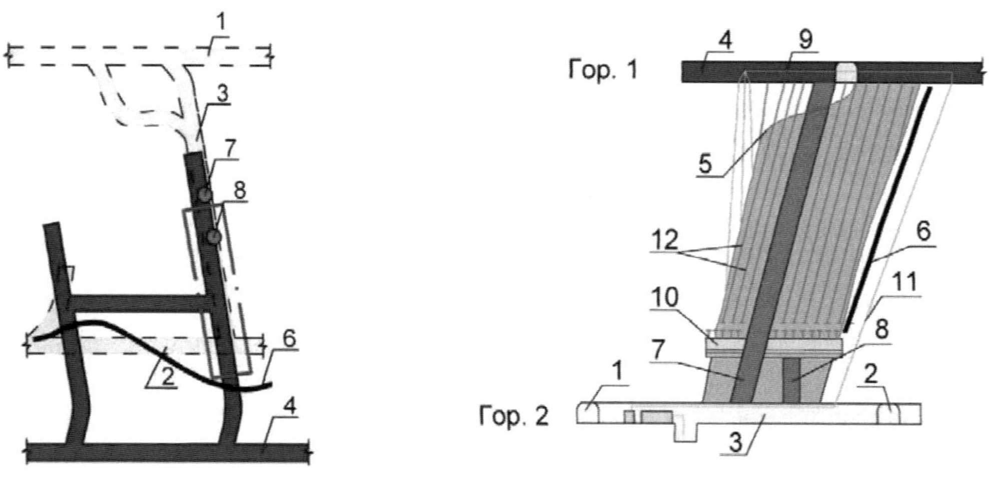


**Figure 2.** Block of underground leaching: 1, 2-drifts; 3-ort; 4-ort; 5-boundary of the ore body; 6-fracture; 7-rising; 8-dimple; 9-installation layer; 10-subsea layer; 11-solutions bore; 12-blast holes.

The reason for the increase in the number of substandard mineral raw materials in the storage is the prevailing trend of gross ore extraction from the subsoil in the expectation of technological progress for their enrichment on the surface, which in practice does not occur.

The possibilities of traditional enrichment technologies are limited by the use of solely mechanical energy. Modernization of traditional enrichment processes is carried out by involving hydrometallurgical and chemical processing operations, that increase efficiency of enrichment through the use of other energy types, but there are no real achievements in this direction yet.

The new direction of metals extraction from substandard metal-containing mineral raw materials is a combined chemical enrichment and activation in working body of disintegrator [6].

Leaching in disintegrator increases a general efficiency indicator-extraction of metals into solution, and processes combination of off-balance ores preparation and tailings in disintegrator and leaching further increases metals extraction into solution.

Elimination of chemical pollution sources radically changes present dangerous impact of mining on the environment and cannot be estimated due to the lack of reliable methods of calculation.

Mechanochemical technology of metals extraction from enrichment tailings is based on the phenomenon of their properties change at the speed of 250 m/s (figure 3).

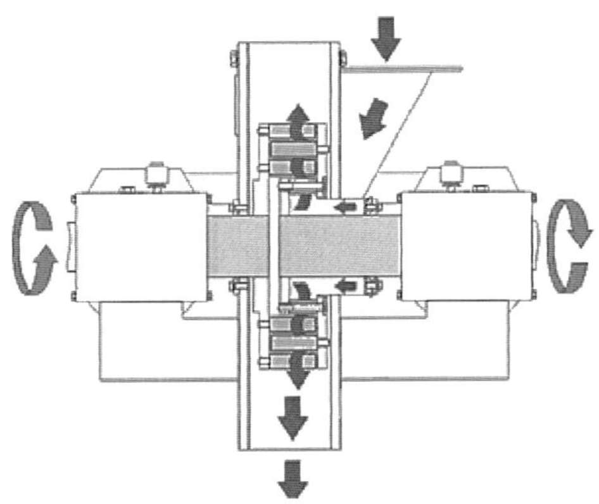

**Figure 3.** Disintegrator work scheme.

When processing substance with an overload of up to four hundred million free-fall accelerations, kinetic energy is accumulated in it, which is used not only for mechanical components activation, but also for chemical processes. Disintegrator plant creates electrically non-equilibrium charged centers in the mineral. Mineral is destroyed on crystals surfaces, processes of phase separation are simplified, and the yield of the activated product increases. The leaching solution is pressed into cracks, and metal is extracted into the solution along with mineral destruction [7].

Experimental lead and zinc extraction from enrichment tailings allows to conclude that leaching of tailings in disintegrator under identical conditions is significantly more effective than agitation leaching. It provides the same extraction, but reduces process duration by 2 orders, and reduction in residual metal content in secondary tailings is reached by increase in the multiplicity of processing.

Metals extraction from solution can be carried out by the option of membrane technology with salts and metals deposition.

Towards leaching technologies realization there is a number of problems: metals extraction from collective leaching solution, effective neutralization of uterine solutions, increase in mechanical and chemical durability of disintegrator working body, etc. constitute the second phase of innovative technology industrial development [8].

The fundamental difference of innovative technology consists in localization of mining products. When using the traditional technology non-conforming on the content of metals, separated from the array, and crushed ores remain in the mined-out space, while with underground leaching, the metals concentration is significantly reduced, less risk of metals migration beyond the mining circuit. The accumulated and current enrichment tailings after leaching are converted into environmentally friendly raw materials for related industries.

Technologies with metals leaching from ores allow extracting from substandard ores up to 70% of metals at the place of occurrence. Rich ores are produce to the surface and processed at the plant, and the rest are leached in blocks, stacks and disintegrators. By-products of the combined technology are metals, construction raw materials, gases, acids and alkalis.

Elimination of chemical pollution sources radically changes present dangerous impact of mining on the environment and cannot be properly assessed due to the lack of reliable methods of calculation [9].

Joint use of mining and processing cluster equipped with modern machinery for hydrometallurgical processes implementation in comparison with separate production of member countries will cut down expenses and will increase the enterprise competitiveness. The technology ensures minimization of load on the environment not by limiting the impact of pollutants at certain production stages, but by eliminating the source of chemical pollution [10].

## 4. Conclusion

Problems of preserving natural environment of transboundary territories in uranium mining can be solved by using leaching technologies that allow to limit the territory of production processes and increase safety of raw radioactive material management.

Common conditions of mineral localization, geographical proximity of deposits and possession of experience in the development of uranium deposits by breakthrough technologies allow to establish a powerful cluster for extraction and processing of chemically uncovered ores on transboundary territories of Russia, China and Mongolia.

## References

[1] Golik V I Bryukhovetsky O S and Gabaraev O Z 2007 *Technologies of Uranium ore Deposits Development* (Moscow: Moscow State Industrial University) p 131

Malatse M and Ndlovu S 2015 The viability of using the Witwatersrand gold mine tailings for brickmaking *J. of the Southern African Institute of Mining and Metallurgy* **115** 321-7

[2] Dmitrak Y V and Kamnev E N 2016 JSC "Leading design and survey and research Institute of industrial technology" - a path of 65 years *Mining J.* **3** 6-12

Babkin V V and Uspensky D D 2015 New strategy Chemistry-2030 *High processing of Raw Materials Clustering Chemicals Used Industry of the Russian Federation* (Moscow: Lika) p 222

[3] Hoevel V I, Razorenov Y I and Polukhin O N 2015 Metal extraction from ore benefication codas by means of lixiviation in a disintegrator *Int. J. of Applied Engineering Research* **1017** 38105-9

Sinclair L and Thompson J 2015 In situ leaching of copper: Challenges and future prospects *Hydrometallurgy* **157** 306-24

[4] Lyashenko V I and Golik V I 2007 Improving the technology and technical means for underground development of uranium deposits *Mining J.* **1** 11-4

[5] Golik V I, Gabaraev O Z, Maslennikov S A, Khasheva Z M and Shulgaty L P 2016 The provision of development conversion perspectives into underground one for russian iron ore deposits development *J. of Soc. Sciences* **1118** 4348-51

[6] Lyashenko V I, Franchuk V P and Kislyi B P 2015 Modernization of technical and technological complex of uranium mining production *Mining J.* **1** 26-32

[7] Golik V I, Khasheva Z M and Shulgaty L P 2015 Economic efficiency of utilization of allied mining enterprises waste *J. of Soc. Sciences* **106** 750-4

Stefanescu D and Foidas I 2017 *Recovery Factor and Replacement Ratio of Reserves, Key Parameters of Monitoring the Natural Gas Process* (Proc. of the 17th Int. Multidisciplinary Scientific Conf.) **17**(14) 843-50

[8] Shemetov P A and Glotov G N 2011 Theoretical foundations of automated systems Geotechnology uranium leaching *Mining J.* **11** 35-40

Karimi N S, Atashpanjeh A and Mollaei F M R 2001 Design considerations of heap leaching at the sarcheshmeh copper open pit mine *IMCET Int. Mining Turkey Cong.* p 513–16

[9] Lyashenko VI 2015 Improving environmental safety in the area of influence of uranium production *Izvestiya Vuzov Geology and Exploration* **1** 43-52

Gryazev M V, Kachurin N M and Stas G V 2017 Radon emission in underground coal mining and forecast of radon hazard cleaning and preparatory coal mine sections *News of the Tula State University Earth Science* **4** 303-11

[10] Krupskaya L T, Golubev D A and Volobueva N D 2016 Assessment of the environmental situation of the territory in the zone of influence of the tailings pond with toxic waste *Modern Trends in Science and Technology* **1-2** 97-100

IOP Conf. Series: Earth and Environmental Science **190** (2018) 012011 doi:10.1088/1755-1315/190/1/012011

# Influence of social and demographic aspects on increase of financial literacy of the population in Russia and China

**T I Klepikova**

Irkutsk National Research Technical University, Irkutsk, Russia

E-mail: klepiktv@gmail.com

**Abstract.** The development of financial literacy in a population helps to manage personal finances, as well as helps to shape the development of financial markets by attracting the public to participate in investment projects. Russia and China belong to a group of states characterized by the initialization of financial education programs for their respective populations. A comparative analysis of the financial literacy level, taking into account financial models of market development, shows China to have an advantage over Russia. The latest prediction models in the field of financial education indicated that China would rise high in the rating on the level of financial literacy of adult population among the G-20 countries. According to results of studies on financial literacy levels of 15-year-olds, conducted in the framework of the Program for International Student Assessment (PISA), Chinese teenagers were shown to be the most knowledgeable due to innovations in Chinese education. The experience of countries leading in the field of financial education of young people is being used to introduce programs in Russia. Target indicators of effectiveness of the financial knowledge of a population should include parameters that take into account both level of financial literacy and changes in people's behavior.

## 1. Introduction

The development of the financial literacy of a population is a vital necessity for every modern state. Individuals need the ability to plan future incomes and expenses, use credit, choose types of insurance, and manage personal finances. International practice shows that governments of industrialized countries are responsible for the development of financial services market. The qualitative growth of financial literacy of the population leads to both an increase in living standards of people and the responsibility of a country's citizens in matters of financial independence. Furthermore financial literacy of a population influences the quality of financial services provided to those citizens, leading to the transparency of financial markets and the otherall growth of the state's economy. As a result, the Organization for Economic Co-operation and Development (OECD) has begun to create a network of financial education programs to deepen and expand awareness of the importance of financial education in the world since 2008 [1].

## 2. Models and Methods

According to the OECD, depending on the participation level of states, states are conditionally divided into three types [2]:

- States with fully developed financial education programs and full government support for these programs;

- States with developed financial education programs and insufficient government support for these programs;
- States with the introduction into the development of financial education programs and certain activities and programs to increase financial literacy by the government.

Russia and China belong to the third group, which have begun to implement programs to ensure the financial education of the population. When considering the advantages and disadvantages of developing financial literacy within the populations of Russia and China, we will use the method of comparative analysis. Russia and China have their own peculiarities connected with economic, social, and regional processes within their respective countries. While both countries have ended complete state socialism, the paths taken towards the creation a market economy have been the same. Russia has been developing on a model marked with the predominance of imputed financial products and the passivity of the financial behavior [3]. This model is characterized by the use of salary cards. Salary cards are used for the calculation of income for more than 47% of Russians. In China, a "technocratic" model of financial market development has been applied. This has resulted in an adoption rate of 35-40% of the population using electronic payments, and 30-36% of the population using mobile financial services [4].

In 2008, the OECD started building an international network of financial education programs in order to deepen and expand awareness of the importance of financial education in the world [5]. In 2011, the OECD / INFE (International Network on Financial Education) published Measuring financial literacy Questionnaire and guidance notes for conducting an internationally comparable survey of financial literacy [6]. For G20 meeting in the summer of 2017, a report based on the comparative international study on financial literacy of the OECD G20 countries was prepared and released. In the final ranking, each participating country could receive a maximum of 21 points, which included a maximum of 7 points for financial knowledge, 5 for attitudes, 9 for behavior. The average financial literacy score for G20 countries is 12.7 points out of the maximum 21. Table 1 represents data of financial knowledge, attitudes and behavior for China and Russia.

**Table 1.** Financial knowledge, attitudes and behavior.

| | Overall score | Knowledge | Behavior | Attitudes |
|---|---|---|---|---|
| **China** | 14.1 | 4.7 | 6.2 | 3.1 |
| **Russian Federation** | 12.2 | 4.1 | 5.1 | 2.91 |

China took fourth place in the ranking, while Russia ranked ninth among the G20 countries.

Based on correct responses of Chinese and Russian respondents, in the analysis of financial knowledge components, the biggest difference in knowledge has been revealed in the estimation of interest plus principal and diversification (Table 2) [2].

**Table 2.** Components of financial knowledge. Percentage correct responses all respondents.

| | Interest plus principal | Diversification |
|---|---|---|
| **China** | 74% | 57% |
| **Russian Federation** | 48% | 41% |

Following the experience of the leading countries, the Russian financial market offers sophisticated and convenient financial products for consumers. Already about 80% of the population use financial services. However, experts argue that 71% of Russians have difficulty in choosing financial services. The greatest difficulties arise for Russians in the field of investment products and loans. Recent polls show that, 55 percent of Russians noted that they need to increase the level of knowledge and skills in the financial sphere [7]. While in China, about 81% of China's surveyed residents said that education in financial services is "very valuable," according to the Xinhua News Agency [8].

Thus, over the last decades of economic reform, financial literacy has become advanced in China. The Chinese believe that the ability to use financial knowledge and skills in the effective management of their finances will allow them avoid depreciating their assets and remain financially aware throughout their lives. The average financial literacy rate for Chinese residents was 63.71% according to the People's Bank of China (PBOC) [8].

However, sharp regional differences in the levels of financial literacy are confirmed both by experts from Russia and China. One of the key factors is the geographical extent of both countries. In Russia, the most vulnerable category of citizens lives in the North Caucasus Federal District, in China they live in Inner Mongolia. The most vulnerable categories of citizens in the field of financial literacy in the world community are consumers with a low level of education, youth, migrant workers, and rural residents. Nevertheless, innovations in the field of financial education make it possible to expand the opportunities for territorial development. At present, the main target groups are children and youth; every second program uses a wide range of tools and channels, among which the Internet becomes the most important. The OECD published results of the 6th international study on the level of financial literacy of 15-year-old schoolchildren from 18 leading countries of the world, conducted as part of the Program for International Student Assessment (PISA) in May of 2017 [9]. The PISA conducts a financial literacy study through a computer test every three years. The results of the survey show the level of financial knowledge and skills of 15-year-old schoolchildren in using such knowledge in making effective decisions to improve financial well-being and participation in the economy of their country. According to the OECD, in terms of awareness of the 15-year-olds in financial matters, the most knowledgeable were teenagers from China (603 points), Russian participants took the 4th place among 18 countries (512 points) [9]. Russia significantly improved its position, rising from 10th place in a similar test conducted in initial 2012 study [10].

## 3. Results and Discussion

The People's Republic of China has made incredible progress in developing the knowledge in financial literacy over the past decade. "As China is building itself into a financial power, the financial literacy level of consumers, is critical to the healthy development of the capital market," said Zhou Baoluo, a chairman of Junior Achievement China, a NGO in the field of financial literacy education [8]. In developing the financial education strategy, China's financial institutions, led by the People's Bank of China, conducted a macro-survey to determine the level of financial knowledge of the population. Based on the conducted studies, a strategic plan was developed to increase the level of financial literacy of the population, which includes the following activities:

- Conducting surveys every two years to adjust the road map and guidelines for the development of best practices at the national level, taking into account real needs;
- Introducing financial literacy into schools from the first class. In math lessons, when learning how to calculate interest, children are taught how to count income from a deposit or assess the terms of a loan;
- Financial institutions participating in trainings and training seminars, as well as other activities on financial literacy of the population;
- Using interactive methods of mastering basic financial knowledge, including comics, films, theater performances, and video games, each targeted at different age groups of the population [11].

The experience of China shows that a family plays a big role in involving young people in finding solutions to financial problems. According to the OECD, 46% of 15-year-old students from Beijing, Shanghai, Jiangsu, and Guangdong have bank accounts. Thirty-six percent of students earn money on holidays or are employed part-time, while 16% of students report having other informal earnings. Furthermore, 78% of students discuss money issues with their parents at least once a month [9]. All this makes it possible to increase the motivation of young people. The Chinese government does not stop at the achieved results, as the level of financial literacy of the adult population of China is still far from the industrialized countries. This inadequacy became quite apparent when in December 2017, the pyramid scheme Qbao.com tricked more than 200 million Chinese. For comparison, the number of people affected in China by Qbao.com is significantly larger than the entire population of Russia [12].

The Russian society faces even more difficult tasks in the field of financial education [13]. Large-scale measures to develop financial literacy by the population of Russia began in 2011:

IOP Conf. Series: Earth and Environmental Science **190** (2018) 012011 doi:10.1088/1755-1315/190/1/012011

- The Ministry of Finance of Russia, together with the World Bank, with the support of the Government of the Russian Federation, developed programs for financial education and consumer protection of financial services.
- The Pension Fund of Russia (FIU) provides training for the working population and youth on pension education, conducts presentations of the continually updated training materials of the FIU on financial literacy, which are freely available on the FIU website [14].
- In Moscow, on February 10-11, 2015, the International Conference on Financial Literacy and Financial Accessibility was held.
- On 9-17 March 2015, a week of financial literacy for children and young people in the Global Money Week Framework was held in partnership with Child and Youth Finance International.

The Ministry of Finance of Russia received the main international award within the framework of the "Global Money Week" for holding the "Week of financial literacy for children and youth" for the best results in 2016 [15].

- On February 13, 2018, the Fourth International Conference on Financial Literacy and Financial Accessibility (Finfin) was held in Moscow, the latest FinTech tools and successful customer service strategies were discussed, to improve the welfare of consumers of the financial market and effective practices of financial education of consumers [12];
- Practical manuals are being published, including textbooks on financial matters for students;
- Seminars are held on the organization of regional methodological centers for financial literacy in schools and secondary vocational education institutions [16].

One of the main criteria of the level of financial literacy is the formation of personal savings. The decline in the savings activity of the Russian population can be traced to the first half of 2017. The share of money incomes that went into savings is estimated at 7.4% of the total revenue, which is lower than in 2016 [17]. The decrease in the share of the population's savings in total income indicates the weakening of the level of work with the population in the formation of additional savings. According to NAFI research, only a quarter of Russians form a "safety pillow."

Credits and investment products as financial services cause the greatest difficulties for Russians. In nominal terms, the total amount of loans to the population at the beginning of May 2017 is equal to 10978.3 billion rubles, the total amount of debt is estimated at 10948.6 billion rubles. The share of savings of the population decreased from 16.1 percent in 2015 to 9.2 percent in 2017 [17], which indicates that the borrowers have not correctly assessed their capabilities and are not fully informed by credit institutions. The illiterate use of loans led to the fact that every fifth respondent with a loan (22%) indicated that payments account for more than 30% of income [16]. Thus, the primary task is to explain the importance of savings and educate the population of basic knowledge on calculating loans using effective technology and international practice.

As a result, the Government of the Russian Federation approved a strategy for increasing financial literacy in the Russian Federation for 2017-2023 in order to expand coverage and improve the quality of education in the field of finance. The strategy will provide citizens with access to reliable information that will help them protect their consumer rights [18].

The development of a joint strategy for the introduction of financial education is being discussed by scientists, experts, governments of the world's leading countries. To this end, international conferences and symposiums are being held. Currently in 2018, Russia is taking an active part in the development of a new concept in the field of financial education of the population.

The symposium Advancing financial literacy globally: implementation and innovation is to be held on the 4-5th of October, 2018 in Moscow. There will be a discussion of innovative implementation of new approaches, such as new digital trends, financial education for vulnerable groups, and consumer protection to improve credit decisions and outcomes. More than 200 international experts are expected to be in attendance.

Following the results and discussion described above, we can conclude that all activities carried out by Russian state structures, financial institutions, and public organizations confirmed the interest of the population in raising financial literacy and readiness to use financial instruments to improve their well-being.

## 4. Conclusion

Involving all segments of the population in the process of financial education requires enormous costs and efforts of interested parties on both sides. The main goal is to increase the financial literacy of young people, who will soon fully enter the difficult financial world. Modern development of innovative programs will accelerate the process of increasing the financial literacy of young people throughout the whole population.

The Russian society faces even more difficult tasks in the field of financial education and steps are taken to move forward. It is necessary to develop new approaches, solutions, and criteria for the success of programs to improve financial literacy and consumer protection. Practice shows that increasing knowledge does not increase the literacy of a person and thus is not enough to change behavior. The main task at the present stage in developing financial literacy, along with increasing financial education, is to change people's behavior and protect their consumer rights. This should be the main criterion for the success of financial literacy programs.

## References

[1] Alifanova E N and Evlakhova Y S 2012 Influence of financial literacy of population on development of financial institution and key segments of the financial market *Terra Economicus* **3** 115-20

[2] G20/OECD INFE report on adult financial literacy in G20 countries 2017 OECD 2017 Income accessed online on 5th May 2018 via http://www.oecd.org/finance/G20-OECD-INFE-report-adult-financial-literacy-in-G20-countries.pdf

[3] Blokhina T K 2012 Russian model of financial market: contradictions and prospects for development *Vestnik of RUDN University Economics Series* **2** 65-72

[4] Market analysis: availability of financial services in Russia 2016 *Banking Technologies* **245**(05) 38-43

[5] Stolyarova A A and Shakhnozaryan G E 2010 Analysis of the world practice of developing financial education and increasing the financial literacy of the population *Finance and Credit* **418**(34) 72-8.

[6] Kuzina O 2015 Financial literacy and Financial capability *Voprosy Ekonomiki* **8** 129-48

[7] The development of payment cards was discussed at the XI International Conference "Bank Cards: Practice and Transformation" 2017 *Banking Technologies* **4** 8-13

[8] Economic Watch: China eyes consumer financial literacy improvement 2018 Income accessed online on 8th February 2018 via http://www.xinhuanet.com/english/2018-02/08/c_136959736.htm

[9] Results from PISA 2015 financial literacy 2017 *The Report of OECD* Income accessed online on 19th April 2017 via https://www.oecd.org/pisa/PISA-2105-Financial-Literacy-Beijing-Shanghai-Jiangsu-Guangdong-China.pdf

[10] Bokarev A A 2015 It is necessary to promote for Russian citizens of reasonable financial behavior *Finance* **9** 3-6

[11] Faiberg T V 2016 Financial literacy increasing of the Russian population: Russian experience and possibilities of cooperation with China Int. *Conf. on Rus-Chin. Relations* (Irkutsk: ISU Press) pp 174-81

[12] Citi fund and Russian microfinancial center have taken the fourth International conference on financial literacy and accessibility in Moscow Issues of credit cooperation 2018 (in Russian) Income accessed online on 16th February 2018 via http://vkk-journal.ru/itogi-finfin-2018/

[13] Sushko N Y 2017 Financial behavior of the population and its determinants *Finance and Credit* **23**(12) 670-82

[14] Klepikova T V 2011 Strategic objectives of microfinance development in Russia *Vestnik IRGTU* **11** 240-4

[15] Bokarev A A 2017 For five years we have basis for system development of financial literacy *Finance* **6** 17-20

[16] Artemyeva S S and Mitrokhin V V 2018 Financial Literacy Assessment of Domestic and Foreign Young Learners and Recommendations for its Enhancement *National Research*

*Mordovia State University* **90**(22) 46-59 doi: 10.15507/1991-9468.090.022.201801.046-059
[17] Andryushin S A 2018 The financial market of Russia in 2019-2020; one step forward, two steps back *Banking* **4** 41-52
[18] Korkin R I 2017 Financial literacy as a factor of state economic security *Theory and Practice of Social Development* **12** 105-7 doi: 10.24158/tipor.2017.12.22

IOP Conf. Series: Earth and Environmental Science **190** (2018) 012012 doi:10.1088/1755-1315/190/1/012012

# The emphasis of Russian regions forest potential realization: efficiency and sustainable development

**S V Makar**

Financial University under the Government of the Russian Federation, Moscow, Russia

E-mail: svetwn@mail.ru

**Abstract**. In the modern world, Russia is the main "environmental donor", ensuring sustainability of the biosphere. Sustainable development of Russia and its regions is connected with the realization of forest potential. From ecological and economic position it is one of the fundamental components of the regional space, regardless of the forest area level of Russian region. Among the modern priorities of the Russian public administration is the efficiency of the timber industry complex (TIC) and the capacity of forestry. These aspects retain their long-term importance at the national level, with the strengthening of regional (subnational) emphasis: the impact of the forest sector on the regional situation and the socio-economic situation of most regions. The current economic practice shows the importance of combining the criterion of efficiency with the criterion of sustainable development for the realization-development of Russian regions forest potential. The author's interpretation of the forest potential as a set of two main functions of the forest, distributed over time - socio-economic and socio-natural, allows achieving this goal. The author's method of constructing an integral indicator of sustainable development of Russian regions forest potential is presented. The analysis of results of opportunities and threats combination is carried out; types of regions are identified. Problems of using the method of integral indices are indicated. The research uses methods of theoretical and empirical analysis, statistical, regional, strategic, content analysis, indicative, index methods, method of grouping, statistical clusters, verbal modeling, typology, logical and content analysis.

## 1. Introduction

Russia has the largest area of forests, the largest area of undisturbed economic activity of the territory, unique ecosystems and biodiversity [1].

The development of Russian territory, which main filling of the natural space is the forest, is a priority of the economic policy of the modern Russian state. This position should be combined with another urgent national economic task- to increase efficiency of the use of national forest potential [2]. Modern parameters of socio-economic development and spatial features of Russia make it necessary to take into account differences in the understanding and calculations of the effectiveness of the forest potential use.

Forest is a reproducible natural resource, a significant part of the national wealth as an intangible non-financial asset according to the system of national accounts, the basis for various types of economic activities and diversification of the regional economy. Forest retains its raw value as a source of timber-raw materials for various industrial activities, but its fuel [3], as well as recreational value increases. Realization of regional forest potential is a factor of organization (transformation) of regional space [4]. The resulting effect not only meets the regional interests (sustainable process of regional reproduction), but also has national and supranational importance. On this basis, it is necessary to take into account not only different types of effects (social, economic, environmental), but also the effectiveness of the realization and development of regional forest potential at different spatial levels – regional, national, supranational [5].

In addition, the effectiveness is complex, it is important to note it as an indicator of the effectiveness of public administration mechanisms - sectoral (forest complex) and territorial (federal and regional levels). According to the author, the main result of the strategy of realization and development of forest potential of the region should be the sustainability of its development as part of regional economy – a subsystem of national level. In this regard, the author developed a method of constructing an integral index of sustainable development of Russian regions forest potential.

## 2. Models and Methods

The transition to sustainable development has become the most important global trend in modern conditions. Modern strategies of development integrate economic, social and environmental approaches, and the welfare is increasingly considered as the summation of consumption, human development and environmental sustainability, taking into account their quality and stability [6]. As a result, socio-natural-ecological-economic space is represented as a system of elements of biological, geographical, economic and social nature. The principle of balance that provides consideration of this space (environment of economic activity) as the whole is put in a basis. Therefore, sustainability assessments are inherently multi-component and require integration of environmental, social, economic and institutional aspects [7].

There are two most developed approaches to building sustainable development indicators. The first is based on the construction of a system of indicators, each of which reflects individual aspects of sustainable development. The following subsystems of indicators are highlighted: economic, environmental, social, and institutional. The strict division of indicators into groups is rather conditional. Some indicators may reflect different aspects of sustainability. In this regard, separate indicators are interpreted as mixed - environmental-economic, environmental-socio - economic, socio-environmental, etc.

The second approach involves construction of an integral (aggregated) indicator, on the basis of which it is possible to judge about the degree of sustainability. Aggregation is usually carried out on the main aspects of development – environmental, economic, social, and institutional. In particular, the economic aspect combines economic structure, production and consumption. Social aspect combines health, education, equality, housing, security, and population. Institutional dimension includes organizations and capacities of social institutions. The step-by-step integration of private indices allows to create heterogeneous blocks of indicators. However, multi-step integration makes it difficult to understand what the resulting index shows, causing the problem of interpreting the result.

Compared to the integrated indicators of sustainability, the approach based on construction of indicator systems is more widespread in the world. However, it should be noted that among many foreign integrated assessments there are practically no measurements corresponding to national definition of "social and economic development", since this concept for western practice sounds very abstract in contrast to the traditional one in our country. Along with the construction of indicator systems, the development of integrated (aggregated) indicators is quite active. The presence of integrated indicators is especially valuable for decision makers. One such indicator could be used to judge about the level of stability of the country (region), the balance of the development trajectory. In that form, this indicator can be an analogue, for example, of GDP, which is now often measured by the level of economic development and wellbeing. The principal issue in the aggregation of information into indicators is the determination of weights of initial indicators to preserve their importance and to eliminate subjectivity as much as possible. The higher the level of information aggregation, the more difficult it is to weigh heterogeneous quantities. The author takes into account formulated [8] basic requirements for indicators:

- be significant for management decisions (policy relevance);
- be easy to understand for a wide audience (simplicity);
- adequately reflect the actual situation (validity);
- based on the available data and not require large costs for data collection (data availability);
- be representative, possibly complex (representativeness);
- adequately reflect changes (sensitivity).

The current ecological and economic situation in the country is to a significant extent the result of extensive development and accumulated structural deformations in the economy of Russia (the dominance of nature-intensive sectors and industries, raw material export orientation, etc.). It should be noted that in Russia the term "sustainable development" is still an alien one. "Green economy" occupies a narrow niche and is associated with unreasonably high costs [9]. The report "National assessment in the transition of the Russian Federation to sustainable development" contains a large number of sustainability indicators. It is emphasized that Russia is characterized by significant regional differentiation. Existing regional differences were the result of historical imbalances in the development of production, focus on the industry of primary processing of natural resources, backwardness and imperfection of the technologies used. This fact determines the need for differentiated regional policy on the basis of a single regulatory framework. Due to the above reasons and the considerable spatial extent of the country, it is reasonable for Russia to assess the progress in sustainable development goals in the regional context, which significantly distinguishes us from the world experience. Foreign systems of indicators are characterized by a small number of regional indices, although it is obvious that many methodological problems of sustainable development aspects are similar for countries and regions.

In countries with economies in transition, there is a clear conflict between economic development and environmental security. For Russia, the issue of developing a sustainable development strategy is relevant, since today's economic growth of extensive type is accompanied by aggravation of environmental situation. It is obvious that economic growth is possible only in conditions of stable level of environmental safety, so as the country's development experience of diversified assessments of sustainable development is becoming more and more popular [10].

Methodology for measuring the sustainability of regional development is only emerging. This is due to a rethinking of basic principles for measuring development, and to statistical constraints. Especially great are the difficulties of constructing integral indices; none of the existing indices can be called "trouble-free".

Building a system of indicators of sustainable development of Russian forest potential is associated with solution of a number of specific tasks. It is known that in Russian forest sector the main problems of the current period are not related to reserves, but to their use. In this regard, the transfer of sectoral indicators, which are typical for most countries with limited forest resources, is inadequate, and it is necessary to substantiate the basic indicators that reflect problems of Russian forest sector.

Based on the indicators recommended by the world community, in terms of sustainable forest management principles, Rosleskhoz prepared "Criteria and indicators of sustainable forest management and forestry of Russia." The criteria and indicators for sustainable forest management for Europe's forests (the pan-European process) and for the world's temperate and boreal forests (the Montreal process) were the basis for their development.

The author's interpretation of the forest potential as a set of two main functions of the forest, distributed over time: socio-economic and socio-natural, in addition to the construction of indicators related to sustainable development of forestry implies also construction of indicators related to the TIC, as well as the impact of socio-economic situation in the region on the level of sustainability.

Within the framework of this study, various approaches to the construction of indicator system were considered. Indicators were constructed in such a way as to give a quantitative description of the identified problems, basing on the database of state Russian statistics. The approach used contains a fairly "compressed" system of key/basic indicators. A number of priority basic indicators and their modifications based on the structure of "problem – indicators" are proposed. Key / basic indicators are selected to reflect the specifics of Russian forest sector, including features of modern period of development. In this regard, it should be noted that the list of indicators cannot be a frozen system and should be adjusted when trends and problems change.

According to the author, the method of constructing integrated index of forest potential sustainability in Russian regions is based on the interrelated assessment of two main sustainability components, which can be represented in coordinates: potential (opportunity) - risk (threat).

In accordance with the above - mentioned logic, a set of private indicators-indices was formed:

- characterizing the prerequisites for sustainable development of forest potential;
- defining the limitations of realizations of forest potential sustainable development.

The former are presented as opportunities for sustainability, the latter, on the contrary, as threats that provoke an increase in instability, characterized by assessment "the more, the worse".

Most often particular indicators are complex, performing interrelated socio-natural (environmental) or socio-economic (economic and social) functions. Often, a particular indicator of threat of this development corresponds to a particular indicator characterizing possibilities of sustainable development.

In general, normalization of private indicators-indices is made by finding a deviation from the average allrussian indicators (standardization of indicators is made by referring the value of each particular indicator for this region to the average value of this indicator for the Russian Federation as a whole). The overall rate of opportunities (or threats) is calculated as arithmetic value of the sum of partial indicators of opportunities/ threats. The composite index summarizes values of particular indicators (indices) in such a way as to take into account the contribution of each from position of stability [11]. Formally, all indicators receive equal weight in the calculation of the final index, since there are no universally recognized priorities in ranking economic, social and environmental problems. At the same time, in order to avoid the subjective approach to the maximum extent, each component is not weighed due to the lack of a representative number of expert assessments. Therefore, the values of all particular indices are taken into account with the same weight, based on equal importance of all the selected factors.

General indicator of opportunities united 6 particular indicators characterizing potential of sustainable development: health (resilience) of the population, production of gross regional product (GRP), forest supply of territory, sufficiency of protected areas (specially protected natural areas - SPNA), intensity of forest use, development of forest industry.

General indicator of threats grouped 6 particular indicators that form risks of forest potential sustainable development: demographic load on forest lands, pollution, forest capacity of TIC production, wood consumption by the population, fire danger, saturation of roads for development/reproduction of forest potential.

As a result, each Russian region is characterized by a quantitative assessment: how great is its potential as an object of sustainable development and how great are threats to sustainable development of forest potential of the region compared to the average Russian. The resulting process is presented as an outcome of positive and negative processes interaction.

## 3. Results and Discussion

The results of calculations of stability level for all regions of the Russian Federation on the basis of official data are presented graphically in the form of distribution of points-regions on coordinate plane, where the horizontal axis characterizes possibilities (arithmetic mean of particular "positive" indicators - indices) of sustainable development, and the vertical – threats (arithmetic mean of particular "negative" indicators - indices) (figure 1). The final stage is the construction of a matrix of regions based on the average values of opportunities and threats of sustainable development of forest potential (table 1). This greatly facilitates the interpretation and practical application of the technique.

IOP Conf. Series: Earth and Environmental Science **190** (2018) 012012 doi:10.1088/1755-1315/190/1/012012

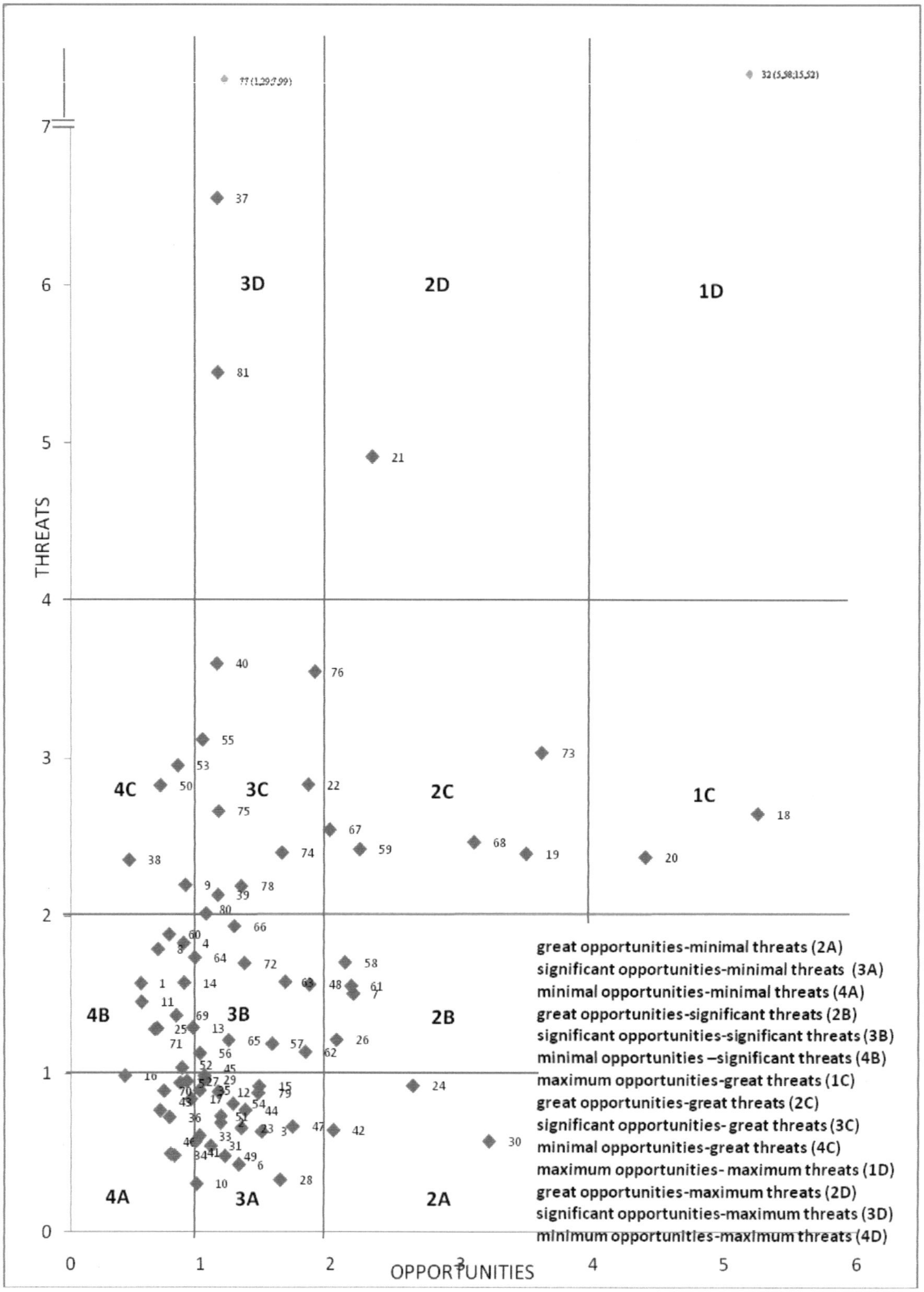


**Figure 1.** Sustainability of Russian forest potential in context of regions.

**Table 1.** Integrated index of sustainable development of Russian regions forest potential in the coordinate system " Opportunities "(O) – "Threats" (T). Regions are represented within each category as the threat indicator increases.

| Point № | Region | Point № | Region |
|---|---|---|---|
| | *great O- minimal T (2A)* | | *minimal O - significant T (4B)* |
| 24 | LeningradRegion | 52 | Samararegion |
| | *significant O - minimal T (3A)* | 56 | Sverdlovskregion |
| 28 | RepublicofAdygea | 65 | AltaiKrai |
| 6 | Kalugaregion | 71 | Omskregion |
| 49 | NizhnyNovgorodRegion | 25 | Murmanskregion |
| 30 | RepublicofIngushetia | 13 | Smolenskregion |
| 3 | Vladimirregion | 69 | Kemerovoregion |
| 42 | RepublicofMariEl | 11 | Oryolregion |
| 23 | Kaliningradregion | 1 | Belgorodregion |
| 47 | PermKrai | 14 | Tambovregion |
| 79 | Sakhalinregion | 64 | RepublicofKhakassia |
| 15 | Tverregion | 8 | Kurskregion |
| | *minimal O - minimal T (4A)* | 4 | Voronezhregion |
| 10 | Moscowregion | 60 | Chelyabinskregion |
| 34 | RepublicofNorthOssetia | | *maximum O - great T (1C)* |
| 46 | ChuvashRepublic | 20 | Arkhangelskregion |
| 31 | Kabardino-BalkarRepublic | 18 | RepublicofKarelia |
| 41 | RepublicofBashkortostan | | *great O - great T (2C)* |
| 33 | Karachay-CherkessRepublic | 19 | KomiRepublic |
| 2 | Bryanskregion | 59 | Yamalo-NenetsAutonomousArea |
| 36 | KrasnodarKrai | 68 | Irkutskregion |
| 51 | Penzaregion | 67 | KrasnoyarskKrai |
| 43 | RepublicofMordovia | 73 | Sakha (Yakutia) Republic |
| 44 | RepublicofTatarstan | | *significant O - great T (3C)* |
| 54 | Ulyanovskregion | 78 | Magadanregion |
| 17 | Yaroslavlregion | 74 | KamchatkaKrai |
| 12 | Ryazanregion | 75 | PrimorskyKrai |
| 70 | Novosibirskregion | 22 | Vologdaregion |
| 35 | ChechenRepublic | 76 | KhabarovskKrai |
| 5 | Ivanovoregion | | *minimalO - great T (4C)* |
| 27 | Pskovregion | 80 | JewishAutonomousRegion |
| 29 | RepublicofDagestan | 39 | Volgogradregion |
| 45 | UdmurtRepublic | 9 | Lipetskregion |
| 16 | Tularegion | 38 | Astrakhanregion |
| | *great O - significant T (2B)* | 50 | Orenburgregion |
| 61 | AltaiRepublic | 53 | Saratovregion |
| 58 | Khanty-MansiAutonomousOkrug | 55 | Kurganregion |
| | *significant O- significant T (3B)* | 40 | Rostovregion |
| 62 | RepublicofBuryatia | | *great O - maximum T (2D)* |
| 57 | Tyumenregion | 21 | NenetsAutonomousOkrug |
| 26 | Novgorodregion | | significant O - maximum T (3D) |
| 7 | Kostromaregion | 81 | ChukotkaAutonomousOkrug |
| 48 | Kirovregion | 77 | Amurregion |
| 63 | RepublicofTyva | | *minimal O - maximum T (4D)* |
| 72 | Tomskregion | 37 | StavropolKrai |
| 6 | ZabaykalskyKrai | 32 | RepublicofKalmykia |

In general, the proposed indicators and their aggregation were informative and applicable for integrated assessment of sustainability, although certain problems (reliability, technical capabilities of calculation, timely receipt of data) still remain.

IOP Conf. Series: Earth and Environmental Science **190** (2018) 012012 doi:10.1088/1755-1315/190/1/012012

## 4. Conclusion

The results allowed us to draw a number of conclusions.

- Calculation of integral index (matrix construction) is quite simple in comparison with other integrated sustainability indices, it includes balanced economic, social and environmental indicators, that allows to use it not only for development of the forest sector decisions, but also for assessment of socio-environmental and economic sustainability. This index is based on a stable statistical database and can be updated regularly.
- Results, based on a one-time observation demonstrate the significance of composite index (matrix). At the same time, they show that regular use of existing statistical base allows to trace the dynamics and predict further development.
- Further improvement of composite index (matrix) is advisable through specification of the list of economic, social and environmental indicators included in it, as well as introduction of weights for particular indicators, depending on the changes in their importance in the conditions of the state policy.
- An important goal is reflecting those aspects of economy that lie outside the monetary circulation. The rule of the market economy- "the good (service) that does not have a price/valuation does not exist in economic reality" distorts indicators on which the decision-making process is based. Therefore, the widespread use of natural indicators with further standardization (the assignment of the value of each particular indicator for the subject of the Federation to the average value of this indicator for the Russian Federation as a whole) in construction of indicators allows to display all aspects.
- One of the typical problems for Russia, which often remains out of sight, is the imbalance in development of individual components. Diagnosis of particular indicators of sustainable development allows to identify the "weak link" and to focus resources to address priorities, not just equalizing.

The problem of applicability of integrated indices, capable of measuring development sustainability is rooted, first of all, in the inconsistency, on the one hand of economic and, on the other, social and environmental development trends, that makes many integrated assessments difficult to interpret. Taking into account the mentioned problems, it is necessary to improve the methodological apparatus of sustainable development monitoring and integrated assessments. In general, we should note that a significant way needs to be done to turn the system of sustainable development indicators into a powerful tool of management, decision-making adjustments, optimization of resources (natural, financial) impact and understanding of the factors, determining dynamics of sustainability processes. The natural capital of our country plays an important role in the world economy, so the formation of environmentally sustainable development in Russia has not only national, but also supranational meaning.

## References

[1] Makar S V and Neshchadin A A 2014 Development of Russian forest potential: environmental priorities *Economy, Taxes and Law* **6** 42-8

[2] Makar S V 2010 Development of regional forest potential in coordinates of the state strategic planning *Regional Economy: the Theory and Practice* **38** 32-41

[3] Fedorenchik A S, Lednitsky A V, Kozhukhov N I and Nikishov V D 2010 *Biofuel from Wood Raw Material* (Moscow: SEI HPE at MSFU) p 384

Makar S V and Yarasheva A V 2017 Development of regional forest potential of Russia in the context of bio-economic trend *Managing Service, Education and Knowledge Management in the Knowledge Economic Era* 13-8

[4] Shamina L K, Makar S V and Kashin V K 2016 Cluster-oriented approach when forming and implementing region's potential *Scientific and Technical Vedomosti of St. Petersburg State Polytechnic University. Economics* **235**(1) 58-66

[5] Vdovin S M Exec. edit. 2017 Theory and practice of harmonization of the interaction of natural,

social and production systems of the region *Proc. of the Int. Scientific and Practical Conf.* **1** 31-9
[6] Thomas B et al 2001 *Quality of Growth* (Moscow: Ves mir) p 352
[7] United Nations CSD 1996 *System of Indicators of a Sustainable Development of the Commission on a Sustainable Development (CSD)* p 315
[8] Bobylev S N, Zubarevich N V, Soloveva S V and Vlasov Y S 2011 *Sustainable Development: Methodology and Measurement Techniques* ed Bobylev S N (Moscow: Economics) p 358
[9] Tkachenko A A 2014 Green economy and its future *Economy Taxes Law* **6** 26-31
Zamyatina M F 2018 Green economy as a basis for sustainable development of the region problems of transformation and regulation of regional socio-economic systems *Collection of Proc. SpB: Saint-Petersburg University of Aerospace Instrumentation* **43** 33-9
[10] Stroev P, Fattakchov R and Makar S 2018 *GIS-technology for Managing Sustainable Spatial Development of Russian Regions* (Moscow: A-project) pp 7-9
[11] Fedotova M A et al 2013 *Economic Development and Regional Diversity of the Russian Federation* (Moscow: Finance University) pp 77-116

# Meteorological potential of the atmosphere: Irkutsk – Beijing – Ulan-Bator

**O P Osipova**

Sochava Institute of Geography SB RAS, Irkutsk, Russia

E-mail: olga@irigs.irk.ru

**Abstract.** The air quality is one of most relevant problem connected with the quality of life for human population and preservation of the Earth's ecosystem. Here we examined and analyzed the atmospheric ability to self-purification according to the method proposed by T.S. Selegay over the cities of Irkutsk, Beijing, and Ulan-Bator. The meteorological potential of the atmosphere (MPA) was chosen as a criterion for estimating of the atmospheric self-purification. In summer season 2017 meteorological conditions over the cities of Irkutsk and Ulan-Bator contributed to the dispersion of impurities in the atmosphere. MPA in Irkutsk in summer was 0.99, in Ulan-Bator 0.78. In summer 2017 and in winter 2017-2018 in Beijing occurred unfavorable meteorological conditions (UMC) which did not contribute to self-purification of the atmosphere from pollution. Summer MPA in Beijing varied from 2.14 to 4.66 with an average of 3.69. During the winter months in all three cities MPA was >1 which indicates the accumulation of impurities in the atmosphere. In Ulan-Bator unfavorable conditions for dispersion were developed and the maximum MPA index (9.5) was observed in December. Calm weather made the biggest contribution to the deterioration of the self-purification ability of the Ulan-Bator atmosphere in winter. The maximum accumulation of pollutants in the atmosphere over Beijing occurs in winter. MPA varied from 15.0 in February to 19.0 in December with average value of 16.7. The fogs made the biggest contribution to the deterioration of the self-purification ability of the Beijing atmosphere in December-February.

## 1. Introduction

The monitoring of atmospheric composition and air quality control in cities is now of great attention around the world. Studies of the meteorological conditions for the dispersion of impurities in the surface layer of atmosphere in urban areas are promising [1]. Spatial and temporal variability of the atmospheric dispersion potential has the greatest significance in industrial centers [2]. In recent decades, the growth of megacities has led to increase in the anthropogenic pressure both on megacities themselves and surrounding regions. Strong air pollution leads to a significant increase in morbidity and premature mortality among the population. Studies carried out in recent years showed that the causes of many diseases were the negative effects of polluted air. In particular, this is applied to Moscow, Beijing, and Ulan-Bator. The level of air pollution depends on the variability of anthropogenic emissions and meteorological conditions of dispersion of impurities in the atmosphere. There are many characteristics and coefficients that describe the influence of meteorological conditions on the dispersion of pollution in the surface layer of the atmosphere [3]. Previously, specialists in atmospheric physics considered the "potential for atmospheric pollution", which took into account the climatic characteristics of surface air and the vertical stratification of the atmosphere

[4]. At present, the spatial frequency of aerological atmospheric sounding has decreased. Therefore, it is not possible to use the information about the vertical temperature distribution with necessary resolution. Specialists began to develop criteria for assessing the accumulation of pollutants in the air, based on surface meteorological observations [5]. Meteorological conditions are the main factors that influence the concentrations of pollutants in the atmosphere [6]. Weak wind, fog, inversions refer to unfavorable meteorological conditions (UMC), under which pollutants accumulate in the surface layer of the atmosphere. When the UMCs are observed, the quality of atmospheric air deteriorates and pollutants exceed the maximum allowable concentration (MAC). Precipitation and strong wind (with speed more than 6 m $s^{-1}$) refer to favorable meteorological conditions, which reduce concentrations of pollutants and contribute to self-cleaning of atmospheric air. The purpose of this study was to calculate the meteorological potential of the atmospheric (MPA) for Irkutsk, Beijing and Ulan-Bator and to assess the possibility of self-cleaning the atmosphere over these cities. In this work we used the MPA index developed by T.S. Selegay [5].

## 2. Study area end methods

The studies were conducted in three cities: Irkutsk, Beijing and Ulan-Bator (figure 1). Irkutsk is located in the Asian part of Russia in the territory of Eastern Siberia in 520 km north from Ulan-Bator, and 1650 km northwest from Beijing. Irkutsk is located on the two banks of the Angara River on the southern edge of the Irkutsk-Cheremkhovo Plain. Geographical coordinates of Irkutsk 52° 17' N, 104° 16' E. The elevation is 420-550 m above sea level. The climate of Irkutsk is sharply continental, with long and frosty winter and warm and rainy summer.

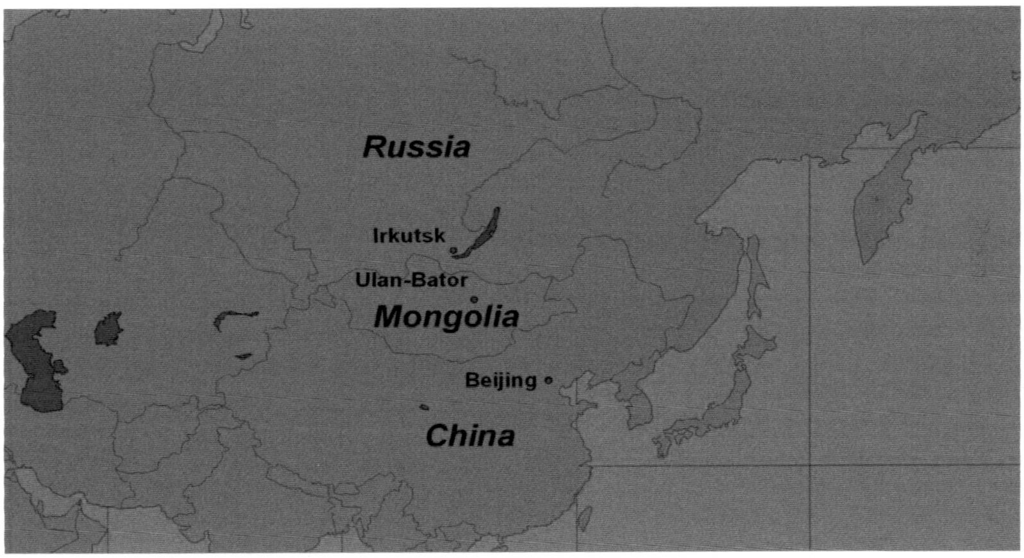


**Figure 1.** Geographical position of the studied cities.

Ulan-Bator is the capital of Mongolia, stretched for 20 km in length along the valley of the Tuul River. The city is surrounded by mountains from all sides. The geographical coordinates of Ulan-Bator are 47° 54 'N, 106° 52' E. The elevation is 1300-1350 m above sea level. The climate of Ulan-Bator is sharply continental with long and frosty and dry winter and cool summer. Among all capitals, Ulan-Bator has the lowest average annual temperature (–0.4 °C). Beijing is the capital of the People's Republic of China. From north and west sides Beijing is surrounded by the mountains of Xishan and Yangshan. Its geographical coordinates are 39° 54' N, 116° 23' E. The elevation is 40-50 m above sea level. The climate of Beijing is wet continental. The East Asian monsoon affects the climate in summer. The summer monsoon brings heat and heavy rains. Siberian anticyclone affects the climate in winter.

The study period was June-August 2017 and December-February 2017-2018. Meteorological data was taken from the site (https://rp5.ru/).

The meteorological potential of the atmosphere (MPA) was calculated according to the method of Selegay [5].

$$MPA = (Fc + Ff)/(Fp + Fv) \quad (1)$$

where Fc is the frequency of wind speed 0-1 m $s^{-1}$ (%), Ff is the frequency of days with fogs (%), Fp is the frequency of days with precipitation (%), and Fv is the frequency of wind speed $\geq 6$ m $s^{-1}$ (%).

The parameters of the formula were calculated from the data of meteorological stations. Repeatability of wind speed was calculated as follows: the number of cases with wind speed of 0-1 m $s^{-1}$ per month was divided by the total number of cases for this month. Repeatability of days with precipitation $\geq$0.5 mm and fog is calculated as follows: the number of days with precipitation and fog was divided by the number of days in a month. The results were expressed as a percentage. The numerator of the formula characterizes the processes in the atmosphere that contribute to the accumulation of impurities, and the denominator, respectively, the processes that contribute to the scattering of impurities. At MPA < 1, the processes favoring dispersion are predominant, in the case of MPA > 1 the impurities are accumulated in the atmosphere.

We used quantitative estimation of meteorological conditions by MPA criterion:

- MPA> 1.25 – unfavorable conditions for scattering;
- 0.8 <MPA ≤ 1.25 – relatively unfavorable conditions for scattering;
- 0.4 <MPA ≤ 0.8 – relatively favorable conditions for scattering;
- 0.25 <MPA ≤ 0.4 – favorable conditions for scattering;
- MPA ≤ 0.25 – extremely favorable conditions for scattering

## 3. Results

According to the formula (1), MPAs of Irkutsk, Ulan-Bator and Beijing were calculated. The results of the MPA calculations for three cities are presented in table 1.

**Table 1.** Monthly and seasonal values of MPA.

| City | Month (Season) | | | | | | | |
|---|---|---|---|---|---|---|---|---|
| | Dec | Jan | Feb | Winter | Jun | Jul | Aug | Summer |
| **Irkutsk** | 0.80 | 1.16 | 1.10 | 1.04 | 0.57 | 0.60 | 1.81 | 0.99 |
| **Beijing** | 19.03 | 16.02 | 15.00 | 16.68 | 4.66 | 4.28 | 2.14 | 3.69 |
| **Ulan-Bator** | 9.50 | 7.00 | 7.00 | 7.83 | 0.60 | 1.66 | 0.09 | 0.78 |

During the winter months in all three cities MPA was > 1, indicating the accumulation of impurities in the atmosphere during this season. In Irkutsk, the MPA changed from 0.80 in December to 1.16 in January. The contribution of calm, precipitation and fog to the formation of MPA for December-February is shown in figure 2. In Irkutsk January and February are characterized by relatively unfavorable conditions for scattering of impurities in the atmosphere, and December relatively favorable conditions for scattering. The contribution of precipitation to the self-purification of the Irkutsk atmosphere in December was 58%. Contribution of calms diminishing the dispersing capacity of the atmosphere in Irkutsk amounted to 32-39% in January-February (figure 2a). The haze was observed 19 times in Irkutsk in winter. In Beijing, during all winter months there were unfavorable conditions for dispersion. In 50% of cases, there was a fog in the city. MPA in December reached a maximum value of 19.03. The fogs made the biggest contribution to the deterioration of the self-cleaning ability of the Beijing atmosphere in December-February. Their contribution was 48-50% (figure 2b). In Ulan-Bator, unfavorable conditions for scattering were also developed throughout the winter months. The maximum index of MPA (9.5) was observed in December. Càlms made the biggest contribution to the deterioration of self-cleaning ability of the Ulan-Bator atmosphere in winter. In January, the contribution of calms was 90%, in December and February 61 and 71%, respectively (figure 2c). The haze was observed in winter every day in this city. The contribution of calm, precipitation and fog to the formation of MPA for December-February in Irkutsk, Beijing and Ulan-Bator is shown in figure 3.

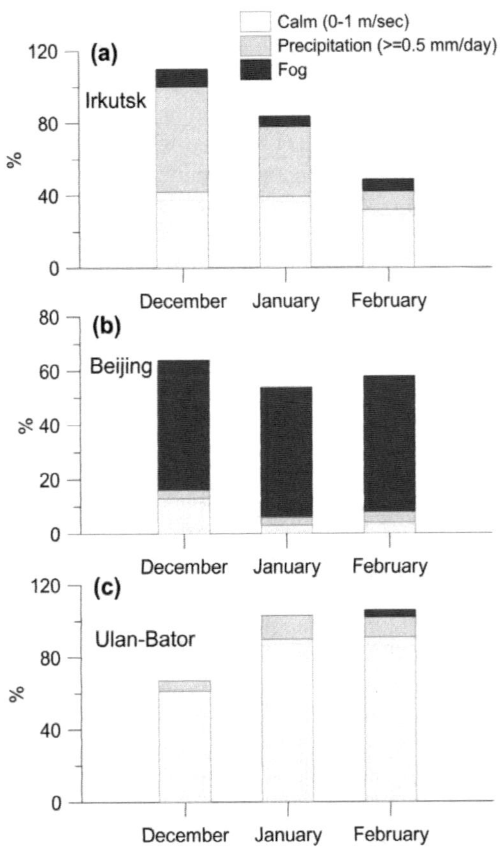


**Figure 2.** The contribution of calm, precipitation and fog to MPA during the winter months in the cities of Irkutsk (a), Beijing (b), Ulan-Bator (c).

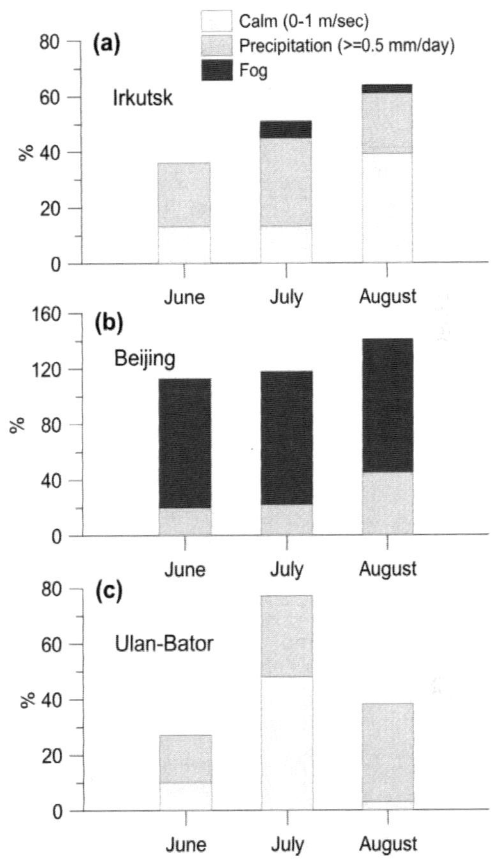


**Figure 3.** The contribution of calm, precipitation and fog to MPA for the summer months in the cities of Irkutsk (a), Beijing (b), Ulan-Bator (c).

During the summer season over the cities of Irkutsk and Ulan-Bator, the processes contributed to the dispersion of impurities were prevailing in the atmosphere. The contribution of precipitation in the self-purification of the Irkutsk atmosphere in June-July amounted to 23-32% (figure 3a). However, not all summer months were favorable for purifying the atmosphere in these cities. Meteorological conditions of August in Irkutsk did not contribute to the dispersion of impurities. Contribution of calm for this month was 39%. In July, unfavorable conditions for scattering were observed in Ulan-Bator (MPA is 1.66). The maximum contribution to the formation of highest MPA was made by calms, up to 48% (figure 3c). In Ulan-Bator extremely favorable conditions for dispersion occurred in August. The contribution of precipitation during this month was 35%. Hazes in summer were observed in 10% of cases, and sandstorms in 5%. In Beijing, during the summer season, the MPA varied from 2.14 in

August to 4.66 in June. In this city the observed UMCs contributed to the accumulation of pollutants in the atmosphere. The contribution of fogs to the deterioration of the self-cleaning ability of the Beijing atmosphere in June-July was 93-96% (figure 3b). The gloom was observed in 60% of cases during the summer season in Beijing. Thus, it can be concluded that during summer periods of 2017 in Irkutsk and Ulan-Bator meteorological conditions reduced the concentration of pollutants in the atmosphere. In Beijing, summer 2017 and winter 2018 were characterized by UMCs, which reduced the self-cleaning ability of the atmosphere.

## 4. Discussion

Atmospheric air pollution is most often considered using the values of air emissions, calculating the maximum allowable concentration (MAC), not taking into account the synoptic conditions under which these emissions occur. Accumulation or dispersion of impurities occurs in different synoptic conditions. Synoptic situations associated with anticyclonic processes contribute to increase in the pollutant concentrations. In winter, the atmospheric circulation over Irkutsk, Ulan-Bator and Beijing is determined by the impact of the Asian maximum and the prevalence of anticyclonic processes. The Asian anticyclone (Mongolian, Siberian) determines the weather conditions of winter months over these cities. A high atmospheric pressure is observed (figure 4).

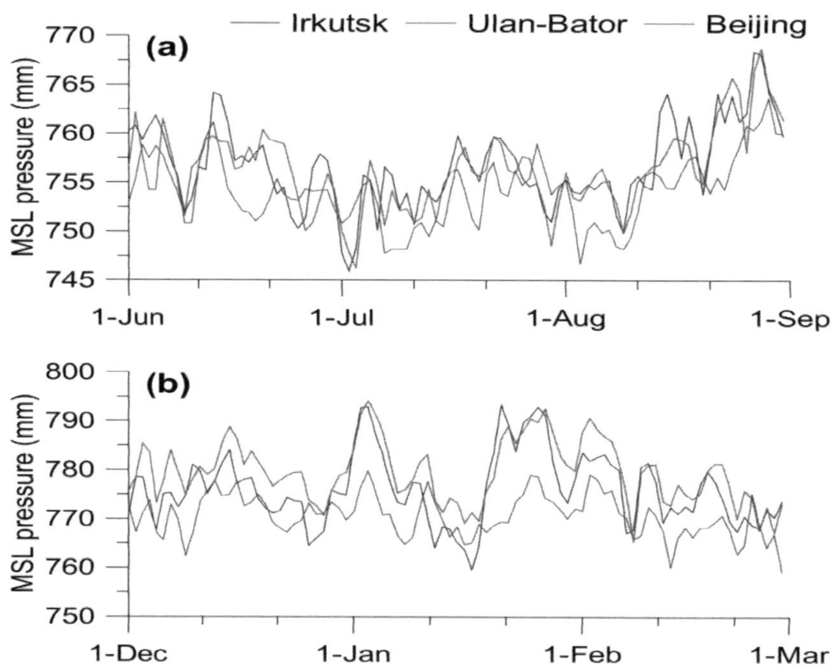


**Figure 4.** Monthly variation of atmospheric pressure, reduced to sea level in Irkutsk, Ulan-Bator and Beijing June-August (a), December-February (b).

In Irkutsk, the mean sea level pressure in winter period varied within 759-792, in Beijing 759-780, and in Ulan-Bator 765-794 mm Hg. In this case, the descending air movements are predominated, with little precipitations and developed calm. Calms hampered the exchange processes in the surface layer of the atmosphere. UMCs were observed, pollutants accumulated in ground layer of the atmosphere of the cities. This was often the cause of smog situations. Three main factors, mountain-hollow relief, Mongolian anticyclone, and coal burning contributed to reducing the dispersing capacity in Ulan-Bator atmosphere in winter [7]. The capital of Mongolia is one of the cities of the world with the dustiest atmosphere, which is confirmed by ecological and chemical research of the last decade [8]. In winter months, there was a maximum of atmospheric pollution in all three cities [9]. The exception was for December 2017 in Irkutsk, with MPA of 0.80 which is relatively favorable for scattering. The monthly precipitation in December in Irkutsk is 19 mm. However in December 2017 about 50 mm was dropped out (263% of the monthly precipitation norm). Thus, the contribution of precipitation to self-purification of the atmosphere of the city was 58%. In summer, there is a weakening of west to east air transfer and at the ground surface the lowered pressure field prevails (figure 4). In summer, the mountain-valley circulation is well expressed in Ulan-Bator which brings clean air from the mountains. In August, cyclonic circulation intensifies and large amount of precipitation falls out. Perhaps, these factors influenced the Ulan-Bator MPA during summer months. In August, the self-cleaning ability of the atmosphere in the city was maximal. The East Asian monsoon affects the weather in Beijing in summer. Synoptic situations associated with cyclonic processes and rapid air transport contributed to increased scattering power of the atmosphere. However, despite the fact that the mechanism of summer monsoon circulation is associated with cyclones, the Beijing MPA in summer was high. Perhaps such high level of air pollution is observed because of geographical location of the city: from the north and west, Beijing is surrounded by the mountains of Xishan and Yangshan. Thus, when the air pressure increases, there is no movement of air masses in the city. The coal industry and transport also contribute to the formation of smog in this city [10]. Beijing is recognized as one of the most environmentally dirty cities in the world.

## 5. Conclusion

The used method of determining the meteorological potential of self-purification of the atmosphere allows to promptly assessing the changing climatic conditions that determine the degree of atmospheric air pollution. The atmospheric processes of June-July 2017 were favorable for a limited dispersion for impurities (MPA=0.57-0.6) in Irkutsk, and in June in Ulan-Bator (MPA=0.6). The most favorable conditions for dispersion occurred in August in Ulan-Bator (MPA=0.09). This was facilitated by precipitation $\geq$0.5 mm, and their contribution was 35%. Precipitation leads to a noticeable purification of air from impurities. Unfavorable conditions for the dispersion over the entire period were observed in Beijing (MPA=2.14-19.03). UMCs contributed to high urban air pollution. The greatest contribution to UMC of Beijing was made by fogs (50-96%). Thus, the peculiarities of relief, synoptic processes, and wind and precipitation regimes have formed the MPA of Irkutsk, Ulan Bator and Beijing during the studied period.

## Acknowledgements

The present study was carried out within the framework of part of the research project (project # 0347-2016-003)

## References

[1] Andreev S S 2006 A short assessment of the bioclimatic resources of the Southern Federal District on the KM index (the coefficient of the climatic potential of self-cleaning of the atmosphere) *Izvestiya Vysshikh Uchebnykh Zavedenii. North-Caucasian region. Series: Natural Sciences* **5** 93–7 (in Russian)
[2] Sankaran S, Murugappan A, Kanakasabai V and Rajendran M 2012 Meteorological potential for air pollutant dispersion in urban and rural areas along the east coast of Tamilnadu *Int. J. Engineering Science and Technology* **4**(6) 2552–9
[3] Perevedentsev Y P and Khabutdinov Y G 2012 Meteorological potential of self-purification and the quality of atmospheric air in Kazan in recent decades *Bulletin of Udmurt University* **3** 23–8 (in Russian)
[4] Bezuglaya E Y 1980 *The Potential Meteorological and Climatic Features of Urban Air Pollution* (Leningrad: Gidrometeoizdat) p 184 (in Russian)
[5] Selegay T S 1989 Meteorological potential of self-purification of the atmosphere of the Siberian economic region *Proc. Zap. Sib. reg. NIGM* (Moscow: Gidrometeoizdat) **86** 84–9 (In Russian)

[6] Lebedeva M G and Krymskaya O V 2003 Air quality in the cities of Central Chernozem region *The Ecology of the Region* (Belgorod: Politerra) Part 1 p 74 (in Russian)

[7] Arguchintsev V K, Ariunsanaa B E and Arguchintseva A V 2013 Distribution of impurities in the atmosphere of Ulan-Bator *Izvestiya Irkutsk State University* **6**(2) 35–46 (in Russian)

[8] Kasimov N S, Kosheleva N E, Sorokina O I, Gunin P D, Bazha S N and Enkh-Amgalan S 2011 An ecological-geochemical assessment of the State of Woody Vegetation in Ulaanbaatar City (Mongolia) *Arid Ecosystems* **1**(4) 201–13 (in Russian)

[9] Sorokina O I and Enkh-Amgalan S 2012 Lead in the landscapes of Ulaanbaatar city *Arid Ecosystems* **2**(1) 61–7

[10] Golitsyn G S et al 2015 Studying the pollution of Moscow and Beijing atmospheres with carbon monoxide and aerosol *Izvestiya Atmospheric and Oceanic Physics* **51**(1) 8–19 (in Russian)

# The natural-resource basis of development of "the river gates" of the Asian Russia

**S A Rodomanskaya**

Far Eastern State Agrarian University, Blagoveshchensk, Russia

E-mail: svetlana_1902@mail.ru

**Abstract.** The article considers one of the largest territorial entities of the Russian Federation – the Amur Region which occupies border location at the south-east of the Asian part of the country. Lying in the geospace of the transport corridors, the region is so called "the river gates" for the Asian Russia and for all countries of the North-Eastern Asia. The territory under study belongs to the most economically developed regions of the Far East that is first of all, stipulated by the specific character of its economic-geographical location, the unique character of natural and climatic conditions, large areas of plain territories with dominating meadow-chernozemic land, very high agricultural development against the general background of agricultural structure of the land-utilization of the Far Eastern economic district. The relevance of the research is indicated by the growth of the regional investment prospects and also by the search of the first choice directions of the perspective use of the natural-resource potential of the Amur Region in the context of the geographical proximity to the most dynamically developing part of the world – countries of the Asian-Pacific region. The regional factor variety of the Amur Region development highlighted by the author – allows to consider it as a powerful and real reserve of the build-up of the food, fuel and energy, mineral-extractive basis of the development not only of the Russian Far East but countries of the Asian Pacific Region in whole.

## 1. Introduction

The current processes of the transformation of the international relations system and border (transboundary) cooperation are accompanied by the consolidation of the economic positions and accelerating development of the Far Eastern region targeting its development as a basis territory with ensuing consequences – the extraction and part time processing of resources that are in high demand in the countries of the Eastern and South-Eastern Asia [1]. The incentive for such interests are high rates of the Asian-Pacific Region (APR) development, and integration of the Far Eastern Federal District (FED) into the North-Eastern Asia is very perspective for Russia as "a transport-logistic corridor" of the Russian products promotion to the most perspective world markets, one of which is Asian Pacific [2]. The administrative-territorial entities of the Far East varying in the degree of the development in view of their geopolitical location and task-oriented regional policy of the Russian Federation are natural guides of Russia's joining APR [3]. For Far Eastern regions such border cooperation is very important in the context of the compensation of business ties weakening with the regions of European Russia, Ural, Western and Eastern Siberia because of the sharp increase of transport rates.

One of the largest territorial entities of the Russian Federation which occupies border position with the longest frontier along the Amur River (almost 1250 km) is the Amur Region [4]. The territory of the region has very important geopolitical location in the south-east of Asian part of the country, lying in the geospace of the transport corridors and is the region of so called "the river gates" for the Asian Russia and for all countries of the North-Eastern Asia. The backbone natural center piece of this cross border territory is the Amur River with its feeders [5]. The territory under study belongs to the most economically developed regions of the Far East that is first of all, stipulated by the specific character of its economic-geographical location, the unique character of natural and climatic conditions, [6] large areas of plain territories with dominating meadow-chernozemic land, very high agricultural development against the general background of agricultural structure of the land-utilization of the Far Eastern economic district. [7]

The relevance of the research is indicated by the growth of the regional investment prospects and also by the search of the first choice directions of the perspective use of the natural-resource potential of the Amur Region in the context of the geographical proximity to the most dynamically developing part of the world – countries of the Asian-Pacific region.

## 2. Models and Methods

The methodological procedures of the research are complex approach in conjunction with the monographic, structural-functional and organizational-economic analysis which supposes the use of the comparative-geographic factor and other general scientific methods. The case study was made on the basis of the analysis of the prevailing scientific-practical beliefs of different skilled specialists including the natural-resource potential assessment and the study of the territorial organization of agriculture.

## 3. Result and Discussion

Considering the Amur region as the contact zone with neighboring economies of the North-Eastern Asia we can distinguish some regional forming factors that are, in our opinion, one of the most important forces of organizing and reconstruction of cross border geo space. A spatial inequality which is revealed through the cooperation of background regional forming factors should be considered the objective condition of the cross border regional formation display.

### *3.1. Physio-geographical factor*

The main factor transforming natural and natural-resource systems in the Russian and Chinese parts of the Amur River basin is *physio-geographical.* Hilly-flat structure of the Amur Region current terrain and sublatitudinal location of the main mountain groups determined by the factors of the geological and geomorphic structure create conditions for the natural diversity and unique climatic conditions [8]. The peculiarities of the physio-geographical location, vast length of the territory determine great spatiotemporal fluidity of climatic variables with the significant rapid change of season temperatures and reduced number of the annual precipitation cause inequality in settlement and land use [9].

The south and central parts of the region are settled more thickly, here the main business activity is concentrated, and they carry the main anthropogenic load in combination with industrial and agricultural use of land. On this territory fuel-energy, production, transport and ameliorative infrastructure, multiple purpose lands of agro ecological potential are concentrated. The north and northern-western parts of the region are areas of new development that are rich in mineral-extractive, forest and hunting resources in combination with mining and wood and paper use of land. A large part of this territory presents areas for traditional management of nature and resettlement of Indigenous Minorities of the Far East. At the same time weak sustainability of natural complexes to anthropogenic influence is observed here which is stipulated by topographic features and climate patterns in the north of the region. Due to the remoteness of the region from the European regions and industrial centers (7-8 thousand kilometers) there is a curb on resources development, the use of new low waste and minimum impact technologies which creates transport problems and high expenses on transport traffic that has a great influence on the competitiveness of the production and natural resources.

IOP Conf. Series: Earth and Environmental Science **190** (2018) 012014 doi:10.1088/1755-1315/190/1/012014

### *3.2. Demographic factor*

As for labor force we can say that as a result of a high migration mobility the region has a tendency to their reduction and correspondingly to their deficit. The region ranks the fourth among regions of the FED and 67th in Russia in the number of inhabitants (805.7 thousand people as of January, 1, 2016), in the density of population (2.2 person per 1 $km^2$) – it ranks 4 th and 78 th correspondingly. According to the regional statistics the level of employment is 54.5% in the region against 60.3% in the Russian Federation and 60.9% in the Far Eastern district. The contemporary state of the number of inhabitants causes, on the one hand, concern from the position of the economic life of the region – the destruction of the mature labor potential and lack of labor resource, [3] on the other – from the position of geostrategic and geopolitical consequences [10].

### *3.3. The factor of natural resource use*

Recently the Amur Region has been seen as the region of particular strategic and economic interests as the natural-economic state of the region is represented by a large variety of resources – energy, mineral, forest, land, water, biological, recreational and ecological.

**Table 1.** The data of natural-resource potential of the Amur Region.

| Data | Volume |
|---|---|
| Mineral raw material stocks, bln. dollars | 400 |
| Land resources, mln. ha | 3.5 |
| including: the area of farm land | 1.5 |
| Stocks of iron-ore deposits, mln. tons | 388 |
| Stocks of titanium-magnesium ores, mln. tons | 271 |
| Stocks of agronomical ore and bioorganics, mln. tons | 600 |
| Stocks of coal, bln. tons | 70 |
| The total capacity of energy generating capacities, Mega Watts | 3722 |
| The total stock of standing timber, mln.cu.m | 1900 |
| The length of general purpose railroads, km. | 3222 |

They create foundation premises of business activities and depending on the number of natural characteristics and peculiarities of operational factors (management, economy, technologies etc.) are suppliers of natural resources to Russian and even world markets. It is especially important from the perspective of cooperation between the government and business not only in projects of development but also effective use of mineral resources of the region. Their implementation allows the region to join more actively the common economic space of Russia, to broaden relations with APR countries and other border territories, to improve the efficiency of the use of the rich natural-resource potential. First of all, gold reserves, reserves of titanium raw materials necessary for the creation of own base of iron industry, energy generating capacities.

One of the most important peculiarities of the region is the unique agricultural potential. The leading industries of agriculture are_soya cultivating, beef and dairy cattle breeding, chicken farming, beekeeping that form agrifood market, food supply and economic security, labor and settlement potential of rural area. The region accounts for 60% of crop land of the Russian Far East. The area of crop land is 38% of all land area in the region, 3.5 mln. ha are in the agricultural use as of 01.01.2016. There is 2.6 hectare of agricultural lands calculated per a person including 1.2 hectare of arable land [5]. In the structure of land under cultivation 76.4 % is accounted for soya, 16 % for grain crops, 5.5% for forage crops, the rest of the area is a cropland with gourds, potatoes and industrial crops.

It is necessary to note that in the FED the region accounts for approximately: 59% of grain production, 32 % of meat, 28% of milk, 25% of potato, 17% of eggs, and 16% of vegetables. So in 2016 the proportion of the Amur Region in the whole share of the FED agricultural production was 28.6% or 423.8 mln.rub (2008 – 20.9%; in 2009 – 19.6%).

Today the Amur Region remains an absolute leader among the regions of the Far East in the index of soya production, for example, the production of soya has increased by 1.3 times since 2012. The whole yield of grains in the region has had mostly positive trends over recent five years. The data of table 2 shows that compared to 2012 there has been a significant increase by 75% and by 35% compared to 2015.

We assumed the analysis of food security showed that 80% of intraregional market of food production is formed at the expense of its own resources. Moreover the region does not only meet domestic needs in grain and potato but can without any prejudice to itself supply them to the other regions of the country and export them to the countries of the APR, export of grains especially soya is growing annually (for example, export of soya has increased by 68% over five years). The majority of grown potato provides the needs of the regional population.

No doubt in this respect the region can be considered as a powerful and real reserve of increasing food base of the development of the Far East and APR countries. Given the opportunity the agricultural sector of the region can ensure food products not only to its population but also cater high quality and competitive production to other regions which are industrial and where agricultural sector cannot be developed due to climatic environment. [11, p.74] In the midterm of the agricultural sector development it is planned to build seed production companies to produce selected high producing seeds of soya and wheat, development of live stock breeding complexes and feedlots to increase the production of meat and milk.

**Table 2.** The output indices movement of main kinds of agricultural production in the households of all categories of the Amur Region.

| Name of parameter | 2012 | 2013 | 2014 | 2015 | 2016 | Rate of increase 2016 г. % to 2012 г. | Rate of increase 2016 г. % to 2015 г. |
|---|---|---|---|---|---|---|---|
| **Grains, th. tones** | 271.3 | 172.3 | 417.7 | 351.0 | 474.6 | 174 | 135 |
| **Soya, th. tones** | 682.4 | 649.7 | 766.3 | 884.9 | 893 | 130 | 100.9 |
| **Meat of live stock and birds for slaughter (light weight basis), th. tones** | 50.3 | 60.3 | 60.5 | 60.6 | 59.0 | 117 | 97.3 |
| **Milk, th. tones** | 161.8 | 165.1 | 143.6 | 148.6 | 147.7 | 91.3 | 99.4 |
| **Eggs, mln. pieces** | 231.0 | 239.4 | 240.1 | 203.1 | 199.6 | 86.4 | 98.3 |
| **Potato, th. tones** | 296.0 | 118.0 | 298.1 | 286.6 | 278.1 | 93.6 | 96.4 |
| **Vegetables, th. tones** | 69.3 | 35.0 | 67.8 | 69.7 | 67.2 | 96.4 | 97.0 |

The total hydroelectric potential of big and average rivers of the Amur Region is estimated at 76 bln.kWh. which amounts to 3.2% of all Russia and 7.5% of the FED. The technological capacity of the region is 51 bln.kWh. or 3% of the similar potential of Russia and 7.5% of the FED. At present the technological hydroelectric potential is used at 50%. The specific degree of territory saturation with hydro energy compared to the other regions is pretty high – 209 megawatt hour/km$^2$.

Another major component of the natural potential of the Amur Region is forest resources. The total area of forest reserves lands and lands of other categories that are occupied with forest is 31733 th.ha as of January 1, 2016 with the largest share of exploitable forest of 86%. The forest cover percent is 65.4% that according to the existing gradation of the forest cover percent factors refers the territory of the region to the richly wooded. However at the immense territory of the region the forest cover percent factors vary enormously in administrative districts from the north to the south and from the east to the west. According to the calculated index of the forested area per person the highest availability is noted in the northern districts (Zeiski, Magdagachinski, Selemdzhinski, Skovorodinski, Tindinski, Shimanovski) from 56 ha/ per person to 436 ha./per person which occupy three fourth of the regional territory with the forest cover percent factor more than 60%. In nine districts of the region the forest cover percent factor is less than 60% and in the main cultivated areas located on the territory of the Zeisko-Bureinskaya plain it is less than 5% and correspondingly the availability of the forestland is low and very low (Tambovski – 0.04 ha/per person, Ivanovski – 0/08 ha/per person, Belogorski – 0.15 ha/per person, Konstantinovski – 0.12 ha/per person and Mikhailovski – 0.7 ha/per person).

### 3.4. *The factor of the regional economy*

The economy of the region is diversified economy. In the structure of the industrial production a specific place is held by energy-extractive trend and high proportion of the major fuel-and-energy companies, nonferrous metals industry and food industry. The region ranks first in the federal district in the power supply and ranks among leaders of the country in gold reserves and production output of gold. In the structure of GRP the greatest proportion is taken by transport and communication services (18%), construction (12%), extraction of mineral resources (13%), wholesaling and retailing (11%).
The foreign trade turnover of the Amur Region in 2016 was 505.3 mln. US dollars (in 2012 – 1206.2), including export – 340.2 mln. US dollars (398.6) and import – 165.4 mln. US dollars. (table 3).

**Table 3.** The foreign trade turnover of the Amur Region (2016).

| Name of parameter | Export | | Import | |
|---|---|---|---|---|
| | mln. dollars | % | mln. dollars | % |
| Food products and agri supplies | 100.4 | 29 | 37.4 | 23 |
| Mineral products | 207 | 61 | 2 | 1.2 |
| Fuel and energy products | 175.5 | 52 | 1.9 | 1.2 |
| Production of chemical industry | 0.73 | 0.22 | 21.9 | 13 |
| Hides, furs | 0.08 | 0.02 | 0.3 | 0.16 |
| Wood substance and pulp and paper production | 25 | 8 | 1.8 | 1 |
| Textiles, textile goods and footwear | 0.2 | 0.06 | 10.4 | 6 |
| Metals and articles made there from | 5.3 | 2 | 12.9 | 8 |
| Cars, machinery and transport | 0.9 | 0.29 | 65.3 | 39 |
| Other goods | 0.08 | 0.003 | 13.2 | 8 |
| **TOTAL** | **340.2** | | **165.4** | |

Among countries-counter parties the largest trading partner of the region traditionally has been China (91.7% of export and 80.9% of import), the region also has external economic links with more than 60 countries of the former Soviet union and non-CIS states. The main share of export in 2016 was accounted for by mineral products (61%), food products and agri supplies (29%), wood substance and pulp and paper production (8%) (in 2015 – 31.2%, 36.4% and 16.2% respectively). In the structure of import the top performers are cars, machinery and transport (39%), food products and agri supplies (23%), production of chemical industry (13%) (in 2015 – 45.4%, 23% and 12.9% respectively).

It should be pointed out that generally there is a stable model of the external turnover of commodities in the Amur Region: Russian party supplies raw materials, Chinese – consumer goods. The foreign trade turnover of border territories is characterized by the inequality of dynamics. The general trend to increase the turnover of commodities is seen. The comparative analysis of the contribution of border Far Eastern territories to the cumulative foreign trade turnover with China shows that the greatest proportion is taken by Primorski Krai – 79%, Sakhalin Region – 24%, Khabarovsk territory – 23%, the Amur Region – 8%. The foreign trade turnover of the Amur Region in the total volume of the FED is 2.1% and ranks the 7$^{th}$ among regions of the Far East. [12]

IOP Conf. Series: Earth and Environmental Science **190** (2018) 012014 doi:10.1088/1755-1315/190/1/012014

Recently the policy of changing the dynamic development model of the region has been pursued. The major components of this process under the guidelines of "The strategy of the social-economic development of the Far East and Baikal Region for the period to 2025" became priority development areas "Belogorsk", "Priamurskaya", and "Svobodni". The major economic projects of the federal importance – Zeiskaya hydro-electric power station (1300 MWt.), Bureiskaya hydro-electric power station (2010 MWt), Nizhne-Bureiskaya hydro-electric power station (302 MWt), the main gasline "Power of Siberia", cosmodrome "Vostochni", modernization of the Baikal-Amur Mainline and Trans-Siberian Railway – are the powerful resource and that way support the interest of the country in the Amur Region. So from 2000 to 2016 the volume of investments increased by 41.5 times from 0.4 to 16.6 bln.ru with the largest share taken by Fuel and Energy Complex (more than 50%).

In this regard it must be stressed that new perspectives of the international export infrastructure development on the territory of the region will be ensured, first of all, on the basis of the largest investment projects implementation that are integrated into the state-run program of the Russian Federation "The strategy of the social-economic development of the Far East and Baikal Region" through federal budget resources, which includes:

- the construction of the cross border bridge crossing the Amur river in the area of Blagoveshchensk – Heihe (China) with the total length of 1080.66 m;
- the development of the metallurgical cluster – opening of Garinski deposit and construction of Garinski mining and smelting enterprise with the capacity of 7.2 mln .tons of iron-ore concentrate, the Chinese party shows great interest to it;
- the construction of the gas chemical complex as the course of the investment project implementation on the construction of the Amur gas processing plant of advanced processing of natural gas in the area of Svobodni;
- the construction of the heavy-duty oil refinery with the capacity of 6 mln. tons of petroleum products with the export of the part of goods to China;
- the construction of the thermal electric power stations with the total capacity of 7920 MWt based on Erkovetski brown coal field in framework of project implementation on large scale export of power from Russia to China.

### *3.5. Tourist-recreational factor*

It is important to note that recently tourism has become one of the most dynamically developing industries of the Amur Region. In 2017 the Amur Region hit top 10 on implementation of Federal Target Program "The development of domestic and inbound tourism of RF" and ranked third in Russia and second in the FED. Today the most popular tours are Blagoveshchensk, Ivanovka, Muravyevski park, Hinganski reserve, Bureiskaya and Zeiskaya hydro-electric power stations, Albazino and Zeiskoe Sea. In the short term of tourist-recreational complex development it is planned to create comfortable conditions for the tourist sector of the economy through the development of the tourist-recreational cluster "AMUR" including tourist-entertaining zone "Golden mile", and it is planned to have mass excursions as part of cosmic, ecological and rural destinations of tourism.

## 4. Conclusion

Summing up it must be stressed that the territory under study is a complex natural anthropogenic system and natural-industrial peculiarities of the Amur Region have striking area-based character. The regional factor variety of the Amur Region development (physio-geographical factor, the factor of natural resource use, demographic factor, the factor of the regional policy and economy etc.) – allows to consider it as a powerful and real reserve of the build-up of the food, fuel and energy, mineral-extractive basis of the development not only of the Russian Far East but countries of the Asian Pacific Region in whole.

## References

[1] Baklanov P Y 2001 The Far Eastern Region of Russia: problems and prospects for the

sustainable development (Vladivostok: Dalnauka) p 143 (in Russian)
[2] Anderson J and O'Dowdl L 1999 Border regions and territoriality: contradictory meanings, changing significance *Regional studies* **33**(7) 593-604
[3] Vardomski L B 2000 The border belt of Russia: problems and trends of the development *Russia and the Modern World* 54-64 (in Russian)
[4] Rickenbach M and Reed A C 2002 Cross-boundary cooperation in a watershed context: the sentiments of private forest landowners *Envir. Manag.* **30**(4) 584-94
[5] Dontsov A V, Rodomanskaya S A and Shirokov V A 2010 *The Regional Aspects of the Erosion of Agricultural Lands and Land Use of the Amur Region* (Blagoveshchensk: Far Eastern State Agrarian University) p 273 (in Russian)
[6] Rodomanskaya S A 2018 Transboundariness as the vector of foreign economic development of the Amur Region *The Journal of Altaic Branch of Russian Geographical society* **48**(1) 67-76 (in Russian)
[7] Baklanov P Y and Ganzei S S 2004 Asymmetry and asynchrony of land use development within the boundaries of cross border geosystem of the Amur River basin *Materials of the Regional Scientific and Practical Conf.* (Vladivostok: FVGU) 28–34 (in Russian)
[8] Ganzei S S 2002 Geoecological research of the international transboundary territories in the South of the Russian Far East *The Newsletter of Far Eastern Division of the Russian Academy of Sciences* **6** 82-91 (in Russian)
[9] Ganzei S S and Mishina N V 2002 International transboundary territories in the South of the Russian Far East and their role in sustainable natural resource use in border regions. *J. Korean Geogr. Soc.* **4** 522–35
[10] Volinchuk A B 2010 The Russian Far East: challenges of transboundary cooperation. *Humanities Research in the Russian Far East.* **4** 29-34
[11] Pashina L L 2010 Food security of the region *The Far Eastern Agrarian Newsletter* **4** (16) 66-74 (in Russian)
[12] Pastushenko S B 2007 Regional specific features of shaping living standards of the population of the Amur Region *The Far Eastern Agrarian Newsletter* **2** 106-10 (in Russian)

# "Our Common Home – Altai": international ecological-economical and humanitarian integration in the Altai region

**I N Rotanova, A V Krotov**

Altai State University, Barnaul, Russia

E-mail: rotanova07@inbox.ru

**Abstract.** Cooperation of the border regions of the four countries, whose territories are located in the Altai mountain system, the so-called "The Great Altai", contributed to the establishment of the International Coordination Council "Our Common Home – Altai" (ICC "Altai") in 2003. The ICC "Altai" includes representatives of the legislative (representative) and executive powers of the Altai region and the Republic of Altai (Russian Federation), East Kazakhstan Region (The Republic of Kazakhstan), Xinjiang Uygur Autonomous Region (People's Republic of China), and Khovd, Bayan-Ulgii aimags (Mongolia). The main objectives of the ICC "Altai" are the development and strengthening of trust, mutual understanding and good neighborliness among representatives of states, cultures and nationalities in the strategically significant Eurasian region of the Great Altai, as well as ensuring sustainable social and economic development of the border areas. In the conditions of geopolitical splits and geoeconomic rivalries that have become more complicated and openly manifested in international relations, the processes of consolidation of border regions and actions to create cross-border economic actors are updated in a new way. The need for joint measures and steps to stimulate the competitiveness of local economies in the territories of the Great Altai makes it necessary to develop common governance and coordination structures.

## 1. Introduction

The challenges of our time in the sphere of international cooperation at the beginning of the third millennium force us to make a choice: to seek new ways of cooperation or to improve the ones already created. In recent years, there has been an increase in changes in international relations related to both the establishment of a polycentric world order and the strengthening of regional and cross-border centers for economic development [1]. Against a backdrop of emerging models of cooperation, such as "Eurasia", "The Great Asia", "One Belt. One Road", "The Silk Road Economic Belt", new prospects for strengthening international relations are opening up, in particular, in the Altai region unique from the natural, geopolitical and geocultural aspects [2].

Since its inception, the ICC "Altai" has really helped to restore the broken links of the Great Altai's regions. It is not a decision-making body, but an advisory one and considers issues of mutual interest, sends recommendations to the ICC "Altai" participants, and also interacts with the central legislative and executive authorities. One of its functions is to initiate processes of convergence of national laws of the four countries in the field of environmental security, taxation, foreign trade and customs regulation.

A significant and important positive feature of the ICC "Altai" activity is the interest of the leaders of the border regions in maintaining friendly formal and informal contacts, which in some cases accelerates the solution of emerging issues not only of a commercial and humanitarian nature, but also interstate interaction in the border areas. Not infrequently these issues are not within the competence of the regions, and can be dealt only at the state level, but joint lobbying at the federal level ensures the achievement of the necessary results [3].

## 2. Highlights

The ICC "Altai" was initially created as a communication platform, as an international regional association for cross-border cooperation to enhance the efficiency and coherence of actions in common interterritorial interests, and build confidence and good-neighborliness in the region. In the framework of its activity, topical issues of trade and economic cooperation, the development of transport, recreational, social infrastructure and others are also discussed and resolved (figure 1).

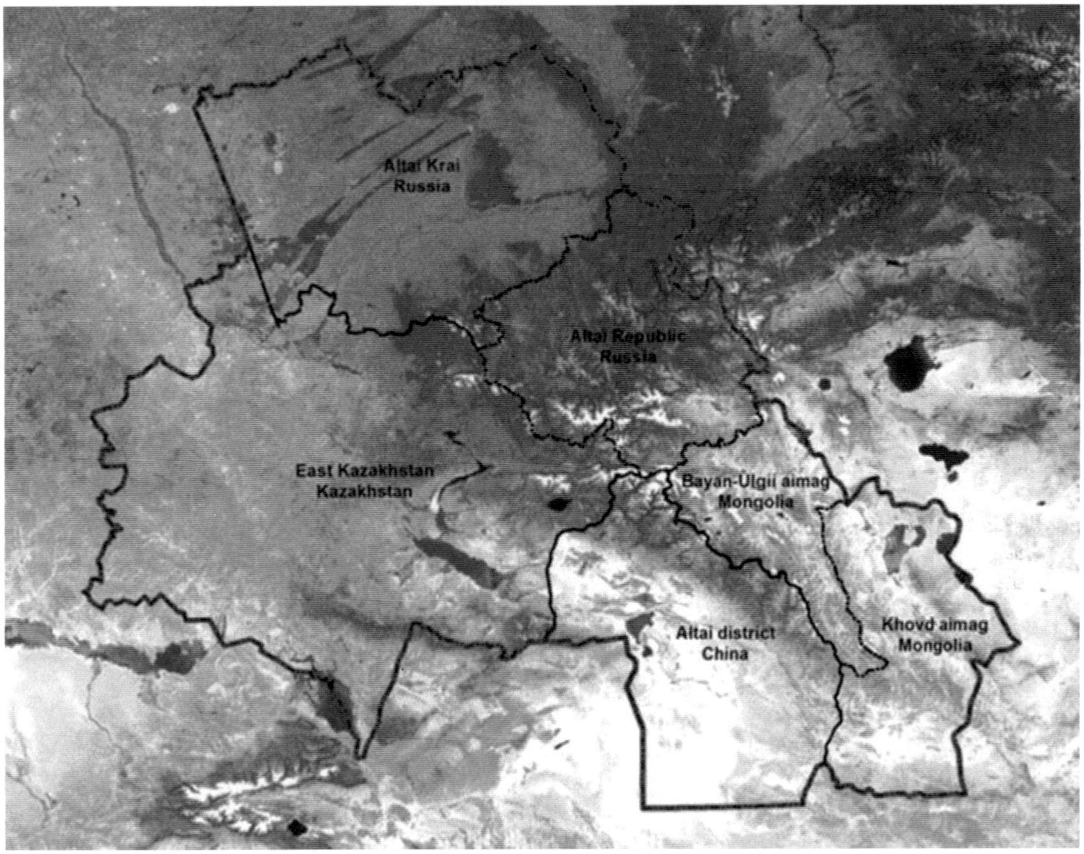


**Figure 1.** "The Great Altai".

The activity of the ICC "Altai" does not substitute for interstate agreements, but promotes their implementation. The issues considered within the framework of the ICC "Altai" activities are most often aimed at assisting to establish inter-human and humanitarian ties, develop direct contacts between people, and overcome often artificial interstate barriers, in scientific discussion and solving problems common to the Altai region: nature protection, environmental safety, preservation of historical heritage, cultural traditions, etc.

Figuratively speaking, the ICC "Altai" is the roof of the Altai house, where the main goal of the participants is to provide comfortable and creative cohabitation of its residents and instill in them a sense of pride and responsibility for their region, with friendship and mutual assistance being objectively beneficial to everybody.

Over the years, multi-year continuing projects have been developed under the auspices of the ICC "Altai": International Conference on Scientific and Technical Cooperation, International Summer School for Students and Young Scientists "Our Common Home - Altai", Children's Public Ecological Expedition "Start From Your Home", and also newer projects at the start-up stage: the Council of Rectors of the Universities of the Great Altai, the Trans-border Tourist Route "Altai - Golden Mountains", and the Tourist-Sports Festival "The Great Altai". The list of the continuing projects of the ICC "Altai" reflects the demand for the development of the humanitarian sphere of international activity and shows that great importance is attached to raising the younger generation, strengthening youth cooperation, education, preservation of historical and cultural heritage and natural values, support to international research, building a favorable image of a neighbor and good-neighborly relations (the "neighborhood belt") [4].

The new mission of the ICC "Altai" is considering it as a regional representative of the activities of the Shanghai Cooperation Organization (SCO). In 2012 and 2014 the Altai territory became the venue for the SCO Youth Forums, in the preparation and conduct of which the ICC "Altai" took an active part, presenting experience of cross-border cooperation in the region of the Great Altai. The SCO Youth Forums held in the Altai region allow to count on the fact that the projects unfolding under its auspices will change the regional status to the global one and attract participants from other regions of Eurasia and the world. So, today the proposal on participation of ICC "Altai" representatives in the activities of the SCO Business Union is already being considered.

Annually the ICC "Altai" meetings are held in one of the border regions of the Great Altai. At the 12th meeting of the ICC "Altai" held in October 2014, Governor of the Altai region A.B. Karlin called "the most significant level of trust between regions" to be the most important achievement of cooperation and drew attention to the fact that the heads of the agencies of foreign affairs of Russia, Kazakhstan, China and Mongolia attach this unique experience great importance, considering it as one of the best practices of interregional interaction. His opinion was supported by Akim of the East Kazakhstan region.

B. Saparbayev, noting in his speech, that through the work of the ICC "Altai" the border regions managed to achieve a great degree of integration on many issues. Among the main goals of cooperation within the ICC "Altai" was the need to intensify trade and economic cooperation.

## 3. Results and prospects

In June 2018, the 16th ICC "Altai" meeting was held in Khovd (Mongolia). The key point in the work of the ICC "Altai" was the consideration and approval of the draft Concept of development of work of the ICC "Altai" until 2025. The ICC Secretary Boris Larin noted that "The Concept defines the development priorities and the main directions of the Council's work. The decision to create a strategic document of international status is driven by the need to update the goals, objectives and key areas of long-term socio-economic and humanitarian cooperation of the member regions of the Great Altai".

The purpose of the ICC Concept of development of work is to strengthen multi-level cooperation and mutual understanding among peoples in the strategically significant Eurasian region of the Great Altai.

The objectives of the Concept are:

- strengthen the ICC "Altai" as an effective regional organization of the Great Altai, in which cooperation is built on mutual trust and friendship of the administrative members of the ICC "Altai";
- develop trade, economic and investment cooperation, environmental and economic security, joint projects in priority areas of cooperation to improve the well-being, the standard of living of the population and the sustainable development of the ICC member regions;
- develop and strengthen trust, mutual understanding and good-neighborliness among representatives of states, preserve national identity of ethnic groups and peoples inhabiting the Altai region;
- coordinate efforts to preserve the natural, cultural and historical heritage of Altai;

- expand socio-cultural and humanitarian ties using different levels of interaction;
- institute common approaches of the ICC member regions to the initiative "One Belt and One Road", aimed at creating favorable conditions for promoting cooperation, including the Great Altai;
- popularize the uniqueness of the territory of the Great Altai in the world space and use its potential for the development of cross-border tourism;
- discuss specific cross-border projects, identify cross-border cooperation issues and prepare recommendations of the ICC "Altai" to intergovernmental commissions.

The main areas of cooperation within the framework of the Concept are:

*Economic and trade development.* Expansion of mutually beneficial trade and economic interaction. Contribution to greater efficiency of existing trade and economic ties. Discussion of the structure of exports and imports of goods, as well as coordination of efforts to optimize them. Facilitation of the improvement of the efficiency of existing mechanisms of trade and economic cooperation with a view to enhancing their effectiveness. Assistance in solving issues of giving international status to border crossing points; cancellation of the restriction on the passage of transit vehicles between Russia and China, the possibility of improving customs administration and trade facilitation, etc. Creation of “The Economic Council of the Great Altai Regions”. Support to business initiatives, implementation of projects in priority areas of cooperation. Encouraging participation in the project activities of government agencies and business leaders. Facilitation of the development of business cooperation in industry and agriculture. Attending to providing financial support to project activities.

Support to the establishment of cross-border economic and technical cooperation centers.

Development of mutually beneficial cooperation in the creation of projects in the energy sector, including wind and solar energy.

Development of the practice of regular holding of joint forums, exhibitions and seminars in order to enhance the interaction of small and medium-sized businesses, as well as research institutions.

Promotion of competitiveness of agriculture and creation of a civilized market of agricultural products, covering the territory of the Great Altai. Support to the implementation of joint projects for the development of the agro-industrial complex, processing of crop production and livestock, with particular emphasis on the development of organic agriculture.

Support to the creation of cross-border production clusters, promotion of cooperation of organizations providing equipment, components, specialized services, and research and educational organizations.

*Development of infrastructure, including transport networks. Facilitation* of the development of transport corridors, creation of favorable conditions for international road transport, cooperation in the field of infrastructure and logistics modernization, support to the development of a network of industrial clusters along the transport highways. The possibility of synchronizing the largest Eurasian transport projects "Trans-Siberian Railway", "NurlyJol" ("Bright Way") and "The Silk Road Economic Belt" with participation of the regions of the Great Altai. Assistance in the creation of a list of priority joint infrastructure projects [5].

*Cultural and humanitarian interaction.* Contribution to the development of joint projects and programs to preserve the traditions and cultural heritage of the nations and nationalities inhabiting the Great Altai.

Support to activities related to joint cultural events in the Great Altai, the exchange of artistic groups, the development of commercial cultural events and the promotion of joint concert programs, holding days of culture of the cooperating regions.

Promotion of the development of cultural and ethnic centers, language learning centers, in particular, by type of "ethnic villages". Activation of cooperation in sports at the level of border areas. Facilitation of the implementation of the project of the International Tourist and Sports Festival "The Great Altai", declared as "Altai Games in the Natural Environment".

*Rational use of natural resources and protection of the environment.* Promotion of the development of environmentally oriented planning and coordination of environmental and ecological and economic activities in the Great Altai. Support to joint projects for the rational use of natural resources in the field of water and land use, energy, etc. Development of interregional cooperation in the field of ecology and environmental protection [6].

Substantiation of new territories of natural and cultural heritage, expansion of UNESCO nominations. Promotion of the development of a unified system of specially protected natural territories in Altai [7].

*Cooperation in the sphere of territorial planning.* The creation of the transport and engineering infrastructure of the Great Altai should be based on the development of coordinated schemes for the territorial development of the border areas [8].

*Science and education.* Revitalization of the Council of Rectors of the Great Altai in the field of educational and scientific and technical cooperation. Development of academic and student exchange. Development of joint educational programs with included education, as well as double diploma programs. Organization of international conferences on scientific, technical, economic and socio-cultural cooperation in the Great Altai. Support to information and publishing projects of the Great Altai (scientific and social almanac "Altai Bulletin", online edition of "Grand Altai Research & Education", etc.) [9].

*Tourism development.* Cross-border tourism in the Great Altai – world brand [10].

*Cooperation in the field of security.* Support to cooperation in the field of prevention of emergencies, natural disasters and catastrophes and combating them.

*Information interaction.* Creation of an international cartographic web resource – the Atlas of the Great Altai. Assistance in the development of the Atlas "The Great Altai: Nature, History, and Culture".

## 4. Conclusion

One of the decisions of the 15th meeting of the ICC was the creation of the Economic Council of the Regions of the Great Altai. Its goal is to coordinate cooperation, promote economic progress and create conditions for further development of the regions.

Interregional integration helps the bordering territories of the Great Altai to more effectively solve urgent problems of development, actively involve representatives of business, scientific and artistic circles and expand the range of partnership relations. Whatever the political situation, the universal issues of international cross-border cooperation remain the priority without losing their relevance. The mission of the International Coordinating Council "Our Common Home - Altai" is regarded to be an important interregional aspect of international cooperation.

**Acknowledgments**
The study was performed with financial support RFBR, research project No. 17-55-53109.

## References

[1] *Border and Cross-Border Territories of Asian Russia and Border Countries (Problems and Prerequisites of Sustainable Development)* 2010 ed P Y Baklanov and A K Tulokhonov (Novosibirsk: Publishing house SB RAS) p 610

[2] *Altai Cross-Border: Ways of International Integration and Sustainable Developmen* 2013 ed M Y Shishin (Moscow: Institute of sustainable development of the Public chamber of the Russian Federation) p 86

[3] Zhuravleva S M and Ivanov A V 2016 The Eurasian civilizational identity as the ideological basis of the new association of Eurasia *Bulletin of Moscow University. Series 12: Political science* **1** 720

[4] Ivanov A V and Kundius V A 2014 Potential and prospects of production of environmentally friendly products in cross-border regions of Altai and Mongolia *Bulletin of ASAU* **11** 149156

[5] Ivanov A V and Fotieva I V 2015 Direct transport and energy corridors to China in the Altai: alternatives and prospects *Bulletin of ASAU* **131**(9) 135138
[6] Mandych A F et al 2012 Biodiversity Conservation in the Russian Portion of the Altai-Sayan Ecoregion under Climate Change *Adaptation Strategy UNDP/ICI Project "Expansion of the Protected Areas Network for the Conservation of the Altai-Sayan Ecoregion"*
[7] Shishin M Y and Engoyan O Z 2015 Ecological framework as the basis for sustainable nature management in Altai region *Polzunov Bulletin* **3** 8690
[8] Ivanov A V et al 2017 Trans-Eurasian transport mega-projects: the design plans *Region: Economics and Sociology* **96**(4) 267284
[9] Korshunov L A and Nikonov N M 2017 Network Interactions in the Great Altai Region *Economy of Region* **13**(4) 11231137 doi 10.17059/2017-4-12
[10] Rotanova I N and Tikunov V S 2018 Approaches of formation of the Altai cluster of international tourism in context of transcontinental transport corridors economic zone of the Silk Road *Science. Innovations. Technologies* **2** 161174

# Chinese migrants in the territory of the Barguzin region in the first half of the 20th century: ethnoeconomic aspects

**L S Tsydypova**

Sochava Institute of Geography SB RAS, Irkutsk, Russia

E-mail: tsidipovaluda@mail.ru

**Abstract.** In this paper we consider the features of the contribution of Chinese migrants to the ethnoeconomic practices of the communities of the Barguzin basin, based on modern approaches to cultural geography and field research. The research area covers an island forest-steppe and steppe, surrounded by mountain-taiga areas. In the valley of the river Barguzin meadow steppes are replaced by forest-steppes and swampy areas. The diversity of natural environment provided a basis for the formation of various systems of ethnic wildlife management. During the establishing their residency the representatives of Evenks, Buryats, Russians, Poles, Jews, Chinese, Tatars, Ukrainians, Byelorussians and others formed the basis of ethnoeconomic practices and traditional economy as a whole complex of socio-natural observations, ideas, mutual interpretation of the traditions of various groups of local population. The key role of cultural traditions and nature management in formation of spatial features of the economic activities of local inhabitants is due to their historical continuity and compliance with both the needs of society and the landscape-geographical environment. The research resulted in a conclusion about the stable idea of the local population about their territorial unity and ethnic tolerance of the Barguzin community. The combination of hunting, nomadic cattle breeding, focal agriculture, farming, lake and river fishing, gathering reveals a connection between people and the natural-geographical and ethno-cultural features of the territory.

## 1. Introduction

The study area has a long and rich history of settlement and development. Different ethnic groups, in accordance with their settling and nature use preferences formed a multicultural settlement network, which was the result of a long period of development. Economic activities, maximally adapted to the natural and geographical landscape conditions are expressed in specific ways of life and nature management. In cultural geography it is established that traditions have a system-forming character in the process of cultural landscape development [1]. At the same time, the tradition is not frozen; it respondsto the challenges of times and changes with its current. Therefore, the research methodology takes into account the features of the convergence of time steps and territorial manifestations of ethnoeconomic transformation in the subjective perception of the Barguzin community. The focus of the study concerns the processes of ethnocultural memory: "site peculiarity," the geographic center of economic landscapes, and the contribution of Chinese migrants to the demographic and economic and cultural characteristics of population. Strongly expressed interethnic communication, mostly, ethno-cultural mosaicism and diversity (differentiation), and the formation of a common regional identity (integration) are typical for this territory.

The territory of the Barguzinskaya depression is located between the Ikatskii and Barguzinskii ranges from the northeast to the southwest, and is one of the largest intermontane depressions of the northeast of Buryatia. The absolute height of the depression ranges from 500 to 700 m, while the height of the surrounding mountain ranges reaches 2000-2700 above sea level. The territory of the depression lies in the mountain-taiga zone, and its middle part is an island of steppe and forest-steppe landscapes located in a closed intermountain depression among the mountain and taiga spaces.

In the floodplain areas and on the terraces of the Barguzin river and its tributaries, the mosaic alternation of meadow steppes, forest-steppes and swampy areas is characteristic. Forest vegetation prevails in foothill areas and on river slopes (up to 1100-1300 m). Forests are rich in berries, mushrooms and nuts. The Barguzinskii range stretches along the north-eastern shore of Lake Baikal. A high and hard-to-reach range, crowned with rocky peaks of golets, rises almost up to 3000 m. The Ikatskiirange borders the Barguzinskaya depression from the northeast. Its crests are wide, flat watershed spaces with gently flattened peaks. The climate of the Barguzinskaya depression is characterized by sharp continentality with comparative severity and aridity. The mountain regions of the Barguzinskii and Ikatskii ranges and the upper reaches of the Barguzin river have more severe climate conditions. The mean annual temperature in the depression is 2-3°, and in mountainous areas - 5-7°. Long winter (more than 6 months) is characterized by severe frosts, dryness, clear sky and wind bull. The mode of air temperature in the area is subject to large fluctuations not only during the year, but also in seasons - in certain months, and even within a day [2]. The administrative structure of the study area has undergone a number of changes. Thus, the Barguzinskii district was consistently considered an administrative unit in the pre-existing Transbaikal region, the Pribaikal region, the Pribaikal province and the Buryat-Mongolian ASSR, where the center was the town of Barguzin. In 1917 the Barguzinskii district was established, which included a compactly residing Russian population. Buryat and Evenk people entered the Barguzinskii aimak, formed within the former Barguzin steppe duma and the Bauntovskii Tungusskaya government. In 1923, after the foundation of the Buryat-Mongolian Autonomous Soviet Socialist Republic (BMASSR), the Barguzinskiidistrict and aimak unite into one administrative unit – the Barguzinskii aimak within the borders of the former Barguzinskii district and are part of the BMASSR.

In 1924, part of the territory occupied by Evenks moved to the Bauntovskiia aimak [3]. In 1925, the Severobaikal'skii aimak has beenfounded and included a part of the Verkhneangarskaya volost' and the territory occupied by the Tungus from the Barguzinskii aimak. By decision of the Presidium of the All-Russian Central Executive Committee of September 26, 1927, "On the new administrative division of the BMSSR", the division of the republic into counties, volosts, and khoshuns was abolished, now the aimaks were divided only into rural (somon) and settlement councils, and the town of Barguzin was renamed into a rural settlement – village [4].

Ethno-economic relations have involved each member of the community in socio-economic and political processes, which have become important stages of personal and family biography. The significant role of local traditions, their transformation and existence in the Soviet period determined the features of regional identity up-to-date.

## 2. Objects, data and methods

The approach presented in the paper reveals the details of local history in the political, social and economic dimensions. The analysis included interviews of residents of the north-eastern part of the Barguzinskaya depression (settlements of Alla, Mogoito, Argada, Garga, Kurumkan, Baragkhan), these are people, born in 1920s and 40s, are descendants of Chinese immigrants, as well as those who directly knew and communicated with the first emigrants of China. Thus, the link between interethnic relations, attitudes towards the habitat, the specifics of natural-economic interactions, values and strategies for space development has been revealed. The ethnoeconomic features of Chinese migrants are considered in the context of the transformation of the Soviet nature use (collectivization and the difficult time after breakdown of traditions), accompanied by crisis and painful changes for all ethnoses of the territory. Constructions of nature use relations, based on the experience of the ethnic group and collected from its significant fragments are sketches that build events in a story, explaining the logic of interaction of the local population with each other and the authorities. The paper also uses unpublished materials from the district archive. The research set out the tasks of studying the territorial arrangement and adaptation of Chinese emigrants, the specifics of their settlements; identification of key events that are significant for the formation of ethno-economic lines; analysis of stories of informants with an emphasis on economic practices and socio-demographic characteristics.

## 3. Results and discussion

The post-revolutionary period was characterized by the establishment in Russia of a state monopoly on gold mining, but the civil war that dragged on in Transbaikalia impededthe state-operated extraction. This was the period of gold industry decline, including the gold mining in the Barguzin taiga. In 1925, 687 workers were registered in the gold mining industry in the Barguzin taiga, about 70% of them were Chinese. Gold was mined mainly by gold diggers, and it was very difficult to keep a record of gold mined by them, and especially by the Chinese, as each sought to hide the true volume of extraction. By 1930, due to the tightening of the tax policy, they all stopped their work [5]. The usual occupation of the Chinese, apart from gold mining, was trade. Thus, M. Novomeiskii recalled: "... Chinese merchants, owners of small shops, got along with the local population and earned its trust. One feature of the Chinese trade was different: it had neither owners nor hired workers, the cooperative principle was dominant. The Chinese learned the spoken Russianto some extent, but somehow they always confused the forms of personal pronouns: "my said," "yours goes," "his has come," the same was noted by local residents: "*... often came from Ikat to Barguzin and sold soap and teato the locals. And then the border has been closed* (the Argada village) "[6]. After the closure of the gold mines, the state demanded the eviction of all migrant workers, explaining this requirement by the need to secure the Russian-Chinese border. During this period, the political leadership of China experienced a crisis [7]. Therefore, the remaining part of migrants at kuituns (northeastern steppe landscapes - L. Ts.) of the Barguzinskaya depression was being gradually attracted to nearby settlements: "*The Chinese arrived even under tsarist Russia. In the 1920s, the Chinese were mostly in Baunta [At the gold mines - L. Ts.]. (the Alla village). Then they appeared in Ulyunkhan, in Alla they were not, and then gradually began to appear and to settle here. Then they settled in Argad, Garga, Kurumkan and further. Those Chinese who dug gold, they, of course, left the Soviet Union, and the simple workers, they remained ... so, step by step. In Kurumkan there were many of them "(the Kurumkan village).* Thus, the chronological period of the resettlement is noted by the informants in the 1920s and 30s, but in the books of economic records Chinese arefirst mentioned in the 1940s, and the place of leaving is Karaftit (Baunta). The main occupations to the 40s of the 20$^{th}$ century are: "truck farmer" and "brickman". Later, workers were attracted to the nearest collective farms, only a few continued to work in a brick factory. Among the additional single types of employment, the old-timers noted the cooking skills of Chinese, as well as sewing and repairing shoes: "*... they cooked masterfully! Men cooked, we had such a chef Ch. The aroma could be heard for many kilometers. And our women did not know how, and we did not eat such dishes before. Later they learned ..."(the Baragkhan village).*

During this period, the first Chinese migrants formed a settlement of compact residing of Sino-Russian and Sino-Buryat families, intermarriage ties were consolidated, and, consequently, acculturation and traditional knowledge exchange have intensified. During 1941-1950 92% of the previously migrated and already settled Chinese joined collective farms and communes. Thus, the population was about 200 people [8]. An important distinctive feature in the Chinese family was education in the spirit of cohesion and diligence [9]. During this period, the places of compact residence of the Chinese in the Buryat settlements were called "Shanghai".

Today, the microtoponymic plan of the locality of the old settlement (remote periphery) is a memorable center, where all the spatio-temporal echoes of political and social changes were manifested. Despite the compact residence in an alien environment, the assimilation process ensured stability and inviolability of residence on the territory and the continuation of the family - the names in Chinese families were preserved in Russian transcription. This process was accompanied by the development of economic and cultural skills; the Chinese studied Buryat and Russianin everyday communication.

Religiously, the descendants honor local traditions. Religious syncretism is observed, feast and funeral rites are taken (95%), traditions of family events (75%), and economic activities (95%). Zhambalova notes: "In the life of the descendants there are currently no special Chinese cultural elements except for one respected family of the Barguzin settlement. Here, the garden is cultivated according to Chinese technology with Chinese tools of labor, on holidays they prepare Chinese dishes; they celebrate the New Year according to the Chinese calendar "[10].

The results of the survey reveal a persistent positive attitude towards migrants in the Soviet period, caused by the inevitable economic difficulties that all residents of the territory experienced. Most families, due to the borrowing the traditional Chinese gardening skills, prevented the problem of hungerto some extent: *"They taught us to conduct irrigation systems,... planted cabbage, I remember such a L., very much saved during the war - gave cabbage, turnips. But the Buryats, they did not know how to plant, and in general they did not eat that way before. Berries were not planted ... "*. The history of the resettlement of Chinese people meant not only a new social environment and cultural traditionsfor local residents, but also the formation of a new land use system, as well as specific ideology of the border regions, which determined the views of local communities on their own place in the local history of the region.

## 4. Conclusion

Under extreme climate and difficult living conditions, the Chinese integration in the ethnocultural landscape of Barguzin Baikal region represents a positive phenomenon of the system of interethnic relations of the region. As a result of the interview, it was established that the descendants of Chinese migrants still represent a close-knit, friendly, industrious collectivein the vision of local residents. Chinese families and migrant groups literally survived due to tolerant inter-ethnic ties of all the ethnic groups of the territory, and in turn, helped to survival to local communities in a difficult economic situation. Buryats, Evenks, Russians, Jews, Chinese and other ethnic groups of the Barguzinskaya depression are united by intermarriages, mutual assistance and support, and the exchange of cultural traditions.

The adaptation of Chinese migrants and their descendants in the predominantly Buryat territories had its own peculiarities. Despite the compactness of living in an alien environment, the processes of assimilation and acculturation provided stability, the opportunity to work and continue their family line. The key to the successful adaptation of migrants was the daily training of the Buryat and Russian languages, the development of economic and cultural skills of Buryats, Evenks and Russians.

The Chinese learned to interact not only with new societies and ethnic groups, but also with the natural environment that is unusual for them and contrasting in landscapes and the result of this interaction can be considered successful. In the materials of the survey of old-timers, a tendency to complement each other's civil and ethnic identity is seen. Traditions, as such, are blurred; religious ideas of descendants of Chinese migrants correspond to the general picture of local communities - Buddhism, shamanism. The ethnic identity, "Chineseness" gives way to other forms of self-identification, like "Barguzinian", and also has a local character: "Argadians, Gargins, Baragkhans" etc. Today the descendants of the Chinese live in the settlements of the Kurumkanskii district (Alla, Argada, Garga, Kurumkan, Baragkhan), and Ulan-Ude.

The obtained results suggest that a consistent analysis of family-generic stories serves as a reliable source providing an understanding of how ethnoeconomic lines of interaction become the basis for mutual assistance and the strength of territorial self-identification of all ethnic groups in the region. As a result, it can be noted that the traditional ethno-economic life-support practices that have been developed over a long period in polyethnic communities contribute to a dynamic and flexible transformation and adaptation to changing socio-economic and sociocultural conditions.

IOP Conf. Series: Earth and Environmental Science **190** (2018) 012016 doi:10.1088/1755-1315/190/1/012016

## References

[1] Barguzin hollow (in the past, present and future) 1993 (Ulan-Ude: BSC Center of the SB RAS) ed V E Vikulov, *Series of "Land Near Baikal"* **5** 157-34

Dashibalov B B 1995 *Archaeological Monuments of Kurykan and Chori* (Ulan-Ude: BSC Center of the SB RAS) p 195

Ragulina M V 2004 *Cultural Geography: Theory, Methods, Regional Synthesis* (Irkutsk: IG of the SB RAS) p 171

Ragulina M V and Savelieva I L 2004 Geographical approach to the allocation of territories of traditional nature use *Geography and Natural Resources* **3** 168-35

[2] Soils of the Barguzin Basin 1983 ed Azmuka T I et al (Novosibirsk: Nauka) p 269

[3] Garmaev S V 2004 *We Are from the Barguzin Valley* (Ulan-Ude: Publishing house of OAO Resp. type) p 268

[4] Tsydypova L S 2014 Socio-demographic characteristics of the Barguzin Valley in the first quarter of the 20th century on the example of local communities with Kharamodun and Barguzin *Izvestia Irkutsk State University. Series of "Earth Sciences"* **9** 140-52

[5] Verkhoturova G A and Zherlov V F 2001 *The Golden Land of Buryatia; Com. natures. Resources for Resp. Buryatia. Administration of the Muysky District* (Ulan-Ude: Rep. type.) p 215

[6] Nagishkin D 1917 *Barguzin Taiga: an Economic Essay* (Verkhneudinsk: Pribaikalsky Union of Cooperatives) p 32

Novomeaysky M 1979 *From Baikal to the Dead Sea* (Jerusalem: Library-Aliya) p 383

[7] Kapitsa M S 1958 *Soviet-Chinese Relations* (Moscow: Gos. publishing house. lit.) p 424

[8] *Kharamodoun. Soviets in Figures and Facts* (1940-1950) 1950 (Kurumkan: District type) p 40

[9] Danilova Z A 2009 *The Chinese Migrants of Ulan-Ude* (Ulan-Ude: Publishing house of BSC SB RAS) p 233

[10] Zhambalova S G 2006 On the typology of intra-ethnic groups of Russian Buryatia *Proc. of Sat. Doc. All-Russia. Scientific-Practical. Conf.* (July 5-6, 2006, Ulan-Ude, Russia) **1** 138-41

# The ecological potential of Baikal region's geosystems

**I N Vladimirov**

Sochava Institute of Geography SB RAS, Irkutsk, Russia

E-mail: garisson@irigs.irk.ru

**Abstract.** The Baikal region is a complex region as regards its natural environment and landscape characteristics; it is endowed with immense reserves of various natural resources, the industrial development of which is steadily progressing, thus posing challenging ecological problems. For assessing the ecological potential of geosystems, the landscape-typological map of the Baikal part of Siberia was generated. The classification of geosystems was carried out on the basis of the taxonomic system of hierarchical units of natural environment as developed in the Sochava Institute of Geography SB RAS. Cartographic analysis determined the most informative units of geosystems of regional dimension, which made it possible to carry out a classification separation of the landscape structure of the Baikal part of Siberia. The investigations into the ecological potential, the factors and the conditions having influence upon its formation used: SRTM data, WorldClim global climate data; data on Net Primary Production; Normalized Difference Vegetation Index, Enhanced Vegetation Index. The main principles of GRID modeling and the Analytic Hierarchy Process technique were used to assess the ecological potential of the Baikal region's geosystems which made it possible to represent the set of natural conditions needed to satisfy the requirements of the population for all the necessary primary means (merely ecological, unassociated with production). The evidence obtained concerning the ecological potential provide a natural scientific basis for a reasonable regional ecological policy, an improvement of the population distribution pattern and the social sphere, a rational organization of labour and recreation, and for the protection of human health.

## 1. Introduction

The Baikal region is directly connected with the world-famous Lake Baikal, and lies in the middle of Asia. The Baikal region includes three subjects of the Russian Federation, united by belonging to the basin of Lake Baikal; they are the Irkutsk region, the Republic of Buryatia and the Zabaikalskii krai.

The relief of the Baikal region is highly differentiated, due to tectonic processes in the Baikal Rift valley, formation of mountain ranges and valley incisions. The Primorskii and Baikalskii Mountain Ranges stretch along the western coast, whereas the Barguzin Range rises in the northeast, and the Khamar-Daban Range limits the Baikal basin in the south. The largest inflowing rivers are the Upper Angara from the north, the Barguzin from the northeast, and the Selenga from the southeast. The Selenga River constitutes the main inflow from a catchment area extending into the trans-Baikal region and northern Mongolia. It forms a large delta area in the southeastern part of Lake Baikal. Numerous minor rivers may also influence sedimentation in the lake.

The regional and altitudinal distribution of landscapes and modern vegetation is controlled by the relief as well as by temperature and precipitation gradients across the region. Generally, Scots pine

(*Pinus sylvestris*) and larch (*Larix*) taiga prevails on dry east-facing mountain slopes, whereas Siberian pine (*Pinus sibirica*), fir (*Abies sibirica*) and spruce (*Picea obovata*) taiga is widespread on moister north- and west-facing slopes of mountain ridges. East of Lake Baikal Larix sibirica is replaced by L. dahurica, and birches (*Betula*) and poplar trees (*Populus tremula*) may replace Scots pine- larch - taiga. Steppe and pine forest-steppe (*Pinus sylvestris*) as well as elm (*Ulmus pumila*) occur within the Selenga catchment area in the trans-Baikal region. Further north steppe areas are limited to the Barguzin River basin, Olkhon Island and the Priolkhon region. However, within the whole region patches of steppe and rock-steppe vegetation occur on east- and south-facing slopes.

The Baikal region is a complex region as regards its natural environment and landscape characteristics; it is endowed with immense reserves of various natural resources, the industrial development of which is steadily progressing, thus posing challenging ecological problems.

The low potential for resistance of this region's landscapes to anthropogenic disturbances, and the retarded self-recovery processes determined by the severity of the natural conditions produce a large number of environmental restrictions on nature management practices, with implications for the environmental quality and ecological safety of the population, for biodiversity preservation across territories as well as for setting the stage for their sustainable ecologo-economic development.

The need for the ecological rationalization of nature management in the Baikal region is also dictated by the fact that most of this region refers to the Baikal Natural Territory for which the ecological regulation of nature management is stipulated by the Russian Federal Law "On the protection of Lake Baikal".

The issue of the ecological rationalization of nature management, and of the prevention or minimization of negative consequences of nature management in the system of human-environment relationships has always been the subject of geographers' attention. Russian academician V.B. Sochava called for "co-creation of Man and Nature", and Academician I.P. Gerasimov defined the scientific and methodological principles of "constructive geography", based on investigations into the problems of systematic transformation of natural environment aimed at effective exploitation of natural resources [1]. The idea of the ecological rationalization of nature management was developed further by the research efforts made by the V.B. Sochava Institute of Geography SB RAS in the field of landscape planning [2].

Nowadays, the principal objective of the ecologization of nature management is to create favorable conditions for harmonious, balanced development of Nature, society and economy. Special emphasis must be placed on recognition of priority of the life-supporting functions of landscapes in the interests of society in comparison with direct use of natural resources provided that the balance of the requirements of the population for the welfare and ecological well-being is maintained. An optimization of nature management involves taking into account the ecological potential of landscapes, which would make it possible to take into consideration the balance of the ecological and economic interests of society, and to create the scientific preconditions for developing the legal environmental regulations.

There are several definitions of the ecological potential but all of them are largely confined to understanding it as the potential for life sustenance, or as the entire set of natural conditions influencing human life and creating a specific local habitat [3].

Such definitions have their origins in the anthropocentric approach leaning upon Man's requirements to the quality of habitat and focusing on the potential usefulness of natural objects for humans and satisfaction of their requirements.

A.G. Isachenko emphasized a fundamental significance of the notion of ecological potential and defined it as the "capacity of the landscape to satisfy the requirements of the population for every kind of primary (i.e. merely ecological, unassociated with production) mans of living : heat, air, water, sources of food products as well as for natural conditions of labor activity, recreation, medical treatment, and spiritual development" [4]. Furthermore, he warned against confusing the ecological potential of the landscape and its production-resource potential.

In our view, to assess the ecological potential of landscapes requires proceeding from multifactor estimation implying an understanding of the essence of natural processes, and a study into object relationships between the components of landscapes. The ecological potential itself is recognized as the set of characteristics of natural environment which have influence upon human health, the state and functioning of landscapes, and upon their resistance to anthropogenic impacts. As a matter of fact, it is the state of natural environment that determines the comfort of human habitat.

A comprehensive characterization of the ecological potential of landscapes requires taking into consideration several tens or even hundreds of indicators, but its comparative assessment must be based on few criteria, namely on leading, or determining, ecological factors. Such factors include ecologically mandatory, i.e. irreplaceable and steadily operating qualities of landscapes, the absence of which reduces the potential to zero, because life is altogether impossible without them.

Among the main criteria in the system of criteria for assessing the ecological potential of landscapes at the regional and local level can be the biota (vegetation in particular) when regarded as a complex, integral, evolutionarily independently developing autotrophic system which is an important and, often, critical component of geographical environment [5]. In the case of vegetation, the ecological potential can be more precisely defined via traditional biospheric ecological functions of vegetation, such as its autotrophicity and intensity of primary productivity, and via the involvement in the formation of atmospheric gas composition or an assessment of its role in the hydrologic and mineral cycle, carbon deposition or ecosystem relationships with the fauna, etc. Of course, such significant ecological functions of vegetation are important, and they determine its role in all global, regional and local ecological processes in the biosphere. Actually, they are assessed in current ecological research as "biosphere and ecosystems services" [6]; as practice shows, however, they are still inadequately taken into consideration in the regulatory/legislative practice of nature management in Russia Federation.

## 2. Materials and Methods

For assessing the ecological potential of landscapes (geosystems), the landscape-typological map of the Baikal region was generated. The map contents were developed having regard to the map "Landscapes of the south of Eastern Siberia" by using high spatial resolution remote sensing data (Landsat 5 TM, 7 ETM+), and Digital Elevation Model (DEM) produced on the basis of Shuttle Radar Topographic Mission (SRTM) data, published data on the state of landscapes as a whole and for separate components as well as research results obtained in different areas of the Baikal region [7].

A classification of landscapes (geosystems) was carried out on the basis of the taxonomic system of hierarchical units of natural environment as developed in the V.B. Sochava Institute of Geography SB RAS: class (system of landscapes) - group of geoms - geom - class of facies - group of facies. Cartographic analysis determined the most informative units of landscapes of regional dimension, which made it possible to carry out a classification separation of the landscape structure of the Baikal region.

The investigations into the ecological potential, the factors and the conditions having influence upon its formation used: the DEM generated on the basis of *SRTM* data, and the set of *WorldClim* global climate data (http://www.worldclim.org); data on *Net Primary Production (NPP)*; the *Normalized Difference Vegetation Index (NDVI)*, and the *Enhanced Vegetation Index (EVI)*.

## 3. Results

For processing and analyzing the resulting value, the *GRID* model with regular steps of 90 arcsec (3 km) was constructed. This *GRID* model was used to calculate in absolute elevation the slope steepness and aspect.

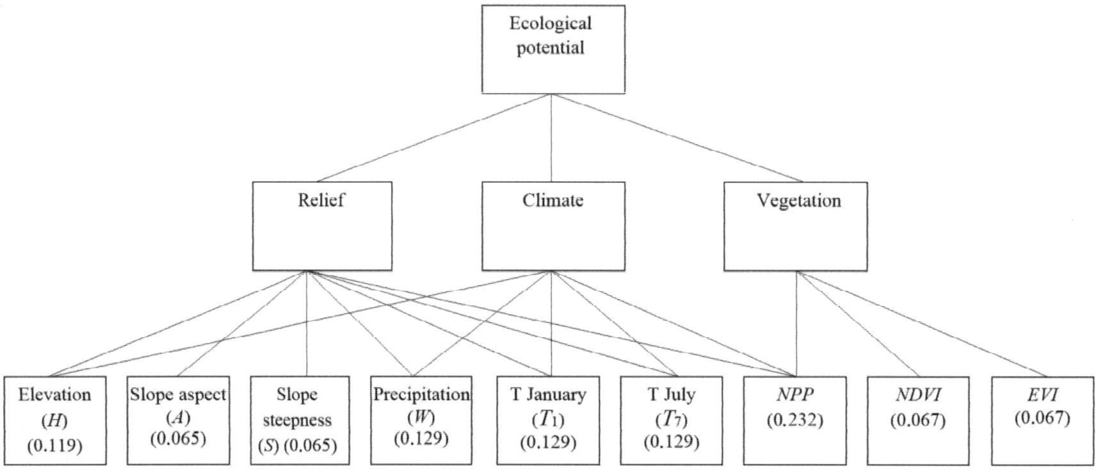


**Figure 1.** The scheme for calculating the ecological potential using the used the Analytic Hierarchy Process

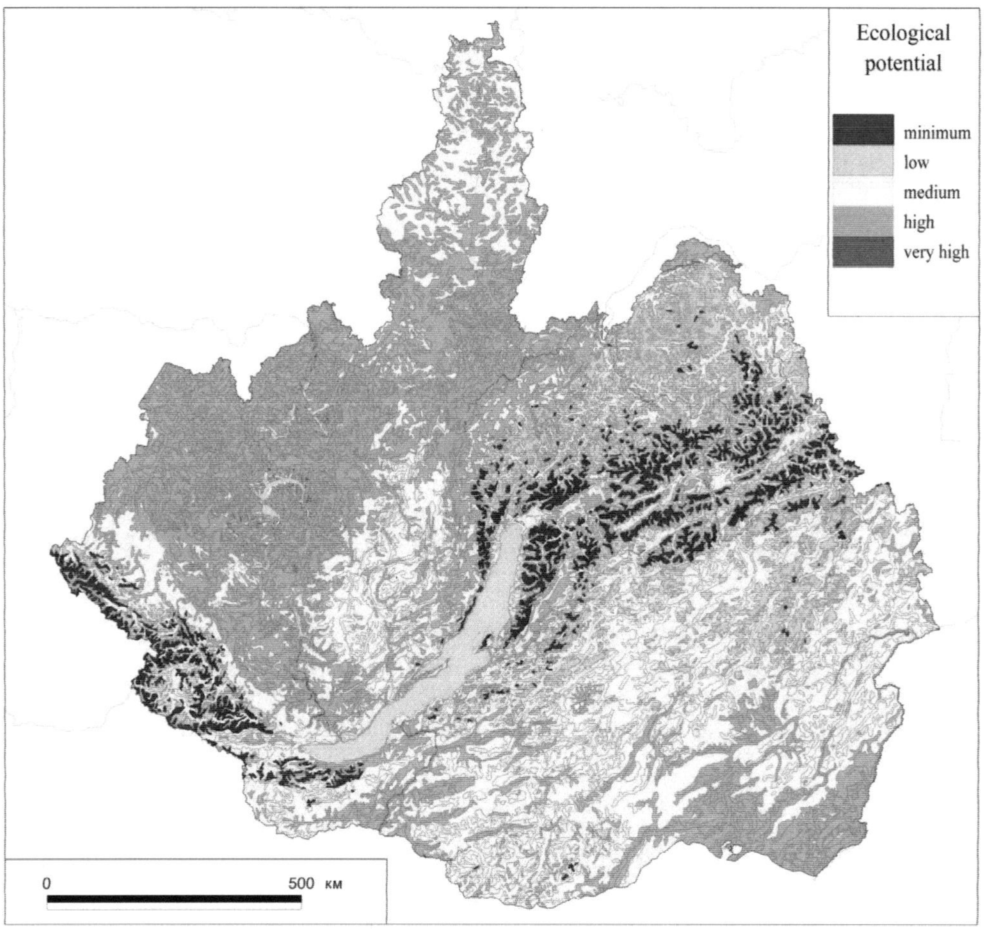


**Figure 2.** The ecological potential of Baikal region's geosystems

The data obtained in the form of regular grids (absolute elevation, slope steepness and aspect, mean annual precipitation amounts, mean January temperature, mean July temperature, the *NDVI*, *EVI* and *NPP* values) were all converted to a vector form, representing a set of regularly occurring points. The indices were all transformed to dimensionless values: ranked into five categories: minimum, low, medium, high and very high values.

To determine the main natural ecological factors and take into account their contribution to the value of the potential used the Analytic Hierarchy Process (AHP) (figure 1) [8].

The formula of this hierarchic structure is $EP = (0.119H + 0.065A + 0.065S) + (0.129W + 0.129T_1 + 0.129T_7) + (0.232NPP + 0.067NDVI + 0.067EVI)$, where *EP* - ecological potential, *H* - elevation (rank), *A* - aspect (rank), *S* - steepness (rank), W - mean annual precipitation (rank), $T_1$ - mean January temperature (rank), $T_7$ - mean July temperature (rank), *NPP* - net primary production (rank), *NDVI* - the Normalized Difference Vegetation Index (rank), and *EVI* - the enhanced vegetation index (rank).

Based on the developed methodology, applying GRID-modeling principles and the hierarchy analysis method using calculated data in the form of regular networks, special evaluation maps were constructed and the weight coefficients of the main formation factors of the ecological potential of Baikal region's geosystems were determined. The result of the calculation of the ecological potential is presented in figure 2.

## 4. Conclusion

The main goal of environmental management at the present time is creation of conditions for harmonious, balanced development of nature, society and economy. Recognizing the priority for society of the life-supporting functions of geosystems before direct use of its resources should be specially emphasized, regarding meeting the balance of the population's needs for welfare and environmental well-being.

In these conditions, the environmental optimization of environmental management based on the ecological potential of geosystems is the main paradigm of environmental policy in the Baikal region, which will allow to take into account the balance of environmental and economic interests of society, to create scientific prerequisites for the development of legal environmental regulations, economic calculations, scientific and information-analytical support of environmental protection and environmental safety.

The main principles of *GRID* modeling and the AHP technique were used to assess the ecological potential of the Baikal region's landscapes which made it possible to represent the set of natural conditions needed to satisfy the requirements of the population for all the necessary primary means (merely ecological, unassociated with production). The evidence obtained concerning the ecological potential provide a natural scientific basis for a reasonable regional ecological policy, an improvement of the population distribution pattern and the social sphere, a rational organization of labor and recreation, and for the protection of human health.

**Acknowledgments**
The reported study was funded by RFBR according to the research project № 17-29-05089

## References

[1] Sochava V B 1986 *The Problems in Physical Geography and Geobotany: Selected Works*. (Novosibirsk: Nauka) (in Russian)
Gerasimov I P 1976 *Soviet Constructive Geography. Problems, Approaches, Results*. (Moscow: Nauka) (in Russian)
[2] Antipov A N et al 2006 *Landscape Planning: Tools and Experience in Implementation* (Bonn – Irkutsk: V. B. Sochava Institute of Geography SB RAS)
Semenov Y M 2017 Landscape planning: The applied branch in complex physical geography *Geography and Natural Resources* **38**(4) 319–23 doi: 10.1134/S1875372817040023
[3] Gavrilenko O P 2008 *Ecogeography of Ukraine* (Kyev: Znannya)

Gusakov V G et al 2008 *Food Security: Terms and Concepts: Wikis* (Minsk: Belarusian Science) (in Russian)

[4] Isachenko A G 2001 *Ecological Geography of Russia* (St. Petersburg: Izd-vo Sankt-Peterb. Univer) (in Russian)

[5] Belov A V and Sokolova L P 2008 Vegetation stability in the system of geobotanical forecasting *Geography and Natural Resources* **29**(2) 124-31 doi: 10.1016/j.gnr.2008.06.016

Belov A V and Sokolova L P 2017 Vegetation as a factor in the system of natural environment preservation of Prebaikalia *Geography and Natural Resources* **38**(4) 341–48 doi: 10.1134/S1875372817040059

Sochava V B 1978 *An Introduction to the Theory of Geosystems* (Novosibirsk: Nauka) (in Russian)

Vladimirov I N and Chudnenko A K 2009 Multilevel modeling of the forest resource dynamics *Mathematical Modelling of Natural Phenomena* **5**(4) 72-88 doi: 10.1051/mmnp /20094505

Vladimirov I N 2014 Predictive mapping of the restoration-age dynamics of taiga forests on the basis of remote sensing data and geographical knowledge *Applied Ecology and Environmental Research* **12**(4) 807-24 doi: 10.15666/aeer/1204_807824

[6] Lal R, Lorenz K, Hüttl R F, Schneider B U and Braun J 2013 *Ecosystem Services and Carbon Sequestration in the Biosphere* (Dordrecht: Springer) doi: 10.1007/978-94-007-6455-2

Tishkov A A 2005 *The Biospheric Functions of Natural Ecosystems of Russia* (Moscow: Nauka) (in Russian)

Smelansky I E and Tishkov A A 2012 The steppe biome in Russia: ecosystem services, conservation status, and actual challenges *Eurasian Steppes. Ecological Problems and Livelihoods in a Changing World. Plant and Vegetation* ed M Werger and M van Staalduinen (Dordrecht: Springer) vol 6 p 45–101

Vanteeva V J and Solodyankina S V 2015 Ecosystem functions of steppe lands capes near lake Baikal *Hacquetia* **14**(1) 65-78 doi: 10.1515/hacq-2015-0016

[7] Mikheyev V S and Ryashin V A 1977 *Landscapes of the South of East Siberia*: *The 1:1 500 000 Map* (Moscow: GUGK) (in Russian)

Farr T G et al 2007 The Shuttle Radar topography mission *Reviews of Geophysics* **45** RG2004 doi:10.1029/2005RG000183

[8] Saaty T L 1980 *The Analytic Hierarchy Process: Planning, Priority Setting, Resource Allocation* (New York: McGraw-Hill)

# Geographical and environmental research of transboundary territories: results and prospects

**I N Vladimirov, V M Plyusnin, A A Sorokovoy**
Sochava Institute of Geography SB RAS, Irkutsk, Russia

E-mail: garisson@irigs.irk.ru

**Abstract**. The study of transboundary territories is an urgent and complex issue of geography since currently there is no well-developed methodology for studying these objects and there are only sectoral approaches to those. The landscape-ecological analysis of transboundary territories with their own characteristic set of relationships between man and nature should be based on a systematic approach that allows the geographical objects with different intersystem connections to be studied. From the point of view of system approach, the object of the study is a set of transboundary systems. In addition, the study of transboundary territories is possible due to the complexity of the study which is based on the landscape theory, the V.B. Sochava's theory of dynamics and evolution of geosystems, methods of ecological zoning.

## 1. Intoroduction

In the current context of transboundary territories, the issues of environmental and economic relations in the sphere of nature use become especially topical, since the growing demand for natural resources in a market economy became a factor in the environmental integrity of natural components.

A powerful factor that is currently responsible for the strategic tendencies in the approaches to the regional policy, is the idea of ecologically sustainable development. "Sustainable development" as a concept and as an imperative has now gained recognition worldwide. And nature and environment conservation and preservation are regarded, along with the economy, as equally important goals. For that reason, in most countries, the quest of science- and reliable data-based solutions aimed at sustainable territorial development is therefore considered a fundamental remit of the politics and management.

The transboundary territories within the economic corridor "China-Mongolia-Russia" are distinguished by a great variety of natural conditions and resources. And yet, natural systems as natural landscapes, habitats of living organisms do not depend on state borders and develop in accordance with natural laws. National boundaries are intersected with integral natural and natural resource systems of various types, for example, the basins of the Selenga river [1], and the Baikal natural territory (figure 1). River systems crossing national borders are a classic example of transboundary systems. Bearing in mind the strategic importance, the Baikal region, being a part of the Russian-Mongolian-Chinese transboundary area, is of great importance for the sustainable social and economic development of Russia.

IOP Conf. Series: Earth and Environmental Science **190** (2018) 012018 doi:10.1088/1755-1315/190/1/012018

## 2. Methods and results

V.B. Sochava Institute of Geography SB RAS has a wide work experience in the transboundary territories. Anthropogenic changes of geosystems in the transboundary territory of Russia and Mongolia are researched [2]. Social interrelations in the border areas have been studied, their population is defined as the most important factor of cross-border migration interactions. Russian-Chinese migration is assessed as an essential element of cross-border ties between Siberia and China, the significance of the tourism component of cross-border migration is justified. The Russian-Mongolian transboundary zone is represented as an integral territorial recreational system characterized by a unique spatial organization, a complex of external and internal interrelations and is part of the international tourist space. The tools of landscape planning in solving social and ecological transboundary conflict situations [3]. It is necessary to recognize that in addition to political and economic tasks, environmental problems are attracting increasing attention lately.

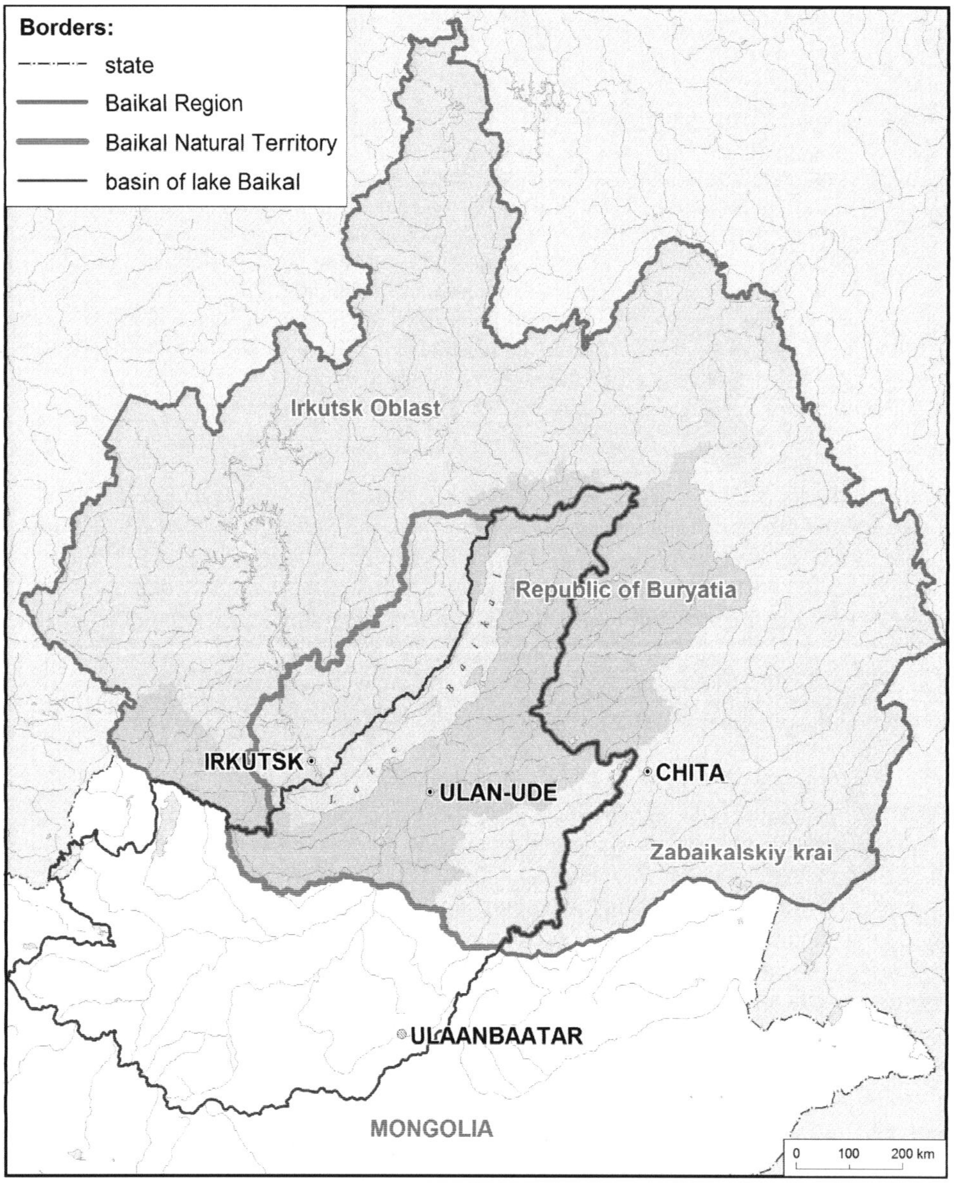


**Figure 1.** Transboundary Baikal region

There are five major directions of anthropogenic impacts that have negative effects on the ecological systems of the transboundary territories.

1. Intensification of natural disasters. The number of human losses and financial damages from natural disasters has increased notably during the last decades. Earthquakes, tsunamis, hurricanes, floods, forest fires, heavy snowfalls, and others do not depend directly on human actions, but the development of periodically inundated floodplains of rivers, deforestation, leading to a change in albedo, as a result cause thermokarst, solifluction, landslides, and occurrence of ravines. Trimming of slopes in the mountains cause avalanches and mudflows.

Floods occupy one of the leading places among natural disasters. They take place during high water and freshets.

Forest fires are one of the major environmental hazards in Siberia. Fires destroy not only the vegetation, but also a significant part of the animal world, and reorganization of ecosystems takes place. The soil cover is damaged, erosion develops, and desertification of a territory is under way. Places of traditional crafts of the population disappear (logging, hunting, gathering of wild plants and crop).

2. Pollution of natural environments. In the atmosphere of cities, where sulfur and nitrogen dioxides, benzopyrene, formaldehyde, suspended solids, and phenol are transferred with the emissions from industrial plants, thermal power plants, and motor vehicles, in the late 20th century the average annual concentrations exceeded the maximum permissible concentration in 185 cities of Russia [4].

3. Depletion of natural resources. Consumption of resources of the biosphere by humanity (nature management) in the world is steadily growing. Non-renewable mineral resources, oil, gas and coal are being reduced; their extraction is getting more expensive. Many mineral resources have been depleted, others are close to it. In their turn, renewable resources, i.e. fresh water, forests, soils, and living organisms are no longer renewed in the same quality and quantity. Drinking water becomes the scarcest resource. The steppe regions are among the areas poor in water because of natural conditions. In the taiga territories, the problem of their quality is of current concern.

4. Land degradation. In the agricultural regions, a high level of plowing under non-compliance with the soil-protecting technologies leads to land degradation, development of processes of water and wind erosion, desertification, salinization, and pollution with heavy metals and toxicants.

Only on the basis of system research that take into account the common interests of the parties we can deal with such challenges. The current situation requires a large-scale and comprehensive study of the border and transboundary territories of Asian Russia in order to develop the scientific foundations for a program for the sustainable development of transboundary territories, taking into account geographical, ecological, socio-economic and geopolitical factors.

The Baikal transboundary region is a complex region as regards its natural environment and landscape characteristics; it is endowed with immense reserves of various natural resources, the industrial development of which is steadily progressing, thus posing challenging ecological problems. Lake Baikal is a unique and fragile ecosystem, which must be protected and preserved. In connection with the assignment of the status of ‘World Heritage Site’, the concept of the Baikal natural territory (BNT) and its ecological zones were legally approved in 2006.

Balanced development of any territory is determined by a combination of socio-economic and natural-ecological factors. Disregarding any one of them leads up to an imbalance and disturbance of the development process. For territories having a high natural value, such as Baikal region, taking into account the ecological requirements is a ‘must’ in connection with what has been said above; at the contemporary stage of development of social and ecological systems, of particular significance is becoming the concept of strategic interrelated ecological and socio-economic territorial planning [5]. This concept is focused on achieving the goal to settle the conflict: economic and social interests – ecological constraints of the territory.

Some explanation is in order as to what is ecological zoning [6].

First, it is a tool for systematization and target-oriented analysis of information about the present status, significance and sensitivity of natural environments and complexes (systematization).

Second, it is a method for assessment of lands in the broad sense, including their geopolitical position, strategic prospects of utilization, and adjustment of the requirements for their utilization in accordance with world standards (comprehensive assessment).

Third, it is a center that assembles - through effective mechanisms of interaction – various agencies and politicians, who take decisions on different levels (management).

Fourth, it is a field for seeking optimal (best-practice) decisions given the competing options of utilizing resources and natural complexes; this is especially true for the conditions of emergent market relations (consideration).

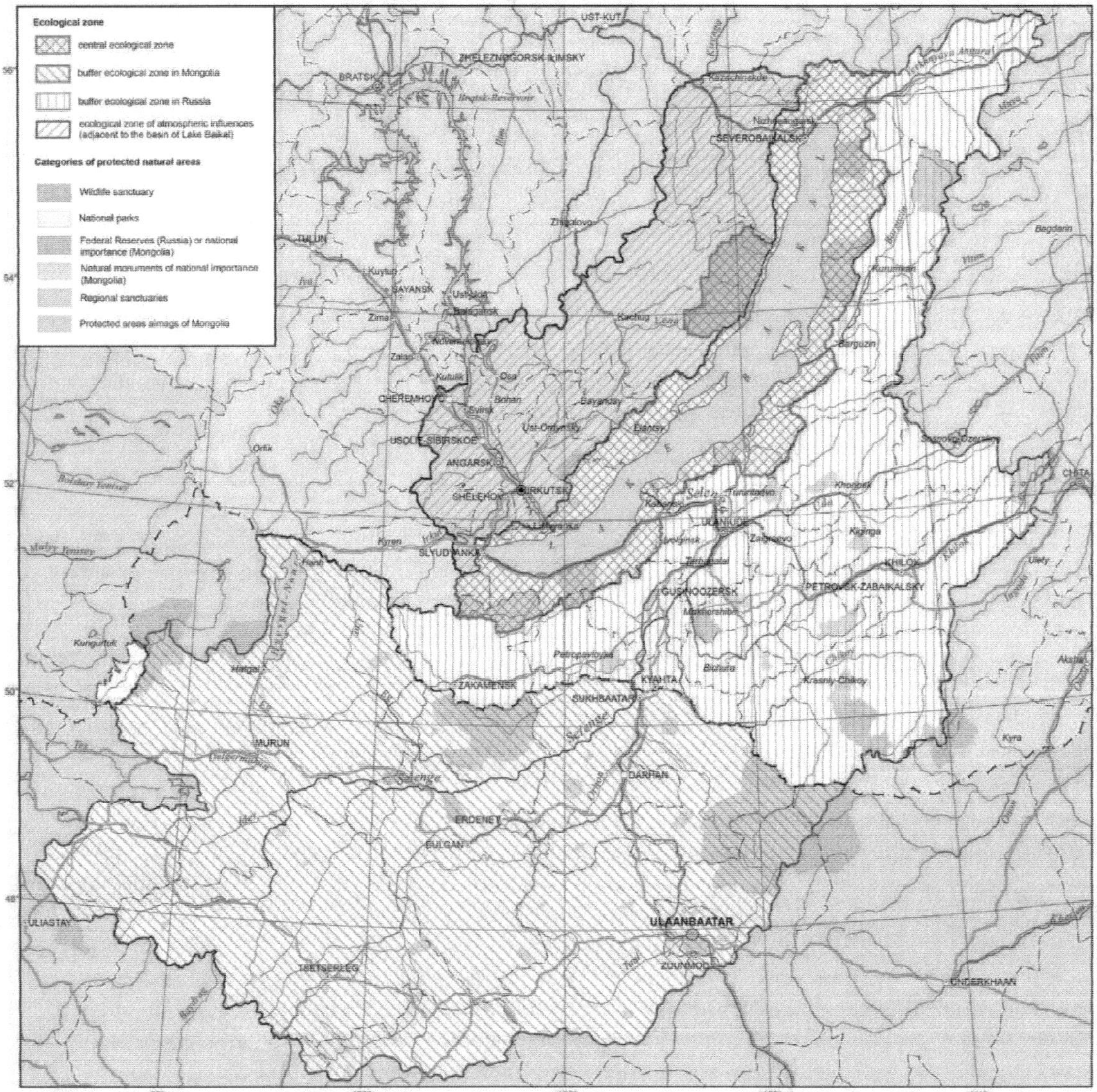


**Figure 2.** Ecological zoning of the Lake Baikal basin

## 3. Conclusion

Differentiation of environmental management regimes for transboundary territories should always be based on a comprehensive assessment of the natural basis and consideration of natural processes. The integration of the goals of landscape, territorial and socio-economic planning for the development of transboundary territories should be the development of the most appropriate strategy for optimizing the territorial organization based on ecological zoning (figure. 2), the main purpose of which is to identify the types of ecological territories that to some extent regulate economic activities. Its achievement is possible with simultaneous solution of two tasks: nature preservation and sustainable socio-economic development of the territory, without infringement of the rights and freedoms living of local people.

## Acknowledgments

The reported study was funded by RFBR according to the research project № 17-29-05089, by V.B. Sochava Institute of Geography SB RAS research program according to the projects № 0347-2016-0001 and 0347-2016-0003.

## References

[1] Chalov S et al 2017 *Reg Environ Change* **17** 2039
[2] Belozertseva I A et al 2017 *Water Resour.* **44** 471
[3] Antipov A and Semenov Y 2006 The Russian school of landscape planning *Environmental Security and Sustainable Land Use - with special reference to Central Asia (NATO Security through Science Series)* ed H Vogtmann and N Dobretsov (Dordrecht: Springer) pp 309-19
[4] Plyusnin V M 2014 *Contemp. Probl. Ecol.* **7** 597
[5] Belov A V, Vladimirov I N and Sokolova L P 2016 *Geogr. Nat. Resour.* **37** 129
[6] Plyusnin V M and Vladimirov I N 2013 *Territorial planning of the Central Ecological Zone of the Baikal Natural Territory* ed I V Bychkov (Novosibirsk: Geo) p 407

# Economic-geographical characteristics of agricultural land use within the Selenga river basin

**O A Ekimovskaya[1], I A Belozertseva[2], S E Amgalan[3], N B Badmaev[4]**

[1]Baikal Institute of Nature Management, Siberian Branch, Russian Academy of Sciences, Ulan-Ude, Russia
[2]Sochava Institute of Geography, Siberian Branch, Russian Academy of Sciences, Irkutsk, Russia
[3]Laboratory of Social and Economic Geography, Mongolian Academy of Sciences, Ulaanbaatar, Mongolia
[4]Institute of General and Experimental Biology, Siberian Branch, Russian Academy of Sciences, Ulan-Ude, Russia

E-mail: oafe@mail.ru

**Abstract.** In the early 1990s, Russia and Mongolia started transformations of the land relations and the system of agricultural land use. This study seeks to identify the distinctive features in agricultural land use in the upper (Mongolian) and lower (Russian) parts of the basin. An analysis is made of the main indicators reflecting the economic-geographical characteristics of agricultural land use, the structure of agricultural lands, the level of agricultural development, the dynamics of the area and efficiency of use of croplands and crop yield of grains. It was found that in spite of the more severe natural and climatic conditions, the Mongolian part of the basin has higher efficiency of agricultural land use. A decrease in the area of the cropland at the beginning of the agrarian transformations was observed throughout the territory of the basin. In the Mongolian aimags within the Selenga basin, the area of the cropland decreased by 27.5%. In the subsequent period, the sown area remained stable, in spite of a significant amplitude of fluctuations during separate years. In the Russian part of the basin, the area of croplands at the beginning of the agrarian transformations decreased by 21.5%. The tendency for a decrease in croplands has persisted to date. The main of the transformation of the agricultural land use system are formulated.

## 1. Introduction

River basins provide the most objective and natural basis for solving many problems related to the organization and governance of rational environmental management [1]. The Selenga river basin is a unified transboundary geosystem which is intensively used in agriculture. The natural-territorial complexes of the Selenga basin are united not only by the arterial hydrological network, geochemical flows of matter, air flows, exogenous processes, etc. but also by agro-economic linkages existing for hundreds of years. The land relations and agriculture in the Russian and Mongolian parts of the basin in the 20th century were evolving under similar socio-economic conditions, and the state monopoly on agricultural lands was prevalent. The Russian and Mongolian parts of the basin are united by historical experience of nomadic livestock husbandry, the long-standing practice of hay production by the sums of Mongolia for settled livestock husbandry farms of Buryatia, the practice of moving livestock from

Mongolia to Buryatia, deliveries of queen bees and honeybee colonies as well as grain crops of elite reproduction, bread and bakery products from Mongolia to Buryatia. Drastic changes in the system of agricultural land use occurred in the mid-1990s. Private ownership over resources became widespread in Mongolia. Up to 48% of agricultural lands are concentrated in private arat farms [2].

The importance of the agricultural lands for agrarian production within the Selenga basin is attested by the following indicators: in the Republic of Buryatia, the districts located in the drainage area account for 73.7% of gross output of agriculture of the republic, and for 83.1% of agricultural lands, including 86.4% of croplands. The aimags located in the Mongolian part of the Selenga basin account for 37.3% of agricultural products of Mongolia. They include 32.2% of natural forage lands and 35.1% of croplands of Mongolia [3].

The goals of this investigation were to identify the distinctive features in agricultural land use and the main directions of transformation of agrarian nature management within the Selenga basin. The findings can be used in developing a unified system of environmental measures on the territory of the transboundary basin. This is particularly relevant in view of the globalization and regionalization of the world economic space. The findings can be used in carrying out the land reform, and in developing programs of development and information support of agriculture in the regions.

## 2. Methods

The information base for this study was provided by material made available by territorial subdivisions of Rosreestr, Rosstat and the State Statistics Committee of Mongolia as well as by material of field expeditions. The method of landscape analysis of the territory was combined with economic assessment methods and with the approaches of comprehensive physical-geographical research. These approaches are based on the theoretical and methodological principles of the Siberian geographical school related to the analysis of the substantial contrasts in the development patterns of agrolandscapes. This research has used the comparative-geographical and economic-mathematical methods as well as the grouping method.

## 3. Results and discussion

Agricultural development of the study territory is highly uneven. The highest degree of agricultural development is characteristic for the southern part of the Selenga middle mountains as well as for the area along the delta of the Selenga. The agricultural lands are continuous and account for 59% of the total area of land use. The croplands are located along broad river valleys and the steppe and forest-steppe intermontane depressions. The proportion of the croplands in the structure of agricultural lands varies from 57% in the lower reaches to 8% in the upper reaches of the basin. The highest degree of plowing is characteristic for the Bichura, Tugnui, Kudara and Borgoi steppes and for the slopes of the Tugnui ouval. The proportion of pastures in the structure of agricultural lands increases along the southward direction of the basin. In the dry-steppe aimags of Mongolia located in the foothill areas of Khentii, the pastures account for up to 92% of the agricultural area. At the beginning of the agrarian transformations, the entire territory saw a drastic reduction of the degree of agricultural development. The chief reason behind this is the disintegration of collective enterprises which are the main land "holders". In the Mongolian aimags, the area of agricultural lands gradually recovered owing to an active participation of arat farms in the land reform which became owners of land parcels. A distinctive feature in land use in the Mongolian part of the basin is a high amplitude of fluctuations of the sown area for separate years. If, however, a long time interval, 10–15 years, is analyzed, the preservation of the sown area as a whole is noticed. In dry years, the area of croplands decreases dramatically but reverts to its former size thereafter. The area of croplands decreased by 21% during 1994–1999, remaining unchanged in subsequent years [2]. No reduction of croplands has been observed since 1999. Characteristic for the Russian part of the basin is a steady decline in the area of croplands. Between 1994 and 1999, the area of croplands decreased by 25% [3].

IOP Conf. Series: Earth and Environmental Science **190** (2018) 012019 doi:10.1088/1755-1315/190/1/012019

The aridity of the climate, the historical traditions and the work skills of the population coupled with a high proportion of natural rangelands on the territory were crucial to the livestock husbandry specialization of the region's agriculture.

Agricultural land use of the Mongolian part of the basin is characterized by integrated use of natural rangelands. Hay is harvested in the river valleys from July to September. In winters and springs, the river valleys are used as pastures. Agriculture has a secondary significance. Most of the croplands are concentrated in the steppe aimags (Selenge, Töv and Khentii) which are endowed with good atmospheric precipitation.

In the Republic of Buryatia, agriculture and livestock husbandry are pursued along the river valleys and in the intermontane depressions as well as gentle slopes of foothill areas. The agricultural lands of the Selenga middle mountains and the delta of the Selenga are represented by a continuous area at 540–720 m above the sea level and occupy the watershed spaces of the Selenga and its main tributaries: the Dzhida, Khilka, Chikoi and Tugnui. Over the course of several centuries this territory was the center of agrarian nature management. A relatively subdued topography of the Selenga middle mountains, the soil-climatic conditions and the historical features in settlement are responsible for the large size of croplands on the territory, including on the Tugnui ouval. Almost all slopes and flat watershed divides are plowed in the valleys of the Tugnui-Sulkhara rivers. A high degree of plowing of the Tugnui-Sulkhara depression is due to the fertile soils (chernozems, and chestnut and dark chestnut soils). Thus, in the farms of the Mukhorshibirskii district, the proportion of chernozems reaches 32.5% of the area of the croplands and 21.4% of the total area of the agricultural lands [4].

The agricultural lands of the Mongolian part of the basin are concentrated in narrow, deeply incised river valleys of the Mongolian Plateau at up to 1500 m above the sea level. The outruns are located throughout the territory, including in the foothill areas of Khentii and Khangai, at 2000 m above the sea level. The natural boundary of the agricultural area in the north-west are the Khan Khukhei, Ulan Taiga and Eastern Sayan mountain ranges which do not form a unified massif; in the middle part, the Khangai range, and in the northeastern part, the Khentii mountain massif.

The main producers of commodity grain are the southern steppe districts of the Selenga middle mountains (Bichurskii, Kyakhtinskii and Mukhorshibirskii). Wheat is the main crop in the structure of grain crops. Most of the grain-sown areas are concentrated in agricultural organizations. In the upper reaches of the Selenga, on the territory of Mongolia, the croplands are distributed evenly among land users of different forms of ownership. Grain crops are cultivated by collective farms as well as by arat farms. They account for 45 and 55% of the sown area, respectively. The structure of sown crops is dominated by grain crops for animal feeds.

In Mongolia, livestock husbandry accounts for up to 80% of gross agricultural produce. Its southern part is dominated by the nomadic system of livestock grazing; therefore, of critical importance are the centuries-long technologies of migrations and natural selection of animals. Different kinds of landscapes are used as seasonal pastures according to the natural and climatic conditions and to the wind regime in particular. In the summertime, "... the northern mountain slopes are a convenient habitat... for livestock", because the prevailing northerly wind in that season "blows round" livestock [4, p. 146]. In winters, the pastoralists move with their herds to higher areas and to lowlands to obtain shelter from the wind.

In the middle and lower parts of the Selenga basin, the main system of livestock husbandry is represented by stall-feeding and grazing which is more particular about animal feeds and the conditions of livestock keeping. With stall-feeding and grazing, of significant importance are the ecological state and productivity of pastures and hayfields. Seasonal pastures in steppes and forest-steppes occupy non-forested interfluve areas, the upper parts of the slopes of depressions, steppizated terraces and high-mountain meadows in places, intermontane depressions and foothill areas. In the Kabanskii district, pastures occupy the area along the delta of the Selenga. The proportion of lands suitable for hay making is very low. The sole exception is the Kabanskii district. The abundance of moisture and subdued topography provide favorable conditions for the growth of vegetation thus providing the farms of the district with rich hayfields. Owing to a lack of natural hayfields, hay making is done in abandoned areas of croplands, grassy glades and fallow lands. Large amounts of forage are produced in field crop rotations, which increases the product cost in livestock husbandry. The area of natural rangelands remains relatively stable. The state of livestock husbandry completely depends on their yield level. A major loss of livestock can occur in dry years. Livestock restocking is governed by an increase in yield level of the pastures.

The sown areas and grain crop yield in the aimags of Mongolia are larger than in the districts of the Republic of Buryatia. Depending on the particular aimag, the absolute size of the croplands varies from 0.599 to 105.8 thou ha. The significant differences in the size of croplands are due to the character of the hydrogeographical network, topographic features and the humidification pattern. In almost half of the aimags (45.6%), the mean long-term crop yield e1xceeds 14 centners/ha [4]. In the Republic of Buryatia, only three districts out of 14 (21%) reach such indicators. The highest yield level of wheat is recorded in the steppe and forest-steppe zones of the Selenga, Tuul and Kharaa valleys dominated by dark chestnut and chernozem soils. In the Orkhon, Bulgan, Selenge and Töv Aimags, the mean long-term indicators, respectively, are 22.9, 15.0, 14.6 and 15.1 centners/ha. Further eastward, atmospheric precipitation amounts are decreasing, and the yield level drops to 13.9 – 7.7 centners/ha. The western aimags in the foothill areas of Khangai are dominated by cryogenic taiga soils of the high mountains. The yield level varies from 0.2 to 22.8 centners/ha.

The distinctive features in the use of natural forage lands are revealed by data on the area of pastures and hayfields per conventional head of livestock (table 1). In most farms of the Selenga middle mountains, 0.8–0.9 ha corresponds to one conventional head. The delta of the Selenga shows a shortage of natural forage lands, and a mere 0.5 ha of hayfields and 0.2 ha of pastures are available for one conventional head. The numerous islands of the delta are used by the population as summer pastures. The number of livestock in Mongolia far exceeds the number on the territory of the Republic of Buryatia; therefore, the availability of natural forage in the Mongolian aimags is lower (table 2).

**Table 1.** Availability of grazing land and hayland in the districts of the Republic of Buryatia*.

| Districts | Total area, ha | | Number of conventional livestock (CLS) | Including per CLS, ha | |
|---|---|---|---|---|---|
| | hay-lands | grazing lands | | Haylands | Grazing lands |
| **Bichurskii** | 9983 | 28574 | 6313 | 1.581 | 4.526 |
| **Dzhidinskii** | 12231 | 124010 | 17216 | 0.710 | 7.203 |
| **Eravinskii** | 14639 | 111038 | 11154 | 1.312 | 9.954 |
| **Zaigraevskii** | 6635 | 16433 | 6104 | 1.087 | 2.692 |
| **Zakamenskii** | 8498 | 16600 | 5662 | 1.501 | 2,932 |
| **Ivolginskii** | 4097 | 12839 | 1673 | 2.445 | 7.674 |
| **Kabanskii** | 3658 | 2016 | 7992 | 0.458 | 0.252 |
| **Kizhinginskii** | 3810 | 6792 | 1965 | 1.939 | 3.456 |
| **Kyakhtinskii** | 8301 | 56395 | 8724 | 0.952 | 6.464 |
| **Mukhorshibirskii** | 10934 | 77197 | 13915 | 0.786 | 5.548 |
| **Pribikal'skii** | 6400 | 8100 | 3972 | 1.611 | 2.039 |
| **Selenginskii** | 16785 | 100031 | 7003 | 2.397 | 14.284 |
| **Tarbagataiskii** | 4277 | 24853 | 5271 | 0.811 | 4.715 |
| **Khorinskii** | 14222 | 66725 | 7954 | 1.788 | 8.389 |

*The table was compiled using data of the Federal State Statistics Service

With the existing traditional system of livestock husbandry, the ecosystems of the Selenga river are in a relatively stable state [5]. The degree of their disturbance corresponds to the number of livestock. With an optimal rate of grazing (up to 2 heads/ha), the pastures show a weak degree of disturbance of the soil cover, which manifests itself in a decrease in productivity of the subsurface vegetation mass by a factor of up to 1.6 and soil compaction within normal parameters (up to 1.1 g/cm3 for the soils of steppe landscapes). The moderate degree of landscape disturbance (grazing pressure from 2 to 4 heads/ha) is characterized by soil compaction up to 1.21 g/cm3, and by a decrease in the mass of plant roots by a factor of up to 4.7. For an intense rate of grazing (more than 4 heads/ha) and, accordingly, a strong degree of landscape disturbance, the study revealed a decrease in root productivity by a factor of up to 22, soil compaction to 1.46 g/cm3 and destruction of the soil sod horizon [5, 6].

**Table 2.** Availability of natural forage lands in the aimags of Mongolia*.

| Aimags | Area of natural forage lands, thou ha | Number of CLS | Natural forage lands, ha, per CLS |
|---|---|---|---|
| **Arkhangai** | 3793.4 | 2912519 | 1.323 |
| **Bulgan** | 2633.0 | 2008110 | 0.137 |
| **Bayankhorgor** | 6100.2 | 2976003 | 2.05 |
| **Darkhan-Uul** | 194.2 | 230406 | 0.871 |
| **Zavkhan** | 6994.2 | 2999903 | 2.311 |
| **Sekenge** | 1771.5 | 1030201 | 1.725 |
| **Orkhon** | 41.1 | 210090 | 0.195 |
| **Töv** | 5434.5 | 2626660 | 2.069 |
| **Övörkhangai** | 5746.2 | 2623304 | 2.190 |
| **Khövsgöl** | 4435.7 | 3425211 | 1.295 |
| **Khentii** | 5404.1 | 2183223 | 2.475 |

*The table was compiled using data of the State Statistics Committee of Mongolia

This work was done under grant 18-45-030039 p_a of the Russian Foundation for Basic Research and within the state assignment of Baikal Institute of Nature Management SB RASH (project no. 0339-2016-0001)

A large portion of the soil cover in Mongolia needs irrigation. However, shallow soils of light particle-size distribution undergo erosion with the removal of mineral and organic compounds and fine earth. In this case, soil optimization requires a package of agrochemical and agrotechnical measures and the protection against deflation. Excessive watering of loamy soils leads to the deterioration of their temperature regime, and enhancement in cryogenic phenomena and the development of salinization processes.

Agricultural land use on the territory of the Selenga basin is concentrated on the Baikalian-type depressions. The existing soil-ecological situation is favorable for traditional livestock husbandry [7].

## 4. Conclusions

In spite of the pre severe natural and climatic conditions, a secondary role of plant-growing, the intensity of development of the grain economy and efficiency of use of croplands in the agrarian economy is higher than in the Russian part of the basin. This implies a faster rehabilitation of the sown areas after dry periods, the absence of abandoned croplands, and a high yield level of grains. The efficiency of plant-growing in Mongolia was positively influence by the privatization of land parcels. The private arat farms engaged in agriculture are distinguished by a higher labor productivity when compared with the collective farms of the Republic of Buryatia.

The Russian and Mongolian parts of the Selenga basin are characterized by an asynchronous development of agrarian nature management. Mongolia has preserved traditional nomadic technologies of grazing. On the other hand, the herd structure forms according to the market conditions. In the Republic of Buryatia, nomadic livestock husbandry had prevailed in almost all regions prior to the collectivization of private farms in the 20th century. Nowadays there is a widespread system of livestock stall-feeding and grazing, with agricultural organizations dominant in land use. A significance importance in the development of the agricultural sectors are the federal and republican programs.

The structure of sown lands existing in the Republic of Buryatia, with a high proportion of feed crops (including grains for feeding purposes), meets the requirements of livestock husbandry and yields a small but stable profit for agricultural enterprises.

The current conditions of economic management demand revival of plant nurseries and seed sorting units in order to develop new high-yield drought-tolerant and early maturing varieties followed by their introduction in both the Buryat and Mongolian parts of the Selenga basin. In general, however, emphasis should be placed on the development of the livestock husbandry component.

To increase productivity of stall-feeding and grazing livestock husbandry requires an enhancement in field and meadow-pasture forage production. Specifically, it is necessary to revive the melioration and fertilization system for hayfields and construct new permanent cultural pastures. In view of the small areas of natural hayfields, it is best to partly use the croplands for the production of hay from annual and perennial grasses.

The districts of the Republic of Buryatia are characterized by a decline in the agrarian production potential in the suburban zone. This implies a decrease in intensity of agricultural land use by collective farms, their total decay in some of the suburban areas, and a reduction of sown areas.

**Acknowledgments**

We are grateful to V.G. Mikhalkovsky for his assistance in translating the manuscript.

**References**

[1] Korytny L M 2001 *The Basin Concept in Nature Management* (Irkutsk: Institut Geografii SO RAN) p 163 (in Russian)

Narozhnaya A G and Kuzmenko Y V 2012 Basin nature management for the environmental protection *Problemy Regionalnoi Ekologii* **2** 12–5 (in Russian)

Smolyaninov V M, Degtyarev S D and Shcherbinina S V 2007 *Ecological-Hydrological Assessment of the State of River Catchments of Voronezh Oblast* (Voronezh: Istoki) p 133 (in Russian)

[2] *Mongolian Statistical Yearbook* 2017 (Ulaanbaatar: State Statistics Committee) (in Russian) Income accessed online on 2th August 2018 via URL: http://1212.mn/tables.aspx?TBL_ID=DT_NSO_1001_021V1

[3] *Characteristic of Agriculture of the Republic in Tables, Diagrams and Groupings* 2017 (Ulan-Ude: Buryatstat) p 74 (in Russian)

[4] Shagdarsuren O 2011 Biology of pasture species of animals and characteristic of nomadic livestock husbandry *Trudy Sovmestnoi Rossiisko-Mongolskoi Kompleksnoi Biologicheskoi Ekspeditsii* vol. LVIII (Moscow: Nauka Publishing) p. 146 (in Russian)

[5] Nogina N A and Dorzhgotov D 1982 Soil geographical zoning of Mongolia *Eurasian Soil Science* **4**(4) 37-42 (in Russian)

[6] Belozertseva I A, Vladimirov I N, Ubugunova V I, Ubugunov V L, Ekimovskaya O A and Bardash A V 2016 Soils of the water protection area of Lake Baikal and their use *Geogr. Prir. Resour.* **5** 47–9 (in Russian)

Vostokova E A and Gunin P D 2005 Degradation of Ecosystems. *Ecosystems of Mongolia, Atlas* (Moscow: Nauka Publishing) p 44

[7] Gunin P D, Vostokova E A, Bazha E A and Bayasgalan D 2005 Ecosystems of the Selenga basin *Trudy Sovmestnoi Rossiisko-Mongolskoi Kompleksnoi Biologicheskoi Ekspeditsii* vol. XLIV (Moscow: Nauka Publishing) p 7 (in Russian)

IOP Conf. Series: Earth and Environmental Science **190** (2018) 012020 doi:10.1088/1755-1315/190/1/012020

# Environmental safety and the "green" economy in the context of the Baikalocentric model of territorial development

**T I Zabortseva**

Sochava Institute of Geography, Siberian Branch, Russian Academy of Sciences, Irkutsk, Russia

E-mail: zabti@irigs.irk.ru

**Abstract.** The results of ecological and socioeconomic scientific research within the central ecological zone of the Baikal natural territory (CEZ of the BNT) are presented, the legislatively fixed function of which is the preservation of the unique ecological system of Lake Baikal. It administratively includes the territory of three districts of the Irkutsk region (Irkutskii, Olkhonskii and Slyudyanskii) and four in the Republic of Buryatia (Barguzinskii, Kabanskii, Severobaikalskii and Pribaikalskii). The research is based on the theoretical and methodological approaches of social geography; we applied the ideology of the "green" economy and the centrality of territorial development. For Lake Baikal as an object of the World Natural heritage (since 1996) on the basis of the adopted federal law "On the protection of Lake Baikal", three federal target programs for the protection of the lake and socioeconomic development of the Baikal natural territory have been developed. Based on the data obtained during the targeted research of the federal and regional level on the territory of the CEZ of the BNT, a matrix is proposed for assessing the location of economic activity objects (e.g. industry of specialization - the recreational sphere). Geographical features of the CES of the BNT and its modern functional purpose predetermine the Baikal-centric vector of economic activity, primarily recreational, and the growing role of the emerging environmental protection infrastructure (GIS). We consider the GIS as part of the environmental infrastructure from the standpoint of the"green" economy as a practical implementation of the principles of sustainable development. Methods of expeditionary research and sociology (questioning, survey) were used.

## 1. Introduction

The "green" economy is interpreted as an economy whereby people's well-being is increased and social justice is provided, environmental risks and degradation are significantly reduced. In the conceptual documents of international organizations, and developed countries, including Russia, the prospects for economic development in the 21st century are associated with the formation of a "green" economy and "green" growth. In the scientific and practical activity over the past few years, according to Bobylev, a consensus has actually been reached on the need to form a new type of economic development, and to develop a new "green" economic course [1]. The concept of environmental safety in the current domestic regulatory and legislative documents is interpreted as a state of protection of the natural environment and vital human interests from the possible negative impact of economic and other activities, natural and man-made emergency situations, and their consequences.

The central ecological zone of the Baikal natural territory is a functionally and legislatively allocated territory with special conditions for ecological, social and economic development. The main function of the central ecological zone of the Baikal natural territory is the preservation of the unique ecological system of Lake Baikal and its coastal environment. Ecosystem functions of the CES of the BNT also include resource (pure drinking water) and recreational (development and restoration of physical and spiritual forces of human being). Formation of a model for the development of the territory in the format of the "CEZ BNT + Baikal ecosystem", consistent and logically aligned, will enable implementation of the principle of sustainable development of the socioeconomic life of the society in relation to the World Heritage site Lake Baikal, to build the matrix of relations in the ideal format, when the population is interested in preserving landscape attractiveness, preventive measures for preserving forests, and demonstrates environmental behavior in practice.

It should be noted that Baikalocentricity as a paradigm of harmonious and inexhaustible nature management is designated, actively promoted in the second half of the last century by academicians G.I. Galazii and V.A. Koptyug, and is fundamentally worked out by the scientific community of the Siberian Branch of the Russian Academy of Sciences [2].

## 2. Models and Methods

The study is based on materials obtained during expeditionary research in the context of municipalities of the low administrative levels.

To conduct a state environmental-oriented policy that could enable a comprehensive analysis of the possible impact of the proposed activity on the environment or facility construction, and also promptly identify the risks of adopting a decision, assess cumulative impacts, develop mechanisms for implementing programs and projects of different status in the CEZ of the BNT (state, corporate, and international), we developed a matrix on the methods of evaluation

Two groups of criteria were used: sociolegal and environmental. Sociolegal group of criteria is based on the basic regulatory and legislative acts in the field of Lake Baikal protection in order to justify the permissibility of placing ecologically oriented types of economic activity. The main characteristics considered in the group of environmental criteria for assessing the permissibility of the location of facilities were as follows: the features of the production infrastructure used to reduce the environmental impact; the main factors of the negative environmental impact; peculiarities of methods for purification of discharges, disposal and utilization of wastes of main production, purification equipment; the probability events with adverse consequences for the environment and caused by the negative impact of economic and other activities, by emergency situations of natural and man-made nature.

## 3. Results and Discussion

The construction of a high-speed highway, the formation of the Eurasian Economic Corridor, that includes the CEZ of the BNT as its part, is a priority for the new powerful integrated development of the Transsib zone and adjacent territories [3]. The author's experience in the implementation of targeted research programs of federal and regional status in the territory of the CEZ of the BNT (1998-2003, 2016-2018, mainly expeditionary research), familiarity with foreign practices of economic development, similar in objects and their function, allow us to recommend a partnership format (co-development) "CEZ of the BNT + Baikal ecosystem". The priority is the "green" economy, which is the practical implementation of the sustainable development goals of this key site of the BNT, which is consistent with other researchers [4]. However, strict prohibitions of the titles on land use in the CEZ of the BNT, which are absolutely necessary concerning significance of Baikal's unique natural site and the "world well" of fresh water, did not achieve the expected result in practice. One of the reasons, the regulation of economic activity, continues to be a problem, without real compensatory mechanisms (financial, organizational and technical, etc.), without updating the whole complex of infrastructure - communal, transport, social, environmental protection, ensuring environmental safety [5].

IOP Conf. Series: Earth and Environmental Science **190** (2018) 012020 doi:10.1088/1755-1315/190/1/012020

The main function of the central ecological zone of the Baikal natural territory is the preservation of the unique ecological system of Lake Baikal and prevention of negative impacts of economic and other activities on its condition. Given the status of federal attention to the problems of Lake Baikal (included in the list of natural World Heritage sites in 1996), as well as the proclamation of the ideology of preserving this Great Lake in the economically well-founded Russia, a peculiar historical analysis of the institutions of interaction between the state, authorities and civil society is highly relevant. Its purpose is to identify and eliminate system failures that prevented, to implement projects of special economic zones of the tourist-recreational type (SEZ TRT) "Gate of Baikal" and "Baikal Harbor", and also the creation of a working scheme for handling waste in the tourist areas on the coast and the development of tourist zoning and the rules of recreational activities in the CES of the BNT in accordance with the Federal Law "On the Protection of Lake Baikal", etc.

The CEZ of the BNT administratively includes the territories of two constituent entities of the Russian Federation (Irkutsk region and the Republic of Buryatia) with seven municipal districts (Irkutskii, Olkhonskii, Slyudyanskii in the region, Barguzinskii, Kabanskii, Pribaikalskii, Severobaikalskii in the Republic) and one city (Severobaikalsk) with an area of about 90 thousand $km^2$, where more than 140 thousand people live (as of 01/01/2018). At the same time, the annual flow of recreants exceeds 2.2 million people [6]. Therefore, placement possibility of various objects of the "hospitality industry" is represented on the example of recreational activities (generalized jointly with O.V. Evstropyeva), with a generalization of a number of known methods mainly in three positions: advantages, disadvantages, selection criteria, supplemented by legal regulations, significant for the actualization of the mentioned table 1.

The location of the CEZ is economically profitable, since it is the developed southern part of Eastern Siberia: along the southern part of the lake there is a section of the Trans-Siberian Railway, in parallel - the federal highway "Baikal", it is located in the zone of influence of the centers of the federation subjects - Irkutsk and Ulan-Ude, along the northern part of the lake the Baikal-Amur Mainline (BAM) is located; During summer water transport connection is provided along Lake Baikal and the Irkutsk reservoir. Geostrategically, this is a local crossroads between the west and east and an international one between the north and south (Siberia-Mongolia-China). A number of the most important factors of the expected qualitative change in the development of the CEZ of the BNT since the late 1980s can be formulated in several positions, under the stable interest of the state, including the regional authorities, municipalities, and civil society.

First, according to our expert assessments, the number of qualitatively developed normative and legislative documents and various proposals of the pre-project, project level, as well as territorial maps for the development of the federal status for the Baikal natural territory (including the CEZ), aimed at protecting nature, organizing economic activities in terms of environmental constraints, exceeds any domestic region for the last quarter of century. Thus, only the "Territorial Complex Scheme for the Protection of the Nature of the Lake Baikal Basin" (hereinafter the Scheme) was presented in the format of 69 volumes (approved by the Presidium of the Council of Ministers of the RSFSR on 14.04.1990), updated in 2009 by the Federal State Unitary Research and Production Enterprise "Russian Federal Geological Fund" commissioned by the Ministry of Natural Resources of Russia. Nevertheless, as a result of the legal inconsistency, the implementation of the Scheme regulating the given vector of the development of the CEZ of the BNT was not carried out [7]. During the reform period, three federal target programs for the protection of Lake Baikal and socioeconomic development of the Baikal natural territory were developed.

Secondly, the strict prohibitive nature of the legal documents (table 1) is absolutely necessary in the context of the importance of the unique natural site of Lake Baikal and the "world well" of fresh water regulating economic activity, has not any real compensatory mechanisms (financial, material, organizational technical, etc.) at the regional and municipal levels. The scale of "different nature" (interests) of government bodies and institutions of civil society causes, in turn, differences in understanding approaches to solving problems in this territory. As a result, the method of development of the CEZ did not work through the mechanism of the two major mega-projects mentioned (the situation requires a separate analysis). Only the self-employment of the population and its entrepreneurial activity, mainly in the organization of various recreational activities, stopped the real increase in unemployment in most settlements, especially after the closure of logging enterprises, fishing enterprises, and the pulp and paper mill in Baikalsk.

**Table 1.** Assessment of placement possibility of various recreational objects within the CEZ of the BNT.

| Methods | Advantages | Disadvantages | Selection criteria | Documents, rules |
|---|---|---|---|---|
| **Special regulated impacts, the load method** | Ratio of anthropogenic load and ecological capacity to local terrestrial systems in the context of local ecosystems. | The application of the standard ratios is possible as a result of detailed research on recreational areas, regular monitoring of recreational loads, the development of functional recreational zoning of the CEZ of the BNT and tourism rules, common for the Irkutsk region and the Republic of Buryatia. | The level of ecological and socio-economic parity | "On approval of standards for exposure limit on the unique ecological system of Lake Baikal" Order of the Ministry of Nature of RF of 05.03.2010 N 63. "On the protection of Lake Baikal. The federal law of 01.05.1999 №94-ФЗ. |
| **Regulatory approach** | General principles of state policy in the field of tourism | Underestimation of regional socio-economic etc. features of territories of different hierarchical status | Priority of central-peripheral development | FL* "On the fundamentals of tourism in the Russian Federation" (as of 24.11.1996) |
| **Departmental regulatory and legal approach** | Regulated general principles of sectoral priorities (in the field of forest management, tourism policy) | Underestimation of regional socio-economic and other features of territories of different hierarchical status | The parity of government-departmental interests | "Standard of the industry OST 56-100-95" Methods and units of recreational load measurements for forest natural complexes "(Order of the Federal Forestry Agency of July 20, 1995 No. 114) Forest management regulations "Pribaikalsky National Park" and others. |

Note: * - Federal Law

In the territory of the CEZ of the BNT, in the light of visible institutional deficiencies, specific ecological and economic problems in the field of the established specialization branch - tourism [8] were also attempted to reveal. Thus, in the field research on the federal target program "Lake Baikal Protection and Social and Economic Development of the Central Ecological Zone of the Baikal Natural Territory" (summer 2016), the heads of local self-government bodies, staff of specially protected territories, as well as representatives of the tourist industry were interviewed. As a result, specific problems related to the recreational activity adjustment in the CEZ of the BNT were revealed. The main part of the unsolved problems is related to the following major deficiencies: lack of cooperation between the tourist industry, local population, local authorities and specially protected territories, as well as infrastructure deficit, primarily in waste management; and a shortage of land for tourism development.

The main environmental problems of the CES of the BNT as a threat to environmental safety are: firstly, the discharge of insufficiently treated sewage into the lake and its tributaries by sewage treatment facilities (STF); and low technical condition of the STF [9]. Secondly, the receipt of pollutants in the form of diffuse and dispersed runoff from recreational, residential, industrial areas and storage sites for production wastes; absence of storm, local and centralized sewerage systems in the majority of settlements and in recreational areas of the CEZ of the BNT. Thirdly, a large amount of biogenic pollutants enters the lake in discharges from the STF and with a dispersed runoff, which contributes to the eutrophication of the lake (at the peak of the summer season on the coast of Lake Baikal, about 20000 tourists can stay at any time, which produce equivalent volumes of solid municipal waste and liquid household waste (SMW and LHW, respectively). The prospects for the development of much-needed environmental protection infrastructure are associated with the creation of a network of waste collection and processing points, with the location of treatment facilities, as well as compliance with environmental regulations of permissible impact standards using environmentally friendly technologies for waste water disposal and utilization.

At Baikal, the processes of "infrastructural disruption" in handling waste have reached an almost catastrophic scale. Over the past decade, the main Baikal theme for obtaining grants is cleaning the coastal areas of Lake Baikal from garbage by domestic and foreign volunteers. One of the reasons is the lack of functioning environmental protective facilities (EPF) (only one polygon in the CEZ is included in the list of state waste disposal facilities); the available special technology for the disposal of consumer waste is only for the local population. Another reason is the increase in the specific accumulation of SMW due to the "packaging boom" and the destroyed system of selective collection of the utilized part of SMW. At the same time, the share of SMW from tourist facilities in the Irkutsk part of the CEZ, according to our calculations, exceeds one third of the total annual SMW flow (8642 and 24811 tons/year, respectively). The highest share of SMW from tourism in the context of municipal districts is in Olkhonskii (52%), and Irkutskii district (32%).

The basic function of the EPF corresponds with the main positions of promoting the 'green' economy: minimizing the impact of the deposited, recyclable and other production and consumption wastes on the environment, as well as involving additional resources in the economic circulation and improving the living conditions of population. The most important task for the CEZ of the BNT is the formation of a mesoregional cluster for waste management. The EPF cluster is a network of geographically localized interrelated economic agents of specialized services, infrastructure enterprises, thematic research institutes, universities and other organizations that complement each other and enhance the competitive advantages of individual territories. A necessary condition for the formation and effective operation of EPF: the creation of an institutional environment (socio-economic, regulatory, legal, cultural, educational, and eco-oriented).

## 4. Conclusion

Analysis of the ecological and socioeconomic conditions of the territory in the CES of the BNT revealed that theoretically there is an ideal situation for the implementation of the principles of sustainable development of the Lake Baikal as a World Heritage site, taking into account the developed strategic documents for the development of the Baikal natural territory, including its central ecological zone in the format promoted by the "green" economy.

The location of the CEZ of the BNT is economically profitable, since it is the mastered southern part of Eastern Siberia. Geostrategically, this is a domestic crossroads between the west and east and the internationally between north and south (Siberia-Mongolia-China). The solution of environmental problems today involves the development and implementation of a new policy, since it is the environmental requirements laid down as principles of the "green" economy that determine the strategic directions of innovative development and modernization of the economic complex of the CEZ of the BNT and further long-term favorable ecological, social and economic development of the territory.

Further development of recreational activities as one of the main branches in the visible perspective "initiates" the formation of environmental protection infrastructure. And the latter one ensures the

IOP Conf. Series: Earth and Environmental Science **190** (2018) 012020 doi:10.1088/1755-1315/190/1/012020

preservation of a favorable environment, including landscapes, attractiveness of the Baikal Territory as a whole.

## References

[1] Bobylev S N 2017 Sustainable development for the benefit of future generations: economic priorities *The World of a New Economy* **3** 90-6
[2] Tulokhonov A K 2010 *The Baikal Problem: History and Documents* (Ulan-Ude: The publishing house "Ecos") p 128
[3] *Transsib Zone as the Eurasian Economic Corridor* 2016 (Irkutsk: IG SB RAS Press) p 251
[4] *Eastern Vector of Russia: a Chance for a Green Economy in Natural Resource Regions* 2016 (Irkutsk: IG SB RAS Press) 247 p
Sysoeva N M and Sysoeva O V 2016 Problems of entrepreneurship development in the coastal zone of Lake Baikal *Geography and Natural Resources* **5** 144-50
Dong S, Li Y, Li Z, Li F, Bilgaev A, Yang Y and Zheng J 2017 Green Development mode of the Lake Baikal region *Baikal as a Site of the World Natural Heritage* (Ulan-Ude: BSU Press) pp 31-8
[5] Tulokhonov A K and Beshentsev A N 2017 The Baikal problem: history and modernity (to the 25th anniversary of the organization of the Governmental Commission for Lake Baikal) *Geography and Natural Resources* **4** 68-75
[6] Evstropyeva O V 2016 Development of the tourist system in the Baikal natural territory *Geography and Natural Resources* **5** 184-95
[7] Korytnyi L M 2011 *Echo of Ecological and Economic Scandals* (Novosibirsk: SB RAS Publishing House) p 328
[8] Zabortseva T I and Evstropyeva OV 2009 Modern socio-ecological problems of recreational development: Pribaikalskii National Park *Geography and Natural Resources* **4** 72-81
[9] *State Report on the State of Lake Baikal and Measures to Protect it in* 2015 (Irkutsk: Siberian branch of Federal State Unitary Enterprise "Rosgeolfond") p 436

IOP Conf. Series: Earth and Environmental Science **190** (2018) 012021 doi:10.1088/1755-1315/190/1/012021

# Assessment of the technogenic impact of the mining industry on the environment of the economic corridors "China - Mongolia – Russia" of the territory of Mongolia and the Baikal region

**A D Abalakov, N B Bazarova**

Sochava Institute of Geography, Siberian Branch, Russian Academy of Sciences, Irkutsk, Russia

E-mail: abalakov@irigs.irk.ru

**Abstract.** The paper considers ecological consequences of the technogenic impact of the development of mineral deposits located in the economic corridors "China - Mongolia - Russia" within Mongolia and the Baikal region including the Irkutsk region, the Republic of Buryatia and the Transbaikal region. Administrative districts along the Transsiberian railway from Ulan-Ude to Ulaanbaatar to the border with China and Karymskoe-Zabaikalsk are being examined. A model that reveals an approach and main directions of the research has been developed. The structure of the model includes blocks containing information on objects and subjects of impact, environmental transformation and its assessment. A mapping method of the ecological situation was adopted as the main one. We created an original mapping language, consisting of a system of mapping signs that reflect the various environmental aspects of mining enterprises. We proposed criteria to assess technogenic impact within the fields and administrative regions. For the local assessment of the technogenic transformation of mining areas, we have taken the type of raw materials, its toxicity, the method of extraction, and the area of disturbed lands as environmental indicators. For spatial assessment, indicators of the environmental state in municipal areas have been adopted. The results of the studies are presented on six environmental maps that provide a visual representation of the mining impact on the environment, including its integrated assessment. A detailed description of the environmental consequences of mining enterprises activity, waste products that have the strongest impact on natural complexes, living conditions, economic activity and public health is given. The problems and prospects of further research are determined.

## 1. Introduction

On June 24, 2016 in Tashkent, at the SCO summit between the Russian Federation, the People's Republic of China and Mongolia, a tripartite agreement was signed to create an economic corridor "Mongolia - Russia - China." This economic corridor will be created, first of all, to increase trade turnover and development of transport infrastructure of the countries-participants of this agreement [1]. It is assumed that the implementation of the Agreement will entail an influx of investments into the adjacent regions, which may lead to an increase in technogenic load on the environment.

Mining industry complex is the most important basic element of national economy. At the same time, it has a significant impact on the environment. Due to the importance of this problem, the assessment of environmental impact of mining enterprises within transport corridors is of great importance to the countries that are parties of this Agreement. This is especially relevant for ecologically vulnerable territories [2], including the Central Ecological Zone of the Baikal Natural Territory [3], national parks and reserves [4].

In the present work, the assessment of the technogenic impact is carried out in relation to the administrative areas along the transit railways running through the territory of Mongolia and the Baikal region. In these areas there are deposits of various minerals, both currently being developed and that have been preserved. The implementation of transport corridor construction projects will improve the transport and economic situation of the adjacent areas. This will give impetus to the further development of settlement systems, to a network of closely interconnected industries, including the mining industry, which will require additional environmental measures.

The research is conducted on the basis of a model, the content of which is disclosed in six associated environmental maps, discussed below.

The present study is a continuation and deepening of previous work on this subject [4, 5]. The maps compiled give a clear idea of the environmental consequences of the impact of mining on mineral deposits and environmental conditions in municipal areas. This information is the basis for planning and protective measures that allow to optimize technogenic load on the environment, which contributes to the conservation of ecosystems in the study region.

## 2. Models and methods

A model has been developed, that we adopted as a structural scheme - a model for solving different research tasks for assessing and mapping the technogenic impact of the mining industry on the environment of administrative regions within transport corridors (figure 1).

The model contains a number of blocks that reveal the ecological effects of the interaction of natural and technogenic systems.

Block "Subjects of impact" means enterprises of the mining complex. Sources of impact on the extraction method: underground (galleries, mines, wells), open (quarries, ditches, pits), and combined. Additionally - Production infrastructure (concentrating factories, tailing dumps, etc.).

Block "Objects of influence" means the environment: natural and socio-economic spheres that experience technogenic impact of the mining industry.

Block "Environmental impacts/local" is presented on the mining deposits, "Areal" within the boundaries of the districts.

Block "Estimated indicators" means qualitative and quantitative indicators, according to which the degree of technogenic impact is calculated. Local - the area of disturbed land in the fields, the toxicity of raw materials. Areal - indicators of the state of the environment in the areas.

Block "Integral indicator of environmental transformation" is estimated by the totality of the indicators on a five-point scale.

To reveal the content of the model, a mapping method was adopted as the main one. The maps are compiled in the ArcMap program based on the interpretation of high-resolution space images (Landsat 2012-2017). The maps of minerals and mining industry were used as the basic ones [6, etc.]. The objects of environmental assessment and mapping are the areas of mining and municipal areas. To indicate the former on the map we used the signs. Considered are currently being developed, or previously developed deposits (except for common minerals), which, depending on the type of extracted raw materials, the method of extraction and the area of disturbed lands, are represented by cartographic signs of different color, shape and size. The method of quantitative background reflects man-made land transformations and other characteristics of the ecological state of the territory within the boundaries of administrative regions. These indicators have relative values and are estimated in points that allow us to compare them with each other and conduct an integral evaluation. The method of pie charts is used to display the percentage of each value in the total amount of indicators for the area. Diagrams provide visualization of data, supplement and improve the reliability of estimates.

IOP Conf. Series: Earth and Environmental Science **190** (2018) 012021 doi:10.1088/1755-1315/190/1/012021

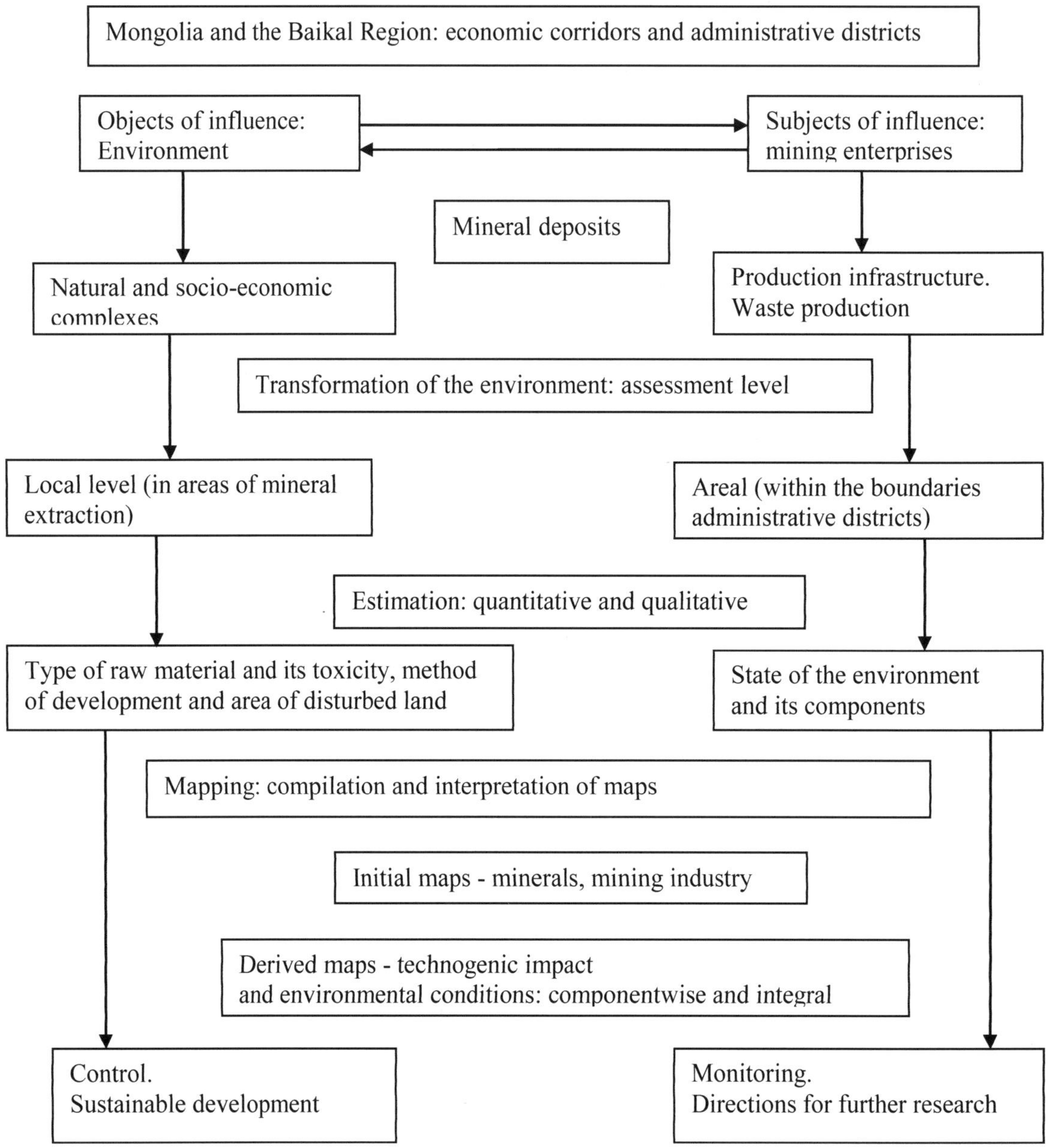


**Figure 1.** Research model of mining industry impact on the environment of Mongolia and the Baikal region within the economic corridors China – Mongolia - Russia

## 3. Results and discussion

The ecological maps are the main result of our research. The following maps are considered.

### *3.1. The technogenic impact of the mining industry on the environment*

This map is basic with the information that is used to compile subsequent maps (figure 2). Non-scale conventional signs reflect the nature of the disturbance, depending on the method of development, the type of extracted raw materials, and the extent of disturbance. As the leading indicator of anthropogenic impact on the environment, the amount of land disturbed in the process of mining has been adopted. For administrative districts, the quantitative background shows the density of disturbance, which is determined by the ratio of the total area of disturbed land in the district to the area of this district.

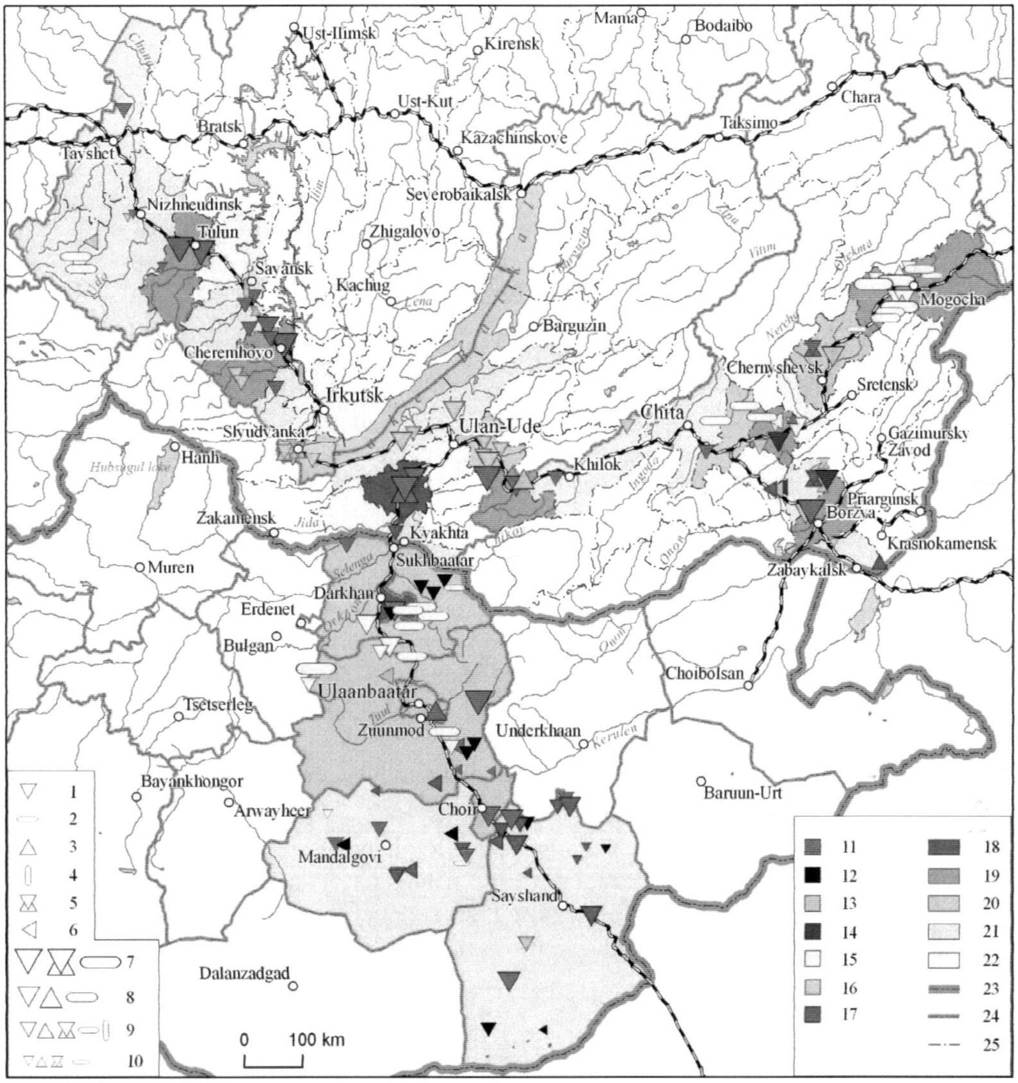


**Figure 2.** Technogenic impact of the mining industry on the environment
Nature of the impact on the environment: 1 - quarries, cuts, dumps (open mining method), 2 - dredge and hydraulic fields (open pit mining), 3 - mines, galleries, dumps (underground mining), 4 - wells (underground mining method), 5 - shafts, galleries, quarries, dumps (combined method), 6 - dumps, ditches, excavations (no data on the development method). The area of disturbed lands, km2: 7 - more than 10, 8 - 1-10, 9 – 0.1-1.0, 10 - less than 0.1. Kind of extracted raw materials: 11 - coal, brown, 12 - ferrous metals, 13 - non-ferrous metals, 14 - rare metals, 15 - gold, 16 - nonmetallic minerals, 17 - fluorite. Density of disturbed lands within municipal districts (aimaks), km2/thou.km2: 18 - more than 5 (5 points), 19 - 1-5 (4 points), 20 – 0.5-1.0 (3 points), 21 – 0.005-0.5 (2 points), 22 - no disturbance were detected (1 point). Borders: 23 - state, 24 - subjects of the Russian Federation, and aimaks, 25 - municipal districts.

The most significant land disturbance was formed as a result of the activities of coal mining enterprises in the Azeisk, Cheremkhovo, Kharanor, Gusinoozersk, Olon-Shibir, Mugun, and Baganur fields. Their development is accompanied by the formation of large quarries and dumps of overburden and enclosing rocks, reaching a volume of more than 200 million cubic meters.

Significant in terms of area, but relatively small in volume of displaced overburden rocks, extended linear sections of disturbed lands occur in river valleys due to the development of placer gold deposits. In the area under consideration, there are about twenty placer gold mining sites. Almost all of them are located in the river valleys in Mogochinskii district, aimaks Selenge and Tuv. The largest areas of disturbed lands (about 40 $km^2$), formed as a result of the work of drags and industrial equipment, were identified in the valleys of the Tuul river and its tributaries [7].

The work is often done right in the riverbeds and is accompanied by frequent discharges of dirty water, which leads to disturbance of aquatic ecosystems. This exacerbates the already complex problem of drinking water supply in Mongolia. In arid regions, under the influence of wind load and increasing gold mining, desertification of the territory also intensifies.

The analysis of the map shows that the territory experiences a significant man-caused strain on the part of mining enterprises, both current and in force in the past (on average, the density of violations of economic corridors is 0.662 $km^2$ / thousand $km^2$). The highest values of this indicator were determined for the Selenginskii district of the Republic of Buryatia, and aimak Darkhan-Uul in Mongolia. Disturbance in the Tulun, Alar, Cheremkhovo districts of the Irkutsk region; Borzinskii, Petrovsk-Zabaikalskii, Mogochinskii, Shilkinskii districts of the Transbaikal Territory are somewhat lower.

*3.2. The structure of disturbed lands as a result of the activities of mining enterprises*

The quantitative background shows the degree of land disturbance within the municipal area relative to the average for all economic corridors (figure 3). The highest values of this indicator are in Tulun, Mogochinskii districts, as well as in the Tuv aimak.

The method of pie charts displays the area of technogenic deformations and the structure of disturbed lands in the districts (table 1).

**Table 1.** The share of lands damaged in the process of mining in economic corridors within the regions, %.

| Regions | Mongolia | Transbaikal region | Republic of Buryatia | Irkutsk region | Total |
|---|---|---|---|---|---|
| | | Type of extracted raw material | | | |
| **Coal stone and brown** | 32 | 30 | 78 | 83 | 46 |
| **Black metals** | 3 | - | - | - | 1 |
| **Non-ferrous metals** | - | 14 | - | - | 5 |
| **Rare metals** | - | 4 | - | - | 2 |
| **Gold ore** | 7 | 5 | - | - | 4 |
| **Gold placer** | 50 | 46 | - | 13 | 36 |
| **Non-metallic raw materials** | - | - | 21 | 4 | 3 |
| **Fluorite** | 8 | 1 | 1 | - | 3 |
| | | Toxicity of extracted raw materials | | | |
| **Low** | | | 21 | 3 | 2 |
| **Moderate** | 50 | 45 | | 13 | 36 |
| **Increased** | 35 | 30 | 78 | 84 | 48 |
| **High** | 15 | 20 | 1 | | 12 |
| **Very high** | | 5 | | | 2 |

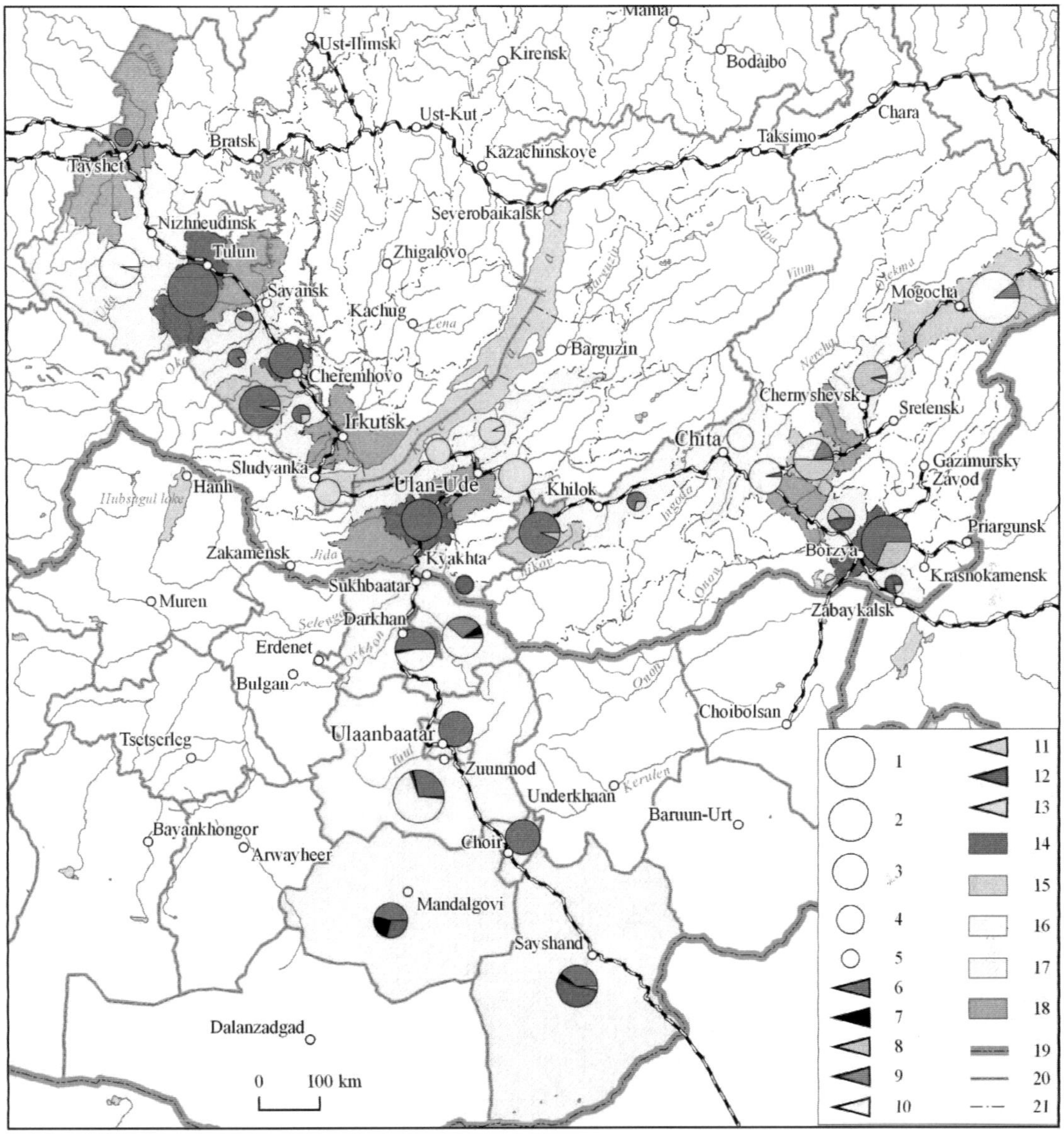


**Figure 3.** The structure of disturbed land as a result of mining enterprises
Total area of disturbed lands in municipal districts (km): 1 - more than 30, 2 - 10-30, 3 - 3-10, 4 - 1-3, 5 - less than 1. Type of produced raw materials: 6 - coal, brown, 7 - ferrous metals, 8 - non-ferrous metals, 9 - gold ore, 10 - gold placer, 11 - rare metals, 12 - fluorite, 13 - non-metallic raw materials. Disturbance in the districts (aimaks) relative to the average along the corridor, km2 / km2: 14 - more than 5 (5 points), 15 - 1.5-5.0 (4 points), 16 - 0.5-1.5 (3 points), 17 - 0.005-0.5 (2 points), 18 - disturbances were not identified (1 point). Borders: 19 - state, 20 - subjects of the Russian Federation, aimaks, 21 - municipal districts.

The size of the diagrams is determined by the total area of disturbance. The maximum values of this indicator were recorded in Mogochinskii (mainly associated with the extraction of alluvial gold and, to a lesser extent, ore), Tulunskii (brown coal mining), Borzinskii (brown coal mining at present and tin in the past) municipal districts, as well as in the Tuv aimak (extraction of brown coal, iron, ore and placer gold, and tungsten). The shape of the diagrams reflects the structure of disturbed lands, which is determined by the percentage of areas affected in the process of mining certain types of minerals, to the total area of disturbed land in the development of all types of minerals within the municipal area. The color of the segment corresponds to the type of extracted raw material.

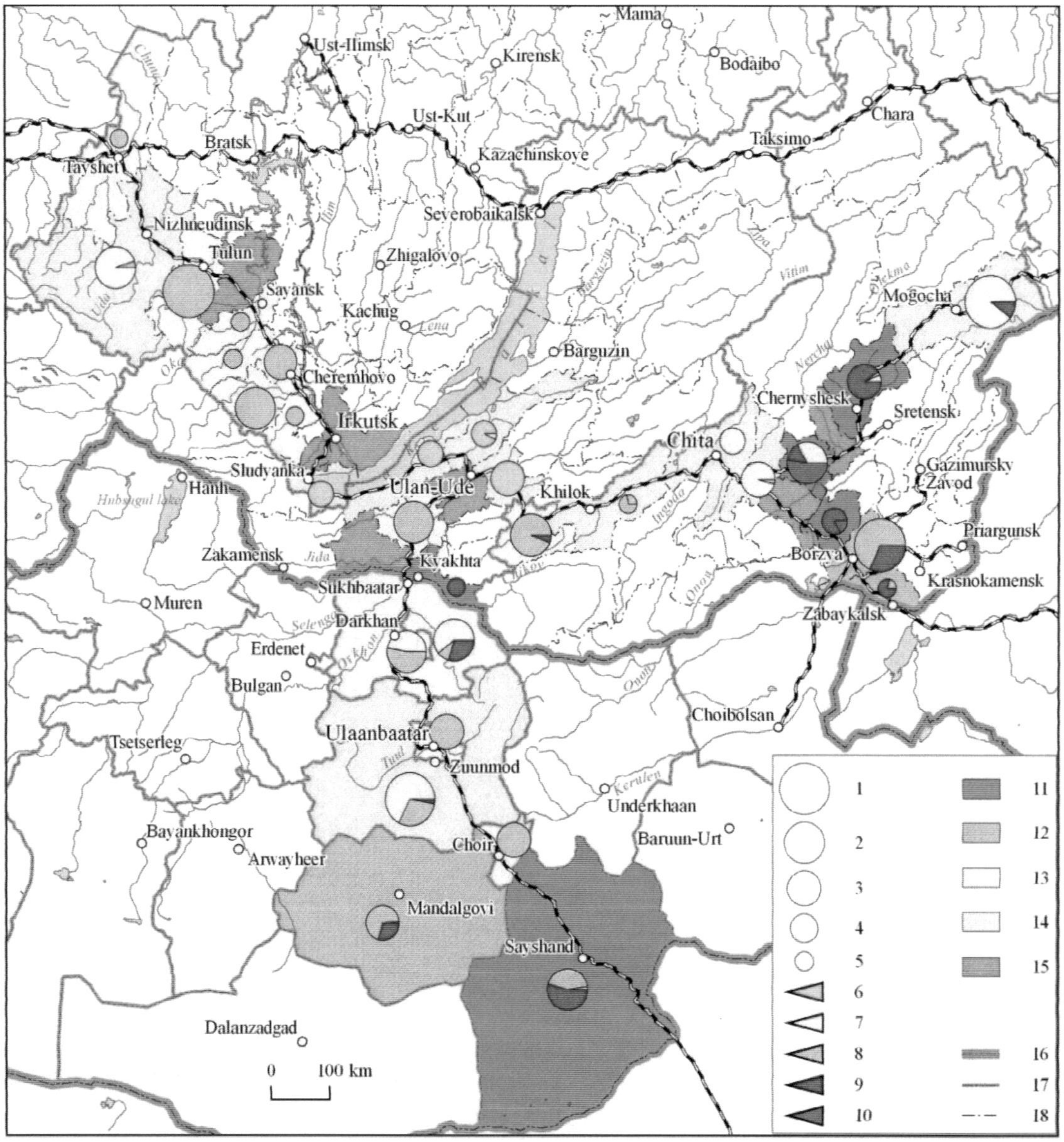


**Figure 4.** Environmental risk assessment of mining industry activity in terms of toxicity of produced raw materials

Total area of disturbed lands in municipal districts (km): 1 - more than 30, 2 - 10-30, 3 - 3-10, 4 - 1-3, 5 - less than 1. Toxicity of the main and accompanying types of raw materials: 6 - low, 7 - moderate, 8 - high, 9 - high, 10 - very high. Toxicity in districts (aimaks), conventional units: 11 - more than 3.5 (5 points), 12 - 3.0-3.5 (4 points), 13 - 2.5-3.0 (3 points), 14 – 1.0-2.5 (2 points), 15 - no disturbance was detected (1 point). Borders: 16 - state, 17 - subjects of the Russian Federation and aimaks, 18 - municipal districts.

The largest disturbed areas were formed during coal mining (Mugunskii, Cheremkhovskii, Azeiskii, Tugnuiskii, Kharanorskii, Baganuur and other sections), the second place is occupied by man-made landscapes that arose during the development of gold-bearing placers (the valleys of the rivers Ithaca, Kia, Kruchina, Davenda, Tuul, Sharyn-Gol, etc.

### *3.3. Assessment of the ecological hazard from mining enterprises activity in terms of the toxicity level of the extracted raw material*

According to the degree of chemical impact on the natural environment, the raw materials and reagents used in its enrichment are divided into five toxicity classes: very high - rare metal and radioactive ores, antimony; high - ores of non-ferrous and precious metals, fluorite; increased - coal stone and brown, oil, iron ore, talc, salt; moderate - natural gas, alluvial gold, tungsten, tin, gypsum; low - non-metallic raw materials (mica, magnesite, perlite, zeolite) [8].

On the map (figure 4), the quantitative background method provides an ecological assessment of the areas by the criterion of toxicity. This assessment was determined by summing up the partial assessments obtained as a result of multiplying the toxicity class of the extracted raw materials and the weight indicators-the shares of the areas disturbed by the extraction of raw materials of this toxicity. The pie chart method represents the disturbance structure according to the indicated toxicity scale (see table 1). The size of the diagrams is determined by the total area of the disturbed territories within the municipal area. The size and color of the segments reflect the areas of man-caused disruptions that have occurred in the development of minerals of certain toxicity.

In the area under consideration, as a result of the extraction of raw materials of a very high toxicity level, 2% of the land was disturbed, and 12% of high toxicity level. Currently, enterprises for the extraction and processing of highly toxic raw materials in the Transbaikal Territory have been mothballed, but the accumulated production waste continues to have a negative impact on the environment, which is observed in Olovyanninskii and Shilkinsky districts.

The enterprises for the extraction and enrichment of raw materials of high toxicity are located in the Transbaikal region and Mongolia (active and mothballed mines for the extraction of molybdenum, tungsten, tin, ore gold, fluorite in Borzinskii, Chernyshevskii, Mogochinskii, Shilkinskii, Olovyanninskii districts, aimaks Selenge, Dundgov' and Dornogov'). A significant contribution to pollution of the territory is made by the dumps and tailing dumps of closed mining enterprises of the Transbaikal Territory. To solve this problem, it is necessary to conduct an assessment of the impact of waste on the environment and the remediation of disturbed land.

### *3.4. The population and the disturbance of lands under the influence of the mining industry*

Environmental damage during mining has a strong negative impact on the health, living conditions and economic activities of the local population (figure 5, 6).

On the first map (figure 5) the quantitative background shows the technogenic load: the ratio of the total area of disturbed lands per 1000 people. The maximum value of this indicator was revealed for Mogochinskii, Petrovsk-Zabaikalskii, and Tulunskii districts. The method of pie charts shows the population in the districts (the size of the diagrams) and its structure (ratio of urban and rural population).

On the second map (figure 6), the quantitative background shows the technogenic load by the criterion of toxicity - the ratio of the total area of the land affected by the extraction of raw materials of high and very high toxicity by 1000 people. According to this indicator, the maximum burden falls on the residents of Mogochinskii and Chernyshevskii districts. Here the pie charts show the population in the regions and the share of the population of these areas in the total population of the Irkutsk region, the Republic of Buryatia, the Transbaikal Territory, and Mongolia within economic corridors.

Close to the settlements with mining and processing enterprises located here, dumps and tailing dumps most strongly affect the sanitary and hygienic conditions of the local population.

This situation is typical for the cities of Tulun and Cheremkhovo, having coal mines and dumps within their territories. The threat to ecological safety for the population of the Sherlovaya Gora settlement was represented by a dusting layer of tail sediments in the tailing dump for ten years. At a distance of 1 km from the tailing dump, a very strong arsenic contamination of 57-300 mg/kg, a weak lead of 28-109 mg/kg and zinc 30-169 mg/kg was detected [9]. In Mongolia, as a result of large-scale mining of alluvial gold in river valleys, water became unsuitable for the swarming of livestock, which led to population migration.

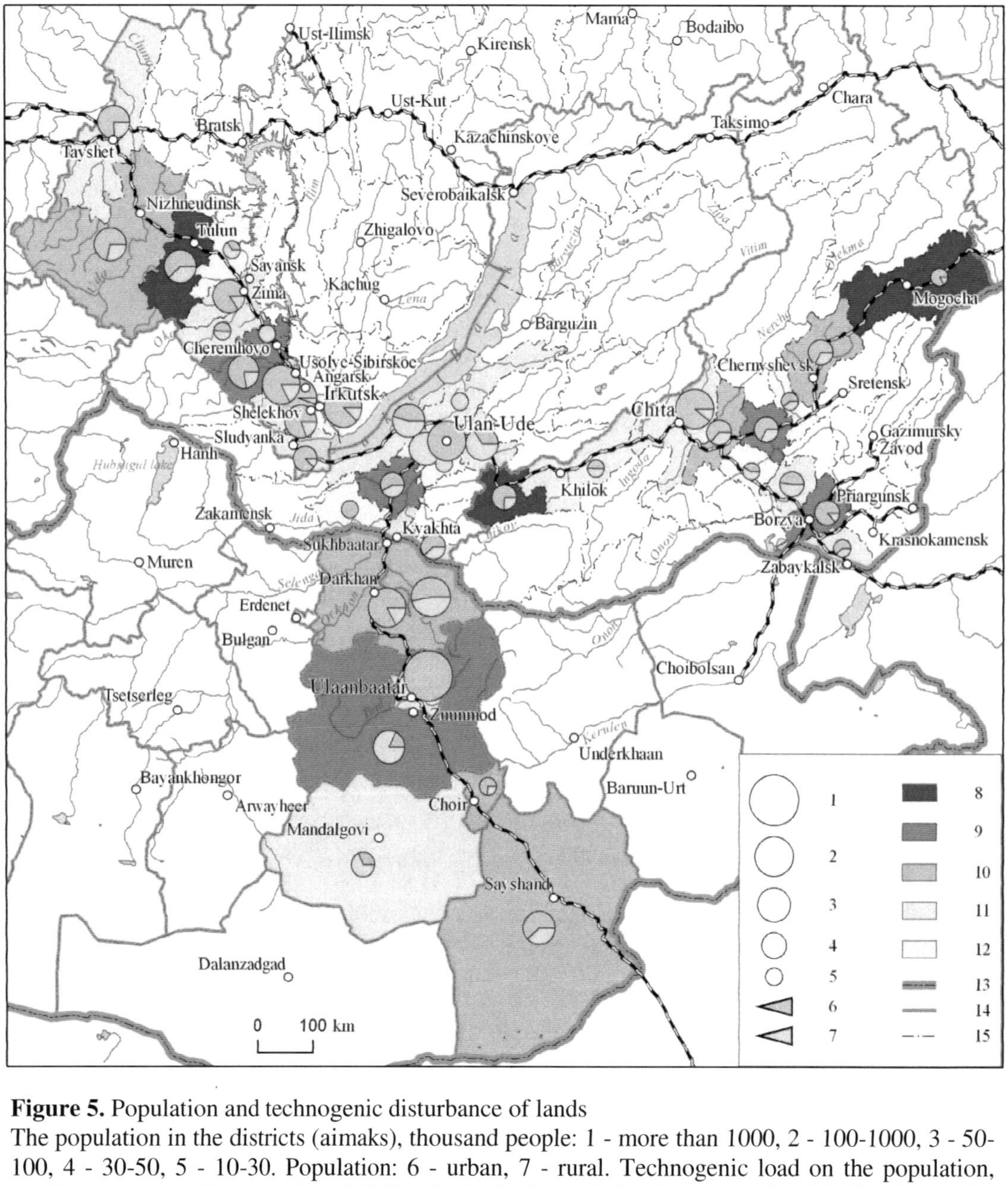


**Figure 5.** Population and technogenic disturbance of lands
The population in the districts (aimaks), thousand people: 1 - more than 1000, 2 - 100-1000, 3 - 50-100, 4 - 30-50, 5 - 10-30. Population: 6 - urban, 7 - rural. Technogenic load on the population, km2/thou. people: 8 - more than 1.5 (5 points), 9 - 0.3-1.5 (4 points), 10 - 0.1-0.3 (3 points), 11 - 0.001-0.1 ( 2 points), 12 - no disturbance was detected (1 point). Borders: 13 - state, 14 - subjects of the Russian Federation and aimaks, 15 - municipal districts

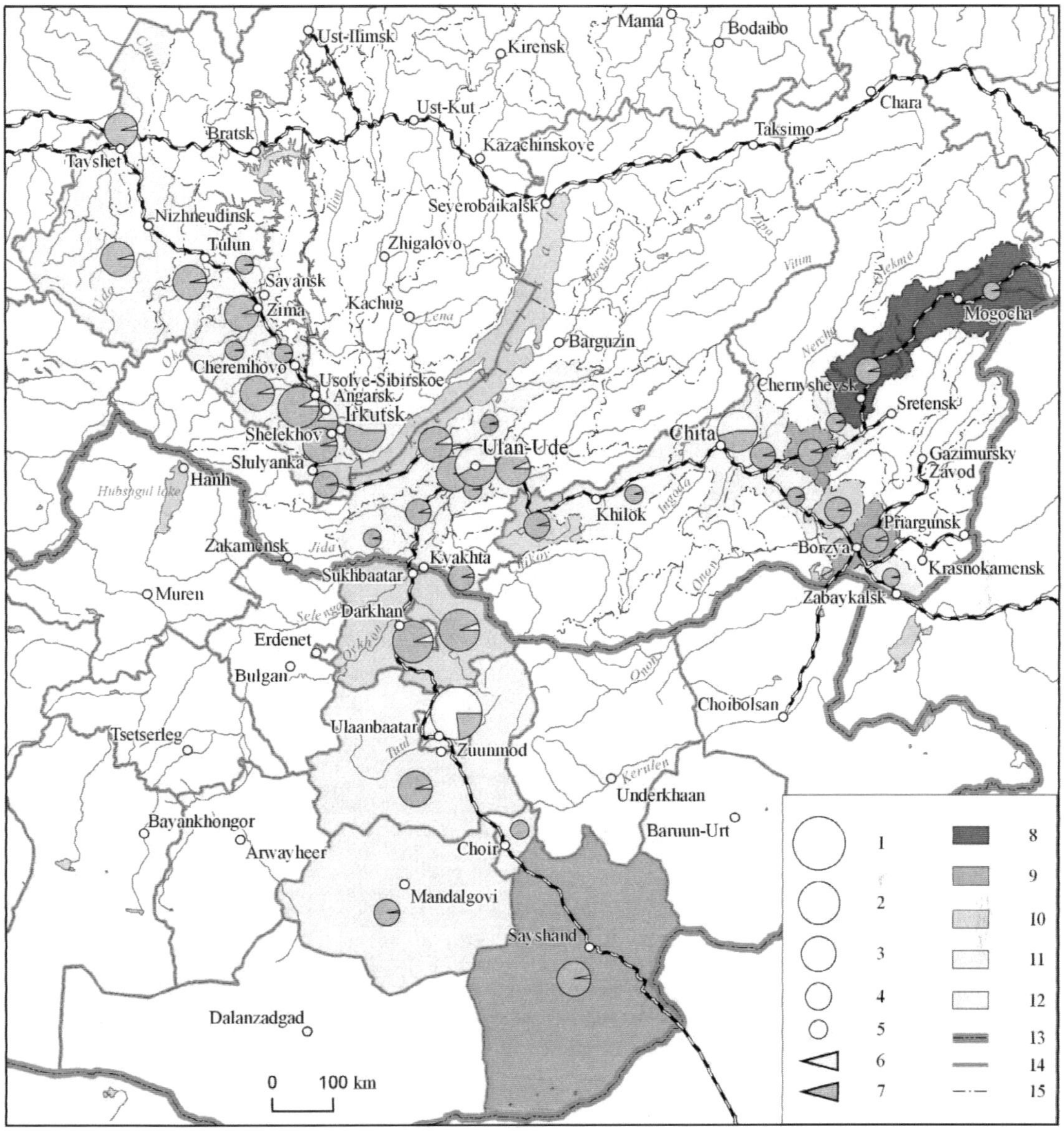


**Figure 6.** Population and technogenic disturbance of lands by the criterion of toxicity
The population in the districts (aimaks), thousand people: 1 - more than 1000, 2 - 100-1000, 3 - 50-100, 4 - 30-50, 5 - 10-30. The ratio of the population of the region (6) and the region within the economic corridor (7). Technogenic load on the population by the criterion of toxicity, km2 / thou. people: 8 - more than 0.25 (5 points), 9 - 0.10-0.25 (4 points), 10 - 0.05-0.10 (3 points), 11 - 0.01-0, 05 (2 points), 12 - no violations were detected (1 point). Borders: 13 - state, 14 - subjects of the Russian Federation and aimaks, 15 - municipal districts.

### *3.5. Integral assessment of technogenic transformation of lands*

The assessment is carried out on the basis of the set of indicators adopted on the maps of various purposes considered above (figure 2-6). Indicators on the degree of environmental impact are assessed on a five-point scale according to the increase in the technogenic impact. Simple scores are summarized and translated into complex scores of the integral score over a range of values - also on a five-point scale (table 2).

The method of quantitative background is used to represent the total evaluation of anthropogenic impact. The method of pie charts is used to represent the spectrum of environmental indicators and their contribution to the integral assessment (figure 7).

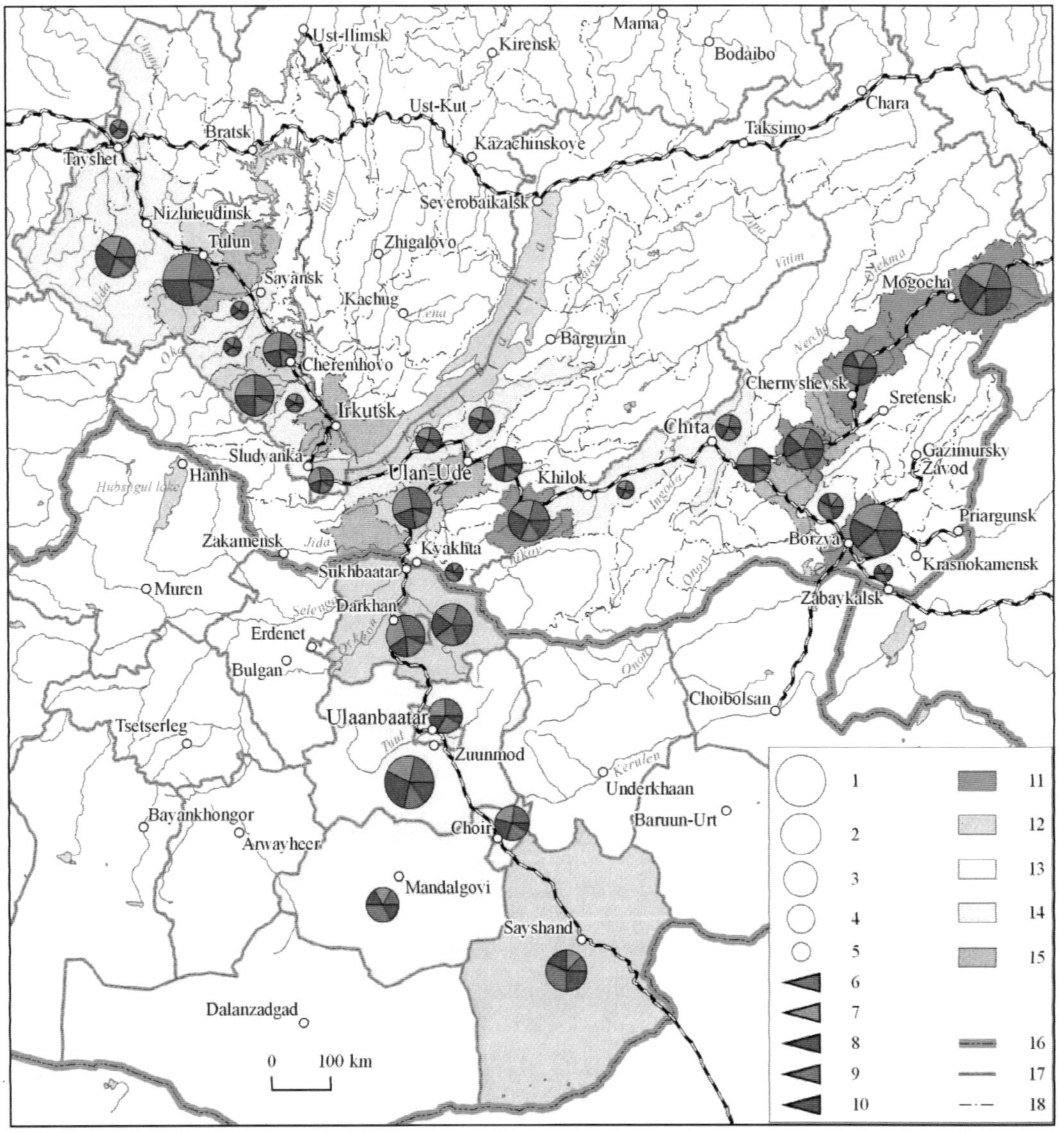


**Figure 7.** Integral assessment of man-made impact
The total area of disturbed lands in municipal districts (km): 1 - more than 30, 2 - 10-30, 3 - 3-10, 4 - 1-3, 5 - less than 1. Indicators of integral assessment of the ecological state: 6 - density 7 - the disturbance density relative to the average along the corridor, 8 - man-caused load per 1000 people, 9 - toxicity in the areas, 10 - man-caused load per 1000 people by the criterion of toxicity. Integral assessment of the technogenic impact, the sum of simple scores: 11 - 19-21 (5 points - very high), 12 -15-18 (4 points high), 13 - 11-14 (3 points moderate), 14 - 9 10 (2 points - low), 15 - 5 (1 point - insignificant - no violations were detected). Borders: 16 - state, 17 - subjects of the Russian Federation and aimaks, 18 - municipal districts.

**Table 2.** Integral assessment of technogenic impact

| Representa-tive municipal districts, aimaks | Estimates, scores | | | | | | |
|---|---|---|---|---|---|---|---|
| | * Dd | Dv | Dt | Tp | Tt | SSP | IA |
| **Irkutsk** | **1 | 1 | 1 | 1 | 1 | 5 | I |
| **Kabanskii** | 2 | 2 | 2 | 2 | 1 | 9 | II |
| **Tuv** | 3 | 3 | 2 | 4 | 2 | 14 | III |
| **Darkhan-Uul** | 5 | 5 | 3 | 3 | 1 | 17 | IV |
| **Borzinskii** | 4 | 5 | 4 | 4 | 4 | 21 | V |

* Dd - density of disturbed lands; Dv – disturbance density relative to the average one along the corridor; Dt - degree of toxicity; Tp - technogenic load on population; Tt - technogenic load on population by the criterion of toxicity; SSP - the sum of simple points; IA - integral assessment
** Assessment of the anthropogenic impact degree: 1 (I) - insignificant, 2 (II) - low, 3 (III) - moderate, 4 (IV) - high, 5 (V) - very high.

Mogochinskii, Borzinskii and Shilkinskii districts of the Transbaikal Territory are highly technogenic disturbed. Irkutskii, Kuitunskii, Tarbagataiskii and other areas are considered environmentally safe, where mining operations (except for common ones) are not conducted.

## 4. Conclusion

The research is based on a model whose formalized image was used to solve problems of assessing the technogenic effect of mining on the environment within the economic corridors of Mongolia and the Baikal region. The model is revealed by a series of conjugated environmental maps of various topics related to the interaction of natural and man-made systems - a local assessment of the negative consequences of the development of specific mineral deposits and an area assessment of the state of the environment in administrative areas. In the series of maps, the first one, containing the information for compiling the subsequent maps is of the greatest importance, and the final one, which gives an integral assessment for the set of indicators.

The implementation of the project for the construction of transport and infrastructure corridors will contribute to strengthening the economic development of the adjacent territories. This will enable additional investments in innovative environmentally oriented technologies, and that is a prerequisite for the sustainable development of the transboundary region. Modernization of transport systems and expansion of mining production can lead not only to additional environmental loads, but also to increased dependence of Mongolia and Russia on raw materials exports, which will further complicate the environmental situation in these countries.

Prospects for further work are associated with the deepening of research to the second and third levels of remote regions, including in the research program all economic corridors, formed on their basis transport-infrastructure and mining complexes.

## References

[1] The program for creating an economic corridor between the Russian Federation, China and Mongolia 2016 *J. Economics and Business* Income accessed online on 19th June 2018 via http://tass.ru/ekonomika/3394629
[2] Miranda M, Burris P and Binggas J F 2003 *Mining and Critical Ecosystem: Mapping the Risks* (Washington: World Resources Institute) p 60
[3] Savel'eva I L 2009 Intraregional resource and environmental factors for the development of the mining industry of the Baikal natural territory *J. Geography and Natural Resources* **3** 109-16
[4] Abalakov A D and Bazarova N B 2018 Assessment of the technogenic impact of the mining industry of the Baikal region on specially protected natural areas: cartographic analysis *J. Geodesy and Cartography* **79**(1) 20-31 https://doi.org/10.22389/0016-7126-2018-931-1-20-31

[5] Abalakov A D and Bazarova N B 2017 The impact of the mining industry on the environment in the Baikal region *J. Engineering Survey* **10** 54-64 https://doi.org/10.25296/1997-8650-2017-1054-64
[6] *National Atlas. People's Republic of Mongolian* 1990 (Ulaanbaatar – Moscow) p 144
*Ecological Atlas of the Lake Baikal Basin* 2015 (Irkutsk: Institute of Geography SB RAS) p 145
*National Atlas of Mongolia* 2009 vol. 2 ed Dorzhgotov D (Ulaanbaatar) p 248
[7] Mongolia. A review of Environmental and Social Impacts in the Mining Sector 2006 *J. Discussion Papers* Income accessed online on 23 th June 2018 via http://siteresources.worldbank.org/INTMONGOLIA/Resources/Mongolia-Mining.pdf
[8] *Ecological-geographical map of the Russian Federation* 1996 (Moscow: State Administration of Geodesy and Cartography Press) p 4
[9] Yurgenson G A, Chechetkin V S, Asoskov V M, Kotelnikov A M and Haritonov Y F 1999 *Geological Studies and Mining Complex of Transbaikalia: History, Current State, Problems, Development Prospects* (Novosibirsk. Nauka. Siberian Publishing Company RAS) p 574

# Stability of landscapes in the areas of creation of economic corridors "China-Mongolia-Russia"

**A D Abalakov, D A Lopatkin, L S Novikova**

Sochava Institute of Geography SB RAS, Irkutsk, Russia

E-mail: abalakov@irigs.irk.ru

**Abstract.** The article considers assessment and mapping of landscape sustainability in the zone of creation of economic corridors "China - Mongolia - Russia". Research was carried out within the borders of Mongolia and the Baikal region of Russia. Various natural and transformed landscapes are presented here. We should expect an increase in their transformation within the economic corridors in the creation of transport and infrastructure complexes. There is a need to determine the permissible level of negative impact of anthropogenic factors. We developed a model that reveals the directions, structure, stages and results of research in their logical sequence. The assessment and mapping of the sustainability of landscapes is based on the interpretation of landscape maps. The research was carried out on regional and topological level. For each level, the taxonomic rank of the assessed landscape, types of impacts, types of sustainability, a system of indicators and criteria are determined. For the calculations the scoring method was used. The characteristic of landscapes and their general stability for Mongolia and the Baikal region is given for all territory, but in the economic corridor China-Mongolia-Russia these characteristics are given more detailed. Prospects and directions of further research work are determined.

## 1. Introduction

The strategic goal of creating economic corridors is to create conditions for the development and expansion of tripartite cooperation between the People's Republic of China, Mongolia and the Russian Federation through joint projects aimed at increasing trade turnover, ensuring product competitiveness, facilitating transboundary transportation, infrastructure development, and environmental protection. The implementation of the Program for the Creation of Economic Corridors and its constituent projects will strengthen the trade and economic cooperation of the three states, and contribute to the creation of transport and infrastructure complex and socio-economic development of the transboundary regions. However, the development of the territory will lead to an increasing anthropogenic impact on the environment and intensification in landscape transformation. Therefore, there is a need in environmental measures, which require information on landscape state and their resistance to anthropogenic pressures. Landscape stability is the most important indicator which can determine the permissible level of anthropogenic impact, followed by irreversible changes in the natural environment. The paper considers research methods and obtained results presented by conjugate landscape stability maps of regional and topological level.

Based on the indices of landscape stability to anthropogenic pressure, standards of permissible impact can be determined. These standards are taken into account in environmental design,

IOP Conf. Series: Earth and Environmental Science **190** (2018) 012022 doi:10.1088/1755-1315/190/1/012022

environmental impact assessment in the construction and operation of enterprises, facilities and other objects.

## 2. Models and methods

A model has been developed that gives an interpretation of the landscapes with respect to the properties of stability. The model simplifies the most significant aspects of such a study. The model presents two series of blocks representing the regional and topological levels of the study. The map is a model of the phenomena under study is presented in this system as the main research tool. The map provides visibility of perception, visual distinguishability of elements and details of the cartographic image, ability to see the main patterns of location and interrelation of objects, the main elements of their structure. The map serves as a tool for understanding the structure and functioning of the landscapes represented on it, their interpretation with respect to their stability.

We used remote methods in creating maps. Their application provides the regularity and fast obtaining reliable information on a large territory, which enables dynamics monitoring. Degree of anthropogenic transformation and the stability of landscapes was determined using scoring method. A promising research is the use of digital terrain models, which allow quantifying the morphometric parameters of the relief and exogenous geological processes.

## 3. Results and Discussion

On the basis of the conducted studies, two conjugate maps of landscape stability of regional and topological levels were compiled.

### *3.1. The map of landscape stability of regional level of Mongolia and the Baikal region*

The initial map information is the landscape maps [1]. Mapping the landscape stability is a continuation and deepening of previously performed works in this direction [2]. This map represents sustainability of regional geographic landscapes (table 1). A geom is the lowest subdivision of geomers of regional dimension, uniting classes and facies groups of a certain zone or belt affiliation [3]. The stability of geoms is determined by the level of natural ecological potential of landscape, based on the index of biological efficiency of climate [4].

The natural ecological potential of landscapes provides population with a favorable habitat. Various components of the landscape participate in the formation of landscape ecological potential. However climatic factors determined by the optimal ratio of heat and moisture are of particular importance.

The integral index of heat and moisture availability is the index of biological effectiveness of climate proposed by N.N. Ivanov. This indicator represents the product of the annual sum of air temperatures above $10^{o}$ C by the wetting factor (the ratio of annual precipitation to annual volatility). Depending on the size of this indicator landscapes are distributed according to the level of ecological potential: 4 or less - very low, 4-8 - low, 8-12 - moderate, 12-16 - relatively high, 16 - 20 - high.

According to the index values of biological effectiveness of climate and ecological potential, landscapes are divided into five ecological groups, which are assigned the appropriate values of stability, ranked on a five-point scale. The sustainability potential is a universal indicator for natural and anthropogenic factors of influence and therefore such stability is considered to be general.

In the landscape structure of the Baikal region and Mongolia, there are mountain-taiga and mountain-hollow steppe landscapes consisting of 16 geoms.

The compiled map represents territorial diversity of these landscapes and degree of their potential sustainability. Mountain and foothill taiga and mountain-hollow steppe geoms of medium-level stability, which occupy most of the territory, dominate in the landscape structure. North-Asian alpine landscapes with the lowest stability and foothill-taiga and south-taiga with the highest one are represented by a small number of geoms, but are relatively wide-spread. Within the economic corridors in the quantitative and spatial sense foothill taiga and mountain taiga landscapes of effective development prevail with a higher value of stability than in the whole territory.

**Table 1.** Landscape stability of the Baikal region and Mongolia.

| North-Asian alpine, taiga and steppe | | | | Central Asian steppe | |
|---|---|---|---|---|---|
| North-Asian alpine and subalpine | Mountain-taiga of Baikal-Dzhugzhur type | Plains-highlands Middle Siberian | Steppes | Mountainous western Siberian and northern Mongolian | High plains of denudation buttes Onon-Argun |
| **I class,** the most unstable, ecological potential is very low, index of climate biological effectiveness - 4 and less | | | | | |
| 1. North-Asian alpine type and mountain tundra | | | | | |
| **II class,** unstable, ecological potential is low, index of climate biological effectiveness 4-8 | | | | | |
| | 2. Mountain-taiga of reduced development<br>3. Intermountain depressions and plains taiga of reduced development | | | 4. Plains and bottoms of depression steppe-meadow cryogenic meadow-boggy | 5. Halophytic |
| **III class**, moderately stable, ecological potential is medium, index of climate biological effectiveness 8-12 | | | | | |
| | 6. Mountain-taiga of reduced development<br>7. Piedmont intermountain depressions and plains taiga of reduced development | | | 8. Mountain steppe<br>9. Piedmont plains and bottoms of depression steppe meadow-bush | 10.Mountain dry steppe |
| **IV class**, stable, ecological potential is relatively high, index of climate biological effectiveness 12-16 | | | | | |
| | 11. Mountain-taiga of effective development | 12. Pine coniferous | 13. Piedmont meadow-steppe | | 14. Plain and valley steppe |
| **V class**, especially stable, ecological potential is high, index of climate biological effectiveness 16-20 | | | | | |
| | 15. Piedmont taiga of effective development and subtaiga | 16. Southern taiga | | | |

The stability of geoms of landscapes of regional level is considered as a background, as these geosystems constitute a regional landscape-ecological background and a subject to the general geographic patterns of landscape-forming processes. Therefore, background stability is taken into account when correcting geosystems of topological level - groups of facies forming a geom. At the regional level it is possible to reduce the stability of the Central Asian steppe landscapes located at the edge of their area, separated by the North Asian alpine-taiga landscapes [5].

### *3.2. Map of the landscape stability of topological level of the key site*

A system of assessing indicators was developed, which formed the basis for the map of landscape stability to anthropogenic impacts on the key site (figure 1, table 2).

IOP Conf. Series: Earth and Environmental Science **190** (2018) 012022 doi:10.1088/1755-1315/190/1/012022

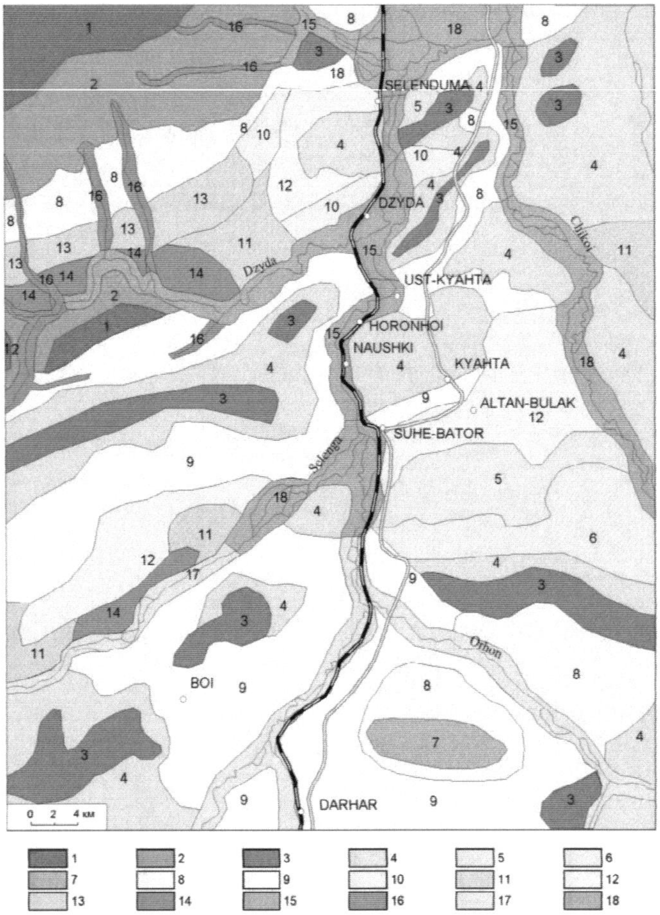


**Figure 1.** Landscapes stability to anthropogenic impact within the economic corridor of the transboundary territory of Buryatia and Mongolia. The stability of landscapes in table 2.

*North Asian alpine-taiga*

Baikal-Dzhugdzhur mountain-taiga

Mountain-taiga larch forests with Siberian pine and Siberian dwarf pine in the undergrowth of limited development: 1. Deplanate and weakly dissected tops of pine-larch with Siberian pine shrubby-moss, with an undergrowth of Siberian dwarf pine on podburs, podzols and cryozems (I- indigenous). 2. Flat and medium steepness of pine-larch slopes with Siberian pine grass-shrub on soddy-podzolic soils, brown soils and podburs (P-pseudoindiginous).

South Siberian mountain-taiga

Mountain-taiga pine with larch with rhododendron in the undergrowth grass-shrub of effective development: 3. Deplanate and weakly dissected tops pine-larch surfaces with Siberian pine grass- shrubs on podburs podzolized (I). 4. Sloping larch-pine grassy slopes with rhododendron on sod podzolized podburs (M). 6. Steep deeply dissected slopes pine-larch steppe and forest-steppe on eroded low-power chestnut soils and lithozems (C-series).

*Central Asian steppe*

West Transbaikalian Daurian type mountain steppe: 7. Deplanate tops steppe on chestnut soils (I). 8. Gentle steppe slopes on chestnut and chernozemic soils (P). 9. Steep, deeply dissected slopes of dry steppe on eroded soils and lithozems (S).

North Mongolian Khangai type hollow-steppe and dry steppe: 10. Plains accumulative dry steppe on chestnut and chernozemic soils (I) 11. Plains eluvial-accumulative halophytic (P). 12. Gentle slopes and terraces of erosion-accumulative steppe on chestnut soils with areas of water and wind erosion (P, S). 13. Elevated gentle-inclined plains erosion-denudation meadow-steppe on low erosion chernozems and chestnut soils (P, S). 14. Denudation buttes in the bottoms of hollows on lithozems (S).

*Valley azonal*

Mountain rivers: 15. Valleys of small and medium rivers with steep slopes and underdeveloped flood plain erosion (S) 16. Antecedent narrow deep valleys of large rivers with steep slopes on the areas of growing uplifts erosional (S).

Plain rivers at the bottom of the hollows: 17. Erosion-accumulative with underdeveloped floodplain (S). 18. Accumulative with a developed floodplain and wide meandering channels (S).

**Table 2.** Landscapes stability to anthropogenic pressure of the key site (fragment).

| Landscapes (groups of facies), landscape numbers correspond to figure 1 | Stability indicators: Geological and geomorphological | | | | | Soil-ecological | | Ecological-climatic | | Landscape and environmental | | | Points: Amount of simple numbers | Points: Integral evaluation |
|---|---|---|---|---|---|---|---|---|---|---|---|---|---|---|
| | Gc | Ss | Rf | Fc | Cs | Hr | Hf | Fh | Eb | Cd | Tt | Gc | | |
| *North Asian alpine-taiga* | | | | | | | | | | | | | | |
| Mountain-taiga of Baikal-Dzhugdzhur type | | | | | | | | | | | | | | |
| Mountain-taiga larch with Siberian pine and Siberian dwarf pine in undergrowth of limited development (area stability – III points*, according table 1) | | | | | | | | | | | | | | |
| 1 Deplanate and flat tops | 5 | 5 | 5 | 4 | 3 | 4 | 3 | 2 | 5 | 4 | 5 | 5 | 45 | V[2] |
| 2 Flat slopes and of mean steepness | 5 | 4 | 4 | 5 | 2 | 4 | 3 | 3 | 5 | 4 | 4 | 5 | 48 | IV[1] |
| South Siberian mountain-taiga | | | | | | | | | | | | | | |
| Mountain-taiga pine-larch rhododendron of effective development (area stability – IV class) | | | | | | | | | | | | | | |
| 3 Deplanate and flat tops | 5 | 5 | 5 | 5 | 3 | 5 | 5 | 4 | 5 | 5 | 5 | 4 | 56 | V[1] |
| 4 Flat slopes and of mean steepness | 5 | 4 | 4 | 5 | 3 | 5 | 5 | 4 | 4 | 5 | 4 | 4 | 52 | V[1] |
| 5 Piedmont plains | 3 | 4 | 5 | 4 | 5 | 3 | 3 | 5 | 4 | 3 | 3 | 4 | 46 | III[4] |
| 6 Steep slopes | 4 | 1 | 1 | 2 | 3 | 1 | 1 | 3 | 2 | 2 | 1 | 4 | 25 | I[6] |
| *Central Asian steppes* | | | | | | | | | | | | | | |
| Transsiberian Daurian type and North Mongolian Khangai type mountain-hollow | | | | | | | | | | | | | | |
| Mountain-steppe (area stability – III class) | | | | | | | | | | | | | | |
| 7 Deplanate and flat tops | 5 | 5 | 5 | 5 | 5 | 3 | 2 | 4 | 2 | 3 | 5 | 3 | 47 | V[2] |
| 8 Flat slopes and of mean steepness | 5 | 4 | 4 | 4 | 5 | 3 | 3 | 3 | 2 | 4 | 3 | 3 | 43 | IV[1] |
| 9 Steep slopes | 4 | 1 | 1 | 1 | 5 | 1 | 1 | 1 | 1 | 1 | 1 | 2 | 20 | I[5] |
| Hollow steppes and dry steppes (area stability – III class) | | | | | | | | | | | | | | |
| 10 Plains accumulative dry steppes | 3 | 5 | 5 | 5 | 5 | 5 | 5 | 4 | 2 | 4 | 5 | 3 | 51 | V[2] |
| 11 Plains alluvial-accumulative halophytic | 2 | 5 | 5 | 4 | 4 | 3 | 3 | 3 | 2 | 2 | 2 | 2 | 33 | III[3] |

**Stability indicators**: geological and geomorphological (Gc - ground conditions, Ss -steepness of the slopes, Rf - vertical fragmentation of the relief, Fc - factor of the territory cover by exogenous geological processes, Gc - geocryological conditions), soil-ecological (Cs - capacity of soils, Hr - humus reserves), ecological-climatic (Hf - hydrothermal factor, Fh - factor of humidifying), ecological-landscape (Eb - biological efficiency, Cd - dynamic categories, Tt - factor of anthropogenous territory transformation).

**Quantitative assessment of stability degree,** points, for individual indicators (1) and in integral expression (I) - 1 (I) - lowest, 2 (II) - low, 3 (III) - medium, 4 (IV) - high, 5 (V) - highest.

**The integral evaluation of stability degree,** determined from the intervals of total values of simple points: I - $\leq$ 25; II - 25-32; III - 32-39; IV - 39-45; V is $\geq$ 45.

***Correction of facies group stability with respect to background stability of geoms:**

[1]it is possible to reduce stability by 1 class;
[2]it is possible to reduce stability by 2 class;
[3]without change;
[4]it is possible to increase stability by 1 class;
[5]it is possible to increase stability by 2 class;
[6]it is possible to increase stability by 3 class.

The key site is located within the economic corridor Ulan-Ude-Ulaanbaatar-Beijing within the transboundary territory of Russia and Mongolia. These are the areas of the Selenginskii (South Buryatia) and Orkhon-Selenginskii (Northern Mongolia) middle mountains, in the zone of development of the Transbaikalian depressions. The depressions and the ridges separating them determine the landscape structure of the territory. The key site is located in the ecotone zone, where penetration and interaction of the North Asian alpine-taiga and Central Asian steppe geosystems occur.

We revealed that the highest stability degree is characteristic of the indigenous mountain taiga and steppe landscapes of deplanate tops and bottoms of hollows. The most unstable are the serial landscapes of steep, deeply dissected mountain slopes.

The stability evaluation carried out at the regional and topological levels is probabilistic. It is necessary to expand the depth of research with a more detailed level of elaboration on the basis of data obtained during engineering surveys [6].

Modernization of transport systems will be accompanied by additional environmental pressures, leading to increased dependence of Mongolia and Russia on raw materials exports [7].

## 4. Conclusion

Landscape stability evaluation, carried out on two levels, reveals different sides and stability properties, complementing each other. Area stability is determined at the regional level, which is taken into account when solving socioeconomic and environmental security issues of the entire transboundary territory. At the topological level landscape stability within the corridor is detailed.

## References

[1] Mikheev V S and Ryashin V A 1977 *Landscapes of the South of Eastern Siberia* (Moscow: State Administration of Geodesy and Cartography Press) pp 1-4

Dash D, Kaydanova O V, Lvov V L, Smirnova E L, Tulgaa H and Fadeeva N V 1990 *Landscapes. National Atlas. People's Republic of Mongolian* (Ulaanbaatar – Moscow) pp 84-5

[2] Abalakov A D and Lopatkin D A 2014 Landscape stability and its mapping *J. The bulletin of Irkutsk state University. Series "Earth Sciences"* **8** 2-14

Kuznetsova T I, Lopatkin D A and Bardash A V 2015 Ecological potential of landscapes (map) *Ecological Atlas of the Baikal Basin* (Irkutsk: Institute of geography Siberian branch of the Russian Academy of Sciences Press) pp 70-1

[3] Sochava V B 1978 *Introduction to the Doctrine of Geosystems* (Novosibirsk: Nauka Pres) p 320

[4] Suvorov E G, Lopatkin D A and Dash D 2015 Geosystems (map) *Ecological Atlas of the Baikal Basin* (Irkutsk: Institute of geography Siberian branch of the Russian Academy of Sciences Press) pp 47-8

Bashnin V N et al 2007 Ecological potential of landscapes *National Atlas of Russia Vol 2 Nature and the Environment* (Moscow: Cartography Press) p 417

[5] Mikheev V S 2001 *Landscape Synthesis of Geographical Knowledge* (Novosibirsk: Nauka Press) p 216

[6] Fomenko I K and Zerkal O V 2016 Assessment of slope stability in engineering surveys: regulatory requirements and problems of their implementation *J. Engineering surveys* **10** 16-21

Shoenemann M R and Pyles M R 1990 Statistical Description of Triaxial Shear Test results *J. Geotechnical Testing* **13** 58-62

[7] Otgonsuren B 2015 Mongolia–China–Russia Economic Corridor Infrastructure Cooperation *J. Erina report (PLUS)* **127** 3-6

# Soil erosion on the agricultural lands in Southern Siberia: current state, risks, soil protection models

**O I Bazhenova[1], E M Tyumentseva[2]**

[1]Sochava Institute of Geography SB RAS, Irkutsk, Russia,
[2]Irkutsk State University, Russia

E-mail: [1]bazhenova@irigs.irk.ru; [2]tumencev@irk.ru

**Abstract.** Territorial distribution of agricultural land erosion of in Siberia is considered. We determine different degradation processes mechanisms from west to east. In the Western Siberia meltwater erosion dominates. In the Middle Siberia soil erosion is caused by meltwater and rainstorm runoff, and in the Eastern Siberia by rainstorm discharge. Agricultural lands in mountains located in inter-mountain basins where concentric zoning of changes in mechanisms and rates of land degradation were analyzed. Predictive estimation of soil loss is carried out. We developed maps of erosion-hazard for agricultural lands. Detailed mapping of soil erosion rates in model basins has been carried out. Obtained are the trends and tendencies of soil erosion development in Siberia and the quantitative indicators of potential soil losses from erosion and deflation. Special attention is paid to soil degradation in the Lake Baikal basin. Positive trend of dust storms in the south part of Eastern Siberia at present day indicate aeolian processes increase under climate warming. In future, desertification of steppe landscapes will continue. The probability of desertification of southern Siberia, caused by the development of soil erosion, is determined. Regional models of soil conservation are presented

## 1. Introduction

Accelerated soil erosion constitutes a very grave consequence of the human irruption into the environment. A complicated ecological situation was also caused by degradation of agricultural lands in the southern of Siberia where the arable land takes more than 25 million ha. Agriculture here has a patchy nature, due to its intermontane location. At the beginning of the 1990s, agricultural lands occupied from 70 to 90 % of this area. Excessive plowing-up of slope areas, combined with large mechanical pressure on soils, gave rise to an intensification of erosion processes. The Khakasia and Buryatia stands out because of its high degree of soil erosion, one of the main criteria for assessing the ecological situation in any region, where this indicator exceeds 80 %, which makes it possible to classify these areas as having an emergency ecological situation (table 1). Active agricultural land use during the 17-20th centuries produced soil degradation. Agricultural lands occupy more than 55 million hectares, and about half of them are plowed. 9.5 million hectares are subject to flushing and deflation in varying degrees (table 1). The attention of researchers to the problems of destruction and degradation of soils in Siberia was attracted in connection with the development of virgin and fallow lands in 1953-1958, accompanied by a sharp activation of erosion processes. In the south of Siberia, comprehensive research was undertaken to protect the soil from erosion and deflation. Presented are regional features of soil erosion, determined are the extent and soil degradation degree in Asian Russia on the basis of generalization of long-term studies of soil erosion.

**Table 1.** Distribution of eroded and deflated soils in some regions of Siberia, thou ha.

| Region | Arable lands | Soils | | | Total degraded soils | |
|---|---|---|---|---|---|---|
| | | Eroded | Deflated | Eroded and deflated | thou ha | % from arable lands |
| **Altaiskii** | 7274.7 | 1099.9 | 2364.5 | no | 3464.4 | 47 |
| **Mountain Altai** | 145.9 | 86.2 | 2.4 | no | 88.6 | 61 |
| **Kemerovskaya** | 1613.0 | 144.8 | 173.6 | 74.6 | 393.0 | 25 |
| **Novosibirskaya** | 3865.4 | 247.0 | 251.6 | 22.0 | 520.6 | 13 |
| **Omskaya** | 4453.5 | 212.9 | 1294.6 | 254.0 | 1761.5 | 40 |
| **Tomskaya** | 667.0 | 6.3 | no | no | 6.3 | 1 |
| **Tyumenskaya** | 1740,0 | no | no | no | no | no |
| **Irkutskaya** | 1612.4 | 252.9 | 177.4 | 57.3 | 487.6 | 30 |
| **Buryatia** | 1018.0 | 272.0 | 308.0 | 250.0 | 830.0 | 81 |
| **Tuva** | 506.0 | 15.6 | 184.6 | no | 200.0 | 40 |
| **Khakasia** | 732.0 | 373.3 | 251.0 | no | 624.3 | 85 |
| **Zabaikalskii** | 2256.0 | 185.6 | 1024.4 | no | 1210.0 | 54 |
| Total | 25883.9 | 2896.5 | 6032.1 | 657.9 | 9586.5 | 37 |

## 2. Data and Methods

To ensure environmental safety in the south of Siberia, a network of complex geographic research stations has been established more than 50 years ago, where soil erosion has been studied as one of the most dangerous processes of degradation within the agricultural lands. For the regions of the south of Siberia we obtained long-term experimental data on the rate of natural erosion. They were created in Institute of Soil Science and Agrochemistry SBRAS, Novosibirsk ( Khairyuzovskii forest-steppe field station (1971), Ust'-Kamenskii forest-steppe field station (1974), Baikalskii mountain-taiga field station (1972), V.N. Sukachev Forest Institute SB RAS, Krasnoyarsk ( Khakasskii field station (1959), V.B. Sochava Institute of Geography SB RAS, Irkutsk ( Berezovskii forest-steppe field station (1981), Novonikolaevskii steppe field station (1970), Kharanorskii steppe field station (1961). The Institute of Soil Science and Agrochemistry SB RAS conducted more than 30 years of observations of snowmelt and of soil loss in arable land in various regions of Western Siberia [1]. Particularly detailed studies were carried out in Presalairye, the Kuznetsk depression and Priobye.

In order to prevent further degradation of land resources in the south of Eastern Siberia, it is necessary to develop land-use policies based on land zoning according to the degree of erosion hazard. We carried out such zoning for all the main agricultural areas of the south of Eastern Siberia (the Nazarovskaya, Kanskskaya and Yuzhno-Minusinskaya depressions, the Irkutsk-Cheremkhovo plain, and the Selenga River basin). Quantitative assessment of soil erosion and mapping of erosion-hazardous lands in the south of East Siberia was carried out using the Universal Soil Erosion Equation, modified by the Laboratory of Soil Erosion and Channel Processes of Moscow State University for the territory of Russia. Applying this empirical dependence to determine the average annual soil loss rate in the southern regions of Siberia is confirmed by the data of full-scale measurements of the slope washout rate. The correlation coefficient of the measured and calculated rates is rather high ($0.86 \pm 0.11$). The long-term dynamics of erosion processes is analyzed based on data on the runoff of suspended sediments in 18 basins. Erosion zoning of Siberia was carried out.

## 3. Results and Discussion

### *3.1. Processes of slope wash*

Here are the data of long-term records of the erosion rate on steppe slopes in undisturbed natural conditions in the south of Siberia. Despite the fact that the measurements were carried out in different regions and by different researchers, the measured rates are close and vary from a few hundredths of a millimeter on gentle slopes to the first millimeters on steep slopes.

According to research by V.A. Khmelev and A.A. Tanasienko [1] from the cultivated slopes of the Presalairskaya plain for 120 years, approximately 9 cm of the humus-accumulative horizon was washed away. Annually lost soil mass from 1 hectare of weakly washed chernozems and dark gray podzolized soils varies within quite narrow limits - 5.7 to 7.1 tons. On the medium-eroded soils, the washing out varies even more widely - 12.9 to 16.5 tons/ha per year. Annual erosion of soil mass on heavily eroded chernozems exceeds 20 tons/ha. Within the Priobye and Presalairye, where the intensive anthropogenic impact on the soil cover is estimated for dozens of years (after the massive development of virgin and fallow lands), the average annual washout of soils does not exceeds 8.1 to - 11.6 tons/ha. Simultaneously with the erosive removal of the soil mass during the intensive land use, there are significant losses of humus soils. From each hectare of weakly washed-out soils with solid runoff products 540 kg of this substance have already been washed out. The humus loss in the medium-eroded soils is 2 times higher (approximately 900-1400 kg/ha), some of the washed-out material accumulates on the downslope aprons, forming soils with high humus content, the proportion of such soils is very small (2-3%). In most river basins in the south of Eastern Siberia, the sediment runoff is usually less than 20 tons per km2, the average density of gully dismemberment does not exceed 0.05 km/km2, and the erosion of arable land averages 5-10 tons/ha per year.

Table 2 presents the distribution of zones with different intensity of potential washout for the main agricultural enclaves of the south of Eastern Siberia.

**Table 2.** Distribution of zones with different intensity of potential soil washout in the main agricultural areas of the south of Eastern Siberia, %.

| Washout intensity, t/ha in year | Nazarovskaya depression | Kanskaya depression | Yuzhno-Minusinskaya depression | Irkutsk-Cheremkhovo plain | Tunka Depression | Kuda basin [2] |
|---|---|---|---|---|---|---|
| ≤ 2 | 27 | 52 | 34 | 19 | 30 | 51 |
| 2-5 | 14 | 23 | 27 | 21 | 25 | 29 |
| 5-10 | 28 | 17 | 12 | 45 | 25 | 15 |
| 10 | 31 | 8 | 27 | 15 | 20 | 5 |

Based on the statistical analysis of long-term observation series of the suspended sediments runoff, the basins are identified, which are merged in four regions with different trends in the behavior of erosion processes. The prevailing trend in the first Ob – Yenisei region with a positive trend of atmospheric humidification is an increase in the intensity of erosion processes. It is characteristic for 80% of the basins in the region. In the structure of erosion processes, the role of flash shower increases, as indicated by the positive trends in the erosion index of precipitation.

The second Angara region is characterized by a complex multidirectional nature of the process intensity change. Against the background of the generally downward trend of erosion processes caused by the negative trend of atmospheric humidification and the decrease in water reserves in snow, in some river basins there is a noticeable activation of erosion activity associated with massive felling of forests. The strongest influence of concentrated felling on erosion processes is manifested in felling not older than 2-5 years. If the deforestation occupies 1-5% of the basin area, the suspended sediment runoff module increases by 50-300%, the turbidity of the water increases it by 100-800%. Currently, the volume of wood harvesting in some areas is increasing. In the third Selenga district, covering the basins of Barguzin and Selenga rivers, the descending trend of suspended sediments runoff predominates. The main reason is an abrupt decline in agricultural activities in this area over the past 20-25 years. In some areas of Buryatia, where practically complete closure of agriculture due to socio-economic reasons is noted, the growth rate of gullies, the rate of flat erosion on slopes and the extent of sediment runoff are decreasing. For the fourth region (the Upper Amur basin), the opposite tendency is characteristic: an increase in the intensity of erosion processes against the background of the progressive growth of atmospheric humidification in the second half of the 20[th] century. Thus, the revealed positive trends of erosion processes are characteristic for the basins with high process intensity (the Yenisei and Amur basins). Due to the climate change, these trends will become even stronger. The most significant increase in sediment runoff should also be expected in areas of new economic development (development of gold, oil, gas, deforestation, etc.). In this regard, in order to reduce the amount of runoff and improve the ecological situation, we need measures of environmental and geomorphological safety, primarily restoration and creation of new anti-erosion protection systems.

Consider the dynamics of soil degradation in the economic land use in the Selenga river basin in the 20th century. It should be emphasized that this river is the main supplier of pollutants to Lake Baikal in solid and dissolved form. For assessing the influence of anthropogenic disturbance to lands upon the soil erosion rate for the period concerned, we carried out a correlative analysis of the changes in the area of eroded lands depending on the degree of land plowing, and on pasture load. It was found that the area of eroded lands increases directly with an increase in anthropogenic impact on landscapes (figure 1).

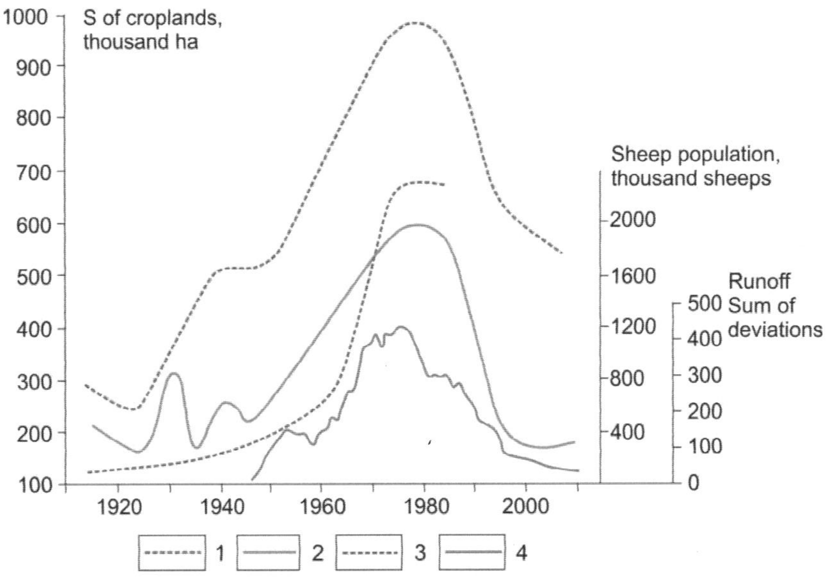


**Figure 1**. Long-term dynamics of changes in the area of arable lands (1), sheep population (2), changes in the area of eroded arable land (3) and in suspended load discharge of the Selenga river represented as an integral difference curve (4).

For the 20[th] century the curve of change in the arable land area represents an increase till the late 1980s with two periods of a slight reduction during the war years (the 1920s and 1940s). Also conspicuous are two periods of an abrupt increase in the proportion of plowed-up lands at the time of collectivization when the area of arable land increased by 40%, and during the opening-up of virgin lands. At the period of collectivization, characterized by merging of small fields, owned by peasant, the plowing-up of boundary paths and formation of relatively large tracts of arable lands, as well as by enhancement in mechanical action on the soil, the soil erosion and deflation intensity has increased considerably. An enhancement in the activity of erosion processes during that period was favored by concentrated forest felling. Virgin and fallow lands in Buryatia began to be developed in 1954 to extend over 20 years. For that period the area of arable land in the Republic increased by almost 40%, i.e. from 646 thou ha in 1953 to 1027 thou ha in 1975. Furthermore, the overall size of plowed-up lands exceeded the increase in arable lands in economic turnover. The post-war period till the end of 1980s showed a considerable increase in sheep population, also contributing to an enhancement of basin erosion, which is testified by an increase in sediment discharge of the Selenga river (figure 1). In the event of plowing up large tracts of lands with low humus content, sand and loamy sands were exposed on the surface, and the first year witnessed the onset of wind and water erosion. For that reason the opening up of virgin lands was often accompanied by the abandoning of old arable lands. At that period, the kolkhozes of the Pribaikal'skii and Zakamenskii districts were abandoned, because of soil degradation, about 90% of arable lands against the area of newly developed lands. The kolkhozes of the Tarbagataiskii district even reduced the overall area of arable lands (they opened up 2.4 thou ha of new lands, but abandoned, as fallow 4 thou ha of plowed fields). During that period, typically with low humidification, deflation was evolving particularly intensely on sandy soils.

An enhancement in deflation was characterized by an expansion of the area occupied by unfixed sands, and by the emergence of new centers of aeolian abrasion. Once again, sands began to encroach onto human settlements, roads, and agricultural lands, which were accompanied by formation of deflation basins. The movement of barkhans averaged 6-8 m/year. Within the Selenga basin, an increased water content and intensive basin erosion were observed in 1959-1973. The subsequent period of high water content, which was observed in 1984-1995, did not show any extreme development of erosion processes, which was reflected on the plot of the difference-integral curve for suspended loads of the Selenga river. (figure 1).

### *3.2. Soil deflation in Siberia*

Active deflation is one of the major problems of soil degradation in southern Siberia. The distribution and intensity of deflation depends on the characteristics of the relief, the wind regime and the presence of the earth's surface without vegetation cover or with sparse vegetation. The flat terrain in Western Kulunda creates favorable aerodynamic conditions for the wind. The presence of wind-impact slopes, gently sloping corridors located along the direction of the prevailing winds in East Kulunda, Minusinskaya depression, on the Irkutsk – Cheremkhovskaya plain, in the Baikal and Transbaikal regions promotes active deflation of plowed surfaces. The area of deflated lands in individual depressions reaches 40%.

Based on experimental research, the intensity of aeolian processes increases from the northern forest-steppe to the steppes and further deserted steppes [3]. With the increase in the deflation intensity, the size of the transferred material increases as well. In the Nazarovskaya forest-steppe, the wind can slightly transfer dust, in the Onon – Argunskaya and Koibal'skaya steppes sand is involved in the transfer. On Lake Baikal in Priol'honye, during the storm winds, gravel and small gravel are moving, periodically by a slip over a distance of 30-40 cm with the strongest gusts of wind.

In the forest-steppe regions aeolian accumulation of substance is dominating, as evidenced by the loess-loamy strata, which are formed by aeolian processes. The rate of aeolian accumulation according to the data of stationary studies varies from 0.01 mm in moderately humid years to 2 mm in extremely dry ones. In the steppes deflation and transport of aeolian material predominate. In the Koibal'skaya, Onon-Argunskaya and Barguzinskaya steppes the deflation rate varies from 0.1 to 2.2 mm/year. In the deserted steppes of Priol'khonye it increases to 3-6 mm/year, in the Peschanka stow to 20-35 mm/year, in the local areas during hurricane winds the catastrophic rates of aeolian denudation reaching 170 mm/year are fixed. The average rate of the areal deflation on sand massifs in the Selenga valley, determined by archaeological data, was 0.6 cm/year for 1000 years, and measured by natural check points 1-8 cm/year [see 3]. In general, the prevailing intensity of deflation in the subarid regions of southern Siberia is 10-50 tons/ha per year. It is close (the same order) with the values of annual soil losses from deflation in the zonal steppes of the East European and West Siberian plains. The dynamics of gully formation is of a pulsating nature, when after several years of almost complete absence of gully peaks growth; their more active growth (1-3 m/year and more) in wet and extremely wet years is noted. It has been established that the most forms of erosion in the last 15-20 years the rate of regressive erosion has decreased to 0.3 m/year. The decrease in the activity of gully formation is explained by the development of a longitudinal equilibrium profile, the use of soil protection measures, and the reduction of arable land. The analysis of the variability of the deflation processes over the past 100 years has revealed a cyclic nature of their course, caused by variations in atmospheric humidity. In the long-term regime of dust storms over the last 30 years of the 20th century, there were opposite trends in their increase in the eastern regions of the south of Siberia and a decrease in the west. They are associated with the change of the atmospheric circulation epochs. Short-period oscillations of deflation processes with small amplitude are characteristic for the western part of the territory. In the eastern regions, longer cycles with larger amplitude are observed, reaching a maximum in the Western Transbaikalia. Against the backdrop of these basically antiphase fluctuations in the intensity of aeolian processes, which are characteristic for individual regions, the general periods of their intensification are marked, associated with severe droughts that spread almost to the entire south of Siberia, when the precipitation was 10-20% or more lower than the norm (in the early 1920s and 1980s). In general, in the south of Siberia, there is a range expansion of deflation by means of forest-steppe regions.

*3.3. Zones of probably aeolian desertification of Subject sites of Siberia*

The results of the study enable ranking aeolian processes according to the degree of their danger to the environment and to divide the semi-arid landscapes into zones with a different level of possible desertification, the aeolian processes. For this purpose, the probability of the development of processes with a weak, moderate, high and very high intensity in each of the regions according to the frequency of dust storms in a year (N) and the complex climate deflation index (C), as well as the module aeolian migration of matter (A). On the basis of these calculations, six levels of desertification (table 3) and the corresponding six possible zones of desertification are identified by the ratio of the repeatability of processes of different intensity (figure 2).

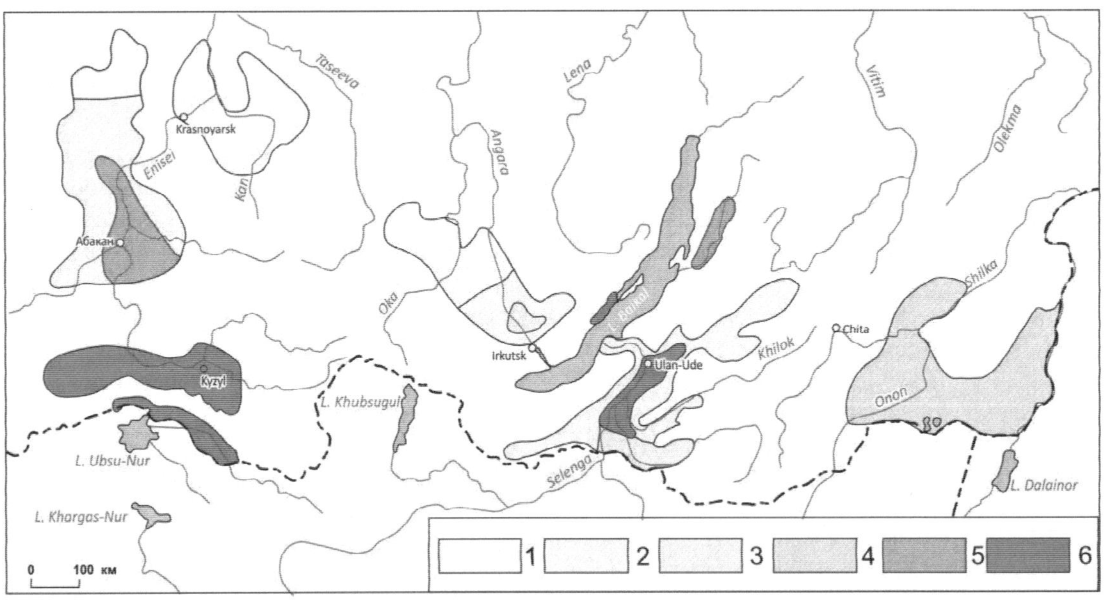


**Figure 2.** Zones of possible desertification of subarid areas of Siberia, associated with the development of aeolian processes: the probality of desertification: 1 – very low, 2 – low, 3 – moderate, 4 – large, 5 – very large, 6 – unusually high.

*3.4. Soil protection models*

Currently there is a further increase in the volume of implementation of agrotechnical erosion control measures, preconditions are created for a significant curbing of erosion rates. Models of soil conservation agriculture in the Siberian region have been developed. Thus, the Altai model of water-defense systems for contour-meliorative agriculture on slopes includes contour-band organization of crop rotation tracts, a complex of agrotechnical erosion measures, curtains of high-stemmed crops, a system of water-retaining and water-guiding hydraulic equipment linked with forest belts and a road network, estuaries and slope storage ponds.

**Table 3.** Probability of desertification of subarid areas of Siberia, calculated on the base of repeatability of dust storms during the year (N), a comprehensive climate index of deflation (C), % and the rate of migration of aeolian matter (A, t/ha/year).

| Desertification degree | Aeolian processes intensity | | | | |
|---|---|---|---|---|---|
| | weak (N < 3.0;C < 1.5) | moderate (N = 3.1–7.0; C = 1.5–3.0) | high (N = 7.1–10.0; C = 3.1–5.0) | very high (N > 10.0; C > 5.0) | A, tons/ha |
| **I** | 85 | 15 | – | – | 0.1–1 |
| **II** | 65 | 25 | 7 | 3 | 1–5 |
| **III** | 60 | 20 | 15 | 5 | 5–10 |
| **IV** | 35 | 35 | 10 | 20 | 10–100 |
| **V** | 30 | 20 | 20 | 30 | 100–1000 |
| **VI** | 30 | 10 | 10 | 50 | more than 1000 |

It has been established that the use of autumn subsurface cultivation with preservation of stubble, in contrast to plowing, is accompanied by significant changes in the processes of snow deposition, by rates of spring snowmelt and the meltwater runoff. The presence of stubble contributes to an earlier and better accumulation, a uniform distribution and preservation of snow on the slopes. The Khakass model of prevention and reliable protection of soils from wind erosion includes the use of complex organizational and economic measures, and forest reclamation and agrotechnical soil protection measures [4]. The most important measure of improving the water properties of soils in arid regions of Siberia is irrigation. Remains of primitive irrigation systems of Khakassia, Tyva and Buryatia are 2–2.5 thou years old. In accordance with the Federal Target Program "Development of land melioration in Russian agricultural purpose for 2014-2020", approved by the Russian government, the reconstruction of Tes-Khemskaya, Barlykskaya, Terezinskaya and Chadanskaya irrigation systems is planned to be carried out on an area of 6394 hectares, for which 7.5 million rubles is allocated [5]. For arid conditions of Khakassia and the Kulundinskaya steppe, the strip placement of crops is still the main soil protection method, without wide application of which it is impossible to effectively conduct agricultural production. The Baikal model of soil protection from joint manifestations of water and wind erosion is based on measures for the proper organization of the territory, such as: strip placement of crops and fallow lands, low- tillage farming, sowing with special anti-erosion seeders, plowing "across the prevailing winds" and "across the slope," erosion crop rotation, grass planting, rationally limited grazing, creation of protective forest belts and landscapes. For more effective land use in the Baikal region, preference is given to the restoration of disturbed agrolandscapes.

## 4. Conclusion

The most urgent priority today for optimizing land use under forced conservation of agricultural lands and some easing of soil erosion should be considered inventory of deflation- and erosion-hazardous lands based on their mapping and quantitative assessment. In order to implement these measures, it is necessary to coordinate and integrate soil erosion investigations carried out in the institutes of the Siberian Branch of the Russian Academy of Sciences, at universities and other institutions, and also training of relevant specialists. The interdepartmental commission for environmental safety of the Russian Federation, in 1993, considered the need for urgent measures aimed at preventing further degradation of the soil cover as a threat to the country's national security [6]. It is finally necessary to adopt a law on soil protection at the federal level and at the level of individual subjects of the

IOP Conf. Series: Earth and Environmental Science **190** (2018) 012023 doi:10.1088/1755-1315/190/1/012023

Federation. To conserve soils and improve the ecological situation, not only legal and financial-economic measures are needed, but also an understanding of the threat of soil and water degradation for the welfare of the population of Russia.

## References

[1] Khmelev V A and Tanasienko A A 2009 *Land Resources of Novosibirsk Region and Ways of Their Rational Use* (Novosibirsk: Izd-vo SO RAN) p 349
[2] Tukhta S A 2017 Quantitative assessment of the intensity of storm shower washout within the Kuda river basin (Leno-Angarskaya forest-steppe) *Geografiya i Prirodnye Resursy* **4** 133-9 doi: 10.21782 /GIPR0206-1619-2017-4(94-104)
[3] Bazhenova O and Tyumentseva E 2015 Contemporary aeolian morphogenesis in semiarid landscapes of the intermountain depressions of southern Siberia *Catena* **134** 50-8 doi:10.1016/j.catena.2015.02.006
[4] On the Federal Target Program "Development of Land Reclamation of Agricultural Lands in Russia for 2014-2020" 2013 *Decree of the Government of the Russian Federation of October* **12** 922
[5] Savostyanov V K 2014 *Use and Protection of Soils of Arid Territories of Siberia* (Abakan: Kooperativ 'Zhurnalist') p 286
[6] Dobrovolskii G V 1997 Quiet Crisis of the Planet *Vestnik RAN* **67**(4) 313-20

# Ecological status of the atmosphere and snow cover of the water area of Lake Baikal and its surrounding territories and its influence on the Lake pollution

**I A Belozertseva, I B Vorobyeva, N V Vlasova, O V Gagarinova, M S Yanchuk, D N Lopatina**

Sochava Institute of Geography SB RAS, Irkutsk, Russia

E-mail: belozia@mail.ru

**Abstract**. The testing and chemical analyses of the snow cover have revealed the anomalies for the following chemical elements and substances: F, Cl, $SO_4$, $NO_2$, $NO_3$, $PO_4$, K, Na, $NH_4$, Mo, Mn, Ba, Al, Pb, Ni, Cu, Be, V, Cr, Fe, Si, Zn, Sr, Ti, Hg and petroleum products which all exceed their background concentrations. Air and snow pollution of the lake's water area has been detected near the littoral residential zones and in the estuary of the Selenga River. The concentrations of $NH_4$, Pb, Be, Hg and petroleum products in the snow water exceed the MAC. According to the research, the current chemical composition of Lake Baikal's water, in terms of its drinking properties, conforms to health and safety standards due to the accumulation of pollutants in soil and alluvial deposits and their delution with a huge amount of lake water. The only exceptions are some local areas of the shoreline. A local increase in the content of Mn, Pb, Ni, Co, Cr, Zn, Cu which all exceed the MAC and the APC values has been detected near the populated areas. High concentrations of F, Hg, Fe, Pb, Zn, Cu, Ni, $PO_4$, $NH_4$ have been detected in the large rivers flowing into the lake. The landscapes of the rivers' estuaries serve as a geochemical barrier for toxic substances.

## 1. Introduction

The uniqueness of the research target is determined by the location of Lake Baikal in its area. Baikal is the deepest and the purest lake in the world as well as the UNESCO World Heritage Site. It is also the largest freshwater lake on the planet. Baikal's enormous freshwater capacity which comprises 20 % of the world's freshwater storage as well as water deficiency problem on a world-wide scale demand the attention of the researchers from all over the globe [1]. The quality of Baikal fresh water that is highly oxygenated and low in mineral salts and organic substances is especially valuable [2]. Unfortunately, the natural area of Baikal is also home to powerful sources of environmental contamination. What threat do they pose to Lake Baikal and its environment?

## 2. Methodical aspects and results of a research

In the winter and spring periods (February-March) of 2015-2017, the authors of the following scientific paper collected snow samples in order to identify the snow cover pollution level which is generally seen as an indicator of the atmosphere pollution in the central zone of the Baikal natural area. The observations and sample selections were carried out according to the key areas system considering the sources of atmosphere pollution and wind patterns (figure 1).

IOP Conf. Series: Earth and Environmental Science **190** (2018) 012024 doi:10.1088/1755-1315/190/1/012024

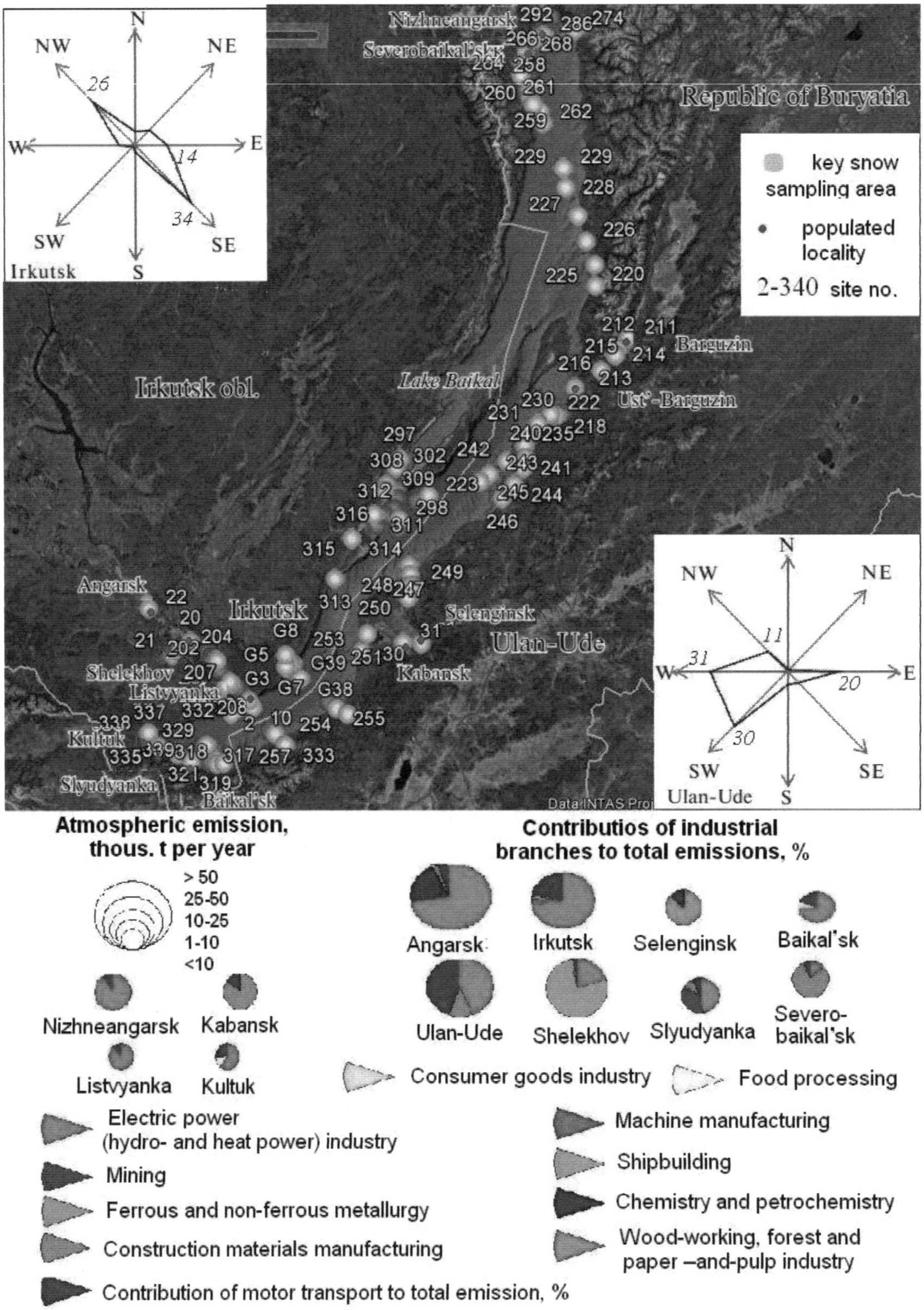


**Figure 1.** Snow sampling points and sources of pollution.

The snow samples were collected using a VS-43 snow gauge throughout the whole depth of the snow cover with the identification of its height and density (weight) [3]. By the time of the sampling, the snow accumulation period had amounted to 131-140 days. More than 300 snow samples have been collected in the water area of Lake Baikal and the adjoining territories. Water samples of Lake Baikal and the rivers flowing into the lake as well as soil samples of its shoreline were also collected at the same period (total over 1000 samples). All the analytical studies have been conducted in the V.B. Sochava Institute of Geography of SB RAS under laboratory conditions using standardized methods and modern analytical equipment. The concentrations of the main anions and cations in the snow water were identified using standardized chemical methods and in accordance with all-union state standards. The input of chemical elements in the water area through atmosphere in the winter period was calculated with the account of the average height and density of the snow cover. Recalculation of the content of chemical elements and substances (mg/dm$^3$ or mg/L) was carried out on the basis of the area index of mg/m$^2$. Metal content was calculated using atomic emission spectroscopy method and Optima 2000DV equipment. Fluorine content was determined using H-120 ion meter and fluorine selective electrode. Petroleum products were detected with the help of a FLUORAT-02-3M fluorescence analyzer, whereas mercury content was identified in the laboratory for the monitoring of ecological and technogenic environments of Irkutsk State Technical University.

## 3. Research Findings

According to the results of the snow survey of 2015-2017 and the scientific papers written by the research workers of various institutes [4, 5, 6, 7, 8, 9, 10] as well as the research carried out by the authors mentioned above [11], it has been concluded that the major sources of air pollution of Lake Baikal are the industrial plants located in the basin of the lake and on its shores, certain segments of the Trans-Siberian and Baikal-Amur railways and airborne pollution caused by the Irkutsk-Cheremkhovo industrial hub. Aerial effluents from the industrial and heating plants of Baikalsk, Sludyanka, Severobaikalsk, Nizhnenagarsk and other small towns situated in the basin of Lake Baikal have the highest chances (from 10 % to 100 %) of penetrating the lake water. According to numerous research [2, 11, 12, 13, 14, 15] conducted in the recent years, it has been found that airborne polluting products of the Irkutsk-Cheremkhovo agglomeration constitute a much smaller part of the total amount of atmospheric pollution due to the remoteness of the source of pollution, windless conditions, and fogs [6, 7, 10, 13].

However, we can observe local pollution taking place under unfavorable weather conditions which altogether leads to a dangerous ecological situation in the towns of the industrial hub mentioned earlier.

Industrial and heating plants of Ulan-Ude, Gusinoozersk, Petrovsk-Zabaikalsk, Kyakhta and other towns located in the buffer zone of Baikal's natural area do not have a major impact on the lake due to their remoteness, specific weather conditions in the area and orographic obstacles [4, 8]. However, they influence the whole ecosystem of the Baikal region [16].

The results of the chemical analysis of the snow cover in the southern kettle of Lake Baikal and the surrounding territory have revealed the anomalies limited to the valley of the Angara River and the populated areas of the lake's shoreline (Listvyanka, Slyudyanka, Kultuk, Baikalsk) with the elevated contrast ratio for the following chemical elements and substances: F – 7.5, Cl – 2.0, $SO_4$ – 9.2, $NO_2$ – 12.8, $NO_3$ – 19.5, $PO_4$ – 18.3, K – 4.3, Na – 8.2, $NH_4$ – 30, Mo – 4.2, Mn – 6.8, Ba – 12.4, Al – 5.4, Pb – 3, Ni – 3, Cu – 3, Be – 4, V – 6, Fe – 7,3, Si – 19.5, Zn – 6.6, Sr – 3.9, Ti – 3.5, Hg – 2.7, petroleum products - 4 (enhanced background) (see figure 2, 3, 4, table 1).

It should be noted that the background concentrations of most of the chemical elements are very low, i.e. are in deficit. The content of $NH_4$, Pb, Be, Hg and petroleum products in the snow water exceeds maximum permissible concentrations (MPC) by a factor of 24, 4, 6, 2 and 3 respectively. Their maximum concentrations are limited to the shoreline zone of the lake's water area.

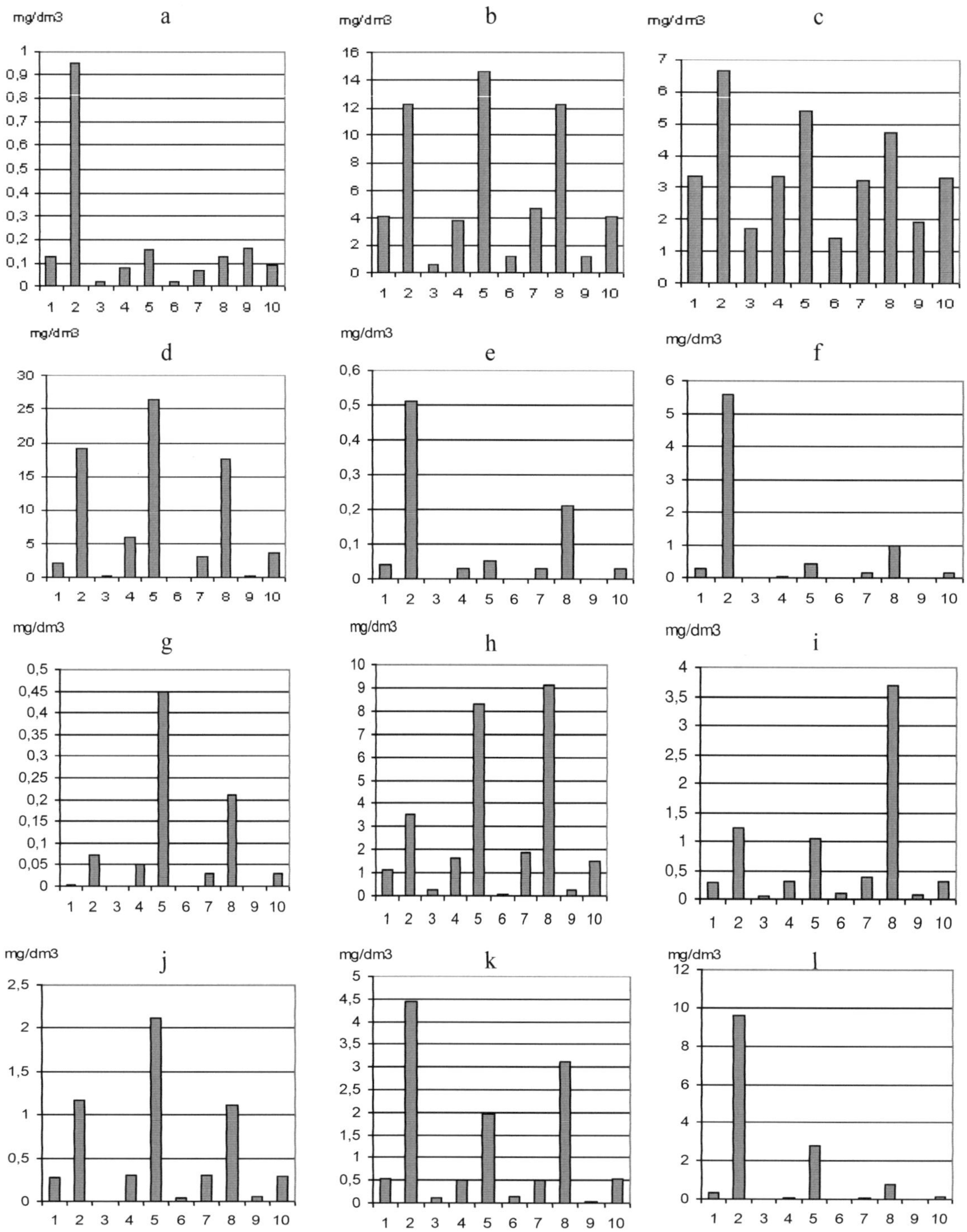


Southern kettle: 1 - average, 2 - maximum, 3 –minimum content; Central kettle: 4 - average, 5 - maximum., 6 –minimum content; Northern kettle: 7 - average, 8 - maximum., 9 –minimum content; Lake Baikal in total: 10 – average content. *MPC, APC: F – 0,7-1,5; Cl – 350; $SO_4$ – 500; $NO_3$ – 130; $PO_4$ – 0,001; Ca – 180; Mg – 50; K – 40-50; Na -120-200; $NH_4$ -0,4, petroleum products – 0,05 mg/dm$^3$*

**Figure 2.** Content of chemical elements and substances a) F, b) $HCO_3$, c) Cl, d) $SO_4$, e) $NO_2$, f) $NO_3$, g) $PO_4$, h) Ca, i) Mg, j) K, k) Na, l) $NH_4$in the snow water of Lake Baikal and the adjoining territory.

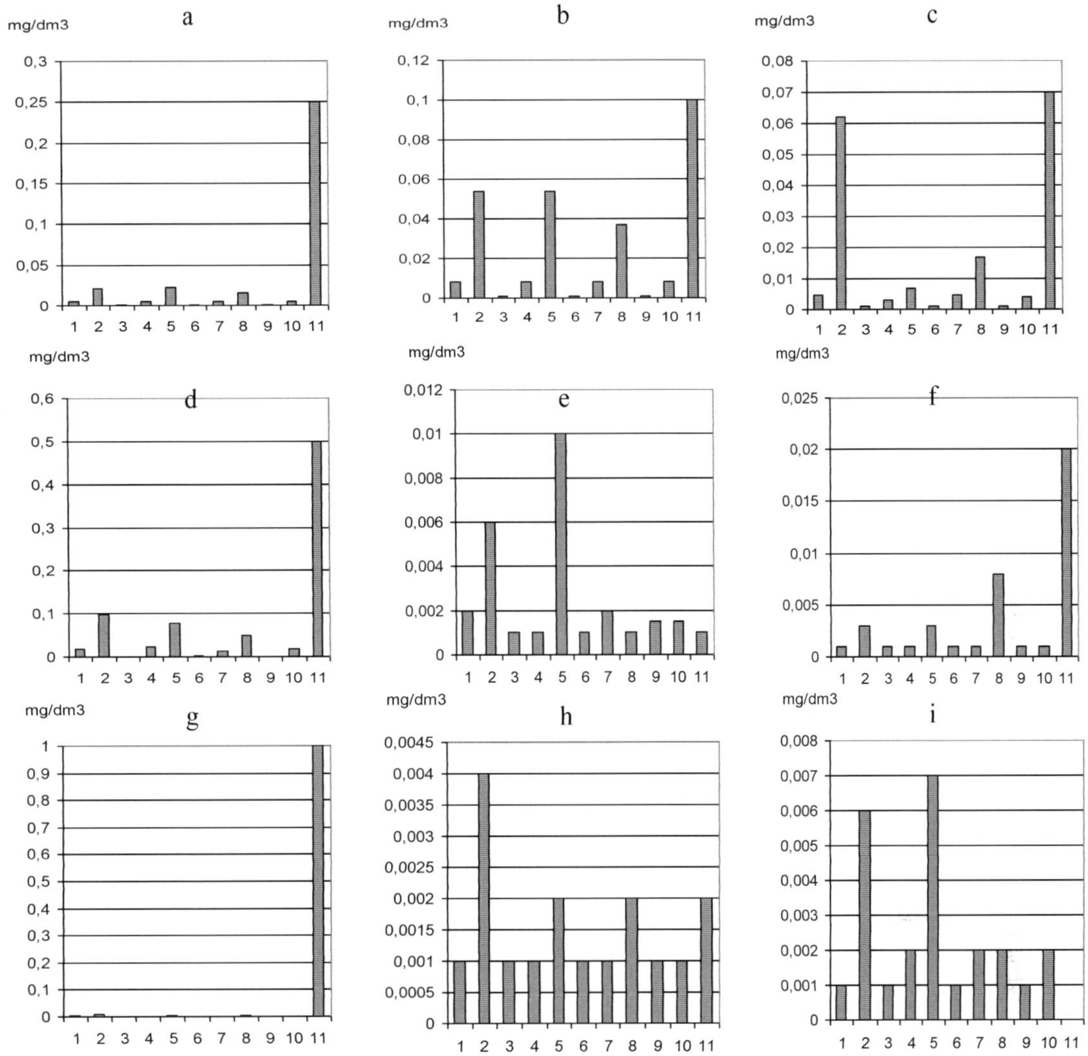


Southern kettle: 1 - average, 2 - maximum, 3 – minimum content; Central kettle: 4 - average, 5 - maximum., 6 - minimum content; Northern kettle: 7 - average, 8 - maximum., 9 - minimum content; Lake Baikal in total: 10 - average content; 11 - Maximum Permissible Concentration and Approximate Permissible Concentration (MPC, APC)

**Figure 3.** Average, maximum and minimum concentrations a) Mo, b) Mn, c) Ba, d) Al, e) Pb, f) Ni, g) Cu, h) Be, i) V in the snow water of Lake Baikal and the adjoining territory.

In terms of their complexity and intensity the anomalies of the most developed South Baikal and Irkutsk-Cheremkhovo industrial hubs refer to the technogenic type and result from the emissions of pollutants from various fuel and energy, chemical and metallurgical industries into the atmosphere [17]. The given data show that hydrogen carbonate ions, sulfate ions and chloride ions comprise a bigger part of the ground mass of the substances penetrating Lake Baikal. Calcium, sodium and potassium cations also have a significant influence on the lake, notably in its central and northern kettles. High concentrations of fluorine ions and some heavy metals have been detected in the snow water near the shoreline village of Listvyanka. They are typical of technogenic emissions of the aluminum industry. However, the concentration of these chemical elements in the liquid phase of snow is dozens and hundreds of times higher (>90 %) near the Irkutsk Aluminum Plant (IAP) rather than in the Baikal shoreline zone of Listvyanka which is 90 km (56 miles) far from the IAP. Therefore, local pollution characteristic of stove heating prevails in the shoreline zone of the village, whereas regional pollution only constitutes 10 %.

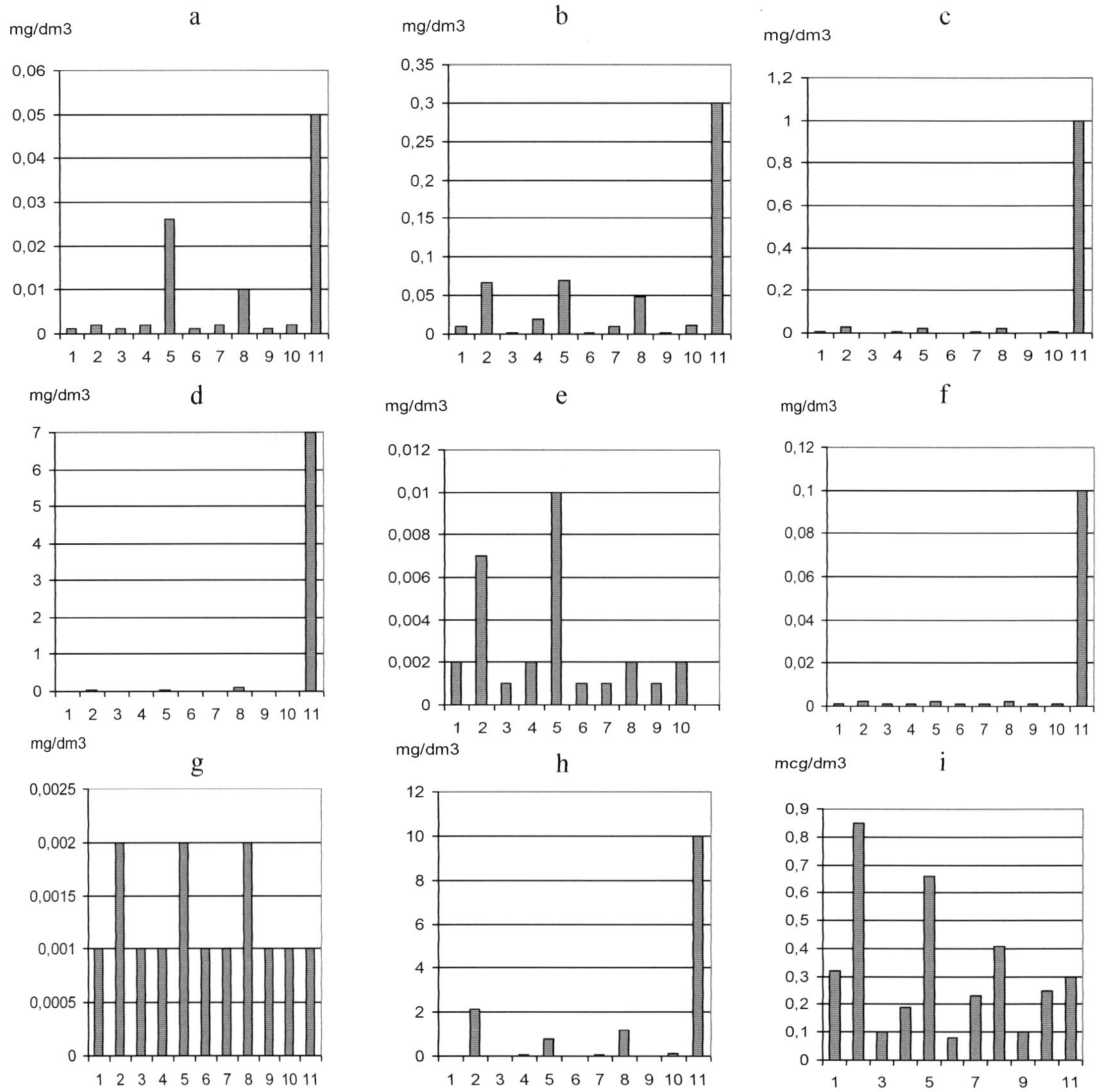


Southern kettle: 1 - average, 2 - maximum, 3 – minimum content; Central kettle: 4 - average, 5 - maximum., 6 - minimum content; Northern kettle: 7 - average, 8 - maximum., 9 - minimum content; Lake Baikal in total: 10 - average content; 11 - Maximum Permissible Concentration and Approximate Permissible Concentration (MPC, APC)

**Figure 4.** Average, maximum and minimum concentrations a) Cr, b) Fe, c) Si, d) Zn, e) Sr, f) Ti, g) Co, h) Cd, i) Hg in the snow water of Lake Baikal and the adjoining territory.

The research carried out in the central kettle of Lake Baikal identified three areas of anomalous concentrations of chemical substances in the snow cover. The first area is limited to the shoreline zone near Olkhon Island, where we have detected a high concentration of chemical substances in the snow cover. In our opinion, it is connected with not only the anthropogenic load near the populated areas and recreational zones but also with the terrigenous dust of the soil-forming materials and roads. The snow in steppificated areas is easily transported to other areas. Bare snowless steppificated spots of the shoreline zone of the Near-Olkhon territory are quite large in area, thus micro and macroparticles of soil and rock typical of the given region are easily carried away to the water zone of the lake by strong winds. This type of pollution is of local and natural origin. Another area of snow cover pollution is limited to the Selenga valley through which technogenic emissions of the Selenga industrial hub are transported. The third area is limited to the villages of Barguzin and Ust-Barguzin.

**Table 1.** Average, maximum and minimum concentration of petroleum products, mineralization, pH in the snow water of Lake Baikal and the adjoining territory. March, 2015

| Region | Content | pH | Suspended Substance, g/dm$^3$ | Mineralization, mg/dm$^3$ | Petroleum products, mg/dm$^3$ |
|---|---|---|---|---|---|
| Southern Kettle of Lake Baikal | average | 6.44 | 0.070 | 10.8 | 0.050 |
| | max | 7.14 | 0.559 | 20.1 | 0.202 |
| | min | 6.06 | 0.001 | 5.84 | 0.008 |
| Central Kettle of Lake Baikal | average | 6.42 | 0.130 | 10.9 | 0.040 |
| | max | 7.39 | 0.690 | 28.3 | 0.140 |
| | min | 6.00 | 0.003 | 5.47 | 0.007 |
| Northern Kettle of Lake Baikal | average | 6.44 | 0.070 | 11.8 | 0.050 |
| | max | 7.03 | 0.192 | 27.2 | 0.117 |
| | min | 5.88 | 0.001 | 5.95 | 0.010 |
| Lake Baikal in total | average | 6.43 | 0.09 | 11.1 | 0.050 |
| MPC, APC, GOST 2874-82 | | | - | - | 0.050 |

MPC – Maximum Permissible Concentration, APC – Approximate Permissible Concentrations; GOST – All-union State Standard; Suspended Substance – solid residue on the filter (dust); mineralization – dry residue (sum of salts)

Maximum ratios of the concentration of chemical substances in the snow cover of the central kettle of Lake Baikal in relation to the background are: F – 2.1, $SO_4$ – 4.4, $NO_3$ – 9, $PO_4$ – 20, K – 7, Na – 3.8, $NH_4$ – 35, Mo – 4.6, Mn – 6.8, Ba – 2.3, Al – 3.5, Pb – 10, Ni – 3, V – 2.3, Cr – 13, Fe – 2.3, Si – 10, Zn – 4.8, Sr – 4.6, Ti – 5, Hg – 3.5, petroleum products – 3,5. The content of NH4, Hg and petroleum products in the snow exceed maximum permissible concentrations by a factor of 7, 3 and 3 respectively. Maximum concentrations of chemical substances are of local nature and are limited to the shoreline zone.

The analysis of the research conducted in the northern kettle of Lake Baikal showed contrast anomalous concentrations of SO4 which are 55 times above background: $NO_3$ – 5.9, $PO_4$ – 11.6, K – 3.7, Na – 6.1, $NH_4$ – 8.9, Mo – 2.7, Mn – 4.6, Ba – 3.4, Al – 4.2, Pb – 3.5, Ni – 8, Cu – 3, V – 2.5, Cr – 5, Fe – 5.4, Si – 14, Sr – 6.1, petroleum products – by a factor of 2.3 near the towns of Nizhneangarsk and Severobaikalsk. The concentrations of $NH_4$, Pb and petroleum products in the snow exceed sanitation and hygiene standards by a factor of 2, 7 and 2 respectively. Construction materials enterprises, heating plants and motor vehicles are the major sources of air pollution. The pollution is of local character.

Comparing the data gathered for the past 6 years [2, 11, 18], we can conclude that we are currently witnessing the increase in the concentration of sulfates, nitrates, fluorine and some heavy metals and a relative decrease in the concentration of calcium, magnesium and potassium in the snow water of the water area of Lake Baikal. It has also been found that regional pollution has slightly increased for the past 6 years (by a factor of 1,1 – 2 depending on a component). Although the data gathered for the past 20 years show that the pollution decreased by a factor of 2 at the end of the 1990s and at the beginning of the 2000s which was caused by the decline in the manufacturing output.

Considering the input of chemical substances through the snow cover (milligram/meter$^2$) during winter period and the total surface of Baikal's water area, we have calculated the input of chemical substances through the atmosphere (relying on the snow cover pollution data) [11]. The content of sulfate sulfur in the southern kettle of Lake Baikal equals 34 million 600 thousand tons [2, 19]. The calculations show that Air emissions equal 1 thousand 540 tons which is literally a thousandth part (1/22000) of the sulfate that is already contained in southern Baikal.

The southern kettle of Lake Baikal has the highest input of nitrates, nitrites, ammonium, phosphates, fluorine, aluminum, sodium, barium, molybdenum, manganese, lead, copper, zinc, strontium, mercury and petroleum products into the water area through the atmosphere despite the fact that it is twice as small as the northern one. However, the northern kettle has a slightly higher input of sulfates, hydrocarbonates, chlorine, calcium, magnesium and potassium. The central kettle has a higher input of beryllium, vanadium, chromium and cobalt due to the activities of the enterprises, heating plants, thermal power stations located in the shoreline zone as well as stove heating, motor vehicle emissions and dust transfer in the poorly snow-covered areas of the central kettle.

The chemical analysis of the waters of Lake Baikal and the soil in the shoreline zones shows that the current chemical composition of Lake Baikal's water, in terms of its drinking properties, conforms to sanitary standards [2, 11, 20] due to the accumulation of pollutants in soil and alluvial deposits and their dilution with a huge amount of lake water. However, high concentrations of lead, phosphates and ammonium nitrogen exceeding sanitation and hygiene norms have been detected in the local areas of the lake's coastal waters (a few meters from the shoreline), notably in such populated areas as Severobaikalsk, Nizhneangrsk, Sarma, Chernorud, Tankhoy, Baikalsk, Buguldeika, Monakhovo, Posolskoye, Dubinino and Ust-Barguzin (see table 2). The concentrations of lead in the coastal waters of Lake Baikal near Chernorud reach 0.009 mg/dm$^3$, Buguldeika – 0.06 mg/dm$^3$, Tankhoy – 0.010 mg/dm$^3$, Ust-Barguzin – 0.020 mg/dm$^3$, Goryachinsk – 0.020 mg/dm$^3$, Nizhneangarsk – 0.023 mg/dm$^3$ (the MPC equals 0.001 mg/dm$^3$ [21]).

Maximum concentrations of $PO_4$ in the lake´s coastal waters near the village of Sarmaare 0.01 mg/dm$^3$, Elantsy – 0.04 mg/dm$^3$, Monakhovo – 0.03 mg/dm$^3$, Goryachinsk – 0.06 mg/dm$^3$, Posolskoyeh – 0.05 mg/dm$^3$, Babushkin – 0.01 mg/dm$^3$, Baikalsk – 0.02 mg/dm$^3$, Sludyanka – 0.02 mg/dm$^3$, Tanghui – 0.01 mg/dm$^3$, Dubinino – 0.02 mg/dm$^3$, lake Kotokel – 0.01 mg/dm$^3$ (the MPC values are 0.001 mg/dm$^3$).

Maximum concentrations of $NH_4$ in the waters of the lake's shoreline near the populated areas of Tankhoy, Baikalsk, Elantsy, Chernorud, Buguldeika are – 0.44, 0.48, 0.55, 0.60, 0.64 mg/dm$^3$ respectively (the MPC values are 0.4 mg/dm$^3$).

High concentrations of copper and zinc exceeding background concentrations have been detected in the shoreline areas near Olkhon Gate and the villages of Kuburlik, Goryachinsk, Vydrino, Bolshie Koty, Babushkin, Tankhoy, Listvyanka, BolshoyeGoloystnoye. All of these concentrations exceed 0.01 mg/dm$^3$. Petroleum concentrations near the populated areas of BolshoyeGoloustnoye, Sludyanka, Maksimovshina, Nizhneangarsk vary from 0.06 to 0.34 mg/dm$^3$ (the MPC values are 0.05 mg/dm$^3$).

High concentrations of F, Hg, Fe, Pb, Zn, Cu, $PO_4$, $NH_4$ have been detected in the rivers Selenga, Barguzin, Utulik, Turka, Kichera, Upper Angara, Tyia, Sarma, Anga, Mysovka, Pereemnaya, Suhaya (see table 3, 4). The content of lead, phosphates, fluorides and petroleum products in these rivers exceed the MPC and APC values.

In waters of the rivers Small Cheremshanka, Big Cheremshanka, Sennushka, Krestovka, Bannaya, Angasolka, Hol, Kultuchnaya, Kurkavochnaya, Snezhnaya, Osinovka, Shamanka, Maksimikha, Babkha, Galutai, Harlakta, Goloustnaya, Pankovka, Enkhalukh, Bolshaya River, Hara-Murin, Kika, Pokhabikha, Haim, Manturikha pollution isn't revealed.

The concentration of fluorine in the waters of the Selenga and the mineral springs of Khakusy reaches 3 mg/dm$^3$. Maximum concentrations of Pb in the waters of the Utulik, Turka, Barguzin, Selenga, Kichera equal 0.011, 0.013, 0.020, 0.020, 0.040 mg/dm$^3$ respectively. High concentrations of $PO_4$ have been identified in the rivers of Turka, Barguzin, Selenga, Upper Angara, Tyia. The figures are 0.01, 0.01, 0.06, 0.10, 0.10 mg/dm$^3$ respectively.

**Table 2.** Content of chemical elements and substances, Mineralization, Suspended Substance, pH of water of Lake Baikal

| Component | Southern Kettle of Lake Baikal | | | Central Kettle of Lake Baikal | | | Northern Kettle of Lake Baikal | | | Coast. | MPC, APC** |
|---|---|---|---|---|---|---|---|---|---|---|---|
| | max | min | aver. | max | min | aver. | max | min | aver. | | |
| pH | 7.80 | 6.97 | 7.35 | 7.81 | 6.78 | 7.33 | 7.90 | 6.66 | 7.47 | 8.37 | - |
| F | 0.650 | 0.060 | 0.270 | 0.210 | 0.004 | 0.122 | 0.652 | 0.091 | 0.443 | 1.060 | 0.7-1.5 |
| $HCO_3$ | 93.33 | 32.12 | 57.39 | 81.74 | 10.01 | 40.13 | 79.30 | 29.70 | 50.27 | 104.07 | - |
| Cl | 2.02 | 0.01 | 0.24 | 0.06 | 0.01 | 0.03 | 0.04 | 0.01 | 0.02 | 3.57 | 350 |
| $SO_4$ | 8.80 | 4.95 | 6.07 | 24.75 | 2.18 | 7.37 | 22.00 | 4.21 | 6.89 | 25.08 | 500 |
| $NO_2$ | 0.05 | 0.01 | 0.03 | 0.40 | 0.01 | 0.06 | 0.04 | 0.01 | 0.02 | 1.260 | - |
| $NO_3$ | 0.02 | 0.01 | 0.01 | 0.01 | 0.01 | 0.01 | 0.02 | 0.01 | 0.01 | 55.00 | 130 |
| $PO_4$ | 0.001 | 0.001 | 0.001 | 0.010 | 0.001 | 0.001 | 0.010 | 0.001 | 0.001 | 0.061 | 0.001 |
| Ca | 30.52 | 10.21 | 17.34 | 18.63 | 10.01 | 15.13 | 28.05 | 10.11 | 14.12 | 41.10 | 180 |
| Mg | 7.48 | 2.01 | 3.88 | 2.92 | 2.01 | 2.78 | 6.23 | 2.01 | 3.16 | 8.78 | 50 |
| K | 1.99 | 1.12 | 1.29 | 1.98 | 1.04 | 1.43 | 1.85 | 1.19 | 1.32 | 2.18 | 40-50 |
| Na | 4.78 | 2.24 | 2.65 | 3.98 | 2.01 | 2.75 | 2.29 | 2.05 | 2.56 | 8.51 | 120-200 |
| $NH_4$ | 0.100 | 0.010 | 0.040 | 0.030 | 0.010 | 0.015 | 0.030 | 0.010 | 0.013 | 0.640 | 0.4 |
| Mo | 0.022 | 0.001 | 0.003 | 0.016 | 0.001 | 0.004 | 0.020 | 0.001 | 0.003 | 0.102 | 0.25 |
| Mn | 0.080 | 0* | 0.012 | 0.013 | 0* | 0.002 | 0.008 | 0* | 0.003 | 0.080 | 0.1 |
| Ba | 0.070 | 0.007 | 0.019 | 0.017 | 0.009 | 0.013 | 0.036 | 0.008 | 0.015 | 0.070 | 0.7 |
| Al | 0.049 | 0.017 | 0.026 | 0.033 | 0.010 | 0.017 | 0.051 | 0.020 | 0.026 | 0.120 | 0.5 |
| Pb | 0.010 | 0* | 0.001 | 0.020 | 0* | 0.001 | 0.023 | 0* | 0.001 | 0.023 | 0.01-0.001 |
| Ni | 0.001 | 0* | 0* | 0.001 | 0* | 0* | 0.001 | 0* | 0* | 0.060 | 0.02 |
| Cu | 0.009 | 0.001 | 0.004 | 0.009 | 0.001 | 0.003 | 0.009 | 0.001 | 0.004 | 0.026 | 1.0 |
| Be | 0.001 | 0* | 0* | 0.001 | 0* | 0.001 | 0.001 | 0* | 0.001 | 0.002 | 0.002 |
| V | 0.001 | 0.001 | 0.001 | 0.001 | 0.001 | 0.001 | 0.001 | 0.001 | 0.001 | 0.007 | - |
| Cr | 0.002 | 0* | 0.001 | 0.001 | 0* | 0.001 | 0.001 | 0* | 0* | 0.010 | 0.05 |
| Fe | 0.022 | 0.001 | 0.006 | 0.040 | 0.001 | 0.006 | 0.160 | 0.002 | 0.028 | 0.165 | 0.3 |
| Si | 7.144 | 0.001 | 1.122 | 1.465 | 0.001 | 0.219 | 4.471 | 1.392 | 2.325 | 6.510 | 10 |
| Zn | 0.015 | 0* | 0.003 | 0.002 | 0* | 0.001 | 0.001 | 0* | 0* | 0.090 | 5-1 |
| Sr | 0.620 | 0.001 | 0.107 | 0.117 | 0.010 | 0.044 | 0.262 | 0.046 | 0.093 | 0.620 | 7 |
| Ti | 0.001 | 0.001 | 0.001 | 0.001 | 0.001 | 0.001 | 0.001 | 0.001 | 0.001 | 0.002 | - |
| Co | 0.002 | 0* | 0.001 | 0.002 | 0* | 0.001 | 0.002 | 0* | 0.001 | 0.002 | 0.1 |
| Cd | 0.001 | 0* | 0.001 | 0.001 | 0* | 0* | 0.001 | 0* | 0.001 | 0.003 | 0.001-0.005 |
| Hg | 0.01 | 0.01 | 0.01 | 0.01 | 0.01 | 0.01 | 0.01 | 0.01 | 0.01 | 0.12 | 0.3-0.5 |
| Petrol. | 0.049 | 0.011 | 0.018 | 0.045 | 0.010 | 0.018 | 0.049 | 0.005 | 0.015 | 0.338 | 0.05 |
| Susp. | 0.0086 | 0.0001 | 0.003C | 0.0262 | 0.0003 | 0.0108 | 0.0290 | 0.001C | 0.006C | 0.0110 | - |
| ∑*** | 135.53 | 63.20 | 89.25 | 113.27 | 37.43 | 69.38 | 120.83 | 53.35 | 82.00 | 160.32 | - |

0* - < 0.001; MPC** – Maximum Permissible Concentration, APC** – Approximate Permissible Concentrations [21, 22]; ∑*** - Mineralization;
aver. – average; Coast. – Water of the coast of the Lake Baikal; Petrol. – Petroleum products; Susp. - Suspended Substance – solid residue on the filter (dust);
F , $HCO_3$, Cl , $SO_4$ , $NO_2$, $NO_3$, $PO_4$, Ca, Mg, K, Na, $NH_4$, Mo, Mn, Ba, Al, Pb, Ni, Cu, Be, V, Cr, Fe, Si, Zn, Sr, Ti, Co, Cd, Petroleum products, Mineralization (mg/dm$^3$); Hg (mkg/dm$^3$); Suspended Substance (g/dm$^3$)

IOP Conf. Series: Earth and Environmental Science **190** (2018) 012024 doi:10.1088/1755-1315/190/1/012024

**Table 3.** Content of chemical elements and substances of water of the Selenga, Barguzin, Upper Angara, Tyia Rivers.

| Component | Selenga | | | Barguzin | | | Upper Angara | | | Tyia | | |
|---|---|---|---|---|---|---|---|---|---|---|---|---|
| | max | min | aver. | max | min | aver. | max | min | aver. | max | min | aver. |
| pH | 9.72 | 6.99 | 7.99 | 8.05 | 7.31 | 7.85 | 8.37 | 7.03 | 7.93 | 8.01 | 7.46 | 7.77 |
| F | 3.140 | 0.059 | 0.162 | 0.223 | 0.085 | 0.164 | 0.234 | 0.105 | 0.195 | 0.162 | 0.041 | 0.105 |
| $HCO_3$ | 125.35 | 18.12 | 81.84 | 90.06 | 36.58 | 65.27 | 58.01 | 40.36 | 48.75 | 76.05 | 37.08 | 57.29 |
| Cl | 3.93 | 1.78 | 2.49 | 18.50 | 1.60 | 4.12 | 2.50 | 0.19 | 0.20 | 2.32 | 1.96 | 2.11 |
| $SO_4$ | 6.80 | 1.00 | 5.50 | 2.10 | 1.00 | 2.70 | 9.37 | 1.00 | 3.56 | 3.80 | 0.10 | 0.70 |
| $NO_2$ | 0.101 | 0.001 | 0.021 | 0.200 | 0.008 | 0.034 | 0.041 | 0.001 | 0.016 | 0.021 | 0.003 | 0.010 |
| $NO_3$ | 63.00 | 0.01 | 12.29 | 11.60 | 0.01 | 1.55 | 0.60 | 0.01 | 0.19 | 3.60 | 0.01 | 1.16 |
| $PO_4$ | 0.061 | 0.001 | 0.017 | 0.077 | 0.001 | 0.012 | 0.100 | 0.001 | 0.005 | 0.100 | 0.001 | 0.006 |
| Ca | 57.40 | 4.46 | 28.92 | 38.85 | 16.80 | 28.95 | 15.88 | 10.06 | 14.32 | 21.79 | 16.01 | 18.13 |
| Mg | 10.19 | 0.83 | 5.96 | 8.55 | 3.39 | 4.33 | 3.55 | 1.64 | 2.82 | 6.62 | 3.21 | 4.46 |
| K | 4.06 | 0.54 | 1.76 | 2.96 | 0.40 | 1.94 | 1.05 | 0.26 | 0.75 | 1.10 | 0.18 | 0.74 |
| Na | 21.02 | 0.19 | 5.55 | 11.37 | 1.90 | 4.86 | 3.58 | 1.30 | 2.69 | 3.65 | 0.01 | 2.11 |
| $NH_4$ | 0.150 | 0.001 | 0.023 | 0.250 | 0.012 | 0.059 | 1.040 | 0.023 | 0.118 | 0.225 | 0.001 | 0.079 |
| Mo | 0.015 | 0.001 | 0.006 | 0.011 | 0.001 | 0.003 | 0.021 | 0.001 | 0.009 | 0.030 | 0.001 | 0.011 |
| Mn | 0.050 | 0.001 | 0.009 | 0.060 | 0.001 | 0.008 | 0.005 | 0.001 | 0.003 | 0.026 | 0.001 | 0.008 |
| Ba | 0.040 | 0.004 | 0.018 | 0.024 | 0.010 | 0.019 | 0.030 | 0.001 | 0.011 | 0.020 | 0.009 | 0.011 |
| Al | 0.060 | 0.001 | 0.036 | 0.050 | 0.001 | 0.020 | 0.030 | 0.001 | 0.015 | 0.325 | 0.020 | 0.065 |
| Pb | 0.020 | 0.001 | 0.003 | 0.009 | 0.001 | 0.002 | 0.003 | 0.001 | 0.002 | 0.003 | 0.001 | 0.002 |
| Ni | 0.050 | 0.001 | 0.004 | 0.001 | 0* | 0.001 | 0.002 | 0* | 0.001 | 0.002 | 0* | 0.001 |
| Cu | 0.020 | 0.001 | 0.006 | 0.004 | 0.001 | 0.001 | 0.008 | 0.001 | 0.002 | 0.005 | 0.001 | 0.003 |
| Be | 0.002 | 0.001 | 0.001 | 0.001 | 0* | 0.001 | 0.002 | 0* | 0.001 | 0.002 | 0* | 0.001 |
| V | 0.005 | 0.001 | 0.001 | 0.003 | 0.001 | 0.001 | 0.008 | 0.001 | 0.004 | 0.010 | 0.001 | 0.004 |
| Cr | 0.010 | 0.001 | 0.004 | 0.001 | 0* | 0.001 | 0.001 | 0* | 0.001 | 0.003 | 0* | 0.002 |
| Fe | 0.236 | 0.001 | 0.061 | 0.186 | 0* | 0.043 | 0.020 | 0.001 | 0.010 | 0.055 | 0.001 | 0.016 |
| Si | 8.490 | 0.221 | 3.513 | 7.640 | 0.280 | 2.691 | 4.390 | 0.001 | 2.020 | 4.955 | 0.330 | 2.540 |
| Zn | 0.220 | 0.001 | 0.028 | 0.002 | 0.001 | 0.001 | 0.005 | 0.001 | 0.002 | 0.042 | 0* | 0.014 |
| Sr | 0.270 | 0.020 | 0.145 | 0.190 | 0.035 | 0.103 | 0.180 | 0.002 | 0.063 | 0.150 | 0.042 | 0.092 |
| Ti | 0.020 | 0.001 | 0.007 | 0.005 | 0.001 | 0.002 | 0.006 | 0.001 | 0.003 | 0.008 | 0.001 | 0.004 |
| Co | 0.010 | 0.001 | 0.003 | 0.002 | 0* | 0.001 | 0.001 | 0.001 | 0.001 | 0.010 | 0.001 | 0.002 |
| Cd | 0.001 | 0.001 | 0.001 | 0.002 | 0* | 0.001 | 0.001 | 0* | 0.001 | 0.001 | 0* | 0.001 |
| Hg | 0.02 | 0.01 | 0.01 | 0.01 | 0.01 | 0.01 | 0.02 | 0.01 | 0.01 | 0.02 | 0.01 | 0.01 |
| Petrol. | 0.129 | 0.001 | 0.016 | 0.082 | 0.002 | 0.004 | 0.080 | 0.001 | 0.018 | 0.068 | 0.003 | 0.007 |
| Miner. | 0.160 | 0.025 | 0.096 | 0.169 | 0.042 | 0.111 | 0.053 | 0.043 | 0.048 | 0.103 | 0.041 | 0.056 |
| ∑*** | 214.95 | 28.05 | 139.58 | 120.23 | 96.51 | 11.56 | 84.28 | 62.11 | 75.43 | 108.92 | 66.45 | 86.29 |

0* and others see in the Table 2

Consumption of river waters: Selenga – 775-2140, Barguzin – 86-175, Upper Angara – 67-128, Tyia – 92-112 liters / second.

The concentrations of zinc and copper in the waters of the Selenga, Kichera, Barguzin, Tyia, Turka do not exceed 0.02 mg/dm$^3$. The concentration of ferrumin the large rivers of the eastern coast of Lake Baikal exceeds background concentrations and equals 0.17-0.37 mg/dm$^3$. High local concentrations of Mn, Pb, Ni, Co, Cr, Sr, Zn, Cu (more than 1500, 60, 80, 20, 150, 350, 100, 130 mg/kg respectively) have been identified in the soil of the shoreline zone of Lake Baikal which exceed the MPC and APC [21, 22] values near the populated areas of Listvyanka, Baikalsk, BolshoyeGoloustnoye, Sarma, Kabansk, Turka, Ust-Barguzin, Severobaykalsk and Nizhneangarsk [23]. Ba, Sr, V and petroleum

IOP Conf. Series: Earth and Environmental Science **190** (2018) 012024 doi:10.1088/1755-1315/190/1/012024

products show maximum concentrations of chemical elements in soil in relation to the background (970, 507, 100 and 242 mg/kg).

**Table 4.** Content of chemical elements and substances of water of the Kichera, Turka, Utulik, Sarma and Anga Rivers.

| Compo-nent | Kichera | | | Turka | | | Utulik | | | Sarma and Anga | | |
|---|---|---|---|---|---|---|---|---|---|---|---|---|
| | max | min | aver. | max | min | aver. | max | min | aver. | max | min | aver. |
| pH | 7.99 | 7.96 | 7.97 | 8.30 | 7.10 | 7.83 | 7.71 | 7.00 | 7.50 | 8.28 | 7.81 | 7.98 |
| F | 0.181 | 0.169 | 0.175 | 0.195 | 0.078 | 0.141 | 0.158 | 0.060 | 0.109 | 0.191 | 0.070 | 0.112 |
| $HCO_3$ | 34.11 | 31.09 | 32.60 | 40.12 | 21.02 | 30.71 | 55.32 | 55.01 | 55.17 | 82.02 | 36.50 | 60.32 |
| Cl | 0.41 | 0.20 | 0.20 | 0.31 | 0.18 | 0.24 | 2.50 | 0.18 | 0.21 | 2.50 | 0.20 | 0.67 |
| $SO_4$ | 4.10 | 0.70 | 1.10 | 8.50 | 0.10 | 1.70 | 22.00 | 0.10 | 4.01 | 20.10 | 0.10 | 5.23 |
| $NO_2$ | 0.022 | 0.022 | 0.022 | 0.030 | 0.008 | 0.019 | 0.650 | 0.008 | 0.029 | 0.050 | 0.004 | 0.020 |
| $NO_3$ | 0.01 | 0.01 | 0.01 | 0.50 | 0.01 | 0.11 | 1.15 | 0.85 | 1.00 | 11.60 | 0.01 | 2.67 |
| $PO_4$ | 0.023 | 0.005 | 0.014 | 0.021 | 0.001 | 0.004 | 0.002 | 0.001 | 0.002 | 0.060 | 0.002 | 0.028 |
| Ca | 15.99 | 10.41 | 13.20 | 15.10 | 9.18 | 12.41 | 31.78 | 13.60 | 17.69 | 37.63 | 16.80 | 28.24 |
| Mg | 3.31 | 2.15 | 2.73 | 3.26 | 1.10 | 2.43 | 7.69 | 4.95 | 3.82 | 8.55 | 3.41 | 6.34 |
| K | 0.90 | 0.47 | 0.69 | 3.10 | 0.73 | 1.43 | 5.03 | 0.98 | 2.10 | 1.89 | 0.31 | 0.75 |
| Na | 3.15 | 1.98 | 2.57 | 8.34 | 1.95 | 3.86 | 5.23 | 4.49 | 4.86 | 3.36 | 0.67 | 1.73 |
| $NH_4$ | 0.550 | 0.020 | 0.119 | 0.300 | 0.002 | 0.055 | 0.025 | 0.015 | 0.020 | 0.036 | 0.005 | 0.019 |
| Mo | 0.005 | 0.001 | 0.003 | 0.014 | 0.001 | 0.004 | 0.009 | 0.001 | 0.005 | 0.009 | 0.001 | 0.005 |
| Mn | 0.002 | 0.001 | 0.002 | 0.054 | 0.001 | 0.009 | 0.010 | 0.001 | 0.007 | 0.010 | 0.001 | 0.007 |
| Ba | 0.020 | 0.010 | 0.013 | 0.013 | 0.006 | 0.009 | 0.019 | 0.009 | 0.016 | 0.019 | 0.009 | 0.016 |
| Al | 0.021 | 0.018 | 0.020 | 0.038 | 0.001 | 0.013 | 0.057 | 0.001 | 0.030 | 0.057 | 0.001 | 0.030 |
| Pb | 0.040 | 0* | 0.002 | 0.013 | 0.001 | 0.003 | 0.011 | 0.001 | 0.002 | 0.011 | 0.001 | 0.002 |
| Ni | 0.002 | 0.001 | 0.002 | 0.005 | 0* | 0.001 | 0.001 | 0.001 | 0.001 | 0.001 | 0.001 | 0.001 |
| Cu | 0.006 | 0.002 | 0.004 | 0.005 | 0.001 | 0.002 | 0.001 | 0.001 | 0.001 | 0.001 | 0.001 | 0.001 |
| Be | 0.002 | 0* | 0.001 | 0.001 | 0* | 0* | 0* | 0* | 0* | 0* | 0* | 0* |
| V | 0.007 | 0* | 0.004 | 0.003 | 0* | 0.001 | 0.004 | 0.001 | 0.002 | 0.004 | 0.001 | 0.002 |
| Cr | 0.002 | 0.001 | 0.001 | 0.002 | 0* | 0.001 | 0* | 0* | 0* | 0* | 0* | 0* |
| Fe | 0.040 | 0.013 | 0.024 | 0.215 | 0* | 0.068 | 0.080 | 0* | 0.028 | 0.080 | 0* | 0.028 |
| Si | 4.150 | 0.790 | 2.016 | 6.984 | 0.015 | 3.472 | 4.610 | 2.537 | 3.769 | 4.610 | 2.536 | 3.769 |
| Zn | 0.002 | 0.001 | 0.002 | 0.004 | 0* | 0.001 | 0.012 | 0.002 | 0.009 | 0.012 | 0.002 | 0.009 |
| Sr | 0.130 | 0.055 | 0.081 | 0.101 | 0.050 | 0.074 | 0.127 | 0.058 | 0.097 | 0.127 | 0.058 | 0.097 |
| Ti | 0.005 | 0.001 | 0.003 | 0.010 | 0* | 0.002 | 0.006 | 0.001 | 0.003 | 0.006 | 0.001 | 0.003 |
| Co | 0.004 | 0* | 0.002 | 0.002 | 0.001 | 0.001 | 0.001 | 0* | 0* | 0.001 | 0* | 0* |
| Cd | 0.002 | 0* | 0.001 | 0.001 | 0* | 0.001 | 0.001 | 0* | 0* | 0.001 | 0* | 0* |
| Hg | 0.01 | 0.01 | 0.01 | 0.02 | 0.01 | 0.01 | 0.01 | 0.01 | 0.01 | 0.01 | 0.01 | 0.01 |
| Petrol. | 0.009 | 0.007 | 0.008 | 0.008 | 0* | 0.005 | 0.012 | 0* | 0.006 | 0.016 | 0.003 | 0.009 |
| Miner. | 0.049 | 0.047 | 0.048 | 0.067 | 0.047 | 0.059 | 0.130 | 0.041 | 0.086 | 0.080 | 0.035 | 0.050 |
| ∑*** | 59.84 | 48.33 | 54.09 | 63.99 | 40.85 | 53.81 | 118.90 | 109.42 | 114.16 | 129.62 | 96.53 | 109.00 |

0* and others see in the Table 2

Consumption of river waters: Kichera – 167-186, Turka – 39-151, Utulik – 19-56, Sarma – 990-1146, Anga – 4687-5690 liters / second.

However, these figures do not exceed sanitation and hygiene standards. High concentrations of organic substances, alkaline and faintly alkaline reactions of the environment contribute to the accumulation of heavy metals in the soil of the shoreline zone, i.e. serve as a deposit environment, a "geochemical barrier" to the input of pollutants to Lake Baikal. However, it should be said that the soils formed on carbonate rocks often have high natural concentrations of macro- and microelements like those in the Near-Olkhon region [24].

The landscapes of the estuary rivers serve as geochemical barriers for toxic substances and accumulate pollutants. The pollution of landscape components leads to the self-purification of the rivers.

However, it is conceivable that, under favourable conditions such as oversaturation or acidification of the en environment, the toxic substances might penetrate the water bodies.

## 4. Conclusions

The examined material points to the significant impact of the manufacturing industries on the environment of Lake Baikal on both regional and local levels. According to the results of the snow survey and chemical analyses, air pollution in the water area of Lake Baikal has been detected near the coastal populated areas of Listvyanka, Kultuk, Baikalsk, Slyudyanka, Ust-Barguzin, Severobaikalsk, Nizhneangarsk and in the estuary of the Selenga River. However, the backgrounds of most of the studied chemical elements and substances in the Baikal region are low, hundreds and thousands of times below the maximum permissible concentrations. Some components are in deficit. Due to the accumulation of pollutants in the soil geochemical barriers and alluvial deposits as well as its dilution, the water of Lake Baikal conforms to sanitary standards. However, the lake has several local areas with elevated concentrations of toxic elements and substances in the coastal waters of Lake Baikal and the soils near the populated areas. High concentrations of pollutants have been identified in the rivers flowing into the lake.

## Acknowledgements

The research has been conducted under the grant of the Russian Foundation for Basic Research № 16-05-00286, federal target program "Protection of Lake Baikal and Social and Economic Development of the Baikal natural area in 2012-2020", and research practice project № 0347-2016-002 and 0347-2016-003

## References

[1] Groundwater Exposed: Editorial 2016 *Nat. Geosci.* **9**(2) 85

[2] Grachev M A 2002 *On the Present State of the Ecological System of Lake Baikal* (Novosibirsk: SO RAN) p 156

Grachev M A 2013 *Lake Baikal Environmental Protection Strategy and Lake Baikal Conservation Law* (Moscow: Nauka) p 36

[3] *GOST 2874-82 1983 Drinking Water. Hygienic Requirements and Quality Control* (Moscow: Standards Publishing House)

[4] Belogolovov V F 1989 *Geochemical Atlas of Ulan-Ude* (Ulan-Ude: Buryatia publishing house) p 52

[5] Vetrov V A and Kuznetsova A I 1997 *Microelements in the Natural Environments of the Lake Baikal Area* (Novosibirsk: Science) p 237

[6] Koval' G P and Belogolova G A 1998 *Anthropogenic Transformation of Natural Geochemical Distributions in Cisbaikalia in Global Changes in Natural Environment* (Novosibirsk: SO RAN) p 248

Koval P V et al 2000 Geochemistry of the Baikal Lakeside environment *Geology and Geophysics* **41**(4) 571-7

Koval P V, Belogolova G A, Burenkov E K and Pampura V D 1993 The Baikal poligon: international and national projects of geochemical mapping and monitoring of the environment *Geology and Geophysics* **34**(10-11) 207-20

[7] Koroleva G P, Gorshkov A G, Vinogradov T P, Butakov E V, Marinaite I I and Khodzher T V 1998 Investigation of the Snow Cover Pollution as Deposit Environment *Chemistry in the Interests of Sustainable Development* **4** 227-37

IOP Conf. Series: Earth and Environmental Science **190** (2018) 012024 doi:10.1088/1755-1315/190/1/012024

[8] Mikheeva A S 2012 *Atmospheric Pollution. Anthropogenic Transformation of Natural Systems and Socioeconomic Consequences in the Basin of the Selenga River* (Ulan-Ude: SB RAS Institute of Environmental Management) pp 124-33
[9] Nechaeva E G and Makarov S A 1996 Snow Cover as the Object of Regional Monitoring of the Living Environment *Geography and Natural Resources* **2** 43–8
[10] Khodzher T V 2005 *Investigation Into Atmospheric Precipitation and Its Influence on Ecosystems of the Baikal Natural Territory Extended* (Moscow: Institute of Geography of the Russian Academy of Sciences) p 305
Khodzher T V and Sorokovikova L M 2007 Evaluation of the Input of Soluble Substances from the Atmosphere to Lake Baikal through the River Flow *Geography and Natural Resources* **3** 185-91
[11] Belozertseva I A 2003 Chemical element composition of the snow cover in the influence zone of the Irkutsk Aluminum Production Plant *Geochemistry International* **41**(6) 614–8
Belozertseva I A, Vorobjeva I B, Vlasova N V, Janchuk M S and Lopatina D N 2017 Chemical Composition of Snow in the Water Area of Lake Baikal and on the Adjacent Territory *Geography and Natural Resources* **1** 68-77
Vorob'eva I B, Naprasnikova E V and Vlasova N V 2010 Study of Hydrocryogenic Components on the Southwest Coast of Baikal (Ecological and Geochemical Aspect) *Led i Sneg* **2**(110) 56–60
[12] Kiselev V Y 1997 Maps of Snow Cover Pollution in Irkutsk Oblast and on Southern Baikal: Inset *Volna* **2**(11) 30
[13] Lomonosov I S, Gapon A Z and Arsentyeva L G 1993 *The Irkutsk Industrial Area: The Cities of Irkutsk, Angarsk and Shelekhov, in Ecogeochemistry of the Cities of East Siberia* (Yakutsk: IMZ SO RAN) pp 25–37
[14] Kuz'min V A 1999 On Chemical Composition of Raised Peat Bogs and Snow of Southern Cisbaikalia (In Connection With the Environmental Pollution Problem) *Geography and Natural Resources* **3** 59–65
[15] Snytko V A and Afonina T E 1993 Technogenic Flows of Hydrocarbon Compounds in Geosystems of the Lake Baikal Drainage Basin *Geography and Natural Resources* **2** 68–72
[16] Tulokhonov A K 1996 *Baikal Region: Problems of Sustainable Development* (Novosibirsk: Science) p 208
[17] Plyusnin V M and Vladimirov I N 2013 *Territorial Planning for the Central Ecological Zone of the Baikal Natural Territory* (Novosibirsk: Geo) p 407
[18] Obolkin V A, Netsvetaeva O G, Golobokova L G, Potyomkin V G, Zimkin E A, Fillipova U G and Khodzher T V 2013 Results of the Long-term Research of Acid Depositions in South Baikal *Geography and Natural Resources* **2** 66-73
[19] Chebanenko B B and Maisuk E P 2002 *Baikal Region: Sustainability Limit* (Novosibirsk: Science) p 160
[20] Vorobyeva I B, Belozertseva I A, Vlasova N V and Janchuk M S 2016 Current State of the Stream Flows in the Estuary Areas of Middle and Southern Baikal *Achievements of Today's Natural Science* **12** 89-93
Vorobyeva I B, Belozertseva I A, Vlasova N V and Janchuk M S 2018 Current State of the Stream Flows in the Estuary Areas of the Eastern Coast of Lake Baikal *Achievements of Today's Natural Science* **1** 86-92
[21] *GN 2.1.5.1315-03 Maximum Permissible Concentration Concentrations (MPC) of Chemicals in the Water of Water Bodies for Household and Recreational Water Use Purposes* 2017 (Moscow: Ministry of Health of the Russian Federation)
[22] *List of fishery standards: the Maximum Concentration Concentrations (MPC) and the Approximate Permissible Concentrations (APC) of harmful substances for water of the water objects having fishery value* 1999 (Moscow: VNIRO All-Russian Research Institute of fishery and oceanography) p 304

IOP Conf. Series: Earth and Environmental Science **190** (2018) 012024 doi:10.1088/1755-1315/190/1/012024

[23] Belozertseva I A, Sorokovoi A A, Beshentsev A H, Pakhakhinova Z Z, Butkhishig O and Oyunchimeg T 2015 *Degradation and Pollution of Soils. Ecological Atlas of the Lake Baikal's basin* (Irkutsk: V.B. Sochava Institute of Geography SB RAS) pp 96-8
[24] Grebenshikova V I, Lustenberg E E, Kitayev N A and Lomonosov I S 2008 *Geochemistry of the Baikal Environment (Baikal geoecological test site)* ed M I Kuzmin (Novosibirsk: Geo) p 234

IOP Conf. Series: Earth and Environmental Science **190** (2018) 012025 doi:10.1088/1755-1315/190/1/012025

# Methodological framework for assessment of anthropogenic impact on the biodiversity of soil-biotic communities of taiga and steppe ecosystems in the Baikal region

**E P Bessolitsyna**

Sochava Institute of Geography SB RAS, Irkutsk, Russia

E-mail: bessol@irigs.irk.ru

**Abstract.** Landscape-ecological analysis of soil invertebrate's communities and their spatial distribution in the Baikal Region was carried out on the local (biogeocenotic), topological (facies) and regional levels. The paper considers Methodological Framework for Assessment of transformation of communities' structure and biodiversity under the influence of natural: phytocenotic (peculiarities of vegetation), edaphical and climatic (hydrothermal regime of the soil) conditions and anthropogenic factors: the removal of the wood as a result of deforestation, fires, irrational use of pastures, technogenic pollution, recreation and others. Peculiarities of structures of invertebrate's communities and their changes due to the climate change and anthropogenic impacts can serve as one of diagnostic criteria of soil conditions and can be used for estimation of an extent of landscape's transformation and for monitoring. In the spectrum of states of the taxonomic diversity of landscape units, three main categories of situations are distinguished: critical, conflict and relatively satisfactory. The main trend of changes in taxonomic diversity of invertebrate's communities is a decrease in the species number in the gradient of an increase of climate aridity, and strengthening of the hypothermal character and anthropogenic pressure.

## 1. Introduction

The problem concerning the study and conservation of biological diversity, the relevance of which was marked by the United Nations Conference on Environment and Development, has one of the central positions in relation to global environmental changes and provision of environmental safety of humans and other living beings.

On the one hand, biological diversity conservation is closely associated with the activity on protection of natural resources and environment, carried out in the system of nature management. On the other hand, it is an independent, specific activity aimed at reducing the impact of the nature management process itself on the state of ecological systems.

Biodiversity as an object for conservation is not only a species composition of organisms with their quantitative parameters, but also a qualitative system of interrelationships between these organisms, which provides for the functional integrity of ecological systems. An effective strategy can be developed only on the basis of specific information about the state of biodiversity of individual regions, causes and trends of its changes, as well as about the distribution of objects within the territory.

IOP Conf. Series: Earth and Environmental Science **190** (2018) 012025 doi:10.1088/1755-1315/190/1/012025

Within the framework of the Convention, the policy of biodiversity conservation was described, a key position of which is to preserve not only the diversity of species itself, but also the natural habitat, where they were formed and acquired their own distinctive features.

## 2. Models and methods

The source of the actual material is the long-term studies of soil invertebrates population on the transect of field stations in the V.B. Sochava Institute of Geography SB RAS, located in the south of the Krasnoyarsk krai. A comparative analysis of the qualitative and quantitative characteristics of zoocenoses was carried out on a wide information basis, from the southern taiga of the Central Siberian plateau to the true steppes of the Minusinskaya depression. In the Baikal region, the study was carried out in four key areas, covering (in the landscape-zonal range) the main regional spectrum of geosystems: from the mountain taiga of the Leno-Angarskoe plateau to the forest-steppe landscapes of the Irkutsko-Cheremkhovskaya plain. This territory is characterized by a multicomponent phytocenosis, mosaicism and a wide variety of derivative communities that arise as a result of human activities.

The collection and processing of material were carried out in a uniform manner using both traditional and modern approaches and methods recommended for ecology-faunistic, soil-zoological, biogeocenological and landscape-ecological studies [1]. To determine the number and biomass of the inhabitants of soil and litter, 6-8 samples with a depth of 25-40 cm (depending on the limiting occurrence of invertebrates) were taken in each area in a checkerboard pattern on a 25x25 cm with a monolithic drill.

## 3. Results and Discussion

High biodiversity of the Baikal region is due to its geographical location and characteristics of the formation of the rift zone, which contributed to the formation of a unique landscape structure. The Baikal Natural Territory is one of the major physico-geographical "nodes" of the Eurasian continent. Here, the imposition of areas of representatives from different biogeographical zones is observed, which leads to a complication of the situation from the standpoint of the distribution of species and faunal complexes, and highlights the region against the neighboring territories. Increased biodiversity of the Pribaikalie, where, along with the taiga forests, there are steppe, wetland and alpine areas, is due to the passage of the northern and southern, eastern and western boundaries of the areas of many species of animals and plants. The peculiarity of the landscape-ecological conditions creates preconditions for local endemism and formation of unique natural sites with a large number of rare species.

The main reason for the impoverishment of communities and species extinction is increasing human impact on the environment. The main factors of anthropogenic disturbance of natural complexes are the impact of industrial enterprises, elimination of tree canopy as a result of felling, fires, overgrazing, plowing of meadows and copses, reclamation, unnormalized use of fertilizers and pesticides, recreational load, and many others, as well as collecting and intense gathering of rare and the most beautiful species. All this leads to the conversion, reduction, and sometimes complete degradation of habitats of animals, and even to the destruction of the least sustainable ecosystems. In most situations, the mechanism of anthropogenic transformations is a change in either hydrothermal regime, or physical and chemical properties of soil, that contribute to corresponding changes of biotic communities.

Impoverishment and trivialization of taxonomic and coenotic diversity of the animal population, and the outbreaks of mass reproduction of insects should be considered as indicators of negative processes in the landscape, i.e. changes in land cover towards its significant xeromorphization and degradation.

As a part of the policy of the environment and natural resources protection, activity on biodiversity conservation is aimed at resource conservation, the mechanism of which should include the following:

- Inventory of biodiversity at the taxonomic and ecosystem levels;

- Identification and assessment of the state of the most valuable, rare and endangered biodiversity objects;
- Classification and systematization of biodiversity objects, and spatial interpretation;
- Predictive assessment of trends of changes.

Biodiversity is in a complex functional relation to the characteristics of an ecosystem, its dimension and stability, and dynamic structure – biota and abiotic environment, representing a whole. An indication of the diversity of species is considered to be the correlation between the number of species and their specific value (numbers, biomass, productivity, frequency of occurrence, etc.) or the ratio of the number of species per unit area [2].

To develop a strategy for the biodiversity conservation of terrestrial ecosystems, bringing together two major biotic units, namely, flora and fauna, an integrated system approach is necessary. One of these approaches can be landscape-ecological or ecosystem one, which according to its content and methodology includes the most important elements of this strategy: from allocation of areas with increased biotic diversity, their inventory, typology and mapping to the development of methods and criteria for assessing the state of biodiversity and monitoring organization.

Landscape-ecological concept of biodiversity conservation involves science-based forms of human activity aimed at the conservation and reproduction of social and ecological functions of a landscape and its biotic potential by limiting the negative impact on the basis of the regulation and standardization of individual forms of economic activity, as well as at the optimization and further development of the system of protected territories.

Key elements in the concept under consideration are:

- identification of the most promising areas in terms of the uniqueness and value of biological and coenotic diversity;
- analysis of the spectrum of natural and anthropogenic factors affecting biotic communities;
- assessment of the current state and trends of changes of biodiversity at the species, coenotic and landscape levels under certain types and forms of human activity;
- determination of the objectives and variants of use of the territory (optimization of existing and potential types of economic activity with a priority for biodiversity conservation);
- development of restrictions for different kinds of impacts on biotic complexes, and environmental and resource-saving measures;
- monitoring of the state of biodiversity and socio-economic and environmental consequences of the anthropogenic impact.

The structure of population, biodiversity and the state of populations of soil invertebrates are the most convenient for biodiagnostics [2, etc.]. The feasibility of using these indicators to assess the state of the communities' diversity of natural and natural-technical systems is determined by the variety of life forms, relatively high abundance, close ties with the soil and sufficiently fast response to changes in environmental conditions. Compared with the inhabitants of mineral soil horizons, gerpetobionts (litter complex) experience direct and indirect effects of anthropogenic loads to a large extent, and can be used to assess the toxicity of the technogenic impact.

The status of individual components of an elementary ecosystem, as well as the degree of change in diversity of biotic communities, can be expressed by different indicators, namely: diversity coefficients, the size of biomass and abundance of individual taxa, their occurrence, indices of similarity of compared objects, and by rating scales, where the deviation of the indication or estimated trait from the background/maximum characteristics is expressed as a percentage or points.

The study into the dynamics of soil mesopopulation under the influence of anthropogenic factors showed that good results for assessing the state of the natural environment are obtained through the use of a comprehensive comparative-geographical approach. When monitoring in the areas of human impact, particular attention should be given to the criteria that characterize the degradation processes of biogeocenoses, namely: in vegetation cover – phytocoenotic rebuilding of communities, and decrease in productivity; and in soil-and-biotic group – restructuring of zoocenoses, destabilization of

IOP Conf. Series: Earth and Environmental Science **190** (2018) 012025 doi:10.1088/1755-1315/190/1/012025

the population level of the background species, reduction of taxonomic diversity, and reduction of biological activity of soil.

Depending on the resistance of animals and their response to the impact of anthropogenic factors, both sensitive (responding positively or negatively) and indifferent taxa that do not have the indication value for a given type of exposure can be distinguished among them.

The following criteria for the degree of transformation of natural complexes and biodiversity state assessment are used as the main ones:

- Changes in species composition and community structure by comparing the taxonomic diversity of natural and disturbed counterparts or according to the gradient of exposure;
- The ratio of dominant life forms and functional-trophic groups of invertebrates;
- Changes in productivity under the influence of environmental factors.

The most common is a five-point scale, where the first category is a deviation of no more than 20% (high diversity), the second one corresponds to the deviation of 21 to 40% (relatively high), the third category is an average with a deviation of 41 to 60% (mean value of diversity), the fourth one is a deviation from 61 to 80% (low diversity), and the fifth category is a deviation in the range of 81 to 100% (very low taxonomic diversity).

For systemization of states and cartographic representation of the obtained data matrix lattices were used, which were constructed taking into account the typological features of elementary ecosystems (and, therefore, biogeocoenoses), the number of taxa, recorded in a landscape unit, their occurrence, and the total abundance and biomass of invertebrates. This approach allows us to estimate the state of taxonomic diversity of a particular biogeocoenose (or landscape unit). Well-ordered spectrum of states is a source of information to create a map of the spatial differentiation of taxonomic diversity.

In the spectrum of states of the taxonomic diversity of landscape units, three main categories of situations are distinguished: critical, conflict and relatively satisfactory (table 1).

The first category includes natural complexes with a high – 4$^{th}$ and 5$^{th}$ – degree of disturbance, and with an almost irreversibly transformed structure, where the state of the environment affects adversely the life of biotic communities. The second category includes reversibly disturbed (3$^{rd}$ degree) natural complexes, where the environmental situation is very dynamic, and the increase in load leads to sharp deterioration of the biota state and to ecological conflicts. Improvement of the situation and recovery of the normal functioning and reproduction of renewable resources are possible by regulating the adverse impacts and by enforcement of environmental protection measures.

**Table 1.** Estimation scale for a degree of transformation of biotic communities

| Scale | | Categories of situations |
|---|---|---|
| in points | in % | |
| 5 | 81-100 | Critical |
| 4 | 61-80 | |
| 3 | 41-60 | Conflict |
| 2 | 21-40 | Relatively satisfactory |
| 1 | <20 | |

A relatively satisfactory situation is characterized by a weak (I and II degree) of disturbance of biotic communities, balanced trophic structure, a certain stability of functioning, and resistance to the effects of anthropogenic factors.

These criteria can be used to identify areas of anthropogenic transformation, i.e. for ecological zoning of territories, involving strengthening of environmental protection regime within the ingredient of exposure from the most stable landscape structures to more sensitive and less stable ones, as well as the control and regulation of the negative impact from the periphery to a protected object.

IOP Conf. Series: Earth and Environmental Science **190** (2018) 012025 doi:10.1088/1755-1315/190/1/012025

An important methodological aspect of the assessment and conservation of biodiversity is an idea about the levels of the spatial dimension of ecosystems. Regularities that take place in the system of general relationships and interdependence within the habitat are not identical in their spatial-and-temporal scales. At the regional level, the structure of communities and quantitative characteristics depend largely on factors of the macro-geographical order, namely: natural zoning, and altitudinal belts, which are manifested in the change of climate, vegetation, soil and topography. In territories small in area, the main factors that determine the structure of biotic communities, their sensitivity to various forms of impact, and the specificity of anthropogenic dynamics, are the location of a biogeocoenosis in the landscape and hydrothermal conditions.

## 4. Conclusion

Changes in biodiversity can be tracked at two levels: at the species level, when there is a reduction in the number of species through some loss of individual species and taxonomic groups, and at the ecosystem level, when under the influence of anthropogenic factors leveling of differences between individual biogeocoenoses (eco- and geosystems) and complete degradation of some of them are observed. The probability of manifestation of adverse effects increases in accordance with the growth of the degree of impact, regarding the ecological parameters of the environment.

Peculiarities of structures of invertebrate's communities and their changes due to the climate change and anthropogenic impacts can serve as one of diagnostic criteria of soil conditions and can be used for estimation of an extent of landscape's transformation and for monitoring.

Landscape-ecological approach ensures the preservation of habitat-forming and regulatory functions of an ecosystem, and increases objectivity of prediction of changes in biodiversity by taking into account the structural-and-dynamic characteristics of natural complexes and related categories of sensitivity, stability, ecological value and uniqueness. Using this approach on a mapping basis provides for typological and spatial certainty of evaluative and predictive definitions, and the possibility to trace the patterns of manifestation of the effects of climate change and various forms of the anthropogenic impact.

## References

[1] Gilyarov M 1975 *Accounting for Large Invertebrates (Mesofauna)* (Moscow: Nauka) pp 12-29
Bjzova J et al 1987 *Quantitative Methods in Soil Zoology* (Moscow: Nauka) p 288 (in Russian)
Bessolitsyna E 2001 *Landscape-Ecological Analysis of the Structure of Soils Zoocenoses in the South of Siberia* (Irkutsk: IG SB RAS) p 166 (in Russian)
Krauklis A 1987 Landschaftsokologische Auswirkungen der anthropogenen Vegetationsdynamik auf die sudliche Taiga in Mittelsibirien *Pet. Georg. Mitt. Jg.* **131** 157-62
Bessolitsyna E, Krauklis A, Kakareka S and Kremer L 1991 *Geosystems of Contact Between Taiga and Steppe: South of Central Siberia* (Novosibirsk: Nauka) p 216 p (in Russian)
Gilyarov M 1965 *Zoological Method for Diagnostics of Soils* (Moscow: Nauka) p 278 (in Russian)
[2] Gere G 1962 Ernahrungsbiologische Untersuchungen an Dipiopoden *Acta Zool. Acad. Sci. Hung. Bd.* **8**(1/2) 25-38
Sohlenius B 1980 Abundance biomass and contribution to energy flow by soil nematodes in terrestrial ecosystems *Oikos* **34**(2) 186–94
Edwards C and Bohlen P 1995 The effects of contaminants on the structure and function of Soil communities *Acta Zool. Fenn.* **1996** 284-89
Bessolitsyna E 2016 *Landscape-Ecological Analysis of Organization of Soil-Biotic Communities in the Geosystems of South Siberia* (Moscow: LAMBERT) p 140 p (in Russian)

IOP Conf. Series: Earth and Environmental Science **190** (2018) 012026 doi:10.1088/1755-1315/190/1/012026

# Geosystem transformations forecast of the Northern Priolkhonye Region

**A Y Bibaeva**

Sochava Institute of Geography SB RAS, Russia, Irkutsk

E-mail: pav_a86@mail.ru

**Abstract**. The current trend of climatic changes in the northern hemisphere is accompanied by an increase in the number of cases of extreme weather events; droughts have become typical for the Baikal region. In the changing conditions of the regional geosystems background impact, the pyrogenic factor acts as a catalyst for the transformation of geosystems. The research is aimed at identifying and mapping the dynamic and structural transformations of the geosystems in the northern Priolkhonie region (the western coast of Lake Baikal), influenced by pyrogenic factor. Spatial analysis of the location of the burnt territories was carried out on the basis of Landsat 8 (OLI) remote sensing in the GIS. Groups of facies in the territory of the northern Priolkhonye region influenced by destructive pyrogenic factor in the period of 2014 – 2015 were designated. Long-term degradation of forest-growing conditions with a long renewal stage, replacement of dark-coniferous forests by light-coniferous forests in a certain part of their range, replacement of larch-pine steppe forests, mainly in the southern slopes, by steppe communities are therefore forecast. In the highland part of the Primorsky Range, the dwarf siberian pine Pinus pumila communities destructed by fire contributes to a wide expansion of stone placers (kurums).

## 1. Introduction

In the last two years, the problem of the preservation of unique geosystems and environmental management within the Central ecological zone of the Baikal Natural territory is being discussed by the Government of the Russian Federation. According to the Federal law № 94-FZ "On protection of Lake Baikal" (The Federal Law "On the Protection of Lake Baikal" dated 01.05.1999 N 94-FZ amended on June 28, 2014), the geosystems of the Baikal region are some of the most important components of the preservation of Lake Baikal as a unique ecological system of the Russian Federation and a world natural heritage site. For this purpose, an expanded system of specially protected natural areas (nature reserves, reserves, national parks) has been developed in the Baikal region.

The current climate warming trend in the Northern hemisphere is accompanied by a number of negative consequences. Atmospheric blocking and disruption of air mass transport in the western direction are detected in certain areas, followed by an increase in the number of extreme weather cases in other areas [1]. There is also an increase in the number of extreme drought cases in the Baikal region, accompanied by rising annual temperatures and a decrease in annual precipitation amounts [2]. In 2014, the water inflow to Lake Baikal was only 67% of the total norm [3], which contributed to a decrease in the lake water level below the established minimum level of 456 m. It has led to a reduction in the water surface area, a decrease in the groundwater level in the lake's watershed area, changes in the microclimates of different areas, etc.

The latter factor aggravates the pyrogenic situation in the region. One of the impacts is a continuous detection of a significant number of areas affected by the forest fires of different intensity. The most devastating consequences affect the unique geosystems of the Olkhon area in the Irkutsk region located entirely in the confines of the Lake Baikal Watershed and its central ecological zone.

Research territory is the Priolkhonie region located the central part of Lake Baikal´s western coast opposite the island of Olkhon. The geosystem structure is determined by the research area located in the Baikal-Dzhugdzhursky physical-geographical region. Climatic conditions are formed mainly under the influence of local physical and geographical factors - orography and lake water mass, - which determine the formation of barrier-shadow, arid-hollow and foothill effects. In accordance with the scheme of fire hazard zoning of The Baikal Natural Territory [4], the Priolkhonie forests refer to the area of potential fire danger. The above-mentioned factors lead to the change in the structural and functional geosystem characteristics of the unique region, loss of their stability, and further radical transformations.

## 2. Models and Methods

Space images of Landsat 8 (OLI) were used to identify the territories subject to the pyrogenic factor and to calculate areas of burnings. The priority was given to the images referring to the autumn season from the end of a fire danger period to the beginning of a snow cover formation (i.e. from the end of September to the beginning of October) and to the images with the cloud cover of less than 20%. If such images were not available, spring season images of the following year in the after-snow cover period (early June) were used.

Forest Resource Management maps with a scale of 1: 25 000 from the Ministry of Forestry of the Irkutsk region and the Pribaikalsky State Natural National Park served as the sources of information about the species composition of burnt forests in the specified period. The information on the geosystems structure of the research area was obtained from the map "Landscapes of the southern part of Eastern Siberia" (1977). The overlay of the burning's contours on the species composition map and on the landscape map was created in the Geoinformation System Quantum GIS.

When recognizing the burnt areas with the help of multizone images, their color tone changes depending on the spectral channel. In this paper, a color RGB image is obtained by the synthesis of the spectral channels 7-6-7 of the multi-zone Landsat 8 OLI image. Using the combination of channels, the visual noise associated with the smoggy atmosphere is isolated, water objects are determined by black color, primary forest vegetation by green color tones, steppe and previously burnt areas by white and light pink color tones respectively, recently burnt areas are identified as heavy purple color tones from bright pink to dark purple.

The identification of burnt areas was carried out in the GIS under a supervised classification method. The maximum likelihood method was used as a classification algorithm. Three class categories of groups of pixels (reference areas) were determined, each relating to water surface, primary vegetation and burnt areas respectively.

Forecast of the Priolkhonie geosystems transformation after the pyrogenic effects was carried out using published materials referring to the research of dynamic changes in the geosystem components: biogeocenoses succession series (A.V. Belov, V.N. Molozhnikov, N.S. Gamova [5], etc.), structural changes of soil cover [6], evolutionary development of geosystems [7], and using the map "Vegetation of the South of Eastern Siberia" edited by A.V. Belov (1971), forest inventory maps from the Ministry of Forestry of the Irkutsk Region and the Pribaikalsky State National Park.

## 3. Results and Discussion

According to the geoinformation analysis of remote sensing data [8], the total burnt areas equaled: in 2013 - 0.29 square kilometers (29 ha), in 2014 - 24.7 square kilometers (2 470 ha), in 2015 - more than 178 square kilometers (17 850 hectares), of which approximately 87 square kilometers (8 665 ha) are the territories of the Pribaikalsky State National Park. In 2016-2017, according to the interpretation of space images Landsat 8 OLI dating to September 24, 2016, and May 22, 2017, no new fire sources were detected in the study area.

The largest burnt areas are confined to the upper zone of the mountains, i.e. geosystems of the dwarf siberian pine (*Pinus pumila*) and dark-coniferous mountain-taiga stone pine (*Pinus sibirica*) forests. Pine, larch and deciduous forests are less affected by the fire. This is due to the spread of the fire from the western macro-slope of the Primorsky ridge to the study area through the watershed. After the fire passes through the area, the geosystem components are subject to various degrees of change. In the case of ground fires, the lower vegetation layers and especially the ground cover are first to be exposed to the destructive impact, and the duration of their restoration period differs depending on a forest type. Thus, in the burnt areas of green moss forest types, mosses and lichens die not only from fire, but from heat exposure of a fire of any intensity [9], and the initial formation of ground cover occurs after 2-3 or more years [10].

The pyrogenic factor causes dynamic transformations of the natural complexes associated with the changes in all the components. In this case, the regenerative dynamics of the geosystems is determined by the processes of self-organization in a changing climate, and for this reason it may also develop in the direction of a radical transformation of geosystems with the change of the invariant (figure 1, table 1).

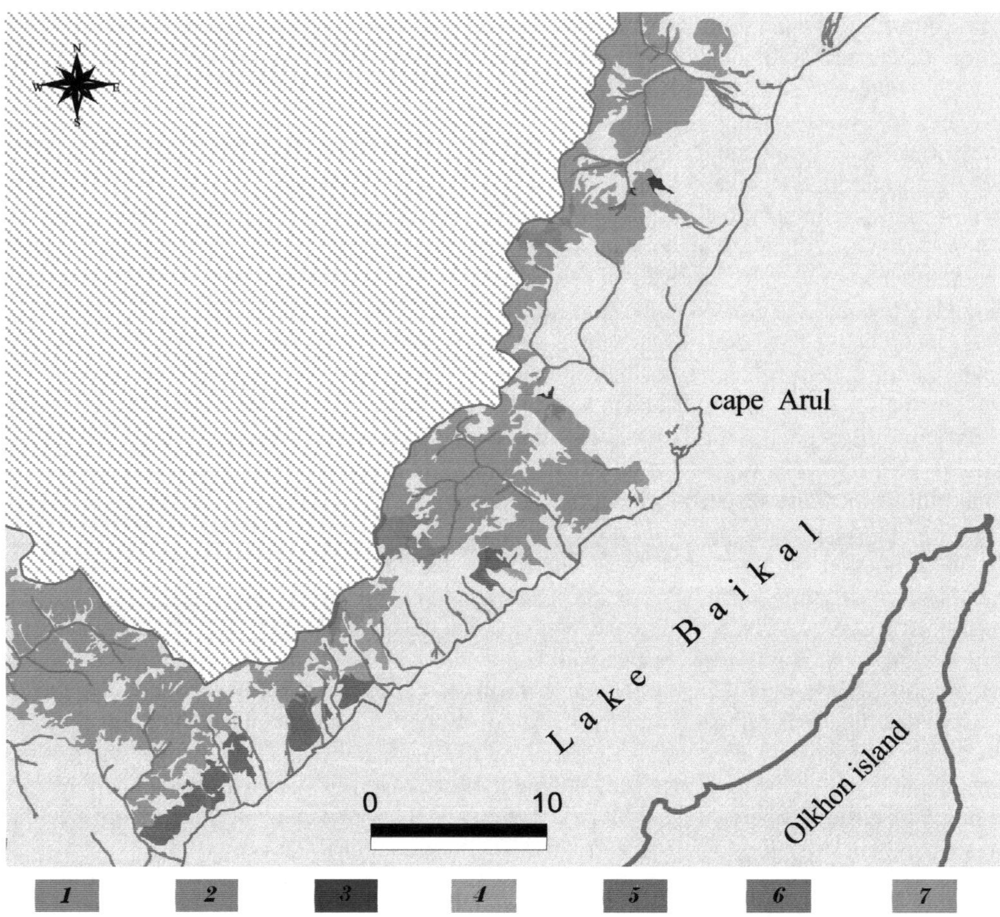


**Figure 1.** Forecasting of the Priolkhonye geosystems transformation after pyrogenic impact in 2014-2015: 1-6 –transformation of geosystems structure (see table); 7 - recovery dynamics of geosystems.

**Table 1.** Forecasting of the Priolkhonye geosystems transformation after pyrogenic impact.

| № | Geosystems subject to the influence of the pyrogenic factor | Eventual geosystems |
|---|---|---|
| 1 | Moss-lichen area in combination with dwarf siberian pine (*Pinus pumila*) in the area of watersheds and gentle slopes | Slow restoration of dwarf Siberian pine (*Pinus pumila*) biogeocenoses (approximately 80-100 years, according to V.N.Molozhnikov [11]). Restoration occurring through *Alnus fruticosa* and *Betula middendorfii*. Substitution by rock fragments is possible a certain part of the area. |
| 2 | Siberian pine *(Pinus sibirica)* shrubby rhododendron aureum green moss forests on slopes | Restoration occurs through Siberian pine, but in a certain part of the area, it may occur through larch and pine forests. |
| 3 | Larch (*Larix sibirica*)-Siberian pine *(Pinus sibirica)* bilberry-small grass green moss forests on slopes | Restoration occurs through the pine-larch blueberry-green moss forests. In the conditions of intense aridity, substitution by larch and pine forests might occur in a certain part of the area. |
| 4 | Larch (*Larix sibirica*)- pine (*Pinus sylvestris*) *rhododendron daurica* steppe forests | Long-term recovery through steppe communities or development of small-grain-grassy lithophilic steppes in a certain part of the area. |
| 5 | Pine-larch rare-growth forests in combination with wormwood low-grass lithophilic and low-bush grass steppes on steep slopes | Long-term recovery through birch communities or substitution by small-grain-grassy lithophilic steppes. |
| 6 | The Siberian pine *(Pinus sibirica)*–fir (*Abies)* bilberry-green-moss forest on low-lying locations | Restoration of the Siberian pine forests of the green-moss group through the succession of small-leaved species (poplar, aspen). |

After passing fires, groups of facies of moss and lichen in combination with dwarf Siberian pine *Pinus pumila* on the watersheds and gentle slopes located in the zone of active frost weathering are replaced by rock fragments in the parts of the study area. Destruction of vegetation cover, mainly dwarf Siberian pine (*Pinus pumila*) which suppresses water erosion, leads to the activation of dwarf soil loss processes and fine earth removal. On the flat surfaces, slow restoration of dwarf Siberian pine (*Pinus pumila*) biogeocenoses (approximately 80-100 years, according to V.N. Molozhnikov [11]) takes place. In the areas of permafrost soils, water-logging processes that impede the renewal of the dwarf Siberian pine (*Pinus pumila*) develop due to an increase in the solar radiation flux to the underlying surface.

The geosystems of dark-coniferous forests are characterized by the shrinking and death of the forest stand. The succession restoration stages of the climax community often occur with a change of tree species.

Fire influence on the facies groups of the Siberian pine (*Pinus sibirica*)–fir (*Abies*) bilberry-green moss forest is characterized by the fir tree death, destruction of the geosystem structures, and formation of the Siberian pine forests of the green-moss group in their areas through the succession of deciduous species (poplar, aspen).

A significant feature of the Siberian pine (*Pinus sibirica*) forests restoration is its close connection with the nutcracker activity. The bird hides the seeds only under the moss cover. For this reason, the restoration of the Siberian pine (*Pinus sibirica*) forest begins only after the moss cover restoration [12]. Restoration of facies groups of Siberian pine (*Pinus sibirica*) shrubby rhododendron aureum green moss forests on slopes occurs without the change of edificatory. However, in the conditions of increasing climate aridity in the region, the Siberian pine (*Pinus sibirica*) forests located the lower border of the distribution range along the southern slopes are replaced by groups of facies of larch and pine forests (see Table). The forest stands thinning leads to the formation of windfall trees that were not affected by the fire due to the "Gornaya" wind. The activation of exogenous processes associated with the dwarf soil losses occurs.

Restoration of the structure of facies groups of larch (*Larix sibirica*)- Siberian pine (*Pinus sibirica*) bilberry-small grass green moss forests on slopes passes through the stage of the formation of the pine-larch green moss forests. In the changing conditions of the background impact of the geosystems at the regional level, this intermediate community in the part of the study area can stay unchanged for a long time. In the upper parts of the slopes along the river valleys, permafrost melting leads to the intensification of bogging processes, degradation of forest-growing conditions, and development of erosion processes develop on the slopes [13].

Geosystems of larch (*Larix sibirica*)- pine (*Pinus sylvestris*) rhododendron daurica steppe forests on the gentle slopes of southern exposures are restored through steppe communities. Transformation of facies groups of pine-larch rare-growth forests in combination with wormwood low-grass lithophilic and low-bush grass steppes on steep slopes in the coastal part of Lake Baikal affected by medium and high-intensity fire is accompanied by the shrinking of the forest stand and the development of small-grain-grassy lithophilic steppes in their area.

## 4. Conclusion

Fire is one of the factors contributing to a sharp change in the dynamics of geosystems. The study area is characterized by arid and semi-arid climatic conditions due to the occurrence of a complex of landscape-forming effects: barrier-shadow, arid-hollow and foothill. In the process of the regional changes in hydrothermal conditions, the dynamic transformations of the Priolkhonie geosystem structures are enhanced by the pyrogenic factor. In the context of the analysis carried out, it should be noted that the area covered by the fires increased, and so did the destructive impact of forest fires. This may be caused by a number of factors, including increasing aridity of the Priolkhonie climate in the recent years, growing popularity of spontaneous tourism in previously inaccessible areas, weak forest protection system that contributes to the spread of the fire to larger areas. The increase of the area of burned moist dark-coniferous forests compared to that of larch-pine forests might be caused by the transition of the fire that has gained strength and significant temperature from the territory of the Kachugsky district through the Primorsky Range under arid and windy weather conditions.

Long-term degradation of forest-growing conditions with a prolonged stage of their restoration, replacement of native dark coniferous forests in some parts of their area by light coniferous forests, replacement of larch-pine steppe forests on the southern slopes by steppe communities, and an increase in the area of rock fragments in mountain-tundra geosystems are therefore forecasted.

## Acknowledgments

The work was done under research project № 16-05-00902 RFBR and research project № VIII.79.2.5 IG SB RAS.

## References

[1] Antokhina O Y, Antokhin P N and Martynova Y V 2016 *Proc. of the International Conference* (Tomsk: Center for Scientific and Technical Information) pp 368-71

[2] Maksyutova E V, Kichigina N V, Voropai N N, Balybina A S and Osipova O P 2012 Tendencies of hydroclimatic changes on the baikal natural territory *Geography and Natural*

*Resources* **33**(4) 304-11

[3] Bychkov I V and Nikitin V M 2015 Water-level regulation of lake Baikal: Problems and possible solutions *Geography and Natural Resources* **36**(3) 215-24

[4] Sofronov M A, Anropov V F and Volokitina A V 1999 Pyrological characteristics of the vegetation of the Lake Baikal basin *Geography and Natural Resources* **2** 52-8

[5] Gamova N S 2014 Post-fire vegetation changes of Central Khamar-Daban (Southern Baikal Region) *Problems of Botany of Southern Siberia and Mongolia* **13** 55-9

[6] Krasnoshchekov Y N 2011 Transformation of Gray-Humus Soils of Pine Forests under the Influence of Fires in Southwestern Lake Baikal Basin *Lesovedenie* **2** 3–12

[7] Bezrukova E V and Letunova P P 2012 Environmental Change in the Middle and Late Holocene Priolkhon Region *Izvestiya of the Irkutsk State University. Series: Geoarchaeology. Ethnology. Anthropology* **1** 91-105

[8] Bibaeva A Y 2016 Analysis of fire-induced influence on the Priolkhonye geosistems by using materials of remote sensing of the Earth *Progress in Modern Natural Science* **12**(2) 347-51

[9] Kovaleva N M, Ivanova G A and Kukavskaya E A 2011 Restoration of the Ground Cover after Surface Fires in Pine Forests of Middle Taiga *Forest Science* **5** 30-5

[10] Sofronov M A and Volokitina A V 2007 *The Method of Pyrological Survey and Description of Forest Areas Passed by Fires* (Krasnoyarsk: V N Sukachev Forest Institute SB RAS) p 71

[11] Molozhnikov V N 1975 *Cedar Stlanik of Mountain Landscapes of the Northern Baikal Region* (Moscow: Nauka) p 203

[12] Popov L V 1982 *Southern Taiga Forests of Central Siberia* (Irkutsk: Irkutsk University Publhing House) p 330

[13] Agafonov B P 1975 Distribution and forecast of physical-geographical processes in the Baikal basin *Dynamics of the Baikal Depression* (Novosibirsk: Science Pub.) pp 59-138

# Geological Information System of environmental and human intervention impact assessment on bodies of water of the Irkutsk region

**I V Bychkov**[1], **A S Gachenko**[2], **A E Hmelnov**[2], **R K Fedorov**[2], **E S Fereferov**[1]

[1]Irkutsk Scientific Center of Siberian Brunch of the Russian Academy of Science, Irkutsk, Russia
[2]Matrosov Institute for System Dynamics and Control Theory of Siberian Branch of Russian Academy of Sciences, Irkutsk, Russia

E-mail: gachenko@icc.ru

**Abstract.** In this paper the authors present a methodology on information support for research on human and environmental impact of Lake Baikal. The geoportal placed on the Internet allows researchers to create a common thematic database (tabular data) and map data (information layers) within the framework of the project, as well as to run Internet services for geoprocessing and complex data analysis problems solving. One of the key features of the engineered geoportal is the ability for users to create the structure of their databases independently (to add, to delete and to modify the tables structure) as well as to configure the run of geoprocessing services. Besides this platform allows you to take on projects on various subjects and with a large number of users simultaneously. They have took on works on 3D modeling of the surface topography of the coastal zone near the village of Listvyanka. Underwater terrain is combined with land relief. As a result catchment basins along which pollutants with groundwater enter the coastal waters of Lake Baikal have been identified. The data on pollution concentration sites is confirmed by practical measurements and coincides with the drainage areas.

## 1. Introduction

The coast of Lake Baikal becomes increasingly attractive place of tourist activity development year by year and remains one of the most relevant places with relation to ecological demands and restrictions as it concerns a unique natural object – Lake Baikal.

A serious impendence of ecological safety of Lake Baikal (one of few natural objects listed in UNESCO immediately according to all accepted criteria for uniqueness) is posed by accompanying economic activity, especially in the absence of engineering constructions [1]. It is explained by the fact that on the coast of Lake Baikal there is a large number of the local sections of mass tourism and rest that are characterized by permanently growth of tourist flow that finally leads to deterioration of the unique ecological system of Lake Baikal status for that reason acceptance of urgent measures for the analysis and assessment of the current state and for the forecast of geosystems dynamics on key sections, that subject to the most powerful anthropogenic loads: Listvyanka, the coast of the Small Sea, the coastal zone in evil Severobaykalsk and Baikalsk, is required. The situation is complicated by the fact that at present no groundwater flow from industrial enterprises, tourism objects, private farms and residential buildings to any surface water body is regulated by any normative documents and, accordingly, there are no uniform methodological approaches to account for the influence of this factor on the ecological system of a surface water body.

Currently the group of authors meet the challenge of a scientific geoinformation system of ecological content creating which is at the initial stage of its formation, although it has a long history of development. Basic and operational geoinformation presentation of materials to ensure the problem solving of environmentally sustainable development of territories, based on new geoinformation approaches and technologies, has become a capstone of scientists' activities [2].

In this paper the authors present a methodology for information support for research on the human impact on the ecology of Lake Baikal [1]. The geoportal (http://atlas.isc.irk.ru/) placed on the Internet allows researchers to create a common under the project thematic database (tabular data) and map data (information layers) as well as to startup Internet services for the purpose of geoprocessing and complex data analysis. One of differential features of the engineered geoportal is the ability for users to create their databases structure [2] by themselves (to add, to delete and to modify tables structure) and also to adjust the geoprocessing services startup. Also, this platform allows you to take on projects on various topics and by a large number of users simultaneously.

## 2. Models and Methods

The subject-matter data of the geoportal being developed (http://atlas.isc.irk.ru/) is presented in the form of a data storage based on the results of the coastal zone of Lake Baikal for 2016-2017 studies containing information in the form of databases (DB) bounded to cartographic layers. Users are able add third-party vector and raster maps. Input and data editing system of the geoportal allows to load any table data in the form of adjustable structures for a particular user. The query subsystem is a powerful dynamically configurable set of filters for any of the database tables. As background maps of the geoportal they connect maps based on the WMS standard from known open GIS: Google Satellite, Yandex Maps, Open Street Map (OSM), 2GIS, Bing.

To present thematic data of research results on a geoportal they created databases that ensure the storage of the following parameters:

- Mineralization of the bottom water on the basis of measurements of Specific Resistance of Soil using electro-tomography methods obtained in the framework of the hydrological survey (using the CTD probe of the Sea Bird Electronics SBE-37 probe).
- Results of vertical electrical sounding.
- The chemical composition of water in rivers, streams, holes in beaches and Lystvenny Bay litters (the basis indexes is the concentration of elements in filtered water samples).
- The content of anthropogenic pollutants in natural environments that are in feedforward together within the framework of landscape-geochemical flows. Quantitative data on a number of pollutants and indicators for each sample is given: microbiological attributes, petroleum derivatives, temperature, suspended substance, mineralization, pH, cations, anions, microelements.
- Sanitary microbiology of the rivers of Listvyanka (the indicators of the base reflect the level of microorganisms that can affect people's health).

The authors created a digital model of the relief of Listvyanka village (figure 1) on the basis of a high resolution (10 meters per pixel) satellite image of WorldDem (Airbus Defense and Space under the common name GEO ElevationServices). The area of the image is 31 650 km$^2$. The global digital model of WorldDEM was created based on the results of interferometric processing of tandem surveys from the TerraSAR-X / TanDEM-X radar satellites). The picture was purchased from Airbus.

For a 3D model construction of the coastline relief they developed the original technology of coupling of the surface-water relief with the underwater relief [2]. The underwater relief was obtained after measurements using echolocation of the bottom relief of the coastline of Listvenichny Bay, field depth measurements were performed up to a 50-meter mark (figure 2).

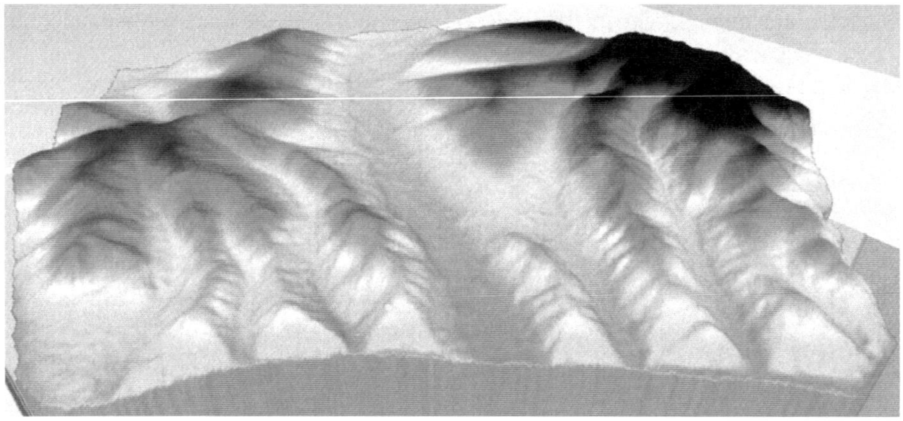

**Figure 1.** Digital model of the relief.

To combine the data on the surface-water and underwater terrain the original algorithm for constructing the Delaunay triangulation [2] with map layer restrictions was tested. While construction of the triangulation the role of every layer is indicated. Every layer can contain information on land relief, standing water level, contours of coastlines, underwater terrain. To work with underwater terrain, information on elevations of water is required. According to the data an auxiliary triangulation is constructed further on this showing information on bench marks is extracted to which the contours of the coastlines are attached and the depths are counted. As a result of these procedures a final agreed combined (surface with underwater) 3D relief model is obtained (figure 2).

**Figure 2.** Combination of surface and underwater relief up to 50 m depth.

In consequence of a high-quality 3D model creation based on current GIS technologies using complex high-precision data on the surface and underwater reliefs, it is possible to evaluate the human impact on the coastal zone of Lake Baikal near Listvyanka village affecting the increase of the ecological crisis.

## 3. Results and Discussion

Analyzing the research of 2015-2017 it was noted that the anthropogenic pollution end of May is increasing in comparison with early May. As an explanation there is a hypothesis that the result of this increase is the growth of the tourist flow and correspondingly the increasing water removal of tourist sites (cafes, hotels) located in the shoreline of lake Baikal. Comparative analysis of geophysical data also showed high reliability of the data obtained by this method on subterranean flows location having subaqual discharge in the reservoir, as well as the stability of the location of underground drains, and in fact the direct dependence of groundwater runoff on the tourist flow.

To confirm the hypothesis it was suggested to analyze the waste water from the objects of the tourist infrastructure using tools of modern GIS. It is known that the groundwater movement conforms with gravity and is carried out in the form of flows through communicating pores or cracks. In this case the groundwater surface hugs in some ways the surface relief and underflow moves from elevated areas (starting from the watershed divide of groundwater) to lower areas (ravines, rivers, lakes, seas) where they are discharged in the form of descending sources (springs) or by a hidden subaqual distributed method (for example, under the waters of river beds, lakes and seas bottom). In view of the surface topography of the village Listvyanka hotels and cafes are located in the river valleys and along the shore of lake Baikal. It is assumed that discharge of waste waters from specific tourist sites will be in the sector of the corresponding watershed divide.

The authors carried out work on the modeling of catch basins in this area (for the catchment basins maps creation freely distributed Quantum GIS was used):

At the first stage you need to get rid of speckles in the data by applying a Gausa filter for smoothing that will help to avoid the formation of small deeps. For this purpose it is necessary to use the Gaussianfilter geo-algorithm from the SAGA GIS collection.

At the second stage they construct catchment basins using the r.watershed tool from the GRASS GIS collection (figure 3).

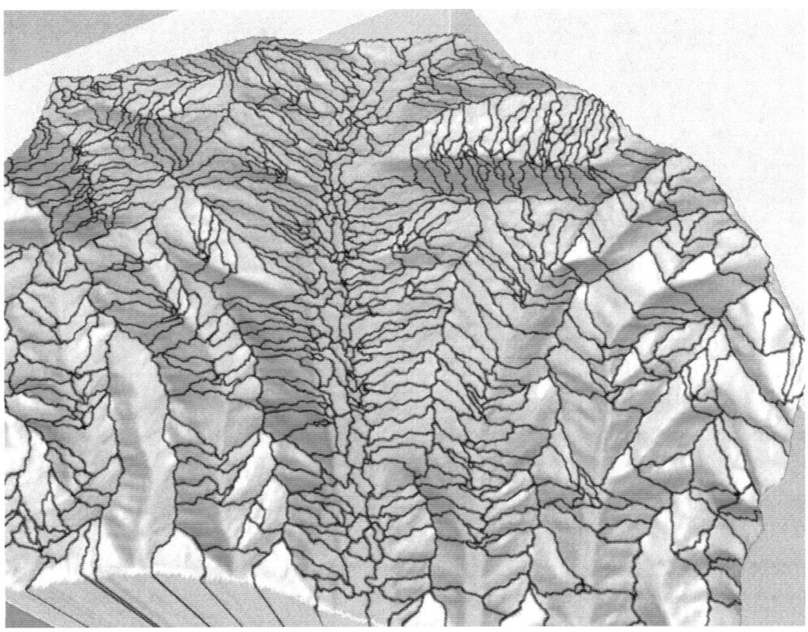

**Figure 3.** The result of catchment basins construction.

At the third stage it is necessary to perform generalization using the v.generalize GRASS GIS feature. But at first you need to transfer the raster into a vector view using the r.to.vect GRASS GIS algorithm. It may also be necessary to combine small catchment basins into larger ones using the QGIS command order: Vector / Geoprocessing / Integration. The result of small basins integration is shown in figure 3.

Further the number of tourist infrastructure objects located in a certain water collection zone was estimated. For this purpose a layer "Hotels and cafes" was obtained based on the data from the parcellary plan and other sources on the number of touristic objects. This layer was put onto the layer of the catchment areas (figure 4).

As a result of the analysis of the obtained data it was discovered that the human pollution indexes in the sub-aquiferous discharge areas of the relevant catchments basins are closely related to the number of tourist infrastructure facilities located in a particular catchment area.

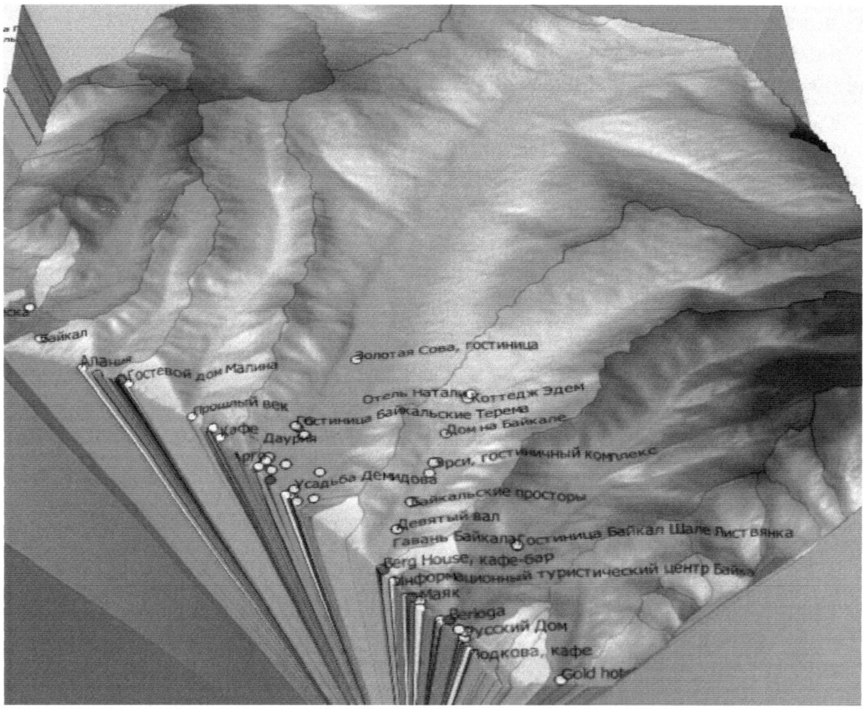


**Figure 4.** Watershed divide with hotels and restaurants.

## 4. Conclusion

As a result of the performed works developed the architecture and created a preproduction model of GIS to support research on the anthropogenic impact on the ecology of the coastal zone of Lake Baikal. The system provides a centralized formation of thematic and cartographic databases of research results for all investigators, monitoring of the ecological situation, a complex analysis of the human impact on the ecology of the coastal zone of Lake Baikal.

A digital elevation model based on a high resolution satellite (10 meters per pixel) WorldDem was created. A fancy technology of coupling between a surface relief with an underwater relief has been developed. Underwater relief was obtained with the help of echolocation survey of the bottom relief of the coastal zone of Listvenichny Bay (depth measurements were carried out up to a 50-meter mark).

The analysis of waste water transfer from the objects of the tourist infrastructure with the help of GIS tools according to the data on relief of drainages and estimation of the number of tourist infrastructure facilities near every basin.

**Acknowledgements**

The presented software is developed using the Shared Computing Center of Integrated Information and Computing Network of Irkutsk Research and Educational Complex (http://net.icc.ru). The study was partially financially supported as part of integration project 4.1: «Scientific substantiation of advanced technologies for organization of integrated monitoring of lake ecological systems».

## References

[1] Grachev M A 2016 Maybe-Bag and Let it Be. Ecological crisis on Baikal: A mystery of the

Century *Science First-Hand* **68** 6–19
Bychkov I V and Nikitin V M 2015 Control of level of Lake Baikal: problems and possible issues *Geography and Environmental Resources* **3** 5–16
Bezrukov L A, Saveliev V A, Nikolskii A F and Podkovalnikov S V 1997 Baikal and hydropower: ecology and economy *Geography and Natural Resources* **4** 158–68
Atutova A A, Pronina N M and Tulohonova A K 1999 *Hydropower and Condition of an Ecosystem of Lake Baikal* (Novosibirsk: publishing house SB RAS) p 280
Itskovich V B, Shigarova A M and Glyzina O 2015 Change in the Hsp70 content in the Biakal endemic sponge Lubomirskia Baicalensis during decolorization and under hypothermia *Aktualnye Problemy Nauki Pribaikalya* 135-8
Vashchenko B 2015 Aliens in Baikal *National Geographic Russia* **146** 6
Kravtsova L S, Izhboldina L A and Khanaev I V 2012 Disturbances of vertical zoning of green algae in the littoral area of the Listvennichnoye Bay, Lake Baikal, as a result of local anthropogenic effect *Doklad RAS* **2** 227-9
Kravtsova L S, Izhboldina L A and Khanaev I V 2014 Nearshore benthic blooms of filamentous green algae in Lake Baikal *J. Great Lakes Res.* **40** 441-8
Timoshkin O A, Samsonov D P and Yamamuro M 2016 Rapid ecological change in the coastal zone of Lake Baikal (East Siberia): Is the site of the world's greatest freshwater biodiversity in danger? *J. Great Lakes Res.* **42** 487-97
[2] Abasov N V, Bolgov M V, Nikitin V M and Osipchuk E N 2017 On regulation of the urovenny mode of lake Baikal *Water Resources* **44** 407-16
Skvortsov A V 2002 *Delaunay Triangulation and its Appliance* (Tomsk: Tomsk University press) p 128

# The use of triangulation processing algorithms for the construction of combined model of the underwater and above-water terrain of the bed of the Bratsk Reservoir

**A S Gachenko[1], A E Hmelnov[2], T A Hmelnova[1], G M Rugnikov[2]**

[1]Irkutsk Scientific Center of Siberian Branch of Russian Academy of Sciences, Irkutsk, Russia
[2]Matrosov Institute for System Dynamics and Control Theory of Siberian Branch of Russian Academy of Sciences, Irkutsk, Russia

E-mail: hmelnov@icc.ru

**Abstract**. The article summarizes the experience gained during processing of the data about the terrain of the Bratsk Reservoir bed. We discuss the available sources of information on the land and underwater terrain and justify the choice of the sources used in the project. To measure the actual depths in the regions of interest an echo-sounder of the consumer grade was applied, and we consider the processing of its measurements in some details. We describe the free, open source and custom software, which we apply for the processing of the data. We consider the use of the several algorithms that we have developed for processing of the terrain points. The main goal of this processing is to obtain the combined terrain model, which summarizes the data from the various information sources: the topographic maps, the pilot charts and the echo sounder measurements. All the algorithms considered build and process the triangulations of irregular sets of points. The algorithms used are: constrained Delaunay triangulation construction; removal of artefacts of triangulation, constructed from the terrain isolines; map morphing to reconcile the terrain models: and triangulation fragment replacement. The resulting combined terrain model contains all the information about the terrain in the area of interest and allows to use the model for the further hydrological calculations.

## 1. Introduction

For several years the Angara-Baikal basin is in the state of severe water shortage. At the same time, the permissible range of Baikal levels is tightly limited by the RF Government decree [1], in spite of which the Baikal level has fallen below the lower limit of the regulated range and forced the Government to issue the new temporary (especially for the period 2018-2020) document [2], which relaxes the restrictions. Due to the water shortage now the Irkutsk HPP is operating at the minimum acceptable level of water discharge, which ensures the minimum permissible level of downstream water. And the only way to control the entire river cascade is to consume the Bratsk Reservoir. As a result, the threat of shutdown of water intakes located around the reservoir emerged in case of a decrease in the water level below the height marks used in their design. In the summer of 2017 an expedition was conducted at the Bratsk Reservoir in order to study the state of all the major water intakes and some other facilities. Our team was responsible for cartographic support and processing of spatial data for the expedition.

To be able to compute the groundwater level for the wells and the water depths near the surface water intakes it is required to obtain the terrain model of the nearby area. In this paper we consider the algorithms that we utilized to process the expedition materials in order to collect the combined terrain models in the vicinity of the water intakes.

To construct the terrain models we use the following sources of information: the terrain isolines from the vector topographic base of scale 1:50000; the depth marks and isobaths, digitized from the Bratsk water reservoir pilot charts; the depths measurements made during the expedition using an echo depth sounder. It was required to reconcile the information from all the data sources to get a realistic result. To improve the quality of the ground terrain model obtained by isolines, the triangulation correction algorithm developed by the authors was applied, which makes it possible to remove the distinctive artefacts of the models resulting from isolines.

The terrain models used were represented by triangulations. All the algorithms considered in this article construct, transform or use triangulations. We utilize our own implementation of triangulation library, which is fast, robust, and can handle large data sets. Some other tasks were performed using the open source QGIS program and the free version of EasyTrace program [3].

## 2. Models and Methods

Let us explain our choice of the sources of information for construction of the combined terrain model, and consider the algorithms, that were used for their processing.

### *2.1. The sources of information about terrain*

The information about ground terrain is relatively easy available in several forms. The SRTM grids exist for the whole area of interest, but the resolution of the SRTM data (3" ≈ 90 m per pixel) is too low to be able to obtain the required precision of water level estimates. The ASTER grids have higher resolution (1" ≈ 30 m per pixel), but the quality of the data in the Irkutsk region is inferior and prohibitive for any practical usage (an interested reader may try to observe the dam of the Irkutsk HPP, which is quite strait in real life, as well as on the SRTM tiles and on the Google Maps` images, but looks like a saw on the ASTER grid). The high-quality (10 m per pixel) elevation model WorldDEM [4] is available commercially, but by the time of the project start the minimum order size (500 $km^2$) and the price per square kilometre would consume a substantial part of the project budget. More importantly, we did not had any practical experience of the WorldDEM data usage by that time to be sure of the actual quality of the data (and the ASTER case shows, that we shouldn`t completely trust the advertising promises).

As a result we settle upon the vector topographic maps, which exist for the whole Bratsk Reservoir area with the scale 1:50000, and for some parts of the region with the scale 1:25000. The vector maps contain a precise enough representation of the strand lines, which allows to use the map morphing technique for aligning the underwater data with the ground terrain (see the section 2.3. ). The main disadvantages of the available topographic maps are high vertical interval (20 m) of all the nonclassified maps and rather old date of survey (circa 2000), but it is still the best available information source about the ground terrain.

In contrast to the ground terrain it is very hard to find any information about the underwater terrain. At the start of the project we had a hope to find some old topographic maps of the Bratsk Reservoir area before its flooding. Then it would be possible to get some information about the underwater terrain, which was on the ground at the time the map was created. Unfortunately, we have recognized later that the water-flooding of the Bratsk Reservoir began in 1961, and the topographic base of the Irkutsk Region was first created in the 1970s, so we were unable to find the maps with the terrain of the reservoir bed before flooding.

As a result, the only a-priory source of information about the underwater terrain that we have found are the pilot charts of the Bratsk Reservoir [5]. The date of survey of the pilot charts is even much

IOP Conf. Series: Earth and Environmental Science **190** (2018) 012028 doi:10.1088/1755-1315/190/1/012028

more old (1974) than that of topographic maps, but that`s all we have. Moreover, the pilot charts are not maps: they a very schematic and cannot be directly combined with the much more accurate contours of the topographic base. So we apply the special technique of map morphing, which we have developed earlier, to be able to use the data.

The actual state of the underwater terrain in the areas of interest was explored during the expedition using a portable echo sounder with its own GNS receiver in conjunction with a professional GNS receiver. The resulting tracks were used for the terrain model update in these areas.

### *2.2. The ground terrain model*

The source topographic maps are represented in the SXF files [6] – the exchange format of the Panorama GIS. The topographic maps of the scale 1:50000 are split into the sheets of the size 10'x15'. The sheets containing the Bratsk Reservoir parts belong to the two neighbor 6-degree zones. For each of the 6-degree zones we have created a temporary map containing all the sheets from the zone (with an object attribute, containing the original sheet name). From the temporary map we have selected the list of all the water objects named "Bratsk Reservoir" to obtain the list of sheets, which contain some part of the Bratsk Reservoir. For each of the zones the sheets from the lists were merged into a single SXF file using the utility SXFUnion, developed earlier by authors, because the Panorama software doesn't provide this kind of operation. The SXF files were opened in the GIS Panorama Mini and their layers of interest (terrain, hydrography, settlements) were exported to SHP files using the Panorama built in converter. The Panorama SHP converter export always converts the coordinates to degrees on the WGS-84 ellipsoid. We have converted them back to the Gauss-Kruger (transverse Mercator) 6-degree zone 18 using the QGIS utility ogr2ogr and then merged using the same utility. As a result we have several layers in SHP files: terrain isolines, point height marks, water polygons and so on for the whole Bratsk Reservoir region.

The layer of water polygons requires further processing: it contains the parts of the Bratsk Reservoir into which it was split by the boundaries of the map sheets. The boundaries between the parts would become the triangulation hard edges (constraints) and spoil the resulting triangulation in its water areas. To join the parts it is not enough to union the polygons: due to the small differences in the coordinates of the border points for the neighbor subobjects from the different sheets the resulting object may still be divided into several parts or have some narrow slots. So we use the following sequence of actions in QGIS to union the parts. At first we compute the "fixed distance buffer" of all the layer objects with the buffer distance of 1m and the dissolve parameter turned on. As a result all the objects will be merged in spite of the small mismatches in the coordinates of the boundary points, but also we will have a side effect: the resulting object will be replaced by its 1m buffer. To almost completely remove the side effect we perform again the operation "fixed distance buffer" but with the buffer distance of -1m and the dissolve parameter turned off. The resulting side effect of this sequence is a minor smoothing of the resulting object border, and it is absolutely acceptable.

### *2.3. The underwater terrain model*

To obtain the vector representation of the pilot charts we scan the paper sheets and vectorize the resulting files using the program EasyTrace 7.99. As a result we have several SHP files for the layers of the vectorized chart sheet. The pilot chart sheets are arbitrary oriented, and we should rotate and shift them to initially align the sheets with the topographic map. To find the correct initial coordinate transform we find some distinctive points, which are present on both maps. We prepare the list of the coordinates of the distinctive point pairs. The list should contain at least two point pairs. We have developed a utility program, which takes the list of coordinate matches, computes the coordinate transform, and performs it for the list of shape files specified in its parameters.

After rotation and shift the match between the topographic map and the pilot chart still won`t be perfect, because the pilot chart is very schematic. Both maps contain a representation of the shore lines and we can compare the representations. The water level of the Bratsk Reservoir object on the map is

402 m. And the water level on the pilot charts is almost the same – 401.65 m. This difference can be neglected – it allows matching the shore lines.

We have developed earlier the technology and software for map morphing, which allows to specify the matches between the line points, and to shift the coordinates of the imprecise map to their correct positions. The field of coordinate shifts is represented here by triangulation. As a result we move the shore line points to the more precise positions. In fact the shifted shore lines themselves are not required at all (because we already have the correct ones), but the coordinate transform obtained allows to smoothly shift all the other pilot chart layers to the positions, that don`t interfere with the topographic map features. For example, the depth marks or isobathic lines will be placed into the water areas, and not on the land. The process of map morphing was discussed in more details in [7].

Only the sheets of the pilot chart, that contain the areas of interest, were processed. Finally the resulting map layers of the same kind (depth marks, isobaths lines) from all the pilot chart sheets were merged into a single layer by ogr2ogr.

### *2.4. The triangulation software*

Although triangulations are widely used in QGIS, there is no special data structure for triangulations in QGIS and its accompanying products (SAGA and GRASS GIS) like that of TIN in the ESRI products. In QGIS using a triangulation in the intermediate algorithm steps we can compute a raster or a polygonal layer of triangles, but can`t store the triangulation itself for a later use.

Our triangulation program *TIN Smith* is based upon the ideas of dynamic hashing from [8] and uses the robust predicates from [9]. It also implements the constrained Delaunay triangulation [10] algorithm and a family of original triangulation flood fill algorithms. In contrast to [10] our implementation of the constrained Delaunay triangulation algorithm is incremental [11] and allows intersection of the hard edges (constraints) [12]. The program uses the data structure called *nodes and triangles* in [8], which is also used in [11]. The algorithms were already applied to very complex practical problems like generalization of the whole map of Irkutsk and have been thoroughly tested during the application. As a result the algorithm works correctly on the very complex data sets, like that with intersections of almost identical hard line segments (which may crash some implementations for example as a result of repeated segment breaking and intersection due to the limited precision of the float coordinate values).

One of the possible data sources for the construction of triangulations in the TIN Smith are the SHP files. The point height may be taken from the Z-coordinates of the shape points or from an attribute field of the map object or from a field of a joined table. When combining several source files they may take different roles: ground terrain, water level marks, underwater terrain.

### *2.5. The combined a-priory terrain model*

Using the layers of isolines, height marks, the Bratsk Reservoir object boundaries from the topographic map, and the depth marks and isobathic lines from the pilot chart we have created the a-priory (before the expedition) terrain model as a triangulation with constraints.

Besides from the low precision of the pilot charts, the main problem with the underwater terrain of the model is that the current water level of the Bratsk Reservoir is far below its levels on the maps and the pilot charts. In particular, the water level of the Bratsk Reservoir object on the map is 402 m, and during the expedition the actual level was slightly above 395 m. The difference is almost 7 m, and all the isobathic lines on the pilot charts have only the two possible depth values: 3 m and 5.5 m. So, all the isobathic lines describe the terrain, which now become the above-water one, and the only source of the a-priory information about the current underwater terrain is the depth marks, which are rather rarely spaced.

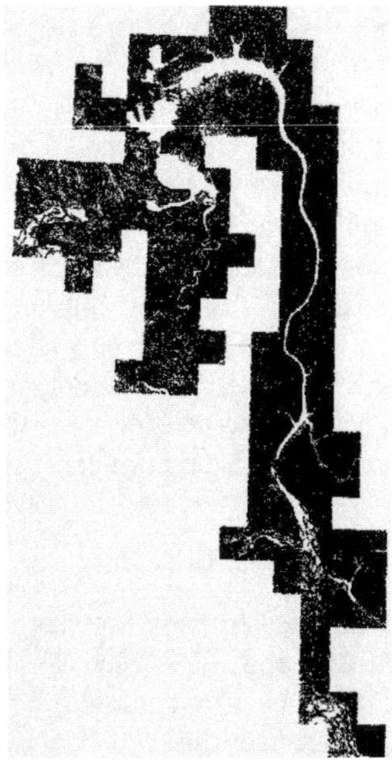

**Figure 1.** The overview of the a-priory terrain model.

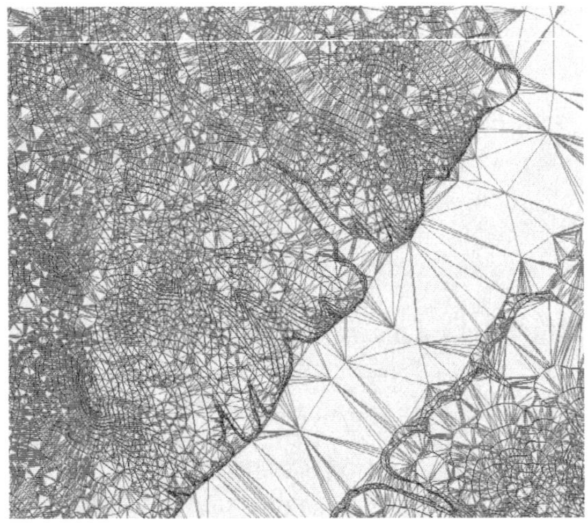

**Figure 2.** A fragment of the a-priory terrain model.

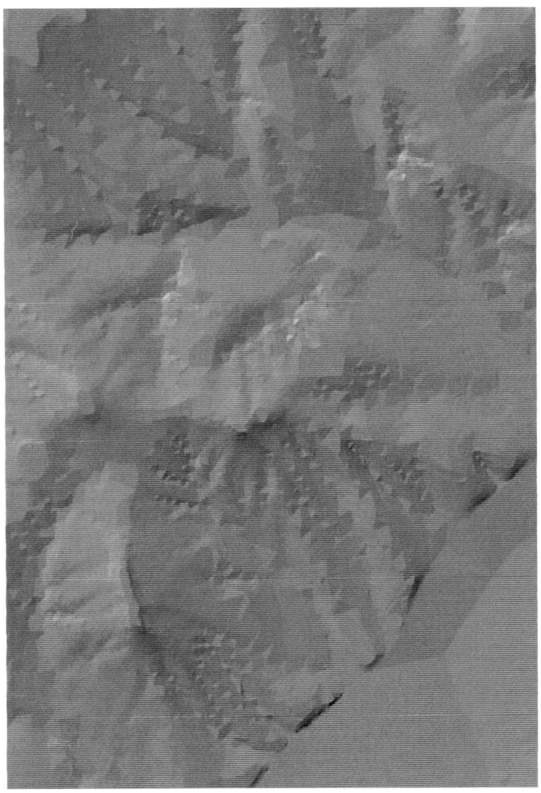

**Figure 3.** The terrain model fragment with artefacts – horizontal steps.

**Figure 4.** The same fragment after fixing the artefacts.

The triangulations obtained from isolines always contain the artefacts, which are clearly visible on the figure 3 – the horizontal steps formed by the points of the isolines of the same height. We have developed the algorithm, which allows to fix the artefacts by adding the hard edges and auxiliary points, which cause removing of the horizontal areas, The results of the algorithm are shown on the figure 4. The fixed triangulation of the Bratsk Reservoir area contains 6610182 points.

The TIN Smith program allows to cut a rectangular fragment from the triangulation and save it into a separate file. All the other processing was performed using this capability: we cut the fragments of interest and add the a-posteriori information to them.

### *2.6. The echo sounder measurements processing*

During the expedition a portable echo sounder with its own GNS receiver was used for the depth measurements. The main disadvantage of the data collected by the echo sounder is the rounding of the measured coordinates to integer values in meters – it produces strong aliasing (steps) of the measured paths. The precision of the GNS receiver of the echo sounder is that of consumer electronics (several meters). That's why together with the echo sounder, in most cases a professional GNS receiver was used: The professional GNS receiver allowed to determine the exact water level markers with respect to which the depth is measured, and in most cases it was placed in the boat along with the echo sounder.

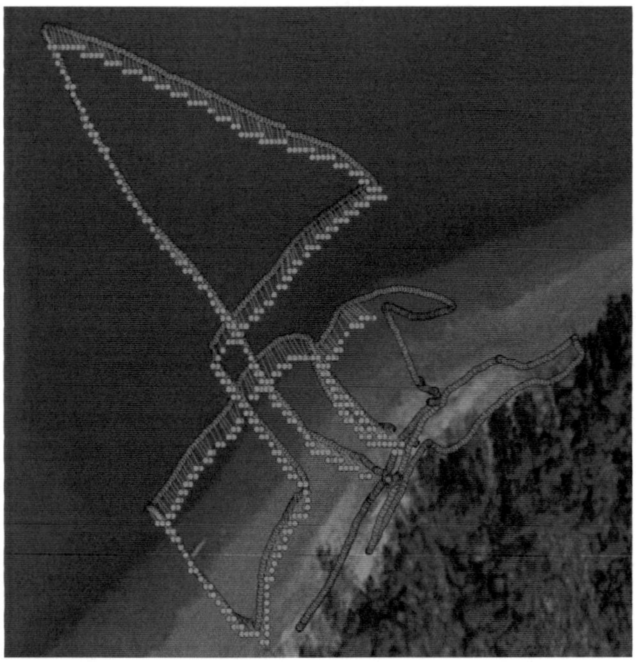

**Figure 5.** The echo sounder track (big green circles), the corresponding professional GNS track (small red circles), and the shifts computed using the time stamps for the echo sounder points (green lines).

Subsequently, the GNS track of the professional receiver was combined with the marks of the depths produced by the echo sounder using the time measurements stamps. Fortunately, both devices record the GPS-time [14], which is very precise for our purposes. The output file of the professional GPS receiver records the position of its rover every second, and the echo sounder records its measurements with a higher frequency. The echo sounder stores the time in milliseconds as an integer value, but the rounding of the horizontal coordinates to meters may place several measurements into the same point. We use the linear interpolation between the professional GNS coordinates to estimate the echo sounder position at the time recorded in its measurement. We have developed a special utility program to perform this task.

The figure 5 presents the results of the hardware test performed before the expedition. Note the aliasing of the echo sounder measurements and their error, which is mostly systematic in its nature.

## 3. Results and Discussion

Since the pilot chart data give a fairly rough idea of the underwater terrain it makes no sense to try to combine it somehow with the more precise data. So, an algorithm was developed to replace the triangulation fragment, which allows to remove all the marks of depth obtained from the pilot chart in the area where the real measurements exist. It removes all the target triangulation points that fall into some triangle of the source (being added) triangulation and then adds the points of the source triangulation.

Thus, to obtain the resulting terrain model in the area of interest we first cut the rectangular fragment from the a-priory terrain model, then create the auxiliary triangulation(s) for the echo sounder measurements in the area and replace the corresponding fragments in the terrain model. An example of the resulting triangulation is shown on the figure 6 (the black lines are the hard edges). Due to the low quality of the a-priory underwater terrain model the replaced fragments usually stand out against the rest of the terrain.

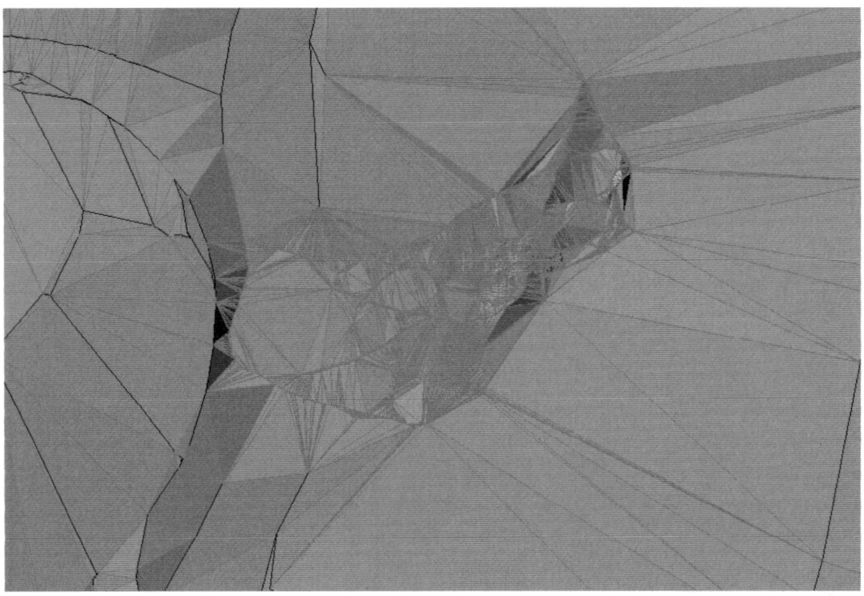

**Figure 6.** The combined terrain model with the fragment replaced by echo sounder measurements.

Using the triangulation computed we can build profiles along any polylines. The profiles are required to perform the hydrological computations for the wells. We can also build isolines of the terrain model, for example, to improve the underwater terrain representation using the echo sounder measurements.

## 4. Conclusion

We have considered the experience gained during preparation and processing of the terrain data around the Bratsk Reservoir. To handle the data we use free and open source software, some custom utilities and our own program TIN Smith, which implements several original algorithms of triangulation processing. We have briefly described those of the algorithms, which were used in the project.

In spite of that our knowledge about the underwater terrain of the Bratsk reservoir (as well as for almost all the other water bodies) is extremely limited and inaccurate, as a result of the application of the algorithms developed it become possible to obtain the terrain model with the accuracy sufficient for performing the further hydrological calculations.

IOP Conf. Series: Earth and Environmental Science **190** (2018) 012028 doi:10.1088/1755-1315/190/1/012028

**References**

[1] *Decree from the 26th of March*, 2001 # 234 On the limit values of the water level in lake Baikal in the conduct of economic and other activities 2001 (in Russian) Income accessed online on 1st August 2018 via http://docs.cntd.ru/document/901783823

[2] *Decree from the 27th of December*, 2017 # 1667 On the maximum and minimum values of the water level in Lake Baikal in 2018-2020 2017 (in Russian) Income accessed online on 1st August 2018 via http://docs.cntd.ru/document/556174677

[3] EasyTrace group 2017 *EasyTrace group free versions Easy Trace Pro 7.99 and 8.65* Income accessed online on 1st August 2018 via http://www.easytrace.com/program/et799_en

[4] Riegler G, Hennig S D and Weber M 2015 Worlddem – a novel global foundation layer, *Int. Arch. Photogramm. Remote Sens. Spatial Inf. Sci.* **XL-3/W2** 183-7 https://doi.org/10.5194/isprsarchives-XL-3-W2-183-2015

[5] *The map of Bratsk Reservoir* 1974 (Irkutsk: State Enterprise Water routs of East Siberian Basin) (in Russian)

[6] KB Panorama. The vector format is SXF. The structure of data is in binary form 2017 PARB.00227-02 99 02 41pp Income accessed online on 1st August 2018 via http://gistoolkit.ru/download/doc/sxf4bin_en.pdf

[7] Bychkov I V, Gachenko A S, Rugnikov G M and Hmelnov A E 2016 3-D modeling of Angara river bed *Mathematical and Information Technologies, MIT-2016 Information Technologies* (Vrnjacka Banja – Budva: CEUR) pp 26–32

[8] Skvortsov A V 2002 *Delaunay Triangulation and its Application* (Tomsk: Tomsk University Press) p 127 (in Russian)

[9] Shewchuk J R 1997 Adaptive precision floating-point arithmetic and fast robust geometric predicates *Discrete & Computational Geometry* **18** 305–63

[10] Chew L P 1987 Constrained Delaunay triangulations. *In Proceedings of the Third Annual Symposium on Computational Geometry (SCG '87)* (New York: ACM) pp 215-22 DOI=http://dx.doi.org/10.1145/41958.41981

[11] Anglada M V 1997 An improved incremental algorithm for constructing restricted delaunay triangulations *Computers & Graphics* **21** 215-23

[12] Nam N M, Kiem H V and Nam N V 2009 A Fast Algorithm for Constructing Constrained Delaunay Triangulation *IEEE-RIVF International Conference on Computing and Communication Technologies* 1-4 doi: 10.1109/RIVF.2009.5174607

[13] Shewchuk J R 1996 Triangle: Engineering a 2D Quality Mesh Generator and Delaunay Triangulator *First Workshop on Applied Computational Geometry* (Philadelphia: ACM) pp 124-133

[14] US Coast Guard 2012 *Navstar GPS Space Segment/Navigation User Segment Interfaces, GPS interface specification* (Global Positioning Systems Directorate, California), IS-GPS-200G Income accessed online on 1st August 2018 via http://www.gps.gov/technical/icwg/IS-GPS-200G.pdf

# Features of the natural environment of the Tea Road corridor in the context of the climate change

**E Z Garmaev, B Z Tsydypov, S G Andreev, A A Ayurzhanaev, Z B Alymbaeva, E A Batotsyrenov, B V Sodnomov, M A Zharnikova**

Baikal Institute of Nature Management SB RAS, Ulan-Ude, Russia

E-mail: bz61@binm.ru

**Abstract.** Low-water period in the basin of Lake Baikal began in the middle 90s of the 20 century. A statistically significant trend in temperature increase and decrease in precipitation is established. The analysis of water inflow in Lake Baikal is performed. It is established that the Baikal level practically directly depends on the water content of the Selenga river. Dendrochronological reconstruction of the Selenga river run-off was carried out. In recent decades, a statistically significant trend towards the descent of the river run-off with the analysis of air temperature and atmospheric precipitation in the whole basin in the Russian part of the basin has been identified. The meridional transect (E 105-107, N 51-44) is chosen for the investigation, covering all arid climatic zones. A set of methods for remote sensing, geoinformation technologies and in-situ surveys of geosystems with the implementation of landscape, geobotanical and soil studies, mapping of vegetation of model polygons and key sites were used. A spatio-temporal assessment of the state of natural-anthropogenic geosystemsis carried out, criteria, indicators and tendencies of their dynamics are determined in connection with the processes of land degradation and desertification.

## 1. Introduction

Lake Baikal is the largest freshwater lake by volume in the world, containing roughly 20 % of the world's unfrozen surface fresh water. Lake Baikal together with the adjacent territories in 1996 was included in the UNESCO World Heritage List. In connection with this, the problem of the protection and conservation of lake water resources is acquired exclusively acute [1].

The Baikal Lake basin is located almost in the center of the vast Asian continent and covers an area of 545 $km^2$ (excluding water area of Lake Baikal), 45 % of which lies within the Russian Federation, the rest is on the territory of Mongolia. About 73 % of river waters are formed on the territory of Russia, 27 % – Mongolia [1].

Because of the global warming in Transbaikalia the temperature of air limit layer increases, precipitation decreases [2]. All this has led to the low water content of the rivers flowing into the Baikal. There has been steady declining trend of the main flow of the Baikal tributary – the Selenga river – due to the lack of rainfall [3, 4].

Land degradation in arid, semi-arid, and dry sub-humid areas – also referred to as dry lands – may result from various factors, including climatic variations and human activities [5]. It includes diverse processes, ranging from changes in plant species composition to soil erosion, and reduces the land's productive potential [6]. Land degradation may diminish the land's resilience, making it more vulnerable and reducing its capacity to recover from disturbances. Furthermore, land degradation can have negative effects on other resources, such as water, soil, flora, and fauna [7].

Causes of land degradation can be natural or man-made. Natural causes include periodic stress from extreme and persistent climatic events, aridity, and droughts. Man-made causes include unsustainable human land use – for example, overgrazing, deforestation, and over-cultivation – as well as indirect socio-economic drivers, such as unstable food market prices and political instability or changes [8].

Land degradation and desertification processes are widespread in Russia and Mongolia. Over 85% of the Mongolian territory is located in dry climate zones: dry sub-humid, semiarid, arid, and extra-arid.

The vegetation is the first experience the impact of desertification [9], and that is why investigation of its changes have a particular importance.

Our research deals with vegetation dynamics of dry climatic zones of Mongolia from NDVI time series analysis and field investigations. The tasks of this study included: 1) field geobotanical studies on testing ranges and classification of vegetation; 2) analysis of the dynamics of the NDVI vegetation index according to satellite data; 3) analysis of the spatial structure of phytocenosis. A number of model monitoring polygons and key sites of contact monitoring has been established in different latitudinal zones according to the longitudinal transect (105-107° E, 51-44°N). These polygons include a wide range of territories with dry climate conditions. This work revealed the main factors, agents, and trends of development for desertification processes in different climate zones.

Two model polygons were considered in Central Mongolia: 1) the Kharaa river downstream basin and the Orkhon river right feeder; 2) the central part of Dundgoviaimag (Middle Gobi). The first polygon is situated in a semiarid climate zone with grassland and bunchgrass steppes. The second is located in an arid climate zone; from botanical and geographical points of view, its main feature is the prevalence of desertified steppes.

## 2. Materials and methods

An analysis of the meteorological series of air temperature and precipitation was held along the whole basin of Lake Baikal based on the initial data of weather stations and a global database of geospatial meteorological parameters Climate Research Unit – CRU (spatial resolution of 0.5×0.5 of angular degree) [10].

The analysis of the hydrometeorological information was held on the basis of historical data and weather gauging stations of the Republic of Buryatia, Transbaikalia region and Mongolia.

According to gauging stations and dendroclimatic stations, a spatio-temporal reconstruction of river flow parameters in the Selenga river basin was conducted [11]. As a result, water flow model reconstruction of the Selenga river and its tributaries – rivers Uda, Khilok, Chikoy, Dzida, Orkhon and Kharaa was obtained. For individual watercourses the time-series until 1666 have been restored. This allowed us to reveal regularities of moisture fluctuations in Baikal Asia in retrospective.

Remote sensing methods, especially satellites, have provided opportunities to organize immediate vegetation monitoring.

Of particular importance in the establishment of a system of remote sensing monitoring is the possibility of organizing completely automated satellite data processing. Over the last years, such technologies were actively being established and developed in the Russian Space Research Institute, Russian Academy of Sciences (IKI RAS). They allowed the creation and actualization of the archive of constant satellite observations on the territory of Russia and adjoining states for the period from 2000 until the present. The basis of automatized technologies established by the IKI RAS is the analysis of NDVI values and comparison with the index variation of the previous years [12].

The temporal dynamics of the vegetation was estimated using archive geoportal data of the IKI RAS for the period of 2001-2017.

To conduct relief morphometric analysis, a number of the corresponding morphometric maps were established and analyzed (along with the data of field observations) on the basis of the Digital Elevation Model SRTM: hypsometry, slopes, and aspects [13].

While identifying vegetation species, Mongolian [14] and Buryatia [15] keys to identification of plants were used.

## 3. Climate change in the Baikal region

On the basis of the annual CRU data a statistically significant trend of increasing temperature and decreasing rainfall in the basin of Lake Baikal was found.

Two humidification periods were allocated: 1980-1998 years – wet period; 1999-2017 years – dry period. The wet period is characterized by positive trends of precipitation totals for almost the entire basin of Lake Baikal, with the highest growth rate observed for Khamar-Daban ridge – 14 mm, and for the area of the Middle Baikal – 10 mm for the period under review.

During the dry season in 1999-2017years there is a negative trend in the amount of precipitation for the basin of Lake Baikal. Extreme reduction in moisture for the dry period under review is marked for Khamar-Daban ridge and amounts to -30 mm. In the Upper Angara and the Barguzin river basins the rate of decreasing reaches -18 mm. Atmospheric precipitation effects the long-term fluctuations in river flow to a greater extent than other elements of the water balance.

From 1885 to 2012 years the average annual air temperature in the Trans-Baikal region increased by 2.0 °C [4] (note that at the same time, the annual average temperature around the globe increased by 0.85 °C) [16]. The temperature increase is observed for the entire catchment area of Lake Baikal, while the areas with low growth rates of surface temperature are interspersed with areas with high growth rates, both in latitudinal and longitudinal directions.

## 4. Peculiarities of Lake Baikal water level regime

It should be noted that since 1996 Lake Baikal is UNESCO World Natural Heritage site. Due to the fact that the Baikal is an artificial reservoir (after the construction of the Irkutsk hydroelectric station in 1956), it can lose this honorary status, which may lead to the risk of inclusion of Lake Baikal to the list of World Heritage in Danger.

The actual water levels of Lake Baikal in vivo (1900-1956 years) ranged from 454.93 m (historical minimum was recorded in 1904) to 457.10 m (maximum was recorded in 1869). In regulated conditions (1960-2015 years) the minimum mark was registered in 1982 – 455.27 m, the maximum in 1988 – 457.42 m (hereinafter used Pacific Heights system).

Average annual useful inflow into the lake during the observation period (1900-2015) is 1872 $m^3/s$, in vivo – 1916 $m^3/s$, minimum average annual inflow into the lake was observed in 1903 – 1106 $m^3/s$, the maximum was in 1932 – 3251 $m^3/s$; over-regulated in the period – 1824 $m^3/s$, the minimum inflow was recorded in 1979 – 1244 $m^3/s$, the maximum – in 1973 – 2848 $m^3/s$ [17].

If the outflow of Lake Baikal will be controlled, it may gradually lose its uniqueness and over the time this can lead to the loss of biodiversity up to the complete disappearance of many species. When the level exceeded 457.0 m, during the high-water years in the middle of the 90-ies of the last century, low-lying coastlines of the east coast were destroyed (coastal forests, recreation zones, beaches and coastal constructions). It caused environmental damage throughout the natural and biological complex of lake system [18].

If the level of Lake Baikal is extremely low, we can observe changes in the groundwater regime and groundwater lowering; violation of the existing mechanisms of Baikal water purification; reduction in water exchange of the sor system with open Baikal, the increase in average temperatures and heavy overgrowth of shallow water; the death of aquatic organisms on the shore of the lake and the coastal system as a result of drying and freezing of the habitats, which are responsible for the processing of organic matter; peat fires in the Selenga river delta [19].

The spring-and-summer period in 2015 was characterized by great losses of forests from fires in the Republic of Buryatia. The dynamics of forest fires on the lands of the State Forest Fund of the Republic for the period from 1936 to 2015 was analyzed. According to the Federal Forestry Agency of the Republic of Buryatia from April to October of 2015 Forest fires caught the area of 750,500 hectares, that exceeds the long-term annual average by 12 times.

Before the construction of the Irkutsk hydroelectric complex (in natural conditions) the level of the lake varied in the range up to 2.17 m. During the period of operation of the Irkutsk hydroelectric station until 2001 the level of the Baikal was 17 times higher than the mark of 457.0 m and 18 times it fell below 456.0 m [20]. After acceptance in March 26, 2001 of the Russian Federation Government Decree № 234 "On the limit values of the water level in Lake Baikal in the implementation of economic and other operations" the level fluctuated in the meter range (456.0-457.0 m), but on 25.02.2015, for the first time since 2001, it fell below the mark of 456.0 m. This is primarily the result of the low water period, which was established in the middle of the 90-ies of the last century and was the highest rate ever recorded. The situation with the low-water level of Lake Baikal basin continues in 2016-2017 years.

Due to the anomalously high air temperatures and almost complete absence of rainfall, inflow of water into the lake in 2015-2017 years was extremely low. The minimum level mark 455.71 m was detected on April 28 to May 06, 2016. The maximum level of Lake Baikal water (after 2014) was 456.27 m (September 7, 2017), which is the lowest value since 2001.

An analysis of the inflow of water into Lake Baikal was carried out. The minimum and maximum water flow, volume of the flow of the major rivers of Lake Baikal basin were studied. It was found that the run-off of the Upper Angara and the Barguzin rivers in the last 20 years has remained within the long-term average rate, the Selenga river run-off was reduced and now stands at 65 % of the normal value. Together these three rivers provide 70 % of the annual inflow of water into Lake Baikal. The level of the lake almost depends on the water content of the Selenga river [21]. A good alignment between the inflow fluctuations in Lake Baikal and the Selenga river run-off was determined, which was confirmed by the high values of correlation of the coefficients between these variables: for the observation period (1934-2014) – 0.85, during the dry periods (1954-1958, 1976-1982, 1996-2014) – 0.68 [17].

The formation of water resources in the catchment area of Mongolia which accounts 67 % of the watershed district of the Selenga river was studied. The dynamics of the Selenga and the Orkhon rivers flow, its main tributary, was studied. High-water period of the Selenga river was observed from 1979 to 1995. In recent years low-water period in the Selenga river basin is observed. The minimum flow during the dry periods, as well as the annual flow, tends to decrease. It is a continuous period of low flow which provided a negative trend of the minimum flow. Thus, the inflow of water into Lake Baikal in recent years is the record-breaking minimum for the whole period of instrumental observations.

## 5. Reconstruction of the water flow of the Selenga river basin

Unfortunately, for a full analysis of the history of the Baikal fluctuations, we have a too short period of instrumental observations – only 115 years. Considering that our object under study is Lake Baikal, which is the oldest lake in the world (20-35 Million years), this short period of observation data does not allow us to get a more or less objective picture of the changes in its level for a longer time. In this regard, only long-term series of observations allow to define changes in moisture regime correctly. An analysis of annual tree rings allows to essentially extend climatic characteristics [22]. For example, the reconstruction of the Baikal level on dendrochronological data is presented in the article of Andreev S.G. [23].

According to gauging and dendroclimatic stations' data, a spatio-temporal reconstruction of river flow parameters in the Selenga river basin was conducted [11]. As a result, water flow model reconstruction of the Selenga river and its tributaries – the rivers Uda, Khilok, Chikoy, Dzida, Orkhon and Kharaa was obtained.

For each river the time-series until 1666 were restored. This allowed us to reveal regularities of moisture fluctuations in Baikal Asia in retrospective.

There is a spatial and temporal coherence of tree growth with water flow dynamics of the Selenga river and its tributaries. This is particularly evident in the low and high water periods. The correlation values for individual tributary basins are analyzed. The maximum value of 0.66 is a generalized chronology obtained through 8 local chronologies and "responsible" for the valley of the Dzida river. The Dzida river valley is an isolated area, surrounded by high mountain ridges. Here, the run-off is formed within the valley. This explains the close relationship between water consumption and growth of trees, depending at some extent on the incoming water from the atmosphere. High correlation can be observed in the small rivers in the steppe and forest steppe zones.

For the Chikoy river connection with tree rings is smaller (R = 0.47). This value can be explained by the fact that the formation of the Chikoy river flow is under the influence of the Pacific air masses coming in the second half of the summer on the east and affecting the upper catchment area. This explains weakening of the connection of dendrochronological series of the stations located in middle and lower parts of the valley.

It should be noted that the reconstruction of the Kharaa river water flow, built on the same stations as for the Chikoy river, shows a similar picture while taking into account that Kharaa basin is located 300 km to the south. This fact is determined by the general laws of the Chikoy and Kharaa rivers run-off formation.

Annual fluctuations in water of the Orkhon river flow showed a close relationship (R = 0.53) with stations located in the Dzida river basin. The values obtained for the correlation between the indices of growth of annual rings and instrumental measurements of river flows on Mongolian part of the Selenga river basin (for the Kharaa river R = 0.47, for the Orkhon river R = 0.53) allow us to expand our network of stations in the basins of these rivers and get more reliable model reconstruction.

20 century in comparison with the previous centuries was provided with water resources. The general trend of the 20 century has the following character: until mid-century water content has increased, reaching a peak in the mid-1940s, and then began to slowly decrease. The modern period, starting in 1950s is characterized by an increase in water content.

## 6. Desertification processes

### *6.1. Semiarid climate zone*

Among the desertification types discovered by FAO-UNEP [24], on the greater part of the Selenga River basin, especially in its Mongolian part, the vegetation cover degradation is the most widely spread.

It is revealed in structural changes of the steppe, forest steppe, and pasture phytocenosis, the successions of their species by synanthropic ones, and decreased projective cover and grass height. In places with the highest development of degradation and land desertification processes caused by a high density of grazing animals on easily deflated sand and sabuloussoils, digression processes of steppes and forest steppe geosystems are increased even more. As a result, soilcover is being destroyed, affected by trampling and subsequent deflation, while the existing types of microrelief are being reformed. We noticed an especially strong manifestation of these processes in the poorly populated pine *Pinussylvestris L.* and drought resistant elms *Ulmuspumila L.* in the Kharaagol sandy area (absolute height values are 735-815 m).

According to the scheme of the landscape and the ecological district division located in the central part of the Selenga River basin, the Kharaagol model polygon is situated in an area of high ecological intensity, caused by both natural and anthropogenic factors. Let us consider the key site of 19.2 $km^2$ situated in the northern part of the Kharaagol polygon, 3 km southeast of the Darkhan city. The biggest part of this site, located on the separated slopes of low hill terrain of the Orkhon-Kharaagol interfluve (maximal height marks are 810-870 m), has steppe caragana-cereal-fringed sagebrush (*Artemisia frigida – Leymus chinesis – Caragana microphylla*) vegetation. As geobotanical studies have shown, the small-leaved *Caragana microphylla (Pall.) Lam.* dominates in the projective vegetation cover (17-20 %). Fringed sagebrush, *Artemisia frigidaWilld* .is also widely spread, as well as Chinese wild rye *Leymus chinensis (Trin.) Tzvelev*, etc. Most of the land is used as pasture.

Using access to the satellite monitoring service "Vega" to monitor the vegetation condition[12], average NDVI curves were built for the Kharaagol key site for the first 17 years of the twenty-first century. Some nonuniformity of the NDVI distribution was observed throughout the years, caused by different climate conditions. It should be pointed out that in 2007, 2008 and 2012 the NDVI values were rather high 0.6, which is related to the higher values of temperature and moisture regime during the summer months of these years. The important feature of the system "Vega" is a possibility of the research with meteorological information. It gives the chance to receive spatial distribution of meteorological parameters for any date of the vegetation period. The average summer value of the NDVI in 2011 was 0.45 for this site, which proves sparse vegetation during this year.

According to the Mongolian State National Statistics Committee [25], a tendency to increase the number of farm animals was observed on the territories of Selenge and Darkhan-Uulaimags in 2000s. Express surveys conducted for arats having a household within the territories of the Kharaagol model polygon showed that nearly half of them immigrated here (to be closer to main market of animal products) with their herds in the 2000s from other periphery Mongolian aimags, the western (mostly) and Gobi.

*6.2. Arid climate zone*

On the territory of Dundgovi aimag (Middle Gobi), moving from north to south, there is a gradual landscape substitute: desertified steppes of semi-deserts are changed for deserts with saxaul (*Haloxylon sp.*).

Studies showed that in conditions of insufficient moisture (an annual sum of atmospheric condensation less than 100 mm on the territory of Ulziitsomon of the southern part of Dundgoviaimag, a maximal value of 150 mm in the north, or a complete absence of fresh surface water), Gobi landscapes are affected by degradation processes easily, especially by the physical weathering, deflation, and degradation of vegetation. Even though the general pasture load is low on the territory of Dundgoviaimag because of the rather low density of the animal base, the soil surface is trodden around bases within a 0.5-0.7 km radius, and there is hardly any vegetation cover (projective cover from 0 to 1 %).

Up north from the aimag center Mandalgovi, southern dry steppe landscapes dominate. They are located on 70-80 % of the territory. Desertified steppes take up 10-20 %, while saline and alkali saline soil comprise under 8 %. South from Mandalgovi, within the limits of the Middle Gobi model polygon, desertified steppe landscapes prevail (65-75 %). The southern steppes take up 20-30 %, and the saline ones comprise less than 9 % of the territory.

The average NDVI value for the last 17 years aggregated for the Middle Gobi model polygon is 0.12, which proves poor vegetation. NDVI values are slightly higher in 2012 and 2013.

Comparison of natural geobotanical studies conducted in 2011-2017 and the results for ecological groups of the Gobi areas of 2010 proves this conclusion, representing the increased share of euxerophyte plants in projective cover for different areas at 12-15 %, indicating continuous invasion of euxeropyte desert species of plants in the north. Thus, the share of euxerophytes has increased in the projective cover from 77 % in August 2010 to 96 % in August 2012 on the key site situated in the lowland, stony, desertified steppes 20 km south of Mandalgovi.

Meteorological data analysis proved that in the 2000s there was more severe drought during the summer periods than in the previous century, which, in our opinion, is the main factor for the digression tendency in the Gobi vegetation.

## 7. Conclusion

Statistically significant trend of increasing temperature and decreasing precipitation with the identification of wet and dry periods is established. In recent years, due to abnormally high air temperatures and almost complete absence of atmospheric precipitation, water flow into Lake Baikal has been extremely low during the whole period of instrumental observations. As a result, there were problems with the level regime of Lake Baikal.

There is high year and seasonal growth in response to the current precipitation in the dendrochronological series for specific areas. Spatio-temporal coherence of the tree growth with water flow dynamics of the Selenga river and its tributaries was revealed. So statistical models of reconstruction of water flow show good correlation (r = 0.47 ... 0.66) between the increase in the annual rings of Scots pine and annual consumption of water. On the example of the Mongolian part of the Selenga river basin, it is shown that the spatial relationship between the water run-off and dendroclimatic stations located on the territory of Russia can be traced for hundreds of kilometers.

It was found that for forest-steppe regions of the Baikal Asia a limiting factor for tree growth is the atmospheric precipitation which form the water flow of the Selenga basin rivers. The data flow rates for a large time interval can serve as a basis for the planning of water resources management. This aspect is particularly relevant for the Selenga river, because of its cross-border location.

Climatic zones of the north of Central Asia experience different trends of desertification processes associated with both climatic variability (mainly aridization) and anthropogenic influence. The performed studies for different channels of optical bands using methods based on the consideration of differences in spectral reflection, together with field natural investigations, prove that the degradation of the vegetation cover has been increasing for the last 25 years.

Cattle overgrazing is observed on the studied areas. It conditions overload for pastures and leads to the digression of their vegetation cover. Therefore, livestock farmers and other land users need to use pasture rationally, changing grazing areas from time to time and thus regulating the grazing rotation.

Remote sensing methods, based on the analysis of spatial and temporal variability of biophysical vegetation parameters, allow quick determination of areas of ecological intensity conditioned by the degradation of vegetation cover and desertification.

Thus, a significant transformation of the natural environment is currently observed in the Tea Road corridor.

**Acknowledgments**
These investigations were partially supported by the Russian Foundation for Basic Research (grants No. 17-05-01059, 17-29-05083).

**References**

[1] Afanasev A N 1976 *Water Resources and Water Balance of the Baikal Lake Basin* (Novosibirsk: Nauka) p 239 (in Russian)

[2] Budyko M I and Izrael Y A 1987 *Antropogenic Variations in Climate* (Leningrad: Gidrometeoizdat) p 406 (in Russian)

[3] Garmaev Y Z 2010 *River Flow of Baikal Lake Basin* (Ulan-Ude: Buryat State University Press) p 272 (in Russian)

[4] Obyazov V A 2013 *Russia. J. Proceedings of the RAS* **450**(6) 713–6 (in Russian)

[5] UNCCD 1994 *United Nations Convention to Combat Desertification in Countries Experiencing Serious Drought And/or Desertification, Particularly in Africa* (Paris: A/AC.241/27)
UNCCD 2018 *Achieving Land Degradation Neutrality* Income accessed online on 12th August 2018 via http://www.unccd.int

[6] Hill J, Megier J and Mehl W 1995 *Rem. Sen. Rev.* **12**(1-2) 107–30.
Reynolds J F 2001 Desertification *Encyclopedia of Biodiversity*, vol 2, ed S Levin (San Diego: Academic Press) 61–8
Reynolds J F et al 2007 *Science* **5826**(316) 847–51

[7] Hennemann R 2001 *Elective on Land Degradation, Assessment, Monitoring and Modelling. Soil Science Division* (The Netherlands: ITC, Enschede)
Hennemann R 2001 *Some Notes About Soil and Land Degradation in the Naivasha Range land Zone. Soil Science Division* (The Netherelands: ITC, Enschede) p 3

[8] NPACD 1997 *National Plan of Action to Combat Desertification in Mongolia* (Ulaanbaatar: Government of Mongolia, Ministry of Nature and Environment) Income accessed online on 12th August 2018 via https://knowledge.unccd.int/sites/default/files/naps/mongolia-eng2000.pdf

UNCCD 2009 *Factsheets aboutthe United Nations Convention to Combat Desertifucation* (Bonn: UNCCD) Income accessed online on 12th August 2018 via http://www.unccd.int/fr/resources/publication/Pages/factsheets.aspx
World Meteorological Organization 2006 *World Day to Combat Desertification, Climate and Land Degradation* (Geneva: WMO) No989
[9] Gunin P D, Miklyaeva I M, Bazha S N, Slemnev N N and Cherdonova V A 2003 *Arid Ecosystems* **9**(19-20) 7–21
[10] Harris I, Jones P, Osborn T and Lister D 2014 *Int. J. Climatol.* **34** 623–42
[11] Garmaev E Z and Andreev S G 2013 *Runoff Reconstruction of Major Rivers of the Selenga River Basin* Certificate of state registration database № 2013621267, Russian Federation. Proposal № 2013620567 Rightholder: Baikal Institute of Nature Management SB RAS; Receipt date: 06.06.2013; Registered in the Registry Database: 26.09.2013 (in Russian)
[12] Lupyan E A, Savin I Y, Bartalev S A, Tolpin V A, Balashov I V and Plotnikov D E 2011 *Russia. J. Modern Problems of Earth's Remote Sensing* **8**(1) 190–8 (in Russian)
[13] Farr T G and Kobrick M 2000 *EOS Trans. AGU* **81** 583–5
[14] Grubov V I 1982 *Key to Identification of Vascular Plants of Mongolia* (Leningrad: Nauka) p 443 (in Russian)
[15] Anenkhonov O A et al 2001 *Key to Identification of the Plants in Buryatia* (Ulan-Ude: Buryat Science Center) p 670 (in Russian)
[16] Kokorin A O 2014 *Climate change: a Review of the Fifth IPCC Assessment Report* (World Wildlife Fund) p 80
[17] Bychkov I V and Nikitin V M 2015 *Russia. J. Geography and Natural Resources* **3** 5–16 (in Russian)
[18] Bezrukov L A and Nikolsky A F 1995 *Russia. J. Geography and Natural Resources* **1** 125–34 (in Russian)
[19] Sinyukovich V N 2016 *Russia. J. Water Sector of Russia: Problems, Technologies, Management* **1** 42–51 (in Russian)
[20] Nikitin V M, Saveliev V A, Berezhnykh T V and Abasov N V 2015 *Russia. J. Region: Economics and Sociology* **3**(87) 273–95 (in Russian)
[21] Garmaev Y Z and Khristophorov A V 2010 *Water Resource of Baikal Lake Basin,* ed A K Tulokhonov (Novosibirsk: Academic Publishing House "Geo") p 231 (in Russian)
[22] Cook E R and Kairiukstis L A 1990 *Methods of Dendrochronology. Applications in the Environmental Sciences* (Dordrecht: Kluwer Academic) p 393
Fritts H C 1991 *Reconstructing Large-Scale Climatic Patterns From Tree-Ring Data* (Tucson: Univ. of Arizona Press) p 286
Shashkin A V and Vaganov E A 1993 *Russia. J. of Ecology* **5** 3–10 (in Russian)
Schweingruber F H 1996 *Tree Rings and Environment: Dendroecology* (Berne: Haupt Publ) p 609
Vaganov E A and Shiyatov S G 1999 *Russia. J.Siberian Journal of Ecology* **2** 111–6 (in Russian)
[23] Andreev S G, Vaganov E A, Naurzbaev M M and Tulokhonov A K 1999 *Russia. J. Proceedings of the RAS* **368**(3) 400–3 (in Russian)
[24] FAO/UNEP 1984 *Provisional Methodology for Assessment and Mapping of Desertification* (Rome: FAO) p 83
[25] Mongolian National Statistical Office 2010 *Mongolian Statistical Yearbook 2009* (Ulaanbaatar) p 448

# Spatio-temporal changes in energetic characteristics of eddy atmospheric circulation in the troposphere of West Siberia in the end of XX – the beginning of XXI centuries

**E V Kharyutkina, S V Loginov**

Institute of Monitoring of Climatic and Ecological Systems SB RAS, Tomsk, Russia

E-mail: kh_ev@mail2000.ru

**Abstract.** In the framework of the study based on ERA-Interim reanalysis dataset the analysis of average values of available potential energy, kinetic energy, their eddy components and parameter of relative vorticity, characterizing cyclonic and anticyclonic types of circulation, was carried out over the period of 1979-2015 for the territory of West Siberia. Also coefficients of their linear trends were calculated. It was revealed that the distribution in positive and negative tendencies of energetic characteristics was significantly changed upon transition from the period of 1979-1998 to the period of 1999-2015 along with considerable increase in their trend values. Their trend areas with negative values were located close to each other and were slightly covered in zone 60°-64°N. The greatest dissipation intensity was observed in the middle troposphere. At the earth's surface eddy circulation generation along the Arctic coast was shifted to East Siberia. The beginning of XXI century was characterized by the significant decrease in eddy circulation activity over West Siberia that was especially pronounced in winter. Furthermore, eddy circulation (cyclonic type) increased just in the certain zones - in the interfluve area of the Ob and the Irtysh rivers. This indicates the possibility of anomalies in meteorological parameters formation there.

## 1. Introduction

Global climate tendencies lead to considerable changes of atmospheric circulation patterns, causing meteorological parameters anomalies in certain regions [1]. This, in particularly, is connected with change in activity of vortices that provide inter-latitude heat and moisture transfer [2]. One of circulation variability effects is in the fact, that anomalous weather phenomena such as waves of heat and cold, heavy rains and snowfalls, storms, floods and droughts became widespread [3]. West Siberia is not an exception here. According to [4], in general warming process in West Siberia was replaced by climate cooling at the beginning of the XXI century, the atmospheric pressure rise was observed together with precipitation amount increase during 1976-1998, and the most statistically significant increase is observed in the northern part of West Siberia in warm season. On the contrary, there was a tendency to negative anomalies of precipitation amount in the south. In the north convective precipitation area expanded on 10% during a warm season.

The tendencies can be related to more frequent events of convective (cumulonimbus) cloudiness development, formation of which are caused by eddy circulation processes. All main aspects of atmospheric processes are expressed through sources and dissipation of distribution of available potential energy, its transport and transformation [5]. Moreover, the important place in research of atmospheric circulation is taken by a question, that deals with the energy of movement (including eddy and wave) of various scales. In [6] components of energy transformation cycle (Lorenz cycle) are estimated for baroclinic eddies. It is shown that only about 10% of eddy energy converted from potential energy to kinetic one, evolved into kinetic energy of main stream, and the rest of 90% are spent for dissipation of eddies relating to processes of surface friction and radiation relaxation.

In [7] it is reported that global "vortex energy" associated with storm, eddy and turbulence generation significantly increased last decades.

Therefore, investigation in energy characteristics of atmospheric circulation and the corresponding analysis allow us to better describe dynamics of separate elements of regional climatic system and to reveal their important features.

Main goal of the study is in estimation of spatio-temporal variability of eddy components of available potential and kinetic energies and, also to obtain parameter of the relative vorticity in the territory of West Siberia over the period of 1979-2015.

## 2. Materials and Methods

Average values of available potential energy (APE), kinetic energy (KE), their eddy components (EAPE and EKE, respectively), parameter of relative vorticity (Vort) and coefficients of their linear trends (here and after - trends) over the period of 1979-2015 were calculated for the territory of West Siberia (50°-70°N, 60°-90°E) and separately to its Arctic part (60°-70°N., 60°-90°E). ERA-Interim reanalysis dataset (http://www.ecmwf.int/en/research/climate-reanalysis/era-interim) were used. Calculations of energetic components were carried out at several levels in the troposphere (1000 hPa and 500 hPa), according to Lorenz's formulas [8]. The term *eddy* was meant as a part of energy that differs from its zonally averaged values.

Used parameter of relative vorticity characterized both anticyclonic (positive values — Vort+) and cyclonic (negative values — Vort-) types of circulation.

Median of cumulative distribution function (CDF) was used as the averaged by the territory characteristic. The statistical significance of the derived values was determined by the two-sided t-test of null hypothesis at a significance level of 0.05.

Calculation of characteristics was carried out separately for two time intervals: during intensive global warming - 1976-1998 and during temperature slowdown - 1999-2015.

In order to define the nature of changes passing from one period to another, for each time interval spatial distribution maps of linear trend estimates were constructed for all calendar months and for all seasons. Spatial distribution maps were presented for the Asian territory of Russia to consider tendencies of these parameters in adjacent regions. Interannual variability was also analyzed. Special attention was paid to the changes in the Arctic part of West Siberia, because there is the highest linear correlation between the Arctic climate presented by the surface air temperature and the meridional energy flux on the border of 70°N [9]. Moreover, the border of 70°N in a considerable degree coincides with the influence area border of annular stationary mode of high-latitude climatic variability - the Arctic Oscillation [10].

## 3. Results and Discussion

Analysis of annual averaged parameters values and their trends allow us to reveal the following features: all considered types of energy have a pronounced annual course with the maximum values in autumn and winter and minimum - in summer.

Kinetic and available potential energy values had inhomogeneous distribution over the territory of West Siberia, especially at the surface, over the period of 1979-1998. APE maxima, characterizing the enhancement of temperature latitudinal contrast, were located in the south of the territory, areas of minima – to the northward and along the western border of the region. The greatest variability of APE was observed in the lower troposphere (up to 3 km). For KE, on the contrary, areas of maxima were located in the Arctic zone, extending into the central part of the territory. In the middle troposphere (500 hPa) there was zonal distribution in KE with decreasing from south to north. In the beginning of XXI century (1999-2015) the APE increase of APE took place at the surface in the southern part of Siberia whereas its almost total disappearance of minima area was in the west. At the same time decrease of KE values was observed in the central part of the region, however, the area of positive values became higher in the north and it was already extended along the whole Arctic coast. At the level of 500 hPa, as well as at 1000 hPa, there was a KE decrease in the central part, and changes in APE were not observed.

Despite the fact that due to great variability of APE and KE values of their trends were statistically insignificant at $\alpha$=0.05 (however, they were significant at $\alpha$=0.1), it was possible to reveal the following tendencies. Firstly, analyzing constructed maps of spatial distribution in trends of the studied parameters, it could be noted that positive and negative tendencies had significantly changed from one time interval to another, and secondly, trend values (figure 1,2) had notably increased.

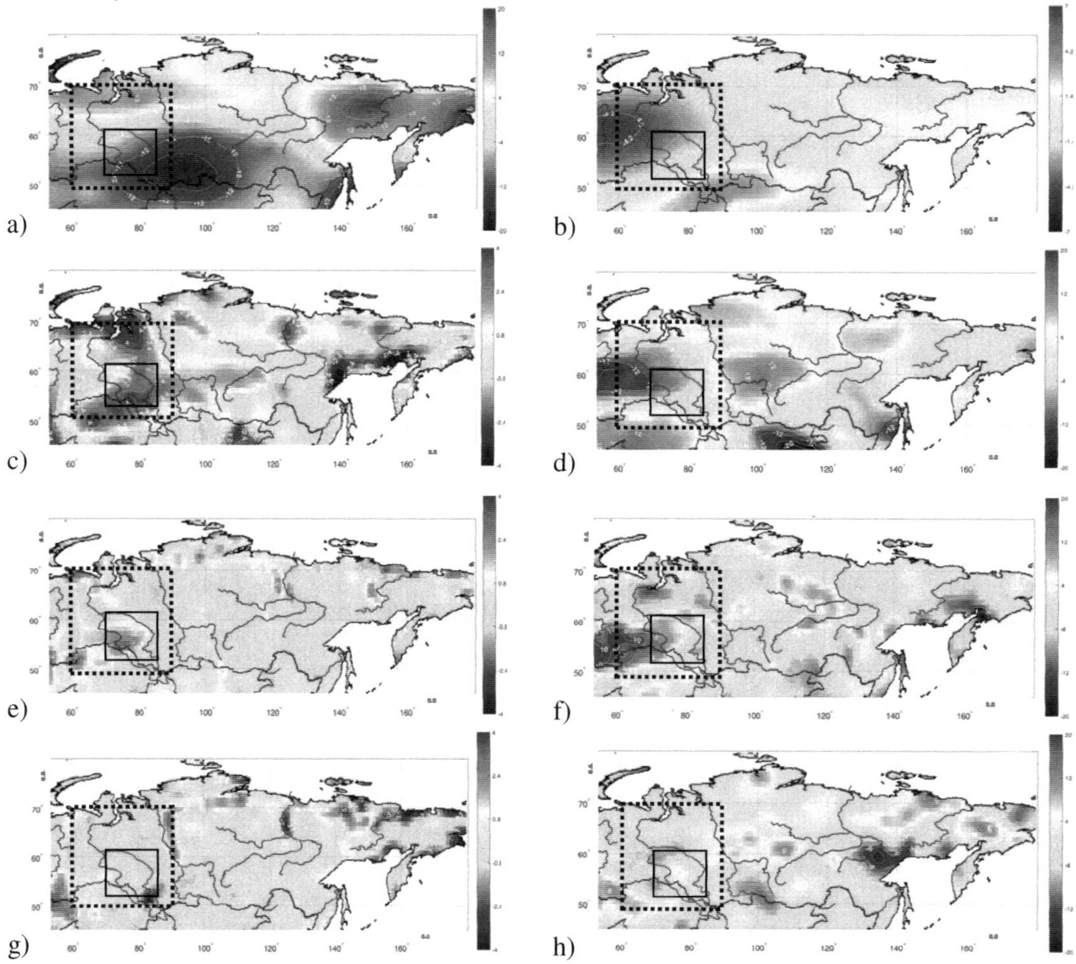


**Figure 1.** Spatial distribution of trend values over the period of 1979-1998: available potential energy, J/m2/decade (a,b), kinetic energy, J/m2/decade (c,d), relative vorticity (s-1/decade) of cyclonic (e,f) and anticyclonic types (g,h). Left panel - 1000 hPa, right panel - 500 hPa. Dash figure – the territory of West Siberia. Solid figure - the interfluve of the Ob and the Irtysh rivers.

IOP Conf. Series: Earth and Environmental Science **190** (2018) 012030 doi:10.1088/1755-1315/190/1/012030

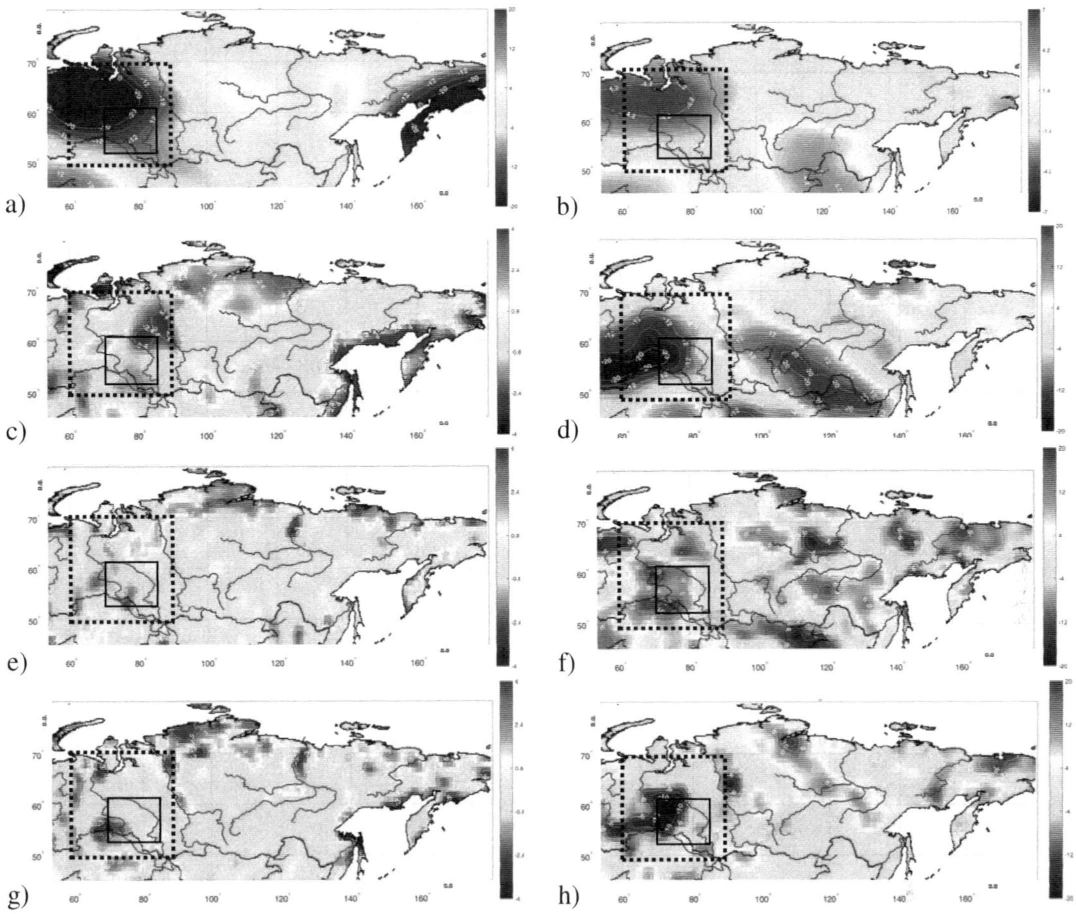


**Figure 2.** Spatial distribution of trend values over the period of 1999-2015: available potential energy, J/m2/decade (a,b), kinetic energy, J/m2/decade (c,d), relative vorticity (s-1/decade) of cyclonic (e,f) and anticyclonic types (g,h). Left panel - 1000 hPa, right panel - 500 hPa. Dash figure - the territory of West Siberia. Solid figure - the interfluve of the Ob and the Irtysh rivers.

Thus, in the beginning of the XXI century a vast range of negative values of APE trends was formed over the Arctic and the central part of the region, that confirmed the decrease in horizontal temperature gradients. The area stretches up to the middle troposphere (figure 2a, b). The same time interval showed almost widespread KE reduction (figure 1, 2) over the whole territory from the surface to the middle troposphere (that indicates the decrease in medium velocity of atmospheric flow), except for existence of the individual centers of positive trends in the center and in the north of the region. While APE negative trends continued to exist in the middle troposphere, a large center of the decrease in air transfer speed was formed for KE almost over all central part of West Siberia. Places where APE and KE trends with the same sign were observed, indicated generation sites (positive) of eddy energy or the sites of its dissipation (negative). Thus, in Siberia APE and KE trend centers with negative values were located under each other (KE – to the southward) and were slightly covered in zone of 60°-64°N (by the isoline corresponded to the half of the maximum value). The greatest dissipation intensity was revealed at 500 hPa. This effect was less pronounced in the boundary layer. It was observed the decrease in eddy activity due to dissipation and radiation relaxation in the energy flux areas.

Areas of eddy circulation generation in the surface layer over West Siberia, reveled along the Arctic coast in beginning of XXI century, moved to East Siberia, whereas no changes occurred in the middle atmosphere – it was located in the north of the Altai.

The CDF showed that in the beginning of XXI century the greatest changes of eddy components parameters of available potential and kinetic energy (EAPE and EKE) occurred at the surface. The second time interval was characterized by significant increase of range in EAPE values, here about 70% of the territory (in comparison with 40% over the first interval) was in the area of negative values (figure 3a). Changes in the Arctic part of the region were not pronounced. EAPE trends were mainly negative in winter and spring and positive in rest of the year. Over the second time interval, positive and statistically significant tendencies were observed in cold season, negative and statistically insignificant - from March to October. In the Arctic part there was a statistically significant decrease in EAPE during the whole year, except for certain months. Noticeable changes for EKE were not determined in the territory of West Siberia (figure 3c). This was confirmed by the distribution of trend estimations: significant tendencies were positive ones (specific for mid-seasons), and their parameters showed the increase over the second time interval. At the same time, in contrast to EAPE, the greatest changes were observed at 500 hPa in the Arctic part of the region. However, there was a tendency to the extending areas of positive values.

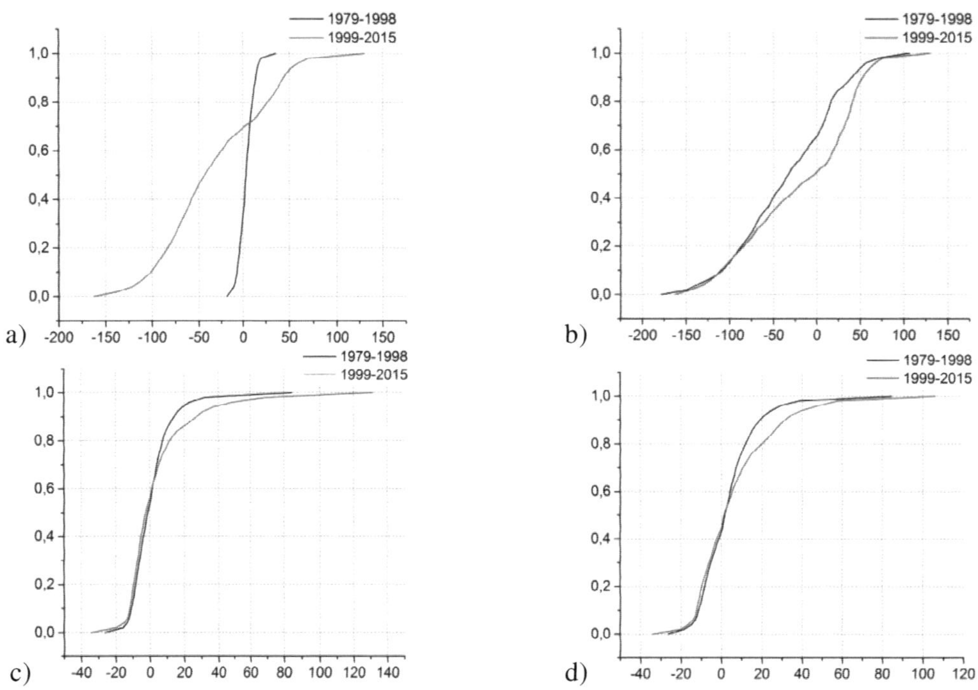


**Figure 3.** CDF of averaged values at 1000 hPa in January in West Siberia: eddy available potential energy (a,b) and eddy kinetic energy (c,d) Right panel (b,d) – for the Arctic part of the region.

In general, the decrease in eddy components existed in winter season over the whole West Siberia during the period of 1999-2015. Transformation process of available potential energy into kinetic one (kinetic energy generation) is caused by baroclinic processes of cold air descending and warm air raising. It is just baroclinic instability or instability of an air flow with a horizontal temperature gradient and, therefore, with vertical gradient of wind speed in the field of Coriolis force, which plays the greatest role in the formation of atmospheric vortices of synoptic scale. If atmosphere stratification is stable, temperature at the fixed altitude increases with air masses descending and vice versa. Consequently, the process would decrease horizontal gradients of temperature until to their total disappearance [11]. Hence, the less horizontal gradients of temperature together and more stability, lower kinetic energy values and, as a result, lower available potential energy values. This was observed in West Siberia in the beginning of XXI century. The revealed tendencies in zonal and eddy energetic characteristics were confirmed by dynamics of relative vorticity (Vort) parameter, describing both cyclonic and anticyclonic processes. Negative sign of values corresponds to the vorticity of cyclonic type (Vort-), positive sign – to vorticity of anticyclonic type (Vort+). It is necessary to consider when analyzing maps that the increase in cyclonic and anticyclonic types of vorticity will be in different colors.

Analysis of relative vorticity variability showed that, in general, average Vort values in the troposphere of West Siberia by 1.5-2 times exceed the relevant estimates at the earth's surface. There was the Vort decrease (though statistically insignificant) from one period to another, whereas it increased in the middle troposphere. Comparison of average values for the northern and the southern parts of West Siberia revealed that the greatest Vort variability was observed in its Arctic part. The decrease of relative vorticity was the most intensive in the north in winter over the early of XXI century.

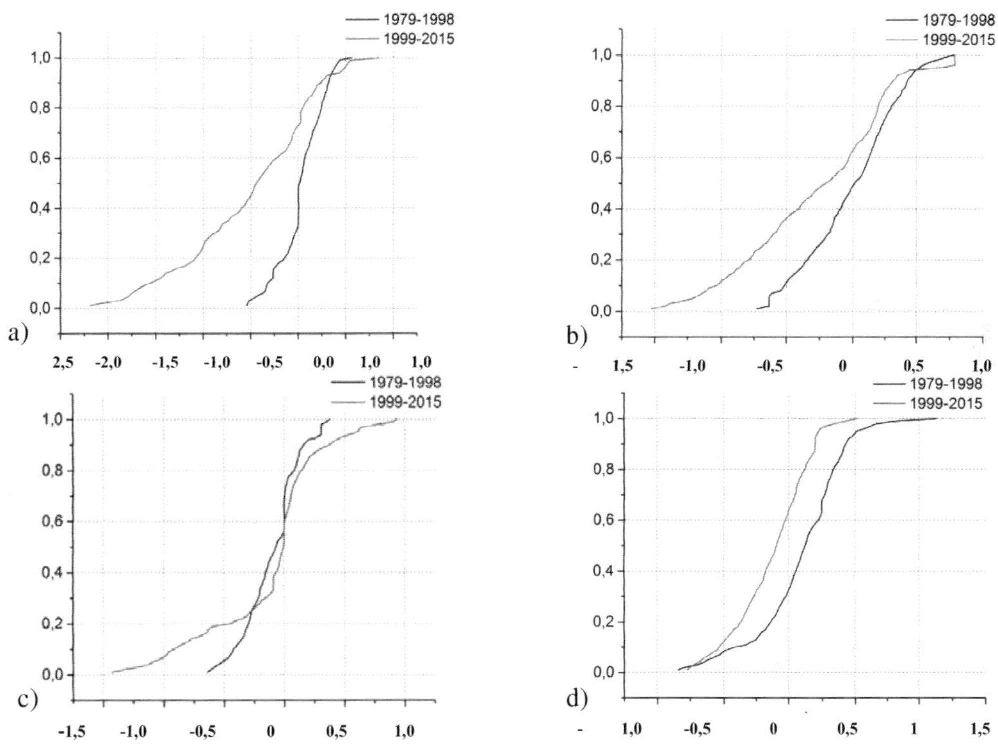


**Figure 4.** CDF of averaged values of relative vorticity at 500 hPa in West Siberia: cyclonic type (a,b) anticyclonic type (c,d). The upper panel (a,b) – for January, the lower panel (c,d) – for July

As for its spatial distribution, over 1999-2015, in comparison with the previous decades, there was cyclonic circulation increase in the surface layer in January that appeared in the central part of West Siberia - interfluve area of the Ob and the Irtysh rivers. The decrease in cyclonic circulation intensity stretched up to the middle troposphere (500 hPa). In West Siberia the intensity of anticyclonic circulation decreased in both surface layer and the middle troposphere, especially near EAPE and EKE dissipation area. It should be noted that those areas where there was a decrease of anticyclonic activity in the Arctic zone, became more extensive and intensive. The tendencies were confirmed also by the CDF, and in January cyclonic activity decrease occured on the most part of the territory, in comparison with anticyclonic one (80% and 60% respectively) (figure 4).

In summer the situation with Vort+ was similar: there was the decrease of тшы parameter over the second interval, the curve of distribution function was shifted to the area of negative values. Concerning Vort-, the general situation from one period to another was not changed. It should be noted that in the beginning of the XXI century the increase in Vort- distribution function values in the range of extreme negative ones (below $10^{th}$ percentiles) was observed both in January and in July. For Vort+ this happened only in January. It means that considerable decrease in eddy activity in the certain areas occured over the whole West Siberia during last decade. This was especially pronounced in winter.

## 4. Conclusion

The conducted research showed that the distribution in positive and negative tendencies of energetic characteristics was significantly changed upon transition from the period of 1979-1998 to the period of 1999-2015 along with considerable increase in their trend values.

In the beginning of XXI century available potential energy values at the earth's surface increased in the south of the territory while almost total disappearance of its minimum area was observed in the western part. Kinetic energy values decreased in the central part of the region, however, positive values area in the north became more intensive and stretched along the entire Arctic coast. In the middle troposphere kinetic energy values also decreased in the central part, and there were no significant changes in available potential energy values.

Moreover, the tendency to decrease in cyclonic circulation intensity was revealed in West Siberia in the beginning of XXI century in the middle troposphere, whereas in the surface layer in January it increased in the central part of West Siberia – in the interfluve area of the Ob and the Irtysh rivers. Anticyclonic circulation intensity in Siberia decreased both in the surface layer, and in the middle troposphere, especially over the dissipation areas of EAPE and EKE. Their trend areas with negative values were located close to each other and were slightly covered in zone 60°-64°N. The greatest dissipation intensity was observed in the middle troposphere and this effect was less pronounced in the surface layer. Here eddy circulation generation along the Arctic coast was shifted to East Siberia. It means that the last decade was characterized by the considerable decrease in eddy circulation activity over West Siberia that was especially pronounced in winter. Furthermore, eddy circulation increased in certain zones, it indicates the possibility of anomalies in meteorological parameters formation there.

Thus, the research in energetic characteristics variability and relative vorticity parameter provides in more detail understanding and description into dynamics of regional climatic system elements and atmospheric circulation processes, in particular.

**Acknowledgements**
The work is supported by RFBR, according to the research project No.16-35-60088 mol_a_dk.

**References**

[1] IPCC 2013 *Climate Change: The Physical Science Basis. Contribution of Working Group I to the Fifth Assessment Report of the Intergovernmental Panel on Climate Change 2013* ed T F Stocker, D Qin et al (Cambridge – New York: Cambridge University Press) p 1535
[2] Akperov M G and Mokhov I I 2013 Estimates of the sensitivity of cyclonic activity in the troposphere of extratropical latitudes to changes in the temperature regime *Izv. Atmos. Ocean. Phys.* **49**(2) 113–20

[3] Pan Y et al 2017 Earth's changing global atmospheric energy cycle in response to climate change. *Nat. Commun.* **8** 14367

Parmesan C 2006 Ecological and evolutionary responses to recent climate change. *Ann. Rev. Ecol. Evol.* **37** 637-69

[4] Kharyutkina E V, Loginov S V and Ippolitov I I 2016 Influence of radiation and circulation factors on climate change in Western Siberia at the end of the 20th century and beginning of the 21st century *Izv. Atmos. Ocean. Phys.* **52**(6) 579–86

[5] Kalinin N A and Sviayzov E M 2008 *Transformation of Kinetic Energy in Midlatitude Cyclones* (Perm: University Publishing) p 116 (in Russian)

[6] Eliseev A V 2017 Linear and nonlinear models of baroclinic waves in atmosphere *Nonlinear Waves 2016* ed A M Sergeev and A V Slunyaev (Nizhniy Novgorod: IAP RAS) pp 129-42

Li L et al 2007 Lorenz energy cycle of the global atmosphere based on reanalysis datasets *Geophys. Res. Lett.* **34**(16) L16813

[7] Pan Y et al 2017 Earth's changing global atmospheric energy cycle in response to climate change *Nat. Commun.* **8** 14367

[8] Lorenz E N 1967 *The Nature and Theory of the General Circulation of the Atmosphere* (Geneva, Switzerland: World Meteorological Organization) p 161

[9] Van der Swaluw E, Drijfhout S and Hazeleger W 2007 Bjerknes compensation at high northern latitudes: The Ocean Forcing the Atmosphere *J. Clim.* **20** 6023–32

[10] Thompson D W J and Wallace J M 1998 The Arctic oscillation signature in the wintertime geopotential height and temperature fields *Geophys. Rev. Lett.* **25** 1297–300

[11] Kalinin N A 1999 *Energetics of Midlatitude Cyclones* (Perm: Perm University Publishing) p 192 (in Russian)

IOP Conf. Series: Earth and Environmental Science **190** (2018) 012031 doi:10.1088/1755-1315/190/1/012031

# Automatic meteorological measuring systems for microclimate monitoring

**M V Kiselev[1], N N Voropay[1,2], E A Dyukarev[1,3], S A Kurakov[1], P S Kurakova[1], E A Makeev[1]**

[1]Institute of Monitoring of Climatic and Ecological Systems SB RAS, Tomsk, Russia
[2]V.B. Sochava Institute of Geography SB RAS, Irkutsk, Russia
[3]Yugra State University, Khanty-Mansiysk, Russia

Email: xplutox@yandex.ru, voropay_nn@mail.ru, dekot@mail.ru, ksa@imces.ru

**Abstract.** The paper presents a description of the atmospheric-soil measuring complex (ASMC) intended for mobile and stationary long-term automatic measurements of the basic parameters of the atmosphere, soil and water. The device developed by the Institute of Monitoring of Climate and Ecological Systems of the SB RAS has a number of differences from analogues. To substantiate the possibility of using the device as a meteorological value meter, the data of measurements of air temperature and humidity, soil temperature, snow cover depth were verified. The data obtained by ASMC were compared with standard meteorological instruments. As a result of the analysis, high correlation coefficients between the time series were obtained. The average error of air temperature in the ASMS measurement is less than 0.2°C, and 91% of the measurements are within ±1°C. The average air humidity error is 5%, 70% of the measurements are within 5% of the error, and 95.5% of the measurements are within 10% of the error. The average error in soil temperature measurements during the warm period is 0.5°C, during the cold period is 1.5°C. The average error in measuring the snow cover depth is 5 cm and 90% of the measurements do not exceed it.

## 1. Introduction

Systematic observation of the climate system is usually carried out by national meteorological centres and other specialised bodies. They take measurements and make observations at standard preset times and places, monitoring atmosphere, ocean and terrestrial systems. Since national monitoring systems all form part of a global network, it is vital that there is as much consistency as possible in the way measurements and observations are made. This includes accuracy, the variables measured and the units they are measured in [1].

The existing network of meteorological stations of Russian Hydrometeorological service has more than 1500 stations, but their density decreases from west to east. Measurements at meteorological stations in Russia began more than a hundred years ago [2], but massive observations began in Russia in 1930-1950 [3]. Current rates and scale of climate change form new conditions for the existence of the natural environment [4]. For their control, it is necessary to have a developed network of climate monitoring stations covering large areas with high spatial resolution.

A unique atmospheric-soil measuring system (ASMS) was developed at the Institute of Monitoring of Climate and Ecological Systems of the SB RAS (Tomsk) [5]. ASMS is designed for mobile and stationary long-term automatic measurements and recording of the main parameters of the atmosphere,

IOP Conf. Series: Earth and Environmental Science **190** (2018) 012031 doi:10.1088/1755-1315/190/1/012031

soil and water. ASMS is different from existing measuring systems with low power consumption, a wide range of connected sensors, an autonomous mode of operation, operability in a wide range of environmental conditions and a low price.

Currently, more than 200 different modifications of ASMS operate on the vast territory of Siberia and the Far East.

## 2. Models and Methods

The atmospheric-soil measuring system can be equipped with soil and air temperature sensors, the depth and temperature of the snow cover, the level of bog or ground waters, the sum of atmospheric precipitation, wind speed and direction, soil and air humidity, solar radiation characteristics, carbon dioxide concentration, etc.

There are two types of ASMS - with the use of concealed installation loggers ("anti-vandal") and stationary non-masked stations.

Vandal proof ASMS designed for hidden installation which determines the inability to use unmasking structures: antennas, solar panels and many types of air sensors. As a result of these limitations, a controller with ultra-low power consumption (battery power) was developed with large internal non-volatile memory for data recording and USB interface, using a special cable with a sealed connector to connect to the logger, and maximum degree of protection of the enclosure IP68 (possible operation below groundwater level). Only underground sensors can be used in vandal proof ASMS: soil temperature, moisture, conductivity, as well as, water level, in some cases, snow depth and air temperature sensors (table 1).

**Table 1.** Main characteristics of vandal proof ASMS.

| Value | Characteristics |
|---|---|
| Size of non-volatile memory | 4 MByte (1 000 000 records) |
| Recording interval | from 2 second to 12 hours |
| Maximum number of connected digital temperature sensors, pcs. | 90 |
| Number of inputs 10-bit for analog-to-digital converter | 7 |
| Current consumption in stand-by mode, mA | 0,07 |
| Average current consumption in active mode, mA | 10 |
| Average battery life, years | 5 |
| Operating temperature range of the recorder unit, °C | from −55 to +65 |
| Interfaces | USB, SDcard, 1-Wire, I2C |
| IP Code | IP67, IP68 |

The microcontroller (ATMEGA1284) is used in the logger, which through the digital interface (1-Wire) interrogates temperature sensors (DS18B20), analog inputs and stores data in non-volatile memory (AT45DB321) with reference to the measurement time, using a clock (PCF8563T). In the memory of the logger, the supply voltage is also fixed (table 2). The USB interface (FT232RL) allows the ASMS to communicate with the computer to configure its operation and read data.

The atmospheric-soil measuring system, with an autonomous controller-registrar (ACR) as a base element, is designed to create monitoring networks [6]. ACR, in contrast to the hidden installation recorder, is supplemented with additional analog and digital interfaces (RS485, RS232 - 2 pcs.), and most importantly, it can connect to GSM, Wi-Fi, and satellite Internet networks through expansion modules. The power of the recorder is provided by the lead –acid battery 12FGHL48, which, through a circuit with a temperature correction of the charge voltage, is powered by a low-power solar battery.

Temperature and humidity sensors located at a level of 2 m are protected from sunlight and atmospheric precipitations by a plastic shield. In the wind speed and direction sensor, a variable resistor and reed switch are used. The total solar radiation sensor (2 channels) is placed on a mast at a height of 1.5 m. The radiation flux, passing through the PTFE protective filter falls on two blackened and two shiny aluminum plates heating them and by attaching them to the back side platinum thermoresistors determine the flux density of solar radiation from the difference in resistances.

**Table 2.** Main sensors used in ASMS.

| № | Component part | Measuring range | Accuracy |
|---|---|---|---|
| 1 | Temperature sensor DS18B20 | from -55 to +55°C | ±0,1°C |
| 2 | Air humidity sensor HIH-5031 | from 0 to 100% | ±3,5% |
| 3 | Soil moisture sensor TRIME-PICO32 | from 0 to 100% | ±2% |
| 4 | Atmospheric pressure sensor MPL3115A2 | from 500 to 1100 hPa | ±4 hPa |
| 5 | Wind speed and direction sensor Davis Anemometer 6410 | from 0,9 to 78 m/s<br>from 0 to 360° | ±5%<br>±7% |
| 6 | Liquid precipitation sensor Davis 7852M | from 0 to 1000 mm/h | ±5% |
| 7 | Pyranometer 2 channel (albedometer) Range from 0,35 to 9 μm | from 0 to 2000 W/m$^2$ | ±10% |
| 8 | The level of groundwater sensor | from 0 to 10,5 m | ±1% |
| 9 | Snow cover depth sensor | from 0 to 2 m | ±0,05 m |
| 10 | Ultrasonic sensor of snow cover depth MB7384 | from 0 to 4 m | ±0,015 m |

According to the sensor-profile of the height of the snow cover the snow level is calculated from the temperature difference at the air-snow interface. Liquid precipitations collected by the inlet section of the precipitation sensor funnel are poured into one of the rocker cups, which, after filling the cup (0.2 mm of precipitation), is tilted, substituting another cup for the funnel, while the magnet fixed to the rocker passes past the reed switch, causing its closure.

The probe of the soil temperature is located at a distance of more than 1 meter from the other sensors and the logger, and contains digital temperature sensors at different levels (from a surface up to 10 m deep into the soil). The probe is made in the form of a three-wire printed circuit board with soldered high-precision digital thermometers, protected by a shrink tube. The soil moisture sensor measures the ratio of the volumetric water content to the volume of the substance by the specific permittivity. The water conductivity is measured by the bridge circuit at the ultra-low frequency of the alternating current. The electrodes are made of stainless steel.

The ASMS software makes it possible to visualize the time series of data. The data is exported to text format. ASMS allows receiving on-line information on the atmospheric and soil climate characteristics, to compile a database for analysis of information.

Due to the use of original plastic mast constructions, the modularity of the constructs and the use of hermetically sealed connectors, it was possible to achieve simplicity in assembling, installing and repairing the complex in remote and inaccessible places (figure 1, figure 2).

## 3. Results and Discussion

Based on the results of parallel synchronous measurements of the atmospheric soil measuring system and the meteorological station Tunka (Republic of Buryatia), a comparative assessment of the data and an analysis of the correctness of ASMS use for autonomous monitoring were carried out. At the moment, the air temperature, soil temperature, air humidity and snow depth sensors have been compared within the period 2012-2018.

According to the Manual on Hydrometeorological Stations and Posts [7], measurements by thermometers and hygrometer are carried out 10 minutes before the end of the period, that is, measurements of the air temperature at the weather station should coincide with the ASMS measurement 15 minutes before the end of the period. With allowance for inertia correction, the smallest difference between measurements was obtained when measured 30 minutes before the end of the period.

**Figure 1.** Hidden vandal proof monitoring station (Tunka National Park, Republic of Buryatia)

**Figure 2.** Open monitoring station (Davsha passage, Barguzin National Park 30 km from Baikal)

Measurements of air temperature by a psychrometric thermometer in a Stevenson screen are made every three hours, and automatic measurements are made every 15 minutes. To understand the reasons for the deviations in the measurements of different instruments, four periods were recorded by ASMS: 15 minutes and 30 minutes before the end of the period, at the end of the observation period (on time) and 15 minutes after the end of the period.

Correlation coefficients between weather station data and data for all four periods of ASMS are more than 99%. The maximum correlation with the data of the weather station in time "for 30 minutes" (99.9%). Mean deviations in all terms were 0.1-0.2°C. The smallest differences were found between weather station data and ASMS data 30 minutes before the end of the observation period at the weather station. The absolute mean deviation is 0.5°C, the amplitude and dispersion are 0.6°C. Analysis of the deviations showed that within ±1°C there were 88% of measurements.

For comparison of air humidity observations, similar studies were conducted. The main method for determining air humidity at a meteorological station is a psychrometric one, which is based on measuring the air temperature and the temperature of a water-moistened thermometer. An additional method for determining air humidity is sorption, or hygrometric, based on the change in the length of the sensing element (skim hair) when the humidity of air is changes. The absolute average error of ASMS when compared with the data obtained at the meteorological station was 5%, the maximum overestimation – 28% and the maximum underestimation – 30%.

For a more accurate comparison, all measurements were divided into two periods: warm (T> 0°C) and cold (T <0°C). When comparing, it was revealed that the deviations between the data were higher in the warm period - an absolute error of 6%, while in the cold – 4%. The maximum overestimation in the warm period is 28%, in the cold period - 19%. The maximum underestimation in the cold period is 30%, in warm - 14%.

Regression equations for different temperature ranges were derived to improve ASMS data. Analysis of the data showed that after correction 54% of measurements fit to the range of deviations of ±3.5%, 70 % of data are within ±5% of error and 95.5% of measurements are in ±10% range of deviations (making the error of the hygrometer at the weather station).

The resulting differences and errors are partly due to poor air exchange in the Stevenson shield, which prevents rapid response to changing conditions in the atmosphere. In automatic equipment, radiation protection is made of plastic. It protects a little worse from direct solar radiation, but the design itself allows the sensor to respond better to atmospheric changes and measure the air temperature and humidity almost without a delay.

Measurements of soil temperature at weather stations by extracting deep soil thermometers are carried out at depths from 80 to 320 cm once a day and, therefore, are not averaged. While the data at depths from 20 to 40 cm in the warm half of the year measured 8 times a day were brought to average daily values. ASMS data on soil temperature for the period from April 1, 2013 to April 1, 2016 with 1 our step were also averaged to daily values.

Correlation coefficients have high values at all depths for each month (from 90% and higher). But at a depth of 240 cm in September, low correlation values were recorded - 66 (2014), 69 (2013), and 83% (2015). In September, the soil temperature, measured with the help of an atmospheric soil measuring system at a depth of 240 cm, is higher by an average of 0.5°C. Such differences were observed in all years. In addition, low values of the correlation coefficient were also observed at a depth of 320 cm and fall in April-May. The values of the correlation coefficient in these months vary from 55% in April 2015 to 6% in April 2014. Comparison of the temperature at these depths during these periods shows a difference of 0.1-0.2°C. Such differences are within the limits of the error (0.2°C) of the extracting thermometers. Thus, we claim that it is possible to neglect the low values of the correlation coefficient, referring to small annual temperature fluctuations at a depth of 320 cm.

So the greatest deviations in soil temperature from ASMS observed at a depth of 20 cm (higher by 1.8-2.0°C) in the cold season. Differences in soil temperature values decrease with the transition from the cold period to the warm one, in summer at a depth of 20 cm they are minimal and average 0.5-0.8°C.

With increasing depth, there is a decrease in deviations between the ASMS and the extracting thermometers. So at a depth of 160 cm, the soil temperature in the ASMS is more by 1.0-1.5°C in spring and summer and less by 0.1-0.3°C in autumn and winter.

Thus, the obtained results show that soil temperature sensors are suitable for monitoring the temperature regime of soil and have small errors at depths of 20 to 320 cm in the autumn, spring and summer periods. Large errors arising in the winter period are partly due to the features of temperature measurement by soil extracting thermometers. At the same time, it is possible to make an additional correction to the temperature in the winter period in order to neutralize the data inaccuracies.

The increase in air temperature has a significant impact on the state of soils in the permafrost zone, which occupies more than 65% of the territory of the Russian Federation. The most important climatic parameters that determine the temperature of the soil and its freezing / thawing, are air temperature and snow cover depth [8].

To solve the problem of automatic measurement of snow cover depth, the method for determining the air-snow boundary was chosen for finding temperature extremes in the vertical temperature profile. The depth of the snow cover is determined by exceeding the preset threshold value of the difference in readings between two neighboring thermometers, starting from the upper one. The measurement period is 1 hour. The number of temperature sensors in the measuring rail is 20, the distance between the sensors is 25 or 50 mm.

To verify the calculation of the height of the snow cover, the meter was equipped with an ultrasonic height sensor, which corresponds to the data of the snow measuring rod. For calculation, the time interval was selected from 09.09.2017 to 31.03.2018.

The correlation coefficients between the sensors for the entire period were more than 90%. At the beginning of the cold period (September-October) ASMS overestimates the depth of the snow cover by 5 cm, which is within the declared error. During the period of active accumulation of snow (October-January), 90% of the measurements are within the limits of the error, the remaining 10% of the ASMS understates the snow cover depth by 6-10 cm. At the end of the cold period (February-March), the ASMS lowers the depth of the snow cover by 5 cm, which is also within the measurement error of the instrument.

Using the methods of mathematical modeling of the measurement of depth and temperature of the snow cover, it is possible to calculate the effective thermal diffusivity of snow and determine its density.

## 4. Conclusion

The atmospheric soil measuring system (ASMS) developed at the Institute for Monitoring of Climate and Ecological Systems of the SB RAS allows to conduct microclimatic and meteorological measurements to be made in difficult-to-access landscapes. The ability to install diverse systems - hidden vandal proof and open, as well as a flexible sensor connection system gives the consumer the opportunity to realize their needs.

In addition to sensors, ASMS can be equipped with modern communication and data transmission systems, and can also be powered by various types of energy carriers (solar panels, batteries). A relatively low price and simple, but high-quality materials allow the consumer to create networks of automatic monitoring in many landscapes, including water bodies [5,9].

Verification of ASMS sensors with standard meteorological data showed that sensors for air temperature, air humidity, soil temperature and snow depth can be used to measure the corresponding meteorological quantities. The series of air temperature data have a correlation of more than 99%. The average deviation in all terms does not exceed 0.2°C. Within ±1°C there are 91% of measurements.

The humidity sensor has an average error of 5%. After correcting the data, 54% of measurements entered the deviation range of ±3.5%, 70% within ±5%, and in ±10% (making the error of the hygrometer at the weather station) have 95.5% of the measurements.

The soil temperature data series have correlations of 90% or more for different soil profiles in winter-spring, the soil temperature according to ASMS is higher by an average of 1.0-2.0°C, in summer and autumn it is lower by an average of 0.1-0.8°C.

Snow depth data have an error of 5 cm, which includes 90% of all values. The maximum error is 10 cm.

## Acknowledgments

The reported study was funded by Russian Foundation for Basic Researches according to the research project №18-05-00306.

## References

[1] WMO *2016 The Global Observing System for Climate: Implementation Needs* p 303
[2] Pavlov A V 2008 *Monitoring of Permafrost Zone* (Novosibirsk: Geo) p 229
[3] Gilichinskii D A, Byhovec S S, Sorokovikova V A, Fedorov-Davidov D G, Barry R G, Chang T, Gavrilova M K and Alekseeva O I 2000 Use of data from meteorological stations to assess long-term trends changes in soil temperature on the territory of the seasonal and permafrost cryolythozone in Russia *Earth's Cryosphere* **4**(3) 59-66
Bugaev V I 2011 *Methodical Guidelines for the Calculation of Runoff from Undrained and Drained Bogs* (Saint-Petersburg: SHI Pub) p 112
Sherstyukov A B 2008 Correlation of soil temperature with air temperature and snow cover depth in Russia *Earth's Cryosphere* **12**(1) 79-87
[4] *The Second Assessment Report of Roshydromet on Climate Change and Their Consequences on the Territory of the Russian Federation* 2014 (Federal service for Hydrometeorology and environmental monitoring) p 605
Gruza G V, Rankova E Y, Rocheva E V and Smirnov V D 2015 Current global warming: geographical and seasonal features *Fundamental and Applied Climatology* **2** 41-62
Ciais P et al 2013 Carbon and Other Biogeochemical Cycles. In: Climate Change *The Physical Science Basis. Contribution of Working Group I to the Fifth Assessment Report of the Intergovernmental Panel on Climate Change* ed Stocker T F and Qin D et al (Cambridge: Cambridge University Press) p 1535

IOP Conf. Series: Earth and Environmental Science **190** (2018) 012031 doi:10.1088/1755-1315/190/1/012031

Anisimov O A and Sherstiukov A B 2016 Evaluating the effect of environmental factors on permafrost in Russia *Earth's Cryosphere* **2** 90-9

Golovatskaya E A and Dyukarev E A 2012 The influence of environmental factors on the $CO_2$ emission from the surface of oligotrophic peat soils in West Siberia *Eurasian Soil Science* **45**(6) 588-97

[5] Kurakov S A 2012 Autonomous Environmental Monitoring System *Sensors and Systems* **4** 29-32

[6] Bazarov A V, Badmaev N B, Kurakov C A, Gonchikov B-M N, Cibenov U B and Kulikov A I 2016 Measuring complex for automatic long-term monitoring of atmospheric and soil climatic parameters *Instruments and Experimental Techniques* **4** 158-9

[7] *Manual for Hydrometeorological Stations and Posts* 1985 (Leningrad: Hydrometeo Pub.) vol 4 chapter 1 p 300

[8] Dyukarev E A 2015 Influences of air temperature and snow cover on the seasonally frozen soil layer characteristics *Earth's Cryosphere* **19**(3) 39-45

[9] Kiselev M V, Voropay N N and Dyukarev E A 2017 The temperature regime of the soil sedge-sphagnum fen in the southern taiga of Western Siberia *Geography and Natural Resources* **1** 110-7

IOP Conf. Series: Earth and Environmental Science **190** (2018) 012032 doi:10.1088/1755-1315/190/1/012032

# The organization of geosystems in southern East Siberia (research and mapping)

**T I Konovalova**

Sochava Institute of Geography, Russian Academy of Sciences, Siberian Branch, Irkutsk, Russia

E-mail: tkonov@mail.ru

**Abstract.** Knowledge of moderngeosystems properties, regularities of their formation, and changes under the influence of natural and anthropogenic factors is the basis for the timely prediction of adverse events occurring during the change of environment. Abstract review of the individual components and static geosystems cannot bring substantive results in solving these problems. The situation is compounded by the need to study and mapping of regions geosystems, which are characterized by tectonic activity and intensive dynamic processes. Currently, when a significant of geophysical, geological data and landscapes virtually no development, dedicated to solving the problem of study geosystems such regions. Specificity of study is complex mapping geosystems, which are characteristic diverse genesis, different stages of development, in the synthesis of space and time into a single whole, the comparison of the current state of geosystems with the natural rhythms and patters of development of the natural environment. These studies are a theoretical model of reality, synthesized information on the functioning, dynamics and evolution of geosystems, the direction of their transformation, as in changing natural and anthropogenic conditions. In development analysis of the future state of geosystems these studies have high scientific and practical value.

## 1. Introduction

One of the major targets of modern geographic research is to find a solution to regional environmental changes evaluating and prognosis and thus to accumulate scientific data background for regional management. That is why research is based on the studies of geosystems composition, functioning, dynamics and evolution. Such studies are vital due to the lack of information about the structure of the regional geosystems. The knowledge of geosystems development regularities, their dynamics and transformation changes that occur due to the influence of natural and anthropogenic factors is the base for prediction of negative phenomena that may occur as a result of both natural and anthropogenic environmental transformation.

## 2. The object for study and methods

The object of this research is the geosystems of the Central Ecological Zone (CEZ) of Lake Baikal and of the extensive territory of the Vitim river basin. The CEZ includes the lake itself, its' islands, adjoining water protection zone and specially protected natural reservations. The CEZ matches the World Natural Heritage site "Lake Baikal". The total area of CEZ is 89165 km$^2$, including the water aria of the lake – 31500 km$^2$ and 24 950 km$^2$ of specially protected natural reservations. The area, occupied by water protection zone outside the natural reservations within the borders of CEZ equals to

20500 km$^2$. Lake Baikal is visited by a huge number of tourists from China and Mongolia and other countries. Based on this study, the most unfavorable areas for recreation and other anthropogenic activities.

The Vitim flows across the Stanovoi and Vitim Uplands and the margin of the Patom Upland. The study area refers to the Sayan-Baikal Stanovoi high mountains whose territory is encompassed by intense modern orogenesis and riftogenesis. The region is characterized by an active reconfiguration of the river network, incessant mountain glaciations since the Pleistocene, and by a high pace and diversity of relief forming processes [1]. The area is a promising source for extraction of gold, uranium, copper and other resources. Also, the landscape structure is experiencing transformation. In addition to natural geosystems, their anthropogenic modifications are of widespread occurrence here. In this area there is a single transport system of Siberia. This is the area where the Transsib and Baikal-Amur railway pass. The Power of Siberia gas pipeline will pass through this territory. Knowledge of the landscape features of the territory will enable the safe development of natural resources of the territory.

Landscape research is based on the results of long-term studies of Siberian geosystems, terrestrial and aerial itinerary studies, interpretation of satellite images and literature data analyses. The work was supported by the Russian Fundamental Research Fund. Methodology geosystem study associated with the implementation of synergetic approach and the further development of the theory of geosystems V.B. Sochava.

## 3. Results and discussion

Most of the region's geosystems are situated within the boundaries of the Vitim, Baikal-Patom and Stanovoi Uplands. The mountain relief of the Vitim basin was transformed by the recent riftogenesis characteristic for Baikal Rift Zone (BRZ). On the other hand, its development influenced the formation of geosystems in the areas adjacent to BRZ. It is thought [2] that the manifestations of deep-seated riftogenic processes in the modern structure and relief of the study region are not limited to the territory of the rift zone bat are spreading from its lateral boundaries westward by 450 and eastward by 300 km. To put it otherwise these structures- and relief-forming processes in an attenuated form continue to be pronounced within the study territory in the Cisbaikalian pre-rift transition zone. The development of the relief, predetermined by the formation of BRZ, led to isolation of the Vitim and Baikal-Patom Uplands which differ in landscape characteristics from the other areas of Transbaikalia. According to the structure of relief and to isolation, they, like the Okinskoe Uplands, may be called Tibet in miniature where the territory with uniform topography is surrounded by high mountains.

### *3.1. The development of geosystems the study area*

The cycle of development of the study area distinctly shows two stages: the pre-rift and rift sages [3]. The early stage is time-coincident with the Oligocene and Neocene when the aftermaths of the riftogenesis manifested themselves for the first time in the region. The locally occurring volcanic activity did not have any substantial influence upon the dominant (in the region) low-mountain deciduous-coniferous (spruce, Tsuga, pine and birch with broad-leaved species) geosystems functioning in conditions similar to subtropical climate.

The end of the Miocene – beginning of the Pliocene was marked by intense volcanism characterizing the second development stage of the southwestern part of the region, which terminated in a relative tectonic calm. The Quaternary tectonic processes had the most dramatic influence upon the areas, which were involved in the uplift and were adjacent to BRZ. On the other hand, most of the Vitim Upland was also evolving subsequently as an autonomous geosystem. The northern part of the tableland in the Eopleistocene, with the riftogenesis encroaching upon the territory of the Stanovoi Upland, had sim-ilar features of development with those of the riftogenic structures of the North-Baikal and Stanovoi sections of the rift zone. At that time, the mountain ranges (Yuzhno-Muiskii, Kalar, and Yankan) were rising slowly along the northern and northeastern margins of the tableland.

IOP Conf. Series: Earth and Environmental Science **190** (2018) 012032 doi:10.1088/1755-1315/190/1/012032

The uplift was time-coincident with the glaciation of the highest mountain ranges. The continental sandy layer (lacustrine, fluvial, aeolian, and water-glacial) was forming in almost all hollows of the region. These deposits absorbed the persisting patches of Miocene landscapes. At that time, the differentiated tectonic movements were accompanied y volcanism. They were most active within the basins of the Bol'shoi Amalat and Dzhilinda rivers to the north-east of the area of Neogene volcanism. The fissuring mechanism operated at the zone of faults. In the Mid- Eopleistocene there emerged an extensive Neogene-Quaternary lava plateau with the area measuring about 4000 km$^2$. However, the southern areas of the Vitim Upland, namely the Uda-Vitim interfluves, and the Konda basin, were only slightly affected by these movements as before.

To date, the upper reaches of the Vitim as well as the area of the Ingur river mouth have retained evidence of volcanic activity of both Miocene (20-2.5 mil. years) and Holocene (10 thou years) age, with covers of basalts and cones of the extinct volcanoes: the Mushketov, Obruchev, Lopatin, Dombrovskii and other volcanoes. The largest basalt cover, Meister Plateau, is situated in the west of the region. It is thought that the basalts of the Vitim Upland are age analogs of the basalts of Eastern Sayan, Khamar-Daban and the Dzhida river basin [4]. Continuous basalt covers introduce some diversity into the region's landscape structure.

In the Mid-Neogene, Siberia saw the start of climate aridization implying a decrease in humidity of air masses caused by the regression of the sea basins in Western Siberia and the Far East [5, 6]. In the Mid-Pliocene, the climate was semihumid and moderately warm. Considerable spaces were occupied by steppes and forests growing on the tops of mountains and hills. Toward the Mid-Pleistocene the conditions of gradual humidification of climate [7] formed sparse pine-birch forests withy fir, and with the inclusion of lime and oak. The grass cover was dominated by xerophytes: grasses, wormwoods, and Chenopodiaceae.

The final formation and spread of the taiga was taking place throughout the entire Quaternary; it was associated with three florogenetic centers: the Southern-Siberian, Angaridian and Okhotsk centers [8, 9]. The Southern-Siberian center developed forests of Siberian larch and Scots pine. Nowadays, the forests of this group occur in the south of the region. The middle and eastern areas are dominated by the communities of the Angaridian center which are younger that the Southern-Siberian communities. The main contribution comes from Arctic dwarf birch, Middendorff's birch, etc. Their development was influenced by the general worsening of the climate at the time of the Late Pleistocene cooling. The forests of this group are dominant in the middle part of the region. Boreal vegetation, with the species of the Beringian complex predominating in its composition, occurs in the north of the region. These include black birch, Erman's birch, and others.

### *3.2. The spatial structure of the region's geosystems*

The spatial structure of the region's taiga geosystems bears evidence of the influence not as much from the latitudinal zonality as the history of BRZ formation. The geosystems are clearly differentiated according to the manifestation of tectonic processes and associated development stages of the region's geosystems. At present, they refer to the Baikal-Dzhugdzhur mountain-taiga region. The study area refers to two provinces: the Vitim taiga-flat-mountain province and the Western-Transbaikalian mountain-taiga-goletz province [10]. We subdivided the region's geosystem into five districts (figure 1).

The Uda-Kondinskii district occurs in the southern part of the territory, which has not experienced any substantial changes through its evolution. The region's oldest morphological relict, the pediplain, has persisted to date. The territory is the home for Lower-Cambrian sandy deposits and Pre- Cambrian granites. The same applies for the Pliocene-Pleistocene relicts, namely the subtaiga light-coniferous forests with patches of meadow steppes, tracts of stony steppes and limestone massifs with populations of relict species, such as *Allium altaicum, Hemerocallis minor*, and others.

The Dzhilinda-Amalatskii district territorially coincides with the ancient basalt plateau, with different-age effusive formations occurring on them. Of widespread occurrence, here are weakly dissected (by erosion) volcanic massifs with step relief and volcanic cones. It is dominated by *Betula pendula* and *Larix gmelinii*, willow-Duschekia taiga-small-grass forests (*Purola incarnate*, and *Maianthemum bifolium*), as well as by tracts of mountain moss-yernik stony tundras.

The Konda-Tsipinskii district lies within the boundaries of the middle and northern parts of the Vitim Upland. It occurs at the denudational, weakly dissected tableland with massifs of farewell rocks and pediments dating back to the Neogene. The region is exemplified by flat or dome-shaped watershed surfaces with the Pliocene red weathering crust and Pleistocene alluvial and lacustrine deposits. The piedmont plains are along-valley pediments. They increase in height from south to north from 20 to 300 m. The oldest relief occupies the summit plane of the region. Its northern part is adjacent to the wing of the Baikal Arch which accelerates its growth and, as a result, was involved together with it in the uplift. At present the district refers to the region of decelerated weakly differentiated recent uplifts. The region's marginal parts are composed largely of solifluction and defluction Quaternary rubble and block-rubble loams. High flat, heavily swamped areas are dominated by *Ledum palustre* larch (*Larix gmelinii*) forests, the moss-lichen cover of which produces a continuous mantle that interrups only on rocky benches of slopes. Green-moss and sedge larch forests occur in topographic lows in conditions of adequate humidification.

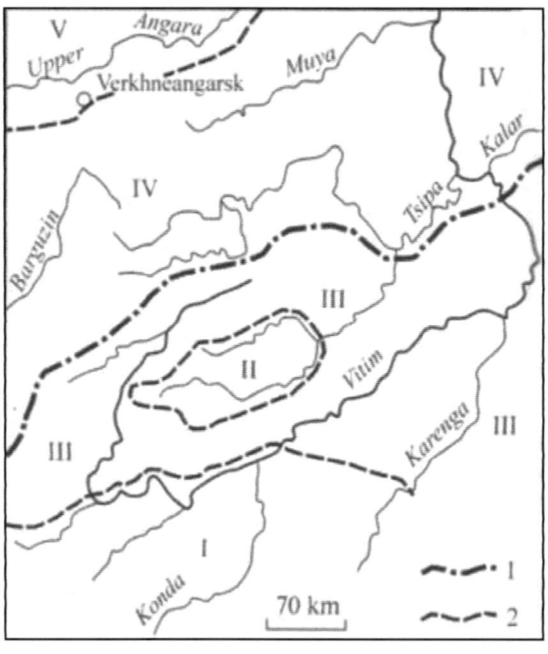


**Figure 1.** Physical-geographical regionalization of the Vitim river basin.
Physical-geographical districts: (1) Uda-Kondinskii, (II) Dzhilinda-Amalatskii, (III) Konda-Tsipinskii, (IV) Norh-Baikal-Stanovoi, (V) Vitimo-Patomskii.
Physical-geographical boundaries: (1) provinces, (2) districts.

The North-Baikal-Stanovoi district, like the next Vitim-Patom district, forms part of the Western-Transbaikalian mountain-taiga-goletz province. It is characterized by active neotectonic processes associated with the development of BRZ, and with the formation of alpine-type mountain topography and Cenozoic hollows of the Baikalian type. Everywhere there occur granites, basites and granodiorites, and the hollows show the presence of lacustrine-alluvial and glacial deposits. It is the area of formation and development of young progressive goletz and subgoletz geosystems. On the mountain slopes, there occur sedge-yernik larch sparse elvin woodlands, having a physiognomic resemblance with northern-taiga geosystems, Siberian mountain dwarf pine subgoletz geosystems, and Erman's birch sparse forests. The North-Baikal depressions are characterized by swampiness of floodplains of the main rivers. Powerful wintertime inversions are favorable for the development (in the lower parts of the intermontane depressions) of larch-taiga sparse yernik cryogenic groups of facies. The elevated trains, terraces and sand massifs in conditions of a different hydrothermal regime provide fragments of subtaiga geosystems on deep-freezing earth materials, alternating with tracts of barchans sands.

The geosystems of the Vitim-Patom district shaped themselves in conditions of the development of differentiated neotectonic movements. The district is characterized by a non-uniform, stepwise lowering of the surface northward thus exposing to the winds from the Arctic Ocean. This determines a dominance of lichen tundra's and oppressed sparse larch forests on its northern margins. On the whole, on the territory of the district there occur taiga sparse Siberian mountain dwarf pine and lichen larch forests with mixed undergrowth. They are genetically related to the development of mountain-tundra forests, at the same time exhibiting evolutionary-dynamical interrelations of the Beringian and Angaridian centers. Examples are provided by forests of Daurian larch with undergrowth of Siberian mountain dwarf pine.

### *3.3. Landscape structure of Lake Baikal and its protection*

Lake Baikal and bordering area are regarded as unique objects of universal importance. The lake itself and surrounding mountain ranges belongs to the area of Baikal continental rift zone. The existence and development of this rift zone are associated with enhanced intensity of the heat flow rate and with the earths' crust warming due to the mantle afflux [11]. The Miocene epoch was marked by massive basalts discharge. Miocene period is associated with the formation of brown soils that nowadays underlay fir and cedar tall grass taiga geosystems of the Khamar-Daban northern macroslope. Intensive tectonic movements of the late Pliocene led to the raising of mountain systems around Lake Baikal and on the east of the Eurasian continent, forming the orographic barriers that influenced the atmospheric circulation [12]. Western air mass disturbance and Asiatic anticyclone greatly influenced the transformation of geosystems.

The growing aridization led to the expansion of the steppe geosystems on the south of the region. This time the precipitous mountain slopes on the shores of Lake Baikal were occupied by dry steppe geosystems with oxytrope, and selaginella. Pliocene –a time of Siberian taiga geosystems and Baikal-Altai forest-steppe complex formation [13]. At the beginning of the Quaternary period due to the Asian and Indian plate's collision new activation of tectonic processes in the region starts. This furthers the raising of the mountain chains and intensive erosional breakdown. The natural environment transformations, growing aridization and climatic cooling, contributed to the dark coniferous taiga geosystems expansion to the south of the region and to the alpine meadows evolvement in the highlands. The south of the region was occupied by mountainous steppe, which can now be found on the western shore of Lake Baikal. Late Pleistocene climatic cooling that coincided with the Baikal rift zone mountains raising led to the development of mountain-valley glaciation. So the mountain tundras and under-goltsy altitudinal belt of cedar and Silver fir woods were formed in the western highlands of the region, while the alpine meadows prevailed in the eastern highlands [14].

In the taiga forest belt was dominated by fir-cedar-ledum and rhododendron groups of slope facies. Stone birch (*Betula ermanii*) penetrated into the region through the mountain highlands during this time. Today this birch is considered to be the northern near-Pacific specie, therefore lake Baikal region the south-western border of its' natural area. The peculiar dark coniferous taiga forests with Siberian cedar and Siberian dwarf pine were formed along the periphery of the glacier at the nearest proximity to the lake. The spread of subsoil glaciation during the Sartan Ice Age, led to the dwarf birch thickets formation [15]. Between the Pleistocene and Holocene ages, the strengthening of climatic continentality and the permafrost development on the northern and eastern parts of the region determinated the formation of new progressive cryosolic taiga geosystems with Dahurian larch. Today the Dahurian larch and Siberian Dwarf pine are both referred as typical representatives of Baikal-Djugdjursky physical geographical subzone.

The historical evolution of Lake Baikal geosystems contributed a lot to the formation of their unicity and diversification. The peculiar features, created by interaction between the atmosphere, the surface area and the water mass of lake Baikal within its' protruded and surrounded by the mountains with highly dissected relief hollow led not only to the formation but also to the preservation of many unique geosystems. The regional factors of landscapes differentiation here are complexities by local geographical peculiarities such as thermal water exits, salination and marshes formation processes, squall wind influence. Landscape contrasts within the Central Ecological Zone of Lake Baikal can be of a very large scale. The Selenga river delta and bordering area can be considered as one of the examples. Here due to the neotectonic movements and the foundering of the crust we can find the combination of marches and dry steppe types of geosystems, which is determined by the groundwater proximity and arid climate. The diversity of the CEZ natural conditions determined the formation of special altitudinal zonality types on the slopes that face the lake. As a result, the geosystems distribution within the CEZdoes not directly correspond with the altitudinal and latitudinal zonality. Instead CEZ of lake Baikal has a peculiar unique combination of mountain taiga and steppe geosystems: the altitudinal zonality in this region is determined by the barrier humid and arid effects that form the continental climate on the western shore and the costal climate on the eastern shore.

Petrologic composition of the area had also influenced the formation of geosystems modifying the latitudinal regularities of their distribution. This influence can especially be observed in the conditions that are far from the ecological optimum. For example, the presence of Proterozoic carbonaceous rocks on the western "continental" shore strengthens the low humifying rate effect, furthering the formation of peculiar under taiga geosystems with combination of grass larch forests and meadow steppes. Landscape contrasts within the Central Ecological Zone of Lake Baikal can be of a very large scale. The Selenga river delta and bordering area can be considered as one of the examples. Here due to the neotectonic movements and the foundering of the crust we can find the combination of marches and dry steppe types of geosystems. This determined by the groundwater proximity and arid climate. Landscape contrasts within the CEZ of Lake Baikal can be of a very large scale. The Selenga river delta and bordering area can be considered as one of the examples. Here due to the neotectonic movements and the foundering of the crust we can find the combination of marches and dry steppe types of geosystems, which is determined by the groundwater proximity and arid climate.

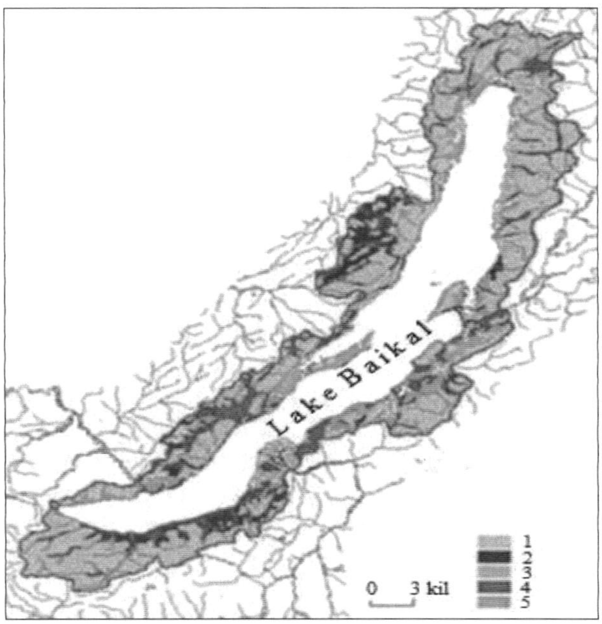


**Figure 2.** Environmentally oriented conservation functions of the geosystems (Explanations are in the text below)

The altitudinal zonality of Barguzinsky and Khamar-Daban mountain ranges differs from the Southern-Siberian mountain zonality type by the presence of Silver fir geosystems along the upper border of the forest belt. The existence of the stone birch steep slopes geosystems is another peculiar feature of Lake Baikal altitudinal zonality. These geosystems are considered as the elements of northern-pacific costal altitudinal zonality types. In the intermountain depressions, valleys of smaller rivers and mountain gorges we can observe the so-called closed depression effect that appears due to the cold air stagnation and extensive permafrost development. These conditions favor the formation of larch-taiga types of geosystems that are genetically associated with ancient periglacial Siberian geosystems. Such landscapes combined with larch-peatmoss bog forests, forestless bogs and dwarf birch thickets, which explained by the Pacific monsoon action. As a result, these geosystems are referred as the fragments of the Far-Eastern nature. The mountain-valley larch and fir geosystems are also referred as the Amur-Sakhalin (Far-Eastern) ones. Another distinctive feature of the Central Ecological Zone of Lake Baikal landscapes is the presence of the so-called false under-goltsy vegetation belt with larch and dwarf-pine. Obviously the reasons for the formation of such geosystems are not only the cooling influences of the lake, but also the temperature inversions that contribute to the continentality of the climate and to the relict character of geosystems; during the glaciation these geosystems "migrated" from the highland to the lakes shore. Dark coniferous fir geosystems with a higher humidity level that are more related to the broadleaf forests are confined to the southern shore of Lake Baikal. Unicity and environmentally oriented conservation functions of the geosystems were classified through the expert assessment. The unique geosystems were classified at 5 categories according to their environmental, water balance and landscape diversity preservation capability (figure 2).

1. Water-control geosystems zone. The geosystems of this category are vital for the regions' water-balance preservation. Changes in their structure will lead to the transformation of the whole regions ecosystem. This category includes goltsy and under goltsy sparse forests, mountain-taiga geosystems of reduced development.

2. Relict geosystems zone. These geosystems reflect the complexity of the regions geosystems formation and development. The examples of such geosystems are poplar and chosenia woods of intermountain depressions,humid fir geosystems that are close to the Pliocene nemoral forests. Due to the limited ecological potential, these relict geosystems are likely to become the subject to destruction.

3. The zone of High geosystems diversity. The existence of such geosystems depends on many local geographical environment processes and conditions such as petrographic composition of the surface, salinization processes, thermal waters discharging, etc. Such types of geosystems can also be found in small river valleys and troughs, where the heat-moisture rate differs from the background ones. Local processes and factors limit the functioning of such geosystems, so any environmental transformation will lead to their elimination.

4. The zone of environment-stabilizing geosystems. Such geosystems developed on the unconsolidated lacustrine-riverine Quaternary deposits of in-termountain lower slopes' parts, valley bottoms and terraces. These pines and larch forests boundary with steppes and negative anthropogenic impact in this place will lead to the soils erosion, forests elimination and the dry out of smaller rivers.

5. The zone of typomorphic geosystems.They geosystems correspond with the contemporary environment conditions.

## 4. Conclusions and Recommendations

The conducted research has identified the low stability of the regions geosystems. It is recommending the implementation of limited industrial exploitation and governmental regulation of any land use in this region. This research may become the ground for the creation of a model for the regional sustainable development. Its' key point lies in the preservation of self-regulated geosystems development by the accurate managing of anthropogenic impacts.

## Acknowledgments

This work was done with financial support from the Russian Foundation for Basic Research (16-05-00902).

## References

[1] Snytko V A and Konovalova T I 2015 Transformation Mechanisms of Taiga Geosystems of Cisbaikalia *J. Geography and Natural Resources* **36**(2) 132-38

[2] Zolotarev A G and Savinskii K A 1978 Pre-rift Structural Zone in Cisbaikalia *J. Geologiya I Geofizika* **8** 60-8 (in Russian)
[3] Florensov N A 1968 *Baikal Rift Zone and Some Problems of Studying Baikal Rift* (Moskow: Nauka) pp 6-19 (in Russian)
[4] Endrikhinskii A S 1974 *The Vitim Upland The Uplands of Cisbaikalia and Transbaikalia* (Moskow: Nauka) pp 6-19 (in Russian)
[5] Logachev N 1986 *Sedimentary and Volcanogenic Formations of Baikal Rift Zone* (Moskow: Nauka) pp 72-101 (in Russian)
[6] Ravskii E A 1972 *Sedimentation and Kkimates of Inner Asia during the Antropogene* (Moskow. Nauka) (in Russian)
[7] Belova V A 1985 *Vegetation and Climate of the Late Cenozoic of the South of East Siberia* (Novosibirsk: Nauka) (in Russian)
[8] Sochava V B 1956 Dark-Coniferous Forests *Vegetation Cover of the USSR*, Explanatorory Note for the Geobotanicak Map of the USSR (Moskow – Leningrad: Izd-vo Akad. Nauk) pp 139-317 (in Russian)
[9] Tolmachev A I 1954 *Toward the History of the Origin and Devekopment of Dark-Conferous Taiga* (Moskow – Leningrad: Izd-vo Akad. Nauk) (in Russian)
[10] Sochava V B and Timofeev D A 1968 Physical-Geographical Regions of North Asia *J. Dokl. In-ta geogr. Sibiri I Dal'nego Vostoka* **9** 3-19 (in Russian)
[11] Priestley K and McKenzie D 2006 The thermal structure of the lithosphere from shear wave velocities *J. Earth and Planetary Science Letters* **244** 285–301 DOI:10.1016/j.epsl. 2006.01.008
[12] Glorie S, Grave J, Buslov M M, Zhimulev F I and Elburg M A 2012 Structural control on Meso-Cenozoic tectonic reactivation and denudation in the Siberian Altai: Insights from multi-method thermochronometry *J.Tectonophysics* **544–545** 75–92
[13] Dumitrashko N V 1956 Origin of lake Baikal and glaciations of Pribaikalie *Proceedings of Geomorphology and Palaeogeography of USSR* **15** 129-46 (in Russian)
[14] Tyulina L N 1950 From the history of vegetation cover of the north-eastern shore of lake Baikal *The Problems of Physical Geography* **15** 62-7
[15] Duchkov A D 2001 Evolution of thermal and phase state of Siberian cryolithic zone *Global Environmental Changes* (Novosibirsk: Nauka) pp 79-104

IOP Conf. Series: Earth and Environmental Science **190** (2018) 012033 doi:10.1088/1755-1315/190/1/012033

# Mercury and its bonding forms in the Soil of the Tunka Depression

**E E Lyapina[1,2], A A Cherkashina[3]**

[1]Institute of Monitoring of Climatic and Ecological System of SB of RAS
[2]Tomsk polytechnic university, Tomsk, Russia
[3]Sochava Institute of geography of Siberian Branch of the Russian Academy of Science, Irkutsk, Russia

E-mail: eeldv@mail.ru

**Abstract.** The work evaluates the general distribution of mercury in the soil cover of the Tunkinskaya depression (the Republic of Buryatia, the national park "Tunkinsky"). For this purpose, its gross contents in natural and agrogenically transformed soils were studied: plowlands, fallow lands, hayfields and pastures. The physico-technical characteristics of soils, the content of organic carbon, the group composition of humus are determined. The method of processing the results included the calculation of the ecological and geochemical parameters: the concentration coefficient relative to the background, MAC, Clark concentration relative to the Earth's crust, the Earth's soils, the identification of the relationship with the physical and technical characteristics of the soil, and the content of C02, CO2 carbonates, fulvic and humic acids. In soils, the largest part of mercury is in free form (>41%). The proportion of physically and chemically related mercury accounts for 9-16% of all mercury in the soil. The least amount of mercury in the soils is in the form of sulfides and chemically bound forms of mercury (<10%). It should be noted that this feature is noted both in native soils and in anthropogenically disturbed soils. The proportion of chemically and strongly bound mercury increases with depth, and the content of free and sulphide forms decreases.

## 1. Introduction

Soil is an important component of biosphere. It has several main functions, such as geological, geoecological, bioecological, bioenergy, hydrological, and meteorological. Soil is an intermedium for regulation of the water and air chemical composition, and for energy and matter exchange in many elements of the global biogeochemical cycles. All these functions determine many interrelations in the global mechanism of geographical environment functioning [1].

Mercury is pervasive, compound, and hydro- and sulfophilous. It is present in all types of ecological environment and has many occurrence forms, which complicate the process of its study. Even in very low concentrations mercury is supertoxic and superpathologic. It is found in the lithosphere in the form of solid compounds, in various gaseous phases and dispersed forms, each of which predominates in concrete physical-chemical conditions, but easily changes from one form to another [2].

## 2. Methods and research objects

This study aims to evaluate the general allocation of mercury in the soil mantle of the Tunka Depression (Republic of Buryatia, Tunkinskiy National Park). For that purpose, a total mercury concentration in natural and agrogenically transformed soil samples was examined [1]. Those samples were taken from arable land, set-asides, hay-fields, and pastures. Soil samples had been collected during the summer periods of 2015 and 2016, using two methods: envelope method according to [3] and method of study of genetic horizons including soil materials. In total, 26 sites were chosen and 77 soil samples were collected. Under laboratory conditions the samples were got to the air-dried state at a room temperature and screened (sieve size is equal to 1 mm).

The Tunka depression is a continuation to the west of the Baikal rift zone. Its length from east to west is 190 km (figure 1). From the south it is bounded by the Khamar-Daban Mountains and the spurs of the Khangarul ridge, from the north by the Tunkinsky loaches, from the west by the southeastern chain of the Great Sayan. In the basin there are six separate basins, separated by narrowings of the bed of the Irkut-Bystrinskaya, Torskaya, Tunkinskaya, Turanskaya, Khoytogolskaya and Mondina.

The sample analysis was conducted in the laboratories of International innovative academic center «Uranium Geology», Geoecology and Geochemistry Department, Tomsk Polytechnic University. Mercury content in the earth samples was detected on the Atomic Absorption Spectrometer RA-915+ with pyrolyzer PYRO-915 (pyrolysis method). Precision is equal to 5 ng/g. Mercury concentration was calculated on 1 gram of dry basis. Forms of mercury in soil samples were determined by thermal desorption using a modification of the attachment to the mercury analyzer RA-915 +. Besides, there were determined physico-technological characteristics of the soil samples, Corg concentration, and humus group composition. Results processing technique included calculation of ecological-geochemical indexes: concentration coefficient with reference to the background, maximum permissible concentration, percentage abundance with reference to the Earth crust and soil, identification of interrelations with physico-technological characteristics of the soil, and also content of Corg, $CO_2$ carbonates, and fulvic and humic acids.

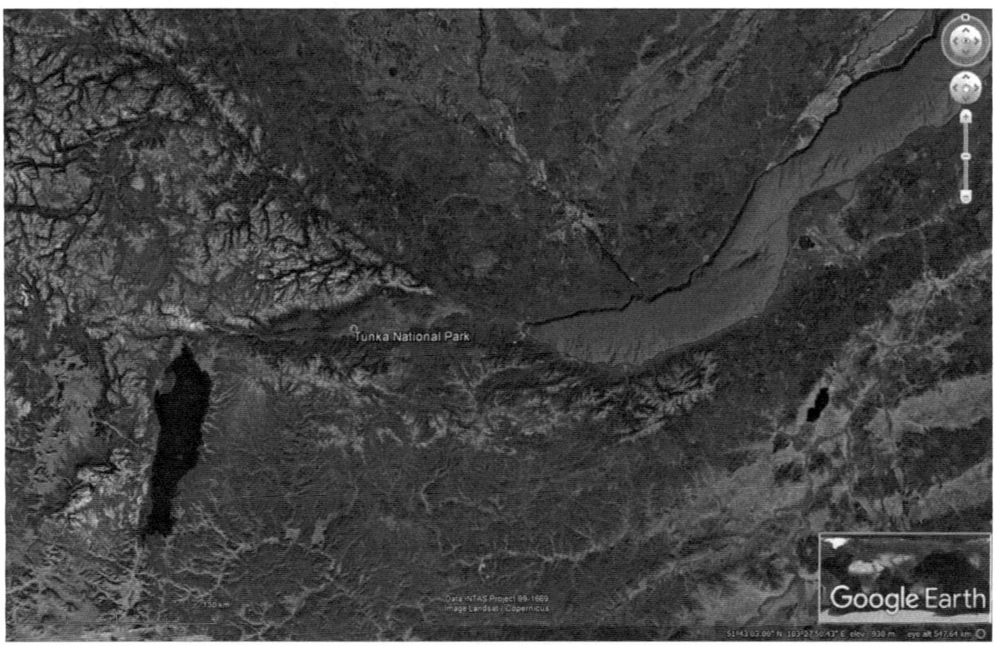


**Figure 1.** Map of the location of the Tunka depression.

## 3. Results and discussion

Average mercury concentration in the soil of the Tunka Depression is equal to 23 ng/g, assuming spread from 1 to 124 ng/g (table 1). All depth profiles of mercury concentration in studied soil cross-sections look similar: maximum values appear in the upper part of the profile with subsequent decrease. Acquired data corresponds to principles of mercury allocation in the soil in many other countries [4]. In addition, it is worth mentioning that the highest mercury concentrations are detected in upper horizons in the interval from 0 to 0,2 meters (figure 2). However, a depth of detection of the highest mercury concentrations is inhomogeneous: 26% of the highest mercury concentrations are found at a depth of 0-5 cm, 21% - at a depth of 10-15 cm, and 19% - at a depth of 5-10 cm.

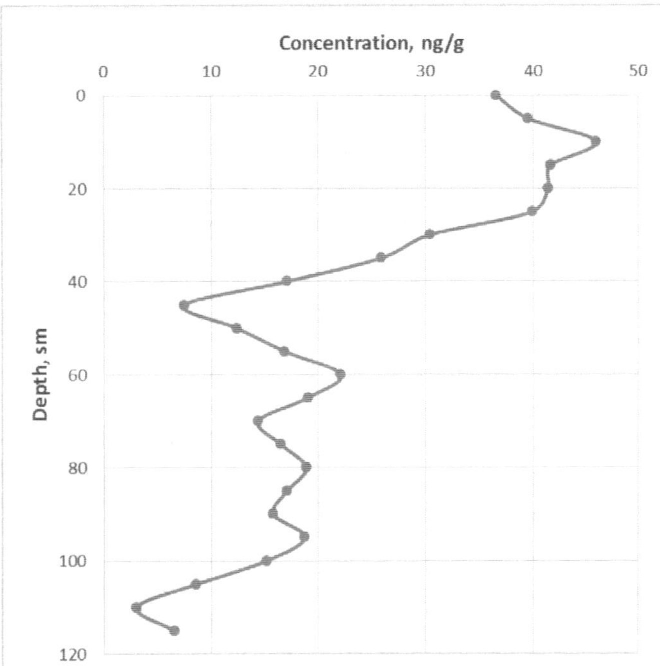


**Figure 2.** Distribution of mercury with depth in the soils of the Tunka depression.

It is found that upper horizons of the soil hold higher total mercury concentrations than lower horizons. Maximum values are detected in the detritus, peat, humus, and topsoil horizons. Relatively lower mercury concentrations are found in the vegetable and agrohumus horizons. In the soils of alfehumus (podbur and sod podbur soil) and texture-differentiated (sod brown podzolic soil) sections, which are widespread on the studied territory, it is clearly observed that there is a certain increase of mercury concentration in accumulation-illuvial horizons (BHF and BT). It can occur due to a downward migration of a mercury in the form of organo-mineral complexes from overlying humus and eluvial horizons [5].

In the studied soil profiles, two kinds of relation between mercury concentration and soil pH are revealed: positive one (r=1 when P=0,02) in transformed soil and negative one (r=-0,9 when P=0,04) in natural soil. There is a negative relation between mercury, Corg and $CO_2$ of carbonates in natural soil (r=-0,76 when P=0,05 for Corg; r=-0,88 when P=0,04 for $CO_2$ of carbonates). Also there is a positive relation between mercury, Corg and $CO_2$ of carbonates in transformed soil (r=0,82 when P=0,05 for Corg; r=0,99 when P=0,03 for $CO_2$ of carbonates). Interrelation between mercury concentration, and fulvic and humic acids on the studied territory is not identified. However, there is a steady interrelation between mercury and undissolved residue (r=0,5 when P=0,02).

Average mercury concentration in natural and agrogenically transformed soils is varied significantly from 20ng/g in transformed soil to 29 ng/g in natural soil. It is worth mentioning here that mercury concentration in transformed soil is lower and conforms to regional geochemical background (20 ng/g) [6]. It can be linked to blending of soil layers when plowing: thin humus horizons and detritus containing the highest mercury concentration are blended with underlying mineral horizons with considerably lower mercury concentrations [7].

Total mercury concentrations in the studied soil samples do not exceed maximum permissible concentration, but are higher than those of the regional background are. Average rate of this excess over regional background is equal to 1,16 (0,05-6,2). Calculated geoecological parameters of the mercury load on the soil of Tunka Depression reveal the rate of excessive mercury concentrations in the upper layer of the soil. It exceeds abundance of mercury in Earth mantle (up to 0,29 in average (0,01-1,55)) and in Earth soils (up to 2,32 (0,10-12,40)).

**Table 1.** The content of mercury, depending on the physical and chemical characteristics of the soils of the Tunka depression.

| Soil | Soil horizons [11] | Capacity, sm | CHg, ng/g | pH | Corg, % |
|---|---|---|---|---|---|
| **Anthropogenic** | P, rr, W, Pw | 0-37 | 20 | 8,0 | 1,21 |
| **Anthropogenic ledge** | AU, W, A0 | 0-30 | 16 | 7,3 | 1,56 |
| **Natural** | AH, AU | 0-13 | 29 | 6,6 | 4,31 |
| **Natural slope** | AY, Aye | 0-17 | 41 | 6,2 | 6,58 |
| **Forest** | T, Th, mr | 0-14 | 43 | 6,4 | 15,64 |
| **Bog** | T1, T2h | 0-12 | 23 | 7,7 | 53,80 |

According to the results of determining the forms of mercury in soils, it should be noted that in most of the soil samples studied, the largest part of mercury is in free form (Table 2), the share of which is over 41%. The proportion of physically and chemically bound mercury accounts for 9 to 16% of all mercury in the soil. This is due to the presence of organic matter in soils in the form of humic and fulvic acids [8], as well as with the activity of a number of microorganisms that transfer mineral forms of Hg into organic forms. These forms of mercury are most accessible to living organisms, and are more toxic. The second place in the investigated soils is the share of physically and firmly bound mercury compounds (5-56%). The lowest amount of mercury in soils is in the form of sulfides and chemically bound forms of mercury, less than 10%. It should be noted that this feature is noted both in native soils and in anthropogenically disturbed soils. The proportion of chemically and strongly bound mercury increases with depth, and the content of free and sulphide forms decreases.

**Table 2.** Forms of mercury thermoforms in soils [9].

| $T^0$, C | Form of mercury | Chemical compound |
|---|---|---|
| 150-160 | Free | HgO |
| 250-290 | Physically sorbed | HgC12 |
| 310-320 | Chemically sorbed | $Hg_2Cl_2$ |
| 350-410 | Sulphide | HgS |
| 500-1000 | Isomorphic | $HgSO_4$ |

## 4. Conclusion

As a result of the study, the total content of mercury in various types of both natural and anthropogenically disturbed soils was determined. At the same time, the average content in natural soils exceeds the concentration in soils experiencing anthropogenic impact. In this case, it is worth noting an increase in the content of mercury in natural soils associated with more acidic soil conditions. In disturbed soils, the concentration of mercury increases in more alkaline conditions. In addition, natural soils are characterized by an increase in mercury content at lower Corg values and $CO_2$ of carbonates. In disturbed soils the picture is the opposite. Calculations of geoecological indicators revealed an excess of mercury in the upper horizons above the background and Clarks.

**Acknowledgements**
This work was partly finacially supported by RFBR 14-05-00183 A (2014-2016) "Technogenesis and forecast of landscape transformation in conditions of global environmental changes, scientific basis for the regulation of nature management".

**Reference**
[1] Yanin E P et al 1999 Mercury. Integrated security system *Collection of Materials of the III-rd Scientific and Technical Conference* p 125
Belousov V M et al 2000 *Physical and Geographical Characteristics and Environmental Issues of the Southwestern Branch of the Baikal Rift Zone* (Irkutsk: IGU) p 152
Kasimov N S et al 2015 Clarks of chemical elements as comparison standards in ecogeochemistry *Bulletin of the Moscow University* **5**(2) 7-17
Sochava V B 1978 *Introduction to the Theory of Geosystems* (Novosibirsk: Science) p 319
Fridland V M 1972 *Soil Combinations and Their Genesis* (Moscow: Science) p 215
[2] Grebenshchikova V I et al 2008 *Geochemistry of the Environment of the Baikal Region (Baikal Geoecological Polygon)* (Novosibirsk: Geo) p 234
*Guidelines for Conducting of Field and Laboratory Researches of Soil and Plants by the Control of Environmental Pollution with Metals* 1981 (Moscow: Gidrometeoizdat) p 108
[3] Ivanov V V 1997 *Ecological Geochemistry of Elements* (Moskow: Ecology) p 576
Fridland V M 1984 *Structure of the soil Cover of the World* (Moscow: M-Press) p 235
[4] Kabata-Pendias A et al 1989 *Microelements in Soils and Plants* (Moscow: Mir) p 439
[5] Cherkashina A A et al 2013 Current state of the landscapes of the Tunka system of Basins *Soils in Space and Time* 125-6
[6] Motuzova G V 2009 *Connections of Microelements in Soils: System Organization, Ecological Significance, Monitoring* (Moskow: LIBROKOM) p 168
[7] Malikova I N et al 2011 Mobile forms of mercury in soils of natural and natural-technogenic landscapes *Geologiya i Fizika* **52**(3) 409-25
[8] Tatsiy Y G 2010 The method of thermal desorption as a method for determining solid-phase forms of mercury. Real opportunities *Mercury in the Biosphere: Ecological Andgeochemical Aspects* 32-7

IOP Conf. Series: Earth and Environmental Science **190** (2018) 012034 doi:10.1088/1755-1315/190/1/012034

# The mapping results of the landscapes of the Upper and Middle Yenisei basin

**G I Lysanova, Y M Semenov**

Sochava Institute of Geography Siberian Branch of the Russian Academy of Sciences, Irkutsk, Russia

E-mail: lysanova@irigs.irk.ru

**Abstract**. The issues of displaying information about diversity, differentiation, integration, dynamics and evolution of conditionally-natural and anthropogenically-altered geosystems on medium-scale maps are considered. The mapping technique is based on the theory of geosystems and the principles of constructing the hierarchical structure of geometers by integrating their indicators. To demonstrate the organization of low-steppe and forest-steppe topogeosystems in southern Siberia on a large scale, a classification was developed based on the principles of the theory of geosystems and the concept of the interdependence of geosystem organization with the differentiation of matter. When compiling maps of geosystems of a smaller scale, the methodological approaches of landscape analysis, synthesis and detailed landscape survey that are applied at the topological level are impossible. The main sources for medium-scale maps are remote sensing materials, topographic, thematic maps only partially controlled by the results of a wide-scale landscape survey and detailed mapping of key areas. The maps of parts of the south of Central Siberia are generalized to create a single medium-scale geosystem map (1: 1000000). When compiling it, the author's own ideas on the classification of geosystems obtained from the experience of landscape mapping of other regions are also used. In future information on geosystems obtained with landscape mapping will be interpreted from the positions of applied geography for the creation of landscape-evaluation maps in order to optimize nature management with the use of landscape planning. To adequately reflect the organization of the geographical envelope from the perspective of the theory of geosystems, it is necessary to develop a modern landscape classification based on correct quantitative data including three taxonomic series of geosystems: typological, horological and dynamic.

## 1. Introduction

Identification of the space-time organization of geosystems remains one of the most important areas of physical geography. Landscape mapping is designed to display and document the results of direct research and generalization of physical and geographical information on diversity, differentiation, integration, dynamics and evolution of conditionally-natural and anthropogenically-impacted geosystems. The scale and detail of the cartographic mapping of the structural and dynamic properties of the geographical envelope are determined by the objectives, research tasks and the hierarchical level of the main mapping units.

Currently some results of landscape mapping of the investigated territory have been summarized. For cartographic support, medium-scale landscape maps and schemes of physical and geographical zoning of the Nazarovo and Minusinsk depressions, maps of administrative areas (soil, vegetation, agricultural and land fund), synthesized satellite imagery, topographic, general geographic, thematic maps of various scales, schemes and materials of route studies and key areas were used.

## 2. Models and methods

The methodology of landscape mapping is based on the basic principles of the theory of geosystems by V.B. Sochava [1]. The classification of geosystems and the creation of a landscape foundation legend are based on the system-hierarchical approach to the identification of subordination of landscape taxons and the evolutionary-dynamic interpretation of the charted units. The structure of the map legend is proposed taking into account the positioning of the territory and the typological spectrum of geosystems in the planetary system [2]. The group of facies was chosen as the lowest charted unit of geomers.

Maps of administrative areas (soil, vegetation, agricultural and land fund), synthesized satellite imagery, topographic, general geographic, thematic maps of various scales, route study materials, key site maps, and previously compiled medium-scale landscape maps and charts, physico-geographical zoning of the Nazarovo and Minusinsk depressions, which were generalized to create a geosystems small-scale maps, were used to map this territory. The results of preliminary decoding of space images were transferred to a cartographic basis, which was a preliminary map layout. Field surveys used profiling in combination with work on key areas selected in regions with the most complex spatial structure of geosystems.

## 3. Results and Discussion

### *3.1. Large-scale mapping*

The development of the classification of lowland steppe and forest-steppe geosystems in the south of Siberia for their large-scale mapping was based on the principles proposed by V.B. Sochava [1] and the concept of the interdependence of geosystem organization with the differentiation of the matter of their components [3]. The method of compiling landscape maps with simultaneous display of geomers and geochors [4] allows us to represent typological and horological units of geosystems on one map, as well as their dynamic state. On these maps, the geohore mosaic includes geomeric parts. If geochors are shown in natural boundaries (taking into account, of course, the map scale capabilities), then the geomers are displayed depending on the scale and level of their generalization, i.e. on the rank.

The differentiation of geosystems is determined by the radial differentiation of matter, with each hierarchical level of geomers of topological level corresponds to certain amplitudes of matter reserve in soils and phytomass, increasing with the rank of the taxon. This enables classification of geomers using the results of landscape-geochemical analysis of geosystems by taking into account the absolute quantities of matter in their components, especially since geomers within geosciences differ in absolute amounts of these or other elements or their radial differentiation [5]. The main mapping unit for large-scale mapping of geomers is usually the facies. A facies combined biogeocenoses with one plant association and one soil variety within the limits of similar locations. By the principle of homogeneity, the facies are combined into subtypes, types and groups of facies. The basic criterion for combining facies into types (eluvial, translevuvial, transaccumulative and superaqual) is the landscape-geochemical conditions for the formation and functioning of typed geosystems, reflected in the classification of locations [6]. Within the facies types subtypes (para-eluvial, ortho-eluvial, neo-eluvial, trans-eluvial steep, very steep, steep and gentle slopes, transcumulative acclivious and gentle slopes, accumulative aligned surfaces), in which the nature and degree of migration of the substance are reflected [4, 7].

Integration of geosystems is caused by lateral flows of matter, and for lower levels of geochoras, unidirectional migration and a single balance of substance differentiation are typical, with all ranks of geochoras topological dimension, corresponding to specific ranks of geomers with certain limits of variability of absolute quantities of matter in components of subordinate geomers whose distribution

IOP Conf. Series: Earth and Environmental Science **190** (2018) 012034 doi:10.1088/1755-1315/190/1/012034

areas are limited by data frames of geochoras [3]. Therefore, it is advisable to carry out geochoric integration stepwise on the basis of landscape-geochemical synthesis of geosystems by taking into account absolute quantities of matter and revealing the balance of migration flows in subordinate geochoras [5].

Introduction of two additional taxa of topological order into the taxonomic system of geomers by V.B.Sochava [1] – a type and subtype of facies - allowed to observe a more complete correspondence between taxa of typological and chorological series. If, with the traditional generalization of the landscape map and zoning schemes, it is possible to intersect geomers areas by the borders of geochoras, then the conjugate mapping of geomers and geochoras makes it possible to generalize geomers in stages, consistent with generalization of geochoras, and therefore on such maps the intersection of geomers contours with the horological borders is absent [5].

Unfortunately, the approaches to landscape synthesis described above are applicable only at the topological level. One example of a medium-scale map constructed in this way is a map of the geosystems of the Nazarovo depression [3]. First, the indices of differentiation of the phytomass substance do not work here due to the large reserves of matter in woody plants, against anthropogenic transformation of native forest vegetation (felling, fires) and succession (spontaneous and regenerative) changes. Secondly, the absolute indices of the differentiation of matter in forest soils do not reflect the specificity of the geosystems: the presence of processes of podzolization, leavage, gleying, etc., which are of decisive importance in determining the classification affiliation of the geosystems of taiga and subtaiga, are not indicated by the total indices of substance differentiation. Besides, with the expansion of the research area in the steppe and forest-steppe regions, serious difficulties were found in interpreting the relationships of substance differentiation and the organization of geosystems. Therefore, in the following years our research was aimed at understanding the role of not the substance as a whole in the formation of the geosystem structure, but in identifying the role of natural and antropogenic flows of specific chemical elements and their associations in the formation of typological diversity, regional and topological patterns of integration of geosystems in mountain taiga and subtaiga.

### *3.2. Medium-scale mapping*
In medium-scale landscape mapping a direct survey with subsequent control of the correctness of the boundaries of all selected contours is, of course, impossible. Therefore, topographic and thematic maps, remote sensing materials, a multi-scale landscape survey and detailed mapping of key areas are used in the mapping of medium-sized geosystems. The mapping of the geosystems of various areas of the Minusinsk depression included the compilation of large-scale maps of the nearest environment of physicogeographical field stations (Berezovskii forest-steppe in the Nazarovo depression, Lena mountain taiga in the foothills of the Western Sayan, and Novonikolaevskii steppe in the Minusinsk depression), and analysis of long-term observations of behavior conditionally-natural and anthropogenically-modified geosystems, the definition of dynamic and evolutionary trends in the development of geosciences and the extrapolation of the results obtained in the course of long-term field investigations, on the adjacent field model. Considerable attention was paid to the problems of determining the landscape-geochemical, landscape-geophysical and biogeocenological indicators of anthropogenic transformation of geosystems under the influence of urbanization, emissions of industrial enterprises, agriculture, and forestry for cartographic mapping of anthropogenic changes in natural formations [5].

On the landscape map of the Nazarovo depression [3] geomers and geochoras are simultaneously displayed, and on the map of the Minusinsk depression [8] - only the geosystems of the typological classification series. The main mapping units on both maps were geomers of the rank of groups of facies, united in the map legend into facies classes and geomes. The map of the geosystems of the Minusinsk depression was compiled using the synthesis of two approaches to the classification of geosystems: as separate families, it shows both conditionally-natural and anthropogenically-modified geosystems - agrolandscapes [8]. At present, these maps of the parts of the Minusinsk depression [3, 8] are being generalized to create a single map of the geosystems of the south of Central Siberia.

Sometimes, the reverse approach is also used, i.e. the detailing of maps of a smaller scale. It is known that when the scale is enlarged, the mosaic and the pattern of contours usually change, and the appearance of additional taxa requires a radical processing of the legend's content. This approach was used in drawing up the basis for landscape planning of the northern regions of the Altai Republic [9], when the map of the Altai landscapes [10] was interpreted taking into account the author's experience in mapping geosystems of adjacent regions and refined based on the own materials of field studies and detailed mapping of key areas.

In recent years the Sochava Institute of Geography is working on the compilation of a landscape-assessment map of the Asian part of Russia for the purposes of territorial planning and the ecological and geographical justification for the creation of large economic facilities. When compiling its sheets, a system-hierarchical approach is used to identify the subordination of landscape taxa and the evolutionary-dynamic interpretation of the charted units, and map legends are constructed taking into account the positioning of the territory and the typological spectrum of regional geosystems in the planetary system [2, 11].

Up-to-date, some results of landscape mapping of the basin of the upper and middle Yenisey (Nazarovo, Kansk, Minusinsk and Tuva basins, and their mountain frames) have been summarized [12-14]. Medium-scale landscape maps and schemes for physiographic division of the Nazarovo and Minusinsk depressions [3, 8], maps of administrative areas, synthesized satellite imagery, topographic, general geographic, thematic maps of various scales, charts and materials of field studies and key sections were used to map this territory. The group of facies was chosen as the lowest mapping unit of geometers. In general, about 200 groups of facies have been allocated on the territory of the region which are united in classes of facies, geomes and groups of geomes. On the small-scale landscape map of the region the differences in the complexity of the horizontal structures of the plain and mountain territories, in their landscape patterns, are first clearly revealed. Plains are characterized by a great homogeneity of the structure, and large areas of landscape contours. The landscape structure of mountainous areas is characterized by considerable complexity and contrast, so the regional specificity of high altitude zoning was especially carefully taken into account.

## 4. Conclusion

Cognition and mapping of the organization of geosystems require the identification of patterns of their differentiation, integration and development, and all these patterns can be obtained only as a result of analysis and interpretation of certain data on properties, both individual components and geosystems in general. Empirical data in geography are obtained in expeditions, at field stations, key areas and profiles. The expedition method allows to collect only a small range of data on the landscape structure and properties of geosystems, and to obtain information on the dynamics and functioning of geosystems, studies at foeld stations and key sites are needed using seasonal or other periodic semi-stationary studies. In key areas of reference landscapes, it is necessary to study typological units that determine the overall epicfacial structuring of knowledge. An analysis of all modifications of the indigenous geosystems should be aimed at identifying the backbone of the studied geochores It is also necessary to develop and unify the methods of description, indicators of the structure and dynamics of geosystems, create inventories of signs, including physiognomic and deciphering for analysis and interpretation of space imagery.

Further development of the theory of geosystems requires the creation of a modern landscape classification that adequately reflects the organization of the geographical envelope, which should include three taxonomic series: typological, horological and dynamic. The classification should be based on correct quantitative data. Hence a task of improving the methodology for collecting, mapping, analyzing and thematic interpretation of geographic information appears. New, geosystem-based, standardized methods of thematic mapping, GIS creation, and aerospace applications are needed here. At the same time, everything should be balanced and work out precisely for the purpose of mapping geosystems.

**References**

[1] Sochava V B 1978 *Introduction to the Theory of Geosystems* (Novosibirsk: Science) p 319
[2] Suvorov E G, Semenov Y M and Novitskaya N I 2009 Landscape-estimation map of the Asian part of Russia: principles and methodical aspects of compilation *Geography and Natural Resources* **4** 5
[3] Semenov Y M 1991 *Landscape-Geochemical Synthesis and Organization of Geosystems* (Novosibirsk: Science) p 145
[4] Snytko V A and Semenov Yu M 1981 Experience in the conjugate mapping of geomers and geochors *Geography and Natural Resources* **4** 28
[5] Semenov Y M and Semenova L N 2006 The role of substance flows in the formation of landscape structure *Landscape Science: Theory, Methods, Regional Studies, Practice*, ed K N D'akonov (Moscow: Moscow State University Press) pp 349-51
[6] Glazovskaya M A 1964 *Geochemical Foundations of the Typology and Methodology of Natural Landscape Studies* (Moscow: Moscow State University Press) p 231
[7] Zagorskaya M V 2004 Landscape structure of Central Priol'honye *Geography and Natural Resources* **4** 58
[8] Lysanova G I 2001 *Landscape Analysis of Agro-Natural Potential of Geosystems* (Irkutsk: IG SB RAS Publishing House) p 188
[9] Semenov Y M and Lysanova G I 2016 Geosystem mapping for landscape planning of the regions of the Altai Republic *Geography and Natural Resources* **4** 66
[10] Chernykh D V and Samoilova G S 2011 *Landscapes of Altai (Altai Republic and Altai Territory), Map* (Novosibirsk: FSUE Cartographic Factory)
[11] Suvorov E G, Semenov Y M and Antipov A N 2007 Concept of landscape information for landscape area, Landscape Analysis for Sustainable Development *Theory and Applications of Landscape Science in Russia*, ed K N D'akonov (Moscow: Alex Publisher) pp 76-80
[12] Lysanova G I, Semenov Y M and Sorokovoi A A 2011 Geosystems of the Upper Yenisei basin *Geography and Natural Resources* **4** 92
[13] Lysanova G I, Semenov Yu M and Sorokovoi A A 2016 Landscape mapping of the Republic of Khakassia *Geodesy and Cartography* **12** 16
[14] Lysanova G I, Semenov Y M, Shekhovtsov A I and Sorokovoi A A 2013 Geosystems of the Republic of Tyva *Geography and Natural Resources* **3** 181

# Self-organization of the geosystems of the southern Cisbaikalia

**M A Nogovitsyna**

Sochava Institute of Geography SB RAS, Russia, Irkutsk

E-mail: 25051204@mail.ru

**Abstract**. The southern Cisbaikalia is located within the Baikal rift zone and the Altai-Sayan folded region. Self-organization of the geosystems in the research area is formed under a specific impact of geodynamically active regions. The main factors of self-organization of geosystems, especially the real-energy exchange, development, internal and external relationships, resonance, stability, are influenced by increased inflow of endogenous heat, Neogene-Quaternary volcanism and high tectonic activity of the territory. Formation and development of the neotectonic structure of the research area determine three energy-carrying elements of the Earth's deep structure - a powerful asthenolite, light blocks of the earth's crust and a vertical channel of the mantle. They in turn affect the formation of components of geosystems and their interrelationships. The paper presents the results of studying the factors of self-organization of geosystems of the southern Cisbaikalia, carried out using the methodology of self-organization process research, field expedition work, and interpretation of space images. The influence of energy-bearing elements on the formation, development and current state of geosystems is shown. It has been observed that within the regions, where elevated values of endogenous heat are recorded, geosystems containing both ancient (relic) and endemic components are concentrated.

## 1. Introduction

The study of spatio-temporal self-organization of geosystems is one of the most important tasks of landscape studies. The concept of self-organization was developed in the 50-60s of the 20th century with the emergence of a new scientific trend - synergetics, which studied self-organizing systems and considered general patterns of different systems development. G. Haken [1] gave a definition of the concepts of organization and self-organization, and carried out a number of works on this topic. Based on the ideas of self-organization, A.D. Armand [2] developed a theory of geographic processes, formulated general principles for the development of geographic systems. The doctrine of geosystems is associated with the name of V.B. Sochava [3], who laid the foundations, contributed subsequently in understanding the factors of self-organization of geosystems. The works of T.I. Konovalova [4] are devoted to complex studies of the self-organization of geosystems in the south of Central Siberia, as well as of V.I. Bulatov and D.V. Chernykh [5] on the organization of the Altai mountains.
Modern directions of geographical research complicate this task in connection with the need to study geosystems developing under climatic and geodynamic changes.

The study of the self-organization of geosystems acquires a special urgency due to the location of the southern Cisbaikalia within the Baikal rift zone (BRZ) and the Sayan-Baikal orogenous region. High tectonic activity, Neogene-Quaternary (basaltic) volcanism, increased values of endogenous heat flow affect the factors of self-organization of geosystems, which in turn affects the formation of new boundaries, preservation of relict and the development of new components of geosystems. The southern Cisbaikalia is characterized by a great variety and contrast of landscapes on closely located sites, which makes the area of research a unique site for solving problems of identifying factors of self-organization of geosystems.

## 2. Objects, Data and Methods

The study aims to identify the features of the geosystem self-organization of the southern Cisbaikalia.

The object of the study is the territory of the southern Cisbaikalia, which consists of Tunkinskaya depression, northern and northeastern macroslopes of the Khamar-Daban Ridge and the south-eastern part of East Sayan.

Currently, there are many definitions of the concept "self-organization", however, all its definitions boil down to the notion that this process that translates an open disequilibrium system that has reached a critical state in its development into a new, stable state with a higher level of orderliness of components and interrelations of the system.

In domestic geographical publications, self-organization is understood as:

- "internal orderliness, coherence, interaction of more or less differentiated and autonomous elements of the whole, conditioned by its structure" [6];
- "the emergence of new structures, evolutionarily more advanced than the previous ones" [2];
- "the formation of qualitatively new structures on a macroscopic scale, when the regularities of the emergence of order from chaos are established by the methods of thermodynamics of nonequilibrium processes" [7];
- "complex forming, preservation and orderly transformation of integrity through internal mechanisms" [4].

In this study, the author adheres to the notion that self-organization is a complex forming, preservation and orderly transformation of integrity through material-energy exchange, internal interrelations, resonance of processes, interrelation with the environment, development, coherence of ongoing processes. The real-energy exchange between geosystems and their components determines their internal unity.

Studies of the self-organization of geosystems of the southern Cisbaikalia were carried out using methods of complex physical-geographical studies, field-based observation and interpretation of space images, cartographic, comparative-geographical methods and GIS. Furthermore, we analyzed literature sources on the landscape structure of the region, works on self-organization, mapping, dynamics and functioning of geosystems, geobotany, tectonic and geological structure, as well as a number of integrated and sectoral maps.

In this work, when analyzing the effect of thermal endogenous fluxes on the geosystems of a district, we used published information on the results of measuring the temperatures in wells and determining the values of geothermal gradients. Geothermal studies in wells were carried out by the Institute of Earth Physics of the Russian Academy of Sciences, the Institute of Petroleum Geology and Geophysics, and the Institute of the Earth's Crust of the SB RAS. Temperature measurements were carried out both in support wells having a depth of more than 2-3 km and located in the basins of the Baikal type and in the Irkutsk amphitheater and in exploration wells with a depth of 300-600 m, drilled in sections of mountain lintels or in thermal water deposits [8].

## 3. Results and Discussion

Factors of self-organization of geosystems of the southern Cisbaikalia change depending on specific impacts of the Baikal rift zone: increased influx of endogenous heat, Neogene-Quaternary volcanism and high tectonic activity of the territory.

According to the published data [9], the formation and development of the neotectonic structure of the research area determine three energy-carrying elements of the Earth's deep structure. They in turn affect the formation of geosystem component and their interrelationships.

1. Powerful asthenolite is a magmatic body, which arose due to the heating of the lithosphere and its thermal expansion, which caused its isostatic movement. This determined the formation of the Mongol-Siberian revived orogenic belt and the maximum elevation within its boundaries of the East Sayan megaanticlinorium. The northern macroslope of the Eastern Sayan coincides with the margin of the asthenolith, as a result of which the mountains stop abruptly toward Baikal.

2. Light blocks of the earth's crust, saturated with granitoid intrusions, which experience prolonged uplifting in the form of arched uplifts. One of these blocks is confined to the Okinskoe plateau, located at an altitude of more than 2000 m and surrounded by high-mountain massifs. From the Miocene time (20-2.5 Ma ago), volcanic activity was intensive here, which ended only in the early Holocene (10 ka). That was illustrated by covers of basalts, cones of volcanoes and lava domes. A specific combination of structural and geodynamic factors defines the plateau as a unique geographical object in the newest structure of the mountainous regions of Eurasia.

3. The vertical channel of the mantle, which is a special element of the newest tectonic structure of Inner Asia. Under its influence in the southern Cisbaikalia, the largest thermal anomaly is recorded among the known thermal maxima outside the regions of active volcanism. In the southern Baikal Cisbaikalia, the heat flux (HF) is 4 times higher than in the neighboring areas of the Siberian platform [9]. The average value of HF is (10.5-11.3) • 10-6 J / ($cm^2$/ s), while for the Siberian platform it is equal to 4.4 • 10-6 J / ($cm^2$/ s). In the Tunka depression, the heat flux averages 5.4-7.5 10-6 J / ($cm^2$/s). Besides, on the general high thermal background in the zones of large young faults within the rift basins, anomalous areas are observed, which are associated with additional removal of deep heat. The greatest values of the heat flux within the territory under study are recorded in the Tunka depression (Arshan, Tunka, Irkut and Irkut), along the Predsayanskii Fault, along straight sections of large river valleys and in the foothills of the Khamar-Daban Ridge (the rivers Snezhnaya, Khara-Murin, Utulik and Zun-Murin). All these areas are associated, first of all, with regional faults and the largest deposits of hot mineral waters.

The impact of energy-bearing elements of the Earth's deep structure led to the formation of diverse and contrasting forms of relief in the region. This glacial erosion mountains with vertical dismemberment from 600 to 1200 m, which are characterized by a steeply slope alpinotypic relief with heights of up to 3284 m (Tunka goletz); gently middle-mountainous north-western spurs of the western part of the Khamar-Daban Ridge, flattened surfaces of Neogene-Quaternary basaltic covers (Khamar-Daban); the depression territory (the system of the Tunka depressions).

The combination of mountain and depressions affects the lithologic-geomorphological conditions and differentiation of climatic parameters. The mean annual air temperature ranges from -0.4 to -7.7 ° C, its highest values are recorded in the coastal part of the territory (from -0.4 to -1° C). The warmest point within the Tunka depression of the basins is Arshan, with a mean annual air temperature of -1.5 ° C, which is two and more times higher than in Mondy (-3 °C), Tunka (-3.5 °C), Kyren (- 3.8°C) [10]. Distribution of precipitation is extremely uneven. The smallest number is recorded in the Tunka depression: from 360 mm in the central part (Mondy) to 550 mm at the foot of the Tunka goletz (Arshan); the largest - on the windward slopes of the Khamar-Daban Ridge - from 520 (Kultuk, Slyudyanka) to 1.450 mm (Snezhnaya, Khamar-Daban) per year [11].

As is known, all processes in the geographic environment occur under the influence of solar energy and inland (endogenous) sources of energy. Into the geographical envelope, most of the internal energy comes in the form of a heat flux [12]. The southern Cisbaikalia is located in the mobile region of the Earth, where the largest thermal anomaly is observed among the known thermal maxima outside the areas of active volcanism. The total solar radiation in this area is on average 4100 mJ /$cm^2$ per year, or 105.3 • 10-6 J / $cm^2$/ s [13], while the values of the heat flux vary from 4.3 to 10.5 • 10-6 J / $cm^2$/ s [8]. In the southern Cisbaikalia, as a percentage, the share of inland energy sources relative to total solar radiation is 4-9%. A comparison of these data shows that solar energy is several times greater than the endogenous energy. However, the value of each type of energy cannot be estimated only quantitatively, since the efficiency of the energy flow depends not only on the power, but also on the form of the intake (concentrated or scattered, to the lower or upper boundary of the geosphere, etc.) [12]. Thus, quantitatively small fluxes of endogenous origin are no less a factor differentiating nature than solar energy [14].

The results of field studies and the study of literature data on the nature of the endogenous flow served as the basis for extrapolating materials to the area of study using space information. For the study, a space image was taken from the Landsat-7 satellite with RGB 4.6,7 channels (0.775-0.9, 10.4-12.5, 2.09-2.35 μm). Comparison of interpretation results, literary data, with topographic link to the territory of representative regions, field studies using a GPS receiver have shown high data comparability.

The effect of endogenous heat is especially evident in the self-organization of geosystems of the Khamar-Daban Ridge and the Tunka depression. Thus, for the depression, within the areas with higher values of heat flow, development of primordial steppe and endemic communities is indicative, such as cryophyte steppe with Festuca kryloviana and sandy with Bromopsis korotkiji. In the Mondy depression within such areas there are cryoxerophilic petrophytic-herbaceous steppes, unique for the Baikal region. Eritrichium sajanense, the Eastern Sayan endemic cryophytic steppe species can be found here, also periglacial relics - Gentiana decumbens, Pulsatilla ambigua, and Eremogone formosa [15].

During the field landscape studies we found that on the same habitats of terraced plots of the Irkut river with a uniform petrological composition, areas of the north-Asian meadow steppes and sub-taiga pine forests are distributed on the left and right bank in regions with smaller values of endogenous heat. Comparison with the literature data and the interpretation results of space images showed that wherever a larger inflow of endogenous heat is recorded, Central Asian dry-step geosystems are represented.

To river valleys of the Khamar- Daban Ridge, which tend to have higher values of heat flux (rivers Utulik, Snezhnaya, Khara-Murin, Zun-Murin), poplar forests are confined, having in its composition tertiary relics such as Oreopteris limbosperma, Fritillaria dagana, Arsenjevia baicalensis, Waldsteinia ternata, Gallium triflorum, Poa ircutica, Polystichum lonchitis, Dryopteris filix-mas, and others.

There are still traces of volcanic activity in the study area. At present, they have a residual character and are noted in the Elovskii spur of the Tunka depression, near the Irkut tributaries (Zamaraiha and B.Taiturka), on the Okinskoe plateau (East Sayan), in the basin of the Zun-Murin river and in the upper reaches of the Utulik, Khara-Murin, Snezhnaya rivers (Khamar-Daban).

At Okinskoe plateau, mainly on Miocene basalts spruce landscapes (with birch and larch), sometimes bushes (Lonicera altaica, Berberis sibirica, Ribes spicatum, Caragana arborescens) with powerful moss lichen cover on thin fragmentary soils of facies group. The overlapping of different age lavas, differences in their chemical composition, peculiarities of the microrelief and microclimatic conditions cause the presence of different-age plant groups here, which provides a variety of landscapes of the basaltic field [16]. On open sections of the basalt flow, steppe geosystems are represented; however, in most of the territory the most common are areas with Aulacomnium palustre, Rhytidium rugosum and Thymus serpyllum. There are also common shrubs and bushes (Lonicera altaica, Berberis sibirica, Empetrum sibiricum), and herbaceous plants (Saxifraga oppositifolia, Bergenia crassifolia, Festuca sphagnicola), as well as larch-spruce undergrowth.

In other areas, where an increased influx of endogenous heat is not recorded, the geosystems, most typical for the zonal background, are represented: Siberian pine within the Khamar-Daban Ridge, steppes (of North Asian type) and larch forests in the Tunka depression.

While analyzing the geosystem self-organization in the region, taking into account natural conditions and the impact of neotectonic processes, it was identified that in regions with an increased inflow of endogenous heat, the relic components of geosystems are still remained; in the areas of development of alpinotype relief forms, young goletz types of geosystems are formed.

## 4. Conclusion

A distinctive feature of the geosystem self-organization in the research area is that the cross-cutting criteria for all hierarchical levels are as follows: the real-energy exchange and development can change as a result of the specific impact of rift factors: high tectonic activity, Neogene-Quaternary volcanism, and increased endogenous heat flux.

The formation and development of the neotectonic structure of the southern Cisbaikalia is determined by three energy-bearing elements of the Earth's deep structure: a powerful asthenolite; light blocks of the earth's crust, saturated with granitoid intrusions; vertical channel of the Earth's mantle. Energy-bearing elements in turn influence the formation of components of geosystems and their interrelations.

The impact of energy-bearing elements of the Earth's deep structure led to the formation of various contrasting forms of relief in the region. These are glacial erosion mountains with vertical dismemberment from 600 to 1200 m, which are characterized by a steeply slope alpinotypic relief with heights of up to 3284 m (Tunka goletz); gently middle-mountainous north-western spurs of the western part of the Khamar-Daban Ridge, flattened surfaces of Neogene-Quaternary basaltic covers (Khamar-Daban) and the depression area (the system of the Tunka depression).

The combination of mountain and depression areas affects the lithologic-geomorphological conditions and differentiation of climatic parameters. Differentiation of the geosystems of the region reflects the features of the formation of their components, occurring under the conditions of a specific impact of endogenous energy sources.

The effect of endogenous heat is especially pronounced in the self-organization of the geosystems of the Tunkinskaya depression and the river valleys of the northern macroslope of the Khamar-Daban ridge where the highest values of heat flow are recorded. Here the dry-step geosystems of the Central Asian type, foothill subtaiga, and steppe geosystems are confined to areas of increased endogenous heat. On the foothills of Khamar-Daban wide-spread fir-forests are common, referring to relicts of the non-moral complex and the connection between fir forests and broad-leaved-coniferous forests of the Pliocene is traced. In fir forests there are still relicts of deciduous forests.

**Acknowledgments**
This work was supported by a grant from RFBR 16-05-00902

**References**

[1] Haken G 1980 *Information and Self-Organization* (Moscow: World Pub) p 224
[2] Armand A D 1988 *Self-Organization and Self-Regulation of Geographic Systems* (Moscow: Science Pub) p 54
Armand A D 1992 *Mechanisms of Geosystem Stability* (Moscow: Science Pub) p 206
[3] Sochava V B 1978 *Introduction to the Theory of Geosystems* (Novosibirsk: Science Pub) p 319
[4] Konovalova T I 2012 *Self-Organization of Geosystems in the South of Central Siberia* (Novosibirsk: Academic Publishing House "GEO") p 148
[5] Chernykh D V and Bulatov V.I. 2002 *Mountain Landscapes: Spatial Organization and Ecological Specificity* (Novosibirsk: GPNTB, IVEP SB RAS Pub) p 83
[6] Preobrazhensky V S 1986 *Organization, Organization Ratio of Landscapes* (Moscow: Institute of Geography Pub) p 33
[7] Reteyum A Y 1988 *Earth Worlds* (Moscow: Thought Pub) p 95
[8] Vilor N V 2002 Satellite monitoring of infrared radiation of geological and structural elements of the Sayan-Baikal-Patom mountainous region and the Baikal rift zone *Research of the Earth from Space* **4** 55-61
Golubev V A 1982 *Geothermy of Lake Baikal* (Novosibirsk: Science Pub) p 150
Golubev V A 2007 *Conducted and Convective Heat Transfer in the Baikal Rift Zone* (Novosibirsk: Academic Publishing House "GEO") p 222
Kiselev A I, Medvedev M E and Golovko G A 1979 *Volcanism of the Baikal Rift Zone and Problems of Deep Magma Formation* (Novosibirsk: Science Pub) p 198
Lysak S V and Zorin Y A 1976 *Geothermal Field of the Baikal Rift Zone* (Moscow: Science Pub) p 91

Lyubimova E A 1968 *Thermal Anomaly in the Region of the Baikal Rift* (Moscow: Science Pub) p 160
[9] Bugaevskii G N 1978 *Seismological Studies of Heterogeneities in the Earth's Mantle* (Kiev: Scientific thought Pub) p 184
Ufimtsev G F 1992 *Morphotectonics of the Baikal Rift Zone* (Novosibirsk: Science Pub) p 216
Shchetnikov A A and Ufimtsev G F 2004 *The Relief Structure and the Latest Tectonics of the Tunkinskii Rift (South-Western Baikal Region)* (Moscow: Scientific world Pub) p 160
Doglioni C and Carminati E 2003 Rift assimetry and continental uplift *Tectonics* **1024**(22) 8-11
[10] *Manuals for Hydrometeorological Stations and Posts* 1966 vol **22**(2) (Leningrad: Hydrometeo Pub) p 360
[11] *Manuals for Hydrometeorological Stations and Posts* 1968 vol **22**(4) (Leningrad: Hydrometeo Pub) p 280
[12] Gerenchuk K I, Bokov V A and Chervanets I G 1984 *General Geoscience* (Moscow: Higher school Pub) p 255
[13] *Baikal Atlas* 1993 (Moscow: Federal Service of Geodesy and Cartography of Russia) p 159
[14] Grigoriev A A 1966 *Regularities in the Structure and Development of the Geographical Environment* (Moscow: Thought Pub) p 382
[15] Namzalov B B and Holboeva S A 1996 The main features of the vegetation cover of the Tunkinskii National Park *Proc. of the Republican Meeting State and Problems of Specially Protected Natural Territories of the Baikal Region* 81-94
Namzalov B B, Holboeva S A and Grishkina T. M. 1998 *The Steppes. General Outline and Description of Steppe Phytocenoses* (Ulan-Ude: Publishing house of BSU) 109-42
[16] Vyrkin V B 2012 The current state of the landscapes of the Oka depression (East Sayan) *Geography and Natural Resources* **4** 98-107

# Mapping of the mountain taiga geosystems of the Baikal Natural Territory

**V M Plyusnin, I N Bilichenko, S A Sedykh**

Sochava Institute of Geography SB RAS, Irkutsk, Russia

E-mail: plyusnin@irigs.irk.ru, bilichenco@irigs.irk.ru, sedykh@irigs.irk.ru

**Abstract.** Results from investigating the geosystems of Baikal Natural Territory in its western part (Primorskii Range) are presented. For studying macroslope of these mountain range (having an easterly aspect for the Primorskii Range), the regional background and the main factors of the landscape differentiation of the study areas were analysed. Structural features of topological geosystems are revealed, and large-scale maps of key areas are compiled (at a scale of 1:50 000). Maps of the landscape-typological structure were created on the basis of the structural-dynamic and facies analysis of landscapes. Based on of V. B. Sochava's theory of geosystems it is meant for use in processing and ordering a large body of information. It is established that the main factors influencing the landscape diversity are the aspect and steepness of the slopes, the composition and structure of rocks, the absolute height, the amount of atmospheric precipitation, and anthropogenic impacts. It is determined that the influence of the lithomorphic factor is widespread throughout the study areas, the hydromorphic factor is also important, the cryomorphic factor occurs additionally in the goletz zone.

## 1. Introduction

The world geographical and ecological literature discusses the issue related to changes in intensity and directedness of many natural processes caused by global and regional climatic shifts as well as by anthropogenic impacts. These processes alter the longstanding functioning of geosystems, their long-term dynamics and eventually the evolution of landscapes. The changes encompass significant areas (thawing of permafrost and northward retreat of the permafrost boundary, desertification of territories, encroachment of forest upon the tundra, etc.). But they also have a local and regional character: the natural characteristics of particular territories leave their imprint on the formation and dynamics of geosystems. As the model region, we used the Baikal natural territory with its mountain-taiga, goletz, subgoletz, mountain-steppe, mountain-depression and aquatic geosystems. Climate warming and humidification has been recorded in recent years, a complex fire-hazardous situation has been emerging, and tourism has been developed intensely. Expedition-based investigations made in the summer seasons of 2012-2018 years sought to determine the spatial regularities in the distribution of geosystems of the topological rank. In accordance with the objective set, we analyzed the regional background and the main factors for the landscape differentiation in the study areas. The analysis revealed the regularities in the structure of geosystems of the topological level, and landscape maps of the key areas were compiled at a large scales of 1:100 000 and 1:50 000.

IOP Conf. Series: Earth and Environmental Science **190** (2018) 012036 doi:10.1088/1755-1315/190/1/012036

## 2. Objects, data and methods

The landscape structure was studied in the key areas of Primorskii Range; the former includes the eastern macroslope of Primorskii Range from Maloe More along the Laninskii brook to the Sarminskii goletz. On the investigated territory of the Primorskii Range, the Sarma bare mountain with the peak of 1658 m rises. The top of the char is composed of proterozoic intrusive rocks of granite-gneiss, shales and coarse-clastic material covered with epiphytic lichens. The precipitation here is from 400 mm to 1000 mm on top of the Range.

Field description in areas 10x10 and 20x20 m in size were carried out by using the techniques developed and tested in landscape investigations [1]. A total of more than 200 areas were documented.

The total area of the study is 120 square kilometers at 1:100 000 scale [2]. The area of the transect of scale 1:50 00 in figure 1 is 30 square kilometers.

To study the current status of the landscapes used the geosystem regional-typological approach of the Siberian geographical school tested in different regions of Siberia [3].

The regional-typological approach forms an important part of V. B. Sochava's theory of geosystems [4] and is meant for use in processing and ordering a large body of information. Therefore, it has gained widespread acceptance in regional geographical research, in landscape mapping, and in other kinds of thematic mapping. When studying and mapping a territory with varied and complicated natural and socio-economic conditions, due regard must be had to the regional and local, individual and typical features of geosystems, the properties of their homogeneity and heterogeneity, and to the chorological and typological aspects of generalization [5]. It is worthwhile making such investigations within the landscape framework using the regional-typological approach where the description of local geosystems is performed in a regional setting with consideration for their generic (typical) and unique (individual) peculiarities. Presentation of elementary geosystems in the structure of hierarchically larger subdivisions of natural environment furnishes a means of assessing their properties with regard to the regional specificity of the landscape sphere. To accomplish this, use is made of the principle of two-row classification of geosystems, identifying geomers and geochores, and homogeneous and heterogeneous spatial systems.

Landscape maps of the key areas at the topological level were compiled in terms of the graphical model of the hierarchy of geosystems for the entire Baikal region [6]. A physical-geographical regionalization has been carried out according to the individual attributes to Legend 1. A mapping of the landscapes was carried out according to the typological attribute by displaying geosystems of topological dimension at the level of groups of facies and separated facies.

The basis for landscape mapping was provided by remote sensing data from Landsat-5 TM, Landsat-7 ETM+, Landsat-8 OLI (resolution of 30 and 15 m, respectively), GeoEye (2.44 m), Kanopus-V (10.5 m), Resource P-1 (Sangur-U scanner, 2 - 0.7 m), STRM 4.1 for creating digital terrain models (in gradation angles for mountainous areas including steep slopes). Field database includes more than 200 points of description for each study area where the structure of landscape units was determined, their boundaries were delineated, and the map legend was updated. Published and archival material, topographic maps and various thematic maps were also used.

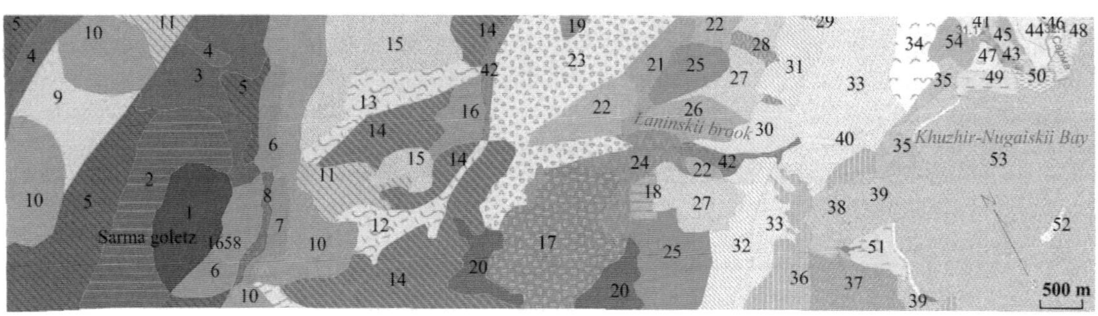


**Figure 1.** Map of geosystems of Primorskii Range.For 1–54, see legend 1.

*2.1. Legend 1 to the 1:50 000 map of geosystems of Primorskii Range.*

*Goletz mountain-tundra geosystems of Primorskii Range on Lower-Proterozoic quartzites, and shale at 1500–1650 m, with physical weathering by cryogenic processes and nivation 1.* Of flattened summits, mountain-tundra detrital lichen on bedrocks (s). 2. Near-the-top lichen with sparse groups of dwarf Siberian pine (pp). 3. Of hilly summits, mountain-tundra lichen, on scattered small-boulders and with sparse dwarf Siberian pine (p). 4. Of gentle slopes, mountain-tundra lichen (pp). 5. Slope mountain-tundra subshrub-moss dwarf Siberian pine (s). 6. Of the nival bench, dwarf Siberian pine bilberries-crowberries with juniper, on detrital coarse-skeletal soils (sf). Subgoletz shrub and sparse-forest on shale, plagiogneisses and amphibolites, at 1350–1500 m, colluvial screes and coarse-skeletal podzols, dominated by cryogenic processes7. Dwarf Siberian pine with dwarf birch, juniper, rowan bilberries-great bilberries (pp). 8. Meadow nival niches with water-filled troughs (sf). 9. Dwarf Siberian pine saddles, bilberries-cowberries, with oppressed young Siberian stone pine trees (pp). 10. Dwarf Siberian pine and yernik moss-lichen on scattered rocks (sf). Mountain-taiga geosystems of reduced development at 1200–1350 m11. Of gentle slopes, sparse-forest Siberian stone pine small grass-lichen, with groups of dwarf Siberian pine, on coarse-skeletal mountain soils (s). 12. Siberian stone pine-spruce subshrub-sedge-sphagnum on peaty-humous soils (s).13 Larch-birch with spruce in underwood, Duschekia/horse-tail on soddy acid soils (v). *Mountain-taiga geosystems of limited development at 900–1200 m.*14. Larch with Siberian stone pine, ledum, with underwood of alder and rowan, with juniper, on forest coarse-humous soils (pp). 15. Birch bilberries-great bilberries with young Siberian stone pine on podzolic soils (s). 16. Siberian stone pine bilberries-great bilberries with alder, rowan and juniper on soddy-podzolic soils (p).17. Larch with birch on screes, on thin podzolized soils (p). 18. Larch-Siberian stone pine subshrub-true moss on peaty-coarse skeletal soils (pp). 19. Larch bilberries-ledum moss on soddy soils (pp). 20. Siberian stone pine-larch cowberries-small grass-moss with rowan on humous soils (ppe). 21. Siberian stone pine subshrub with rock remnants and screes (ppe, s). 22. Larch bilberries-ledum-sphagnum on peaty-humous soils (p). 23. Larch-birch on humous-gley pyrogenic soils (s).24. Siberian stone pine-spruce in topographic lows horse-tail/small grass-moss on illuvial humous soils (s). 25. Larch with birch grass-true moss, with underwood of Duschekia and rowan, on grey thin soils (pp). 26. Larch with pine bilberries-ledum, on weakly podzolized soils (pp). 27. Larch grass with birch, aspen and alder in underwood, on pyrogenic podzolic (s). 28. Larch with spruce small grass-true moss, on humous-gley soils (sf). 29. Of drainage lows, larch-Siberian stone pine sedge-moss, on illuvial humous soils (pp). *Mountain-steppe geosystems of the Western Transbaikalian type on gravelites and conglomerates at 600–900 m.*30. Slope mountain-steppe forb-grass (s). 31. Thyme-Leymus slope on stony thin soils (me, sf). 32. Wormwood-grass slope with Caragana on thin petrozems (me, sf). 33. Thyme-Leymus and wormwood-fescue on petrozems (sf). 34. Meadow-grass and feather-grass/wheatgrass on chestnut soils (p). *Submontane subtaiga light-coniferous xerophytic geosystems on eluvial sands and deluvial loams at 500–600 m.* 35. Sparse larch summit rhododendron-grass on petrozems (s). 36. Sparse larch slope sedge-grass on petrozems (s). 37. Near-the-top larch grass with hard rock remnants (s). 38. Slope larch shrub-grass on petrozems (pp). 39. Open woodland, larch grass on slopes disturbed by recreational activity (a). *Valley meadow and bog geosystems on Quaternary alluvial and deluvial slope deposits40. Floodplain willow-shrub forb-meadow on soddy soils* (sl). 41. Slope willow-shrub forb on alluvial soddy soils (s). 42. Spruce-larch with willow, sedge-grass-moss on alluvial soils. 43. Sedge-forb on peat soils (sf). 44. Forb-grass on alluvial meadow soils (s). 45. Meadow sedge-grass waterlogged (sf). 46. Sedge-small grass on alluvial meadow-boggy soils (sf). 47. Boggy floodplain sedge (sf). 48. Subshrub-grass with pools on peat-bog soils (sf). 49. Sedge-small grass-sphagnum on peat soils (sf). 50. Sedge-small grass near mouths (s). Coastal inclined plains and subaquatic geosystems on modern alluvial deposits51. Coastal meadow-steppe grass-forb on eroded meadow soils (sl, a). 52. Sedge-small grass on outwash sandy-pebble alluvium (s). 53. Shallow hydrophytic coastal and bottom pebble-sandy (s). Anthropogenic geosystems54. Residential (a). Dynamical categories of groups of facies: p – primary stablest, pp – pseudo-primary, s – serial, ppe – pseudo-primary extra-regional, sf – serial factorial, sl – stably long-lasting secondary, a – anthropogenically transformed.

## 3. Results and discussion

The study area in Priol'khonie encompasses the southeastern (facing Baikal) macroslope of Primorskii Range. The topography of the slope is complicated by an extended bench (Obruchev fault) with a height difference of up to 1000 m. Archean and Lower Proterozoic rocks include granites, diorites, quartzites, shale and gneisses. Below the bench, an inclined peneplain of the Priol'khonskoe plateau stretches, which is composed of gneisses, amphibolites and quartzites of the Chernorud Archean formation, with the mantle of Mid- and Upper-Quaternary eluvial sands and deluvial loams. The climate of the territory is distinguished by winters with little snowfall, the anticyclonic winter regime with insufficient atmospheric humidification (100–400 mm), and by a short growing period (4–4.5 months). The mean January and July air temperatures, respectively, are –17.3 and 14.4°C, and the mean annual temperature is 0.7°C [7].

The geosystems of this key area (figure 1, legend 1) belong to the Baikal-Dzhugdzhur mountain-taiga region of the Cisbaikalian goletz-mountain-taiga and depression province. The territory covers the Primorskii goletz-mountain-taiga and Ol'khonskii mountain-subtaiga and submontane-steppe physical-geographical districts [8]. The specific character of the altitudinal-belt landscape structure including an almost entire set of geosystems of the mountains surrounding Baikal is governed by the manifestation of the barrier-shadow, arid-depression and submontane effects complicated by the influence of Baikal's water mass [9]. As a result, there emerges a combination of contrasting (in natural conditions) landscapes united into the Anga-Sarma middle-mountain taiga, largely light-coniferous area and the Olkhon mountain-subtaiga and submontane-steppe area. Noteworthy among the main exogenous processes are frost weathering, rill and sheet wash, linear erosion, cryogenic processes on talus aprons, karst and suffosion in topographic lows [10]. Coastal landscapes are actively used for recreation and undergo a significant anthropogenic load. The study territory is dominated by the Sarma goletz with its summit at 1658 m. This is composed of Proterozoic intrusive rocks, namely, granitogneisses, shale and coarse detritus covered by epiphytic lichens. The atmospheric precipitation amount is 400–500 mm. The facies of the goletz planation surface are represented by mountain-tundra lichen geosystems (1-2, see numbers at figure 1 and in the legend 1) dominated by *Cladonia rangiferina, Tephroseris porphyrantha and Gasparrinia elegans* of the cryomorphic-lithomorphic series. Under the influence of cryogenic and erosion processes there occur scattered granitogneiss small-boulders, small spot medallions of frost heaving of detrital earth materials. The gentle goletz (4) slopes are occupied by shrub groups of *Pinus pumila*, *Betula rotundifolia*, *Betula nana* and singly occurring *Juniperus communis*. The steep slopes have the eastward and south-eastward aspect (5). The upper part of the slopes is home for nival processes at snow patches (6); the middle and lower parts (7) include the subgoletz belt of closed dwarf Siberian pine with *B. rotundifolia* on coarse-skeletal soils. The massif of the shrub belt is sparse because of a presence of lichen waste areas with stone runs. The lows show outcrops of groundwater, with small subalpine meadows of the forb/horse-tail series (8). The upper part of the mountain-taiga belt (14-29) is represented by a sparse subshrub-moss-lichen Siberian stone pine forest with groups of dwarf Siberian pine on a thin layer of podzols, often with outcrops of rocks with an imprint of cryogenic processes. The middle mountains are occupied by larch-Siberian stone pine subshrub-true moss forests.

On flattened watershed divides there occur screes consolidated by vegetation of larch and birch trees as well as weakly consolidated, sometimes with groups of dwarf Siberian pine. Hard-rock ridges are seen to include low rock remnants. The precipitation amount here is smaller, 300–400 mm. The lower part of the mountain-taiga belt (30-34) is the transition zone to the steep slopes of Primorskii Range. The great bilberries-ledum forests with pine are replaced by mixed aspen-birch-pine shrub-grass, Spiraea-grass (with *Duschekia fruticosa* and *Sorbus sibirica*) forests and birch-larch grass-true moss (often pyrogenically disturbed) forests. Pine forb forests with underwood of *Rhododendron dauricum* are observed on the southern slopes of moderate steepness. There occur soddy-podzolic, soddy-weakly podzolized soils. The degree of pyrogenic disturbance caused by frequent ground fires is high. The first signs of steppization emerge here. The lower part of the slopes, from the border o forest down the slope, includes forb meadow steppes, followed by mountain dry petrophytic steppes. The steepness of the slopes of the mountain range reaches 40–45$^0$ (32), and erosion-gravitational processes are taking place. Thyme-Leymus, wormwood-steppes on the slopes reside on thin layers of coarse-skeletal soils. There occur *Caragana pygmaea* and *Cotoneaster melanocarpus*. Sparse larch trees emerge nearer to the foot of the slopes. The submontane subtaiga landscapes along the foot of the mountain range are represented by dry sparse larch ledum grass and pine shrub forb forests which, nearer to Baikal, are replaced by sparse larch forests. This area is characterized by minimal precipitation amounts, 200–300 mm. The deeply incised valley (42) of the Laninskii brook in Primorskii Range, and also th depressions with ouvals and balkas have a V-shaped profile where there is an enhanced contribution from small-leaved species and communities of different species of willows: *Salix jenisseensis*, *S. hastate* and others, and in places from *Populus tremula*. Along the narrow gullies of the streams there occur groups of shrubs with willows, in places with larch trees. The geosystems of the second key area on Barguzin Range also belong to the Baikal-Dzhugdzhur mountain-taiga region, Cisbaikalian goletz-mountain-taiga and depression province, and to the Barguzin high-mountain/goletz and Middle Baikal mountain-Taiga and subtaiga-depression districts. The trail along which the investigations

## 4. Conclusions

The landscape regularities underlying the classification of geosystems and used for the map legends were revealed in large-scale mapping of the key areas. Regional characteristics of the position of facies and their groups were identified in the system of dynamical categories. The landscape diversity of the mountain territories used in this study is associated primarily with the influence of the lithomorphic factor, although a considerable role is played by the hydromorphic factor, the cryomorphic factor at elevations above 1500 m, and by the xeromorphic factor in Priol'khonie. The main differentiation factors for the geosystems include the exposure differences in humidification and insolation, the bedding characteristics of rocks, their fissuring and composition, the degree of stoniness of the slopes, the absolute height, the steepness of the slopes, and consolidation of the slopes by vegetation. In the process of studying the mountain geosystems of the Baikal natural territory, differences of the macroslopes facing Baikal were revealed. Primorskii Range is separated from Baikal by 9 km, the precipitation amount on the watershed divide and at the foot of the mountain range, respectively, is 400–500 and 200–300 mm, and the elevation of the watershed divide averages 1400–1600 m. Landscape structure: 1) goletz round-summit mountain-tundra lichen with screes; 2) subgoletz Siberian stone pine mountain-taiga larch-Siberian stone pine with dwarf Siberian pine, subshrub-true moss with sparse rock remnants and screes; 3) steep-slope mountain dry-steppe (Obruchev fault at 700–900 m), and 4) submontane subtaiga sparse larch forests, dry and forbs.

## Acknowledgements

The work was partially supported by SB RAS (Fundamental Research Program project 17-05-00400).

## References

[1] Vidina A A 1962 *Methodological Guidelines for Large-Scale Field Landscape Investigations (for Purposes of Agricultural Production)* ed N.A. Solntsev (Moscow: Mosk. Univ.) p 120

[2] Bilichenko I N and Sedykh S A 2016 Mapping of the landscapes of the Western Baikal region *Geodesy and cartography* **9** 29-38

[3] Geographical studies of Siberia *Structure and Dynamics of Geosystems* 2007 ed Y M Semenov and A V Belov (Novosibirsk: Acad. publishing house "Geo") p 413

Mikheev V S 1987 *Landscape-Geographical Providing of Complex Problems of Siberia* (Novosibirsk: Science) p 207

Plyusnin V M 2003 *Landscape Analysis of Mountain Territories* (Irkutsk: Publishing house of the Institute of Geography of the SB RAS) p 257

[4] Sochava V B 1978 *An Introduction to the Theory of Geosystems* (Novosibirsk: Nauka) p 320

[5] Abalakov A D and Sedykh S A 2010 Regional-typological study and mapping of geosystems: analysis of the implementation *Geography and Natural resources* **4** 21-27
[6] Mikheev VS 1988 *Landscape-Geographical Support of TerKSOP of the Lake Baikal Drainage Basin* (Irkutsk: Inst. Geogr. of Siberia and the Far East, Acad. Sci. USSR) p 63
Imetkhenov O A 2011 *Modern Landscapes of Buryatia. Methodological Approaches, Spatial Organization* (Ulan-Ude: VSGTU) p 260
[7] *Atlas of the Irkutsk Region: Environmental Conditions of Development* 2004 ed V V Vorobiev et al (Irkutsk: Publishing house of the Institute of Geography of the SB RAS; Moscow: Roscartography) p 90
[8] Plyusnin V M and Sorokovoi A A 2013 *Geoinformation Analysis of the Landscape Structure of the Baikal Natural Territory* ed V A Snytko (Novosibirsk: Geo) p 187
[9] Mikheev V S 1993 Landscape-structural analysis *The Man at Baikal: the Ecological Analysis of an Inhabitancy* (Novosibirsk: Science) 8-39
[10] Danko L V, Sizykh A P and Kuz'min S B 2005 Spatial Structure of Geosystems in Priol'khonie (Exemplified by a Model Transect) *Structure, Functioning and Evolution of Mountain Landscapes in Western Cisbaikalia* ed V A Snytko (Irkutsk: Publishing house of the Institute of Geography of the SB RAS) pp 11–19

# Landscape-typological mapping of the Baikal region (within the boundaries of the Irkutsk region)

**E A Rasputina (Istomina), M V Tsygankova, E E Korolkova**

Sochava Institute of Geography, Siberian Branch, Russian Academy of Sciences, Irkutsk, Russia

E-mail: elenaistoma@gmail.com, zagorsk@irigs.irk.ru, elainefisher@yandex.ru

**Abstract.** The aim of the work is to create a landscape-typological map of the territory to study the territorial differentiation of recreational activities. The variety of landscapes of the Baikal region is determined by a large number of landscape-forming factors, as well as the heterogeneity of the conditions in which geocomplexes are formed. A landscape-typological map of the territory was created on a scale of 1: 500 000. As an input data for creating the map, electronic topographic maps of the territory, space images of Landsat 5, 7, 8 for different seasons and years (including mosaic of MrSID for 2000 and Hansen mosaic for 2016), digital elevation models, landscape, geological, soil and geobotanical maps of different scale were used. A total of 56 landscape units were identified: 13 highland (goltzy, sub-mountain, reduced development mountain-taiga), 6 medium-mountain (limited development taiga), 14 lowland (optimal development mountain-taiga, steppe), 13 foothill (limited development taiga, optimal development taiga, subtaiga, steppe), 10 - intermontane depressions and valleys (limited development taiga, optimal development taiga, steppe). The conclusions drawn about the landscape structure and recreational properties of the territory are generalized. More detailed landscape research is needed, including comprehensive field work to clarify the landscape map and assess the recreational significance of landscapes.

## 1. Introduction

The territory of the Baikal region, which is, on the one hand, promising for the development of various types of economic activity, and on the other hand, unique and highly valuable from an ecological point of view, needs scientifically based planning of nature management. The information basis for such planning should be a reliable medium-scale landscape map reflecting the potential and current state of the landscapes of the territory. Landscape maps of different scale on the territory of the Baikal region, created by employees of various laboratories of the Institute of Geography SB RAS in different years, are mostly fragmentary and require generalization and reduction to a single one [1].

The aim of the work is to create a landscape-typological map of the territory to study the territorial differentiation of recreational activities. Different landscapes with certain properties (vegetation, relief, humidification conditions, aesthetic appeal, comfort, resistance to recreational loads) are appropriate for the development of one or another type of recreational activities.

The themes of planning recreational activities on a landscape basis, sustainable tourism development, assessment and mapping of cultural ecosystem services are widely represented both in Russian and foreign studies [2].

IOP Conf. Series: Earth and Environmental Science **190** (2018) 012037 doi:10.1088/1755-1315/190/1/012037

## 2. Objects

As the object of the study, the territory of the Baikal region within the boundaries of the central ecological zone of the Baikal natural territory within the Irkutsk region was chosen.

The variety of landscapes of the Baikal region is determined by a large number of landscape-forming factors. The geological structure of the territory is very contrasting. Precambrian acid magmatic and metamorphic rocks (granites, gneisses, stannites, argillites and conglomerates) prevail in the area. On the east coast, granitoids with the metamorphic (crystalline schists, biotite-garnet plagioclases, amphibolites) and sedimentary (sandstone, gravel) rocks predominate, and on the western side - Proterozoic sedimentary and metamorphic deposits (sandstones, siltstones, various shales, gneisses and tuffs), as well as igneous rocks (granites, granosyenites, porphyries, granodiorites) [3].

The climate of the Baikal region is determined by the location of the territory in the belt of temperate latitudes, considerable distance from the oceans, and also by the features of the mountain-hollow relief. Average annual air temperatures are negative throughout the territory. The lowest temperatures are recorded in January (-16 above the water area of Lake Baikal, down to -28 on the periphery), the highest - in July (from 8 above the water area of Lake Baikal, to 20 in the basins of the Baikal region). The western transport of air masses with a possible chill of cold air from the north, and a warm moist one - from the south prevails. The annual amount of precipitation varies from 200 mm or less in Priolkhonye to 1000 and more on Khamar-Daban ridge [4].

The complexity of the geological history and modern conditions of the Baikal region determine the contrast of the soil of the territory. In the goltsy belt - stony, low-power organogenic-gravelly, substructures and podzols, in the forest zone - soddy taiga with accompanying podzolic, soddy-carbonate and brown soils, in the steppes of Priolkhonye - chestnut, sod steppe and chestnut soils predominate. Meadow, meadow-marsh, and marshsoils prevail in valleys, and saline soils - near saline lakes in Priolkhonye under saline meadows [3].

The distribution of vegetation in the area is determined by high-altitude zones, complicated by the bolson effect, aspect, lithology, microclimate of the lake Baikal, latitudinal and meridional zoning. Ehe alpine, mountain-tundra, subalpine, subgoltsy, light coniferous taiga, dark coniferous taiga, steppe and shrub-meadow-marsh plant complexes are distinguished in the area [5].

The landscape structure of the Baikal region is characterized by high complexity and contrast. Two large regions of the North Asia - Baikal-Dzhugdzhursky and South-Siberian physico-geographical regions - are in contact here; three types of natural environment are combined: tundra, taiga and steppe; a wide range of landscapes: goltsy, subgoltsy, mountain-taiga, mountain-forest, mountain-forest-steppe (subtaiga) and mountain-steppe [6].

## 3. Data, methods, results and discussion

A landscape-typological map (figure 1) of the territory was created on a scale of 1: 500 000. As an input data for creating the map, electronic topographic maps of the territory, space images of Landsat 5, 7, 8 for different seasons and years (including mosaic of MrSID for 2000 and Hansen mosaic for 2016), digital elevation models, landscape, geological, soil and geobotanical maps of different scale were used. [7]. With the use of a digital elevation model, maps of slope and aspect were constructed. When creating the map, methods of automatic processing of space images were used. Using the methods of visual interpretation of Landsat images of different seasons, as well as using maps of slope and aspect, the territory was differentiated into homogeneous landscape units and their typological affiliation was determined. In the legend (table 1) of the landscape map landscapes are grouped into clusters of high-altitude belts (high-mountain, middle-mountain, low-mountain, piedmont, intermountain depressions and valleys), developmental conditions (optimal, reduced and limited), features of mesorelief and vegetation.

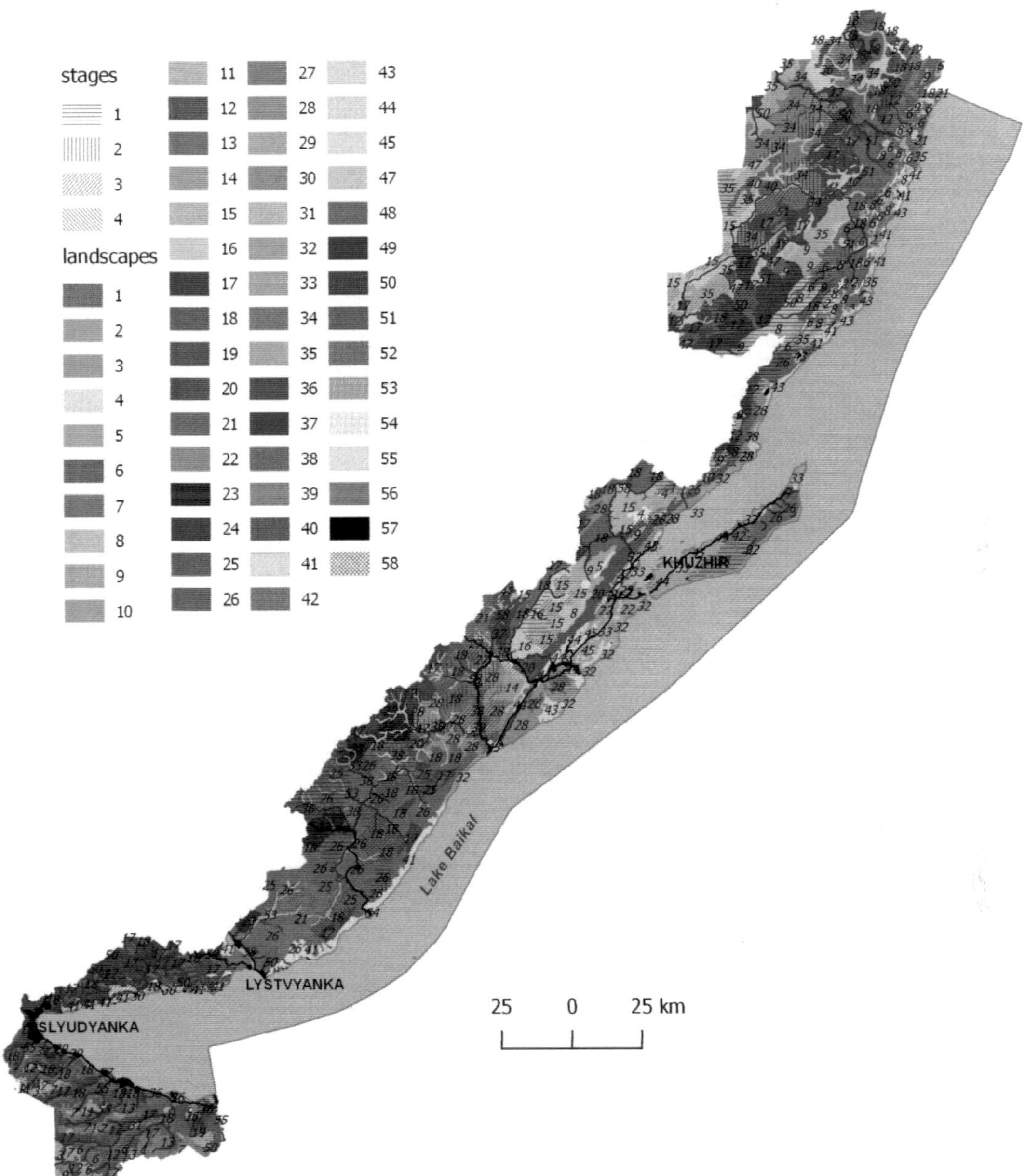


**Figure 1.** Landscape map of the area. Stages: 1 – new (0-3 years) burnt areas, 2 - regenerating burnt areas, 3 - small-leaved forests in place of old burnt areas, 4 - light coniferous stages of restoration of dark coniferous forests. Landscapes: see table 1.

A total of 56 landscape units were identified: 13 highland (goltzy, sub-mountain, reduced development mountain-taiga), 6 medium-mountain (limited development taiga), 14 lowland (optimal development mountain-taiga, steppe), 13 foothill (limited development taiga, optimal development taiga, subtaiga, steppe), 10 - intermontane depressions and valleys (limited development taiga, optimal development taiga, steppe).

IOP Conf. Series: Earth and Environmental Science **190** (2018) 012037 doi:10.1088/1755-1315/190/1/012037

**Table 1.** Legend of the landscape map.

| Topographic feature | Landscape | Number on the map |
|---|---|---|
| | *High-mountain, Goltsy, Alpinotype* | |
| Sharp ridge watersheds, rocky and steep rockfall-talus slopes | With sparse vegetation cover (fragments of lichen tundras) | 1 |
| | *Subalpinotype* | |
| Of watersheds and slopes | Herbaceous with areas of alpine meadows | 2 |
| | *Tundra* | |
| Goltsy planate surfaces, adjacent gentle slopes | Lichen tundra | 3 |
| Of rounded tops и adjacen gentle slopes, detrital-block | Moorland | 4 |
| Slopes | Gravity-solifluction lichen with sparse thickets of Siberian dwarf pine | 5 |
| | *Subgoltsy, Shrub* | |
| Of configured surfaces and gentle slopes | With Siberian dwarf pine moss-lichen | 6 |
| | *Sparse wood* | |
| Of steep slopes | With Siberian dwarf pine | 7 |
| Of top surfaces and slopes | Larch with Siberian pine, fir and birch sparse wood subshrub-lichen combined with rare thickets of Siberian dwarf pine | 8 |
| Of watersheds and slopes | Siberian pine and fir sparse wood with yernik subshrub-true moss and grass-subshrub-lichen | 9 |
| Of slopes | Siberian pine sparse wood with yernik or Siberian dwarf pine | 10 |
| Of trough bottom | With Siberian pine and firry sparse wood herbaceous, grass-subshrub and subshrub | 11 |
| | *Mountain-taiga of reduced development* | |
| Planate surfaces of watersheds, upper and lower slopes | Siberian pine and fir-Siberian pine (with larch on nothern slopes) sparse wood subshrub-true mosses with short grass (with Siberian dwarf pine and rhododendron golden – on nothern slopes, with bergenia on steepe slopes, with rhododendron golden – on gentle slopes and planate surfaces) | 12 |
| Planate surfaces of watersheds, upper and lower slopes | Fir и Siberian pine-fir (with spruce and Siberian pine on nothern slopes) subshrub-true mosses (with bergenia on steepe slope, with Siberian dwarf pine on northern slopes) in combination of subshrub-herbaceous and herbaceous on southern slopes | 13 |
| | *Middle-mountain, Mountain-taiga of limited development* | |
| Of flat surfaces | Larch with admixture Siberian pine cowberry-bergenia | 14 |
| Slopes | Larch with admixture Siberian pine, spruce with mixed undergrowth | 15 |
| | Larch with pine | 16 |
| Flat watersheds and southern slopes | Fir-Siberian pine and Siberian pine-fir subshrub-short grass-true mosses (short grass-subshrub- true mosses in places) | 17 |
| Largely gentle slopes | Siberian pine и fir-Siberian pine (with admixture of spruce and larch in places) subshrub-true mosses with short grass (subshrub- true mosses with marsh tea on northern gentle slopes) | 18 |
| Slopes of different light aspects and steepness | Fir and spruce-fir (on steep slopes - Siberian pine- fir) subshrub- true mosses (with bergenia on steep slopes, with marsh tea – on northern gentle slopes, short grass-large grass-small reed and cowberry-forb with bracken – on southern gentle slopes, with sphagnum moss on the lower parts of slopes) | 19 |
| | *Low-mountain, Mountain-taiga of optimal development, Baikal-Dzhugdzhurski* | |
| Slopes | Larch with shrub undergrowth | 20 |
| | Larch с участием pine and pine-larch forb and cowberry-forb and grass-mosses | 21 |
| Gentle slopes | Larch rare forests with rare shrub undergrowth gramineous-forb combined with steppes | 22 |
| | *Southern-Sibeirian* | |
| Slopes | Siberian pine-fir bilberry-true mosses | 23 |
| | Polydominant Siberian pine-spruce | 24 |
| Of flat surfaces | Pine with rhododendron undergrowth (Rhododendron dauricum) | 25 |
| Slopes | Pine and larch-pine with mixed undergrowth | 26 |
| Slopes of low mountains and high | Pine with additive of larch grass-cowberry | 27 |

| Topographic feature | Landscape | Number on the map |
|---|---|---|
| plains | | |
| | Pine grass, with rhododendron undergrowth (Rhododendron dauricum) steppificated in places | 28 |
| Steep northern slopes, facing the lake Baikal | Siberian pine, with Siberian dwarf pine undergrowth | 29 |
| Steep precipitous slopes facing the lake Baikal | Pine and larch rare forests steppificated lithophile | 30 |
| Slopes and watersheds | Deforested forb-small reed shrub-dominated with washed away the soil and spots of stony outcrop | 31 |
| | *Steppe* | |
| Steep precipitous slopes | Forb-fescue combined with small forb and sagebrush lithophile | 32 |
| Gentle slopes with spots of stony placers | Small sod-gramineous and forb-fescue combined with small forb and sagebrush litophile | 33 |
| | Submountain, Taiga of limited development | |
| Of piedmont high plains | Fir-Siberian pine subshrub-small grass-true mosses | 34 |
| Bottom of depressions | Siberian pine-larch and larch with addition of spruce, with shrub undergrowth subshrub-mosses in places | 35 |
| | *Taiga of optimal development* | |
| Of piedmont high plains | Siberian pine-fir bilberry-grass-moss with tall grass | 36 |
| Of submountain plains | Fir-Siberian pine moss-grass | 37 |
| | *Subtaiga* | |
| Of depressions bottoms, submountain plains | Pine with shrub undergrowth grass, steppificated in places | 38 |
| Submountain plains | Gramineous-forb meadows of anthropogenic genesis, combined with birch grove and sedge lowland swamps | 39 |
| Of piedmont high plains | Siberian pine-larch subshrub | 40 |
| Slopes facing the lake Baikal except for north steep slopes) | Light coniferous and small-leaved, usually rarefied forests small reed-forb, forb-bracken | 41 |
| Bottoms of gullyes | Pine sparse psammophyte | 42 |
| | *Steppe* | |
| Terraces | Small sod-gramineous lithophile | 43 |
| Bottoms of depressions | Large gramineous feather grass-wheatgrass | 44 |
| Bottoms of gullyes | Small sod-grass spotted combined with halophytic meadow | 45 |
| | Psammophyte plant aggregation | 46 |
| *Intermontain hollows and valleys, Taiga of limited development, Baikal-Dzhugdzhurski* | | |
| Valleys | Yernik | 47 |
| | Taiga of optimal development, Baikal-Dzhugdzhurski | |
| Valleys | Meadows with gramineous cover steppificated in places | 48 |
| | Taiga of limited development, Southern-Sibeirian | |
| Valleys | Fir-Siberian pine with spruce horsetail-short grass-true mosses | 49 |
| | Fir and larch-spruce mixed shrub grass-true mosses and forb-small reed | 50 |
| | Siberian pine-larch and larch with fir and spruce forest with yernik, willow | 51 |
| | *Taiga of optimal development, Southern-Sibeirian* | |
| Valleys | Fir-spruce and Siberian pine-spruce large grass | 52 |
| Valleys | Grass and grass-moss swamps sparse wood larch with spruce | 53 |
| | Oligotrophic moors and mesotrophic bogs of submountain plains sedge-sphagnum and subshrub (marsh tea, cranberry, leatherleaf)-sedge-sphagnum with sparse stand of Siberian pine, spruce and birch | 54 |
| | Ligth coniferous and dark coniferous with poplar forb- moist grass shrub-dominated | 55 |
| | *Steppe* | |
| Valleys | Sedge-gramineous meadow-swamp solonetzic | 56 |
| | *Anthropogenic* | |
| Settlements and industrial areas | | 57 |
| Agricultural (arable lands) | | 58 |

The study area has a high landscape diversity, which makes it possible to develop a wide range of tourism activities [8].

Landscapes of the goltsy belt (3% of the research area) are represented by the near-watershed, slope, rocky and rockfall-talus moorlands and with lichen tundra complexes and alpine meadows. In the subgoltsy belt (11% of the study area) moss-lichen, yernik, dwarf Siberian pine and shrub-lichen tundra, as well as larch, Siberian pine and fir sparse forest are widespread. They are confined mainly to the peaks and slopes of the Khamar-Daban and the Baikal ridges, to a lesser extent, the Primorsky ridge.

Mountain-taiga landscapes (62% of the investigated territory) of Khamar-Daban ridge are mainly represented by dark coniferous Siberian pine and spruce-fir- Siberian pine forests. The mountain taiga belt of the Primorsky range is represented mainly by larch and pine, as well as Siberian pine with an admixture of spruce and larch forests. The Baikal ridge is dominated by fir- Siberian pine forests. On the Olkha plateau, fir- Siberian pine forests, as well as areas of their light coniferous and small-leaved succession stages predominante.

Submountain and subtaiga complexes (9% of the study area) are characterized by a moderately roughness of relief and a high species diversity. On Olkha plateau they are mainly represented by light coniferous and small-leaved forests on the slopes facing Lake Baikal, on the Khamar-Daban ridge – by Siberian pine-fir forests of foothills, on the Baikal ridge - by larch with Siberian pine forests, on the Primorsky ridge, in Priolhonye and on the Olkhon island – by subtayga pine and larch forests.

Steppe landscapes of Priolkhonye and Olkhon island, occupying piedmont locations and located in the "rain shadow" of the Primorsky ridge, are spatially differentiated depending on the relief and lithomorphic factor and characterized by the predominance of large-grained feather-grass on the bottoms of the basins (9 %) and small-grained lithotrophic on terraces (6 %).

About 30% of the Baikal Range, 13% of the Primorsky Range, 9% of Priolkhonye and Olkhon, 7% of the Olkha plateau within the central ecological zone of the lake Baikal are occupied by new and regenerating burnt areas.

## 4. Conclusions

Thus, within the territory under study, a number of zones (goltsy, subgoltsy, mountain-taiga, piedmont and subtaiga and steppes) are identified, each of which has unique conditions for the tourism development.

The conclusions drawn about the landscape structure of the territory are generalized. More detailed landscape research is needed, including comprehensive field work to clarify the landscape map and assess the recreational significance of landscapes.

## Acknowledgments

The work was carried out within the framework of the program of research of the V.B. Sochava Institute of Geography SB RAS with partial support of the Russian Foundation for Basic Research (project No. 16-05-00902 and No. 17-05-00588).

## References

[1] Istomina E A, Solodyankina S V, Vanteeva Y V, Konovalova T I, Bibaeva A Y, Frolov A A and Tsygankova M V 2018 Results of landscape-cartographic studies in the Baikal region *Geodesy and cartography* **79**(2) 36-47

[2] Bezrukikh V A, Antonenko O V, Kostrenko O V, Avdeeva E V and Korotkov A A 2016 Mountain-taiga forests of eastern Sayan as a potential for nature-oriented tourism (within the Krasnoyarsk Territory) *Coniferous of Boreal Zone* **37** 266-9

Bezuglova M S, Sharova I S and Sulejmanov A R 2013 Geoecological approaches in studying the tourist and recreational potential of the territory *Geology, Geography and Global Energy* **51**(4) 132-9

Timoshenko N V 2010 Mapping and study of landscapes for the purpose of creating a scheme for the development of recreation and tourism *Izvestiya of Smolensk State University* **2** 30-9

Fadafan F K, Danehkar A and Pourebrahim S 2018 Developing a non- compensatory approach to identify suitable zones for intensive tourism in an environmentally sensitive landscape *Ecological Indicators* **87** 152-66

Mann C, Garcia-Martin M, Raymond C M, Shaw B J and Plieninger T 2018 The potential for integrated landscape management to fulfil Europe's commitments to the Sustainable Development Goals *Landscape and Urban Planning* **177** 75-82

Woźniak E, Kulczyk S and Derek M 2018 From intrinsic to service potential: An approach to assess tourism landscape potential *Landscape and Urban Planning* **170** 209-20

[3] Kuzmin V A 2002 *Soils of the Central Zone of the Baikal Natural Territory (Ecology-Geochemical Approach)* (Irkutsk: IG SO RAN) p 166

[4] *Baikal. Atlas* 1993 (Moscow: Roskartografiya) p 160

[5] Belov A V 1973 *Map of Vegetation in the South of Eastern Siberia. Principles and Methods of Compilation. Geobotanical Mapping* (Leningrad: Nauka) pp 16-30

[6] Mikheev V S 1995 Landscapes of the Baikal region: structure, state estimation, problems *Geography and Natural Resources* **3** 68-78

[7] Frolov A A 2015 Geoinformation mapping of landscape variability (on the example of the southern Baikal region) *Geography and Natural Resources* **1** 156-66

Korol'kova E E 2015 Medium-scale geobotanical mapping of the North-Western Baikal region taking into account the evolutionary-dynamic features of vegetation *Geobotanical Mapping* (St.Petersburg: Boston-Spectr) pp 42-61

Pljusnin V M and Bilichenko I N 2001 Remote and quantitative methods of studying the landscape structure (on the example of the Khamar-Daban ridge) *Geography and Natural Resources* **2** 127-36

Zagorskaja M V 2004 Landscape structure of Central Priolhonya *Geography and Natural Resources* **4** 58-68

Suvorov E G and Titaev D B 1999 Structure of the landscapes of the Southern Pribaikalye *Geography and Natural Resources* **4** 20-30

Suvorov E G, Antipov A N, Semenov YU M 2002 *Ecologically-Oriented Land Use Planning in the Baikal Region: Sludyanka Region* (Irkutsk: IG SO RAN) p 141

[8] Evstropeva O V 2009 *Cross-Border Tourism in Adjacent Regions of Russia and Mongolia* (Irkutsk: IG SB RAS) p 143

# Landscape-interpretation mapping of the air temperature field of the Mondy depression

**O V Vasilenko, E A Istomina**

Sochava Institute of Geography SB RAS, Irkutsk, Russia

E-mail: oksa_na85@mail.ru, elenaistoma@gmail.com

**Abstract.** Air temperature is one of the most important climate characteristics, which is used for climate research, investigation and modeling of vegetation, hydrological parameters and other landscape components as well as geosystems entirely. The goal of this research is investigation and mapping of the seasonal and daily air temperature regime of different landscapes of Mondy valley. For the research we use landscape map of the area on the level of group of facies as well as thermochron devices. For the analysis of air temperature regime of different landscapes we used year-round time air temperature data for 2013 from 12 key areas. For each class of facies of the landscape map monthly air temperature characteristics were calculated, and maps of mean monthly air temperature were made. According to the annual mean air temperature values landscapes of the bottom of the valley are colder than landscapes of slopes. Temperature regimes of landscapes in summer and in winter are different. In winter landscapes of the central part of the valley are characterized by the lowest values of air temperature, and the landscapes of the ridges – by the highest temperature due to winter temperature inversions.

## 1. Introduction

Air temperature is one of the most important climatic characteristics used for studying climate, research and modeling of vegetation, hydrological parameters, various components of landscapes, and geosystems in general. The study of the change in air temperature has become particularly relevant in recent decades due to the observed global warming.

When mapping the air temperature field in mountainous regions, extrapolation of ground-based measurements (in most cases they are the data of meteorological stations) the consideration of a large number of factors requires [1], therefore, additional information such as space images and modeling results are used.

The aim of the research is to study and map the features of the seasonal variation of air temperature in different landscapes of the Mondy depression based on the data of micrometeorological measurements of the surface air temperature and the landscape map of the territory.

## 2. Objects, data and methods

The object of research is the territory of the Mondy intermountain bolson. The basin is located along the Irkut river near the border with Mongolia. It is the most highly located and compact of the basins of the Tunka branch and closes it in the west, and represents a lake-like expansion of the Irkut valley. Its length is about 15 km, width is 6-7 km. The absolute height is 1344 meters [2]. In the south the bolson is bounded by the Khamar-Daban ridge, and in the north by the Tunka Goletz ridge (Eastern Sayans). The territory belongs to the zone of sporadic permafrost [3, 4] and is characterized by strongly continental climate [5]. The annual precipitation in the bottom of the basin is relatively small and increases toward its sides [6, 7]. The study area is characterized by a variety of landscape conditions, forms of relief and soil-forming rocks, and a long history of economic development [7, 8]. All this allows to explore a wide variety of landscapes (the goletz, mountain taiga and mountain steppe landscapes) and their temperature features, as well as the bolson effects on a compact territory.

In 2007 we started interdisciplinary physical-geographical studies in the Tunka valley [4, 9-11]. A GIS dataset of the research area was created, containing topographic, small-scale geological and landscape maps, a digital elevation model (SRTM), Landsat-5, -7 and SPOT-4 space images. In previous studies, the authors validated the data of Landsat satellite measurements in the far infrared band based on ground-based measurements of air temperature using autonomous thermographs DS-1922 [12]. Based on the remote sensing data an analysis of the space-time changes in the surface temperature field in the territory of the Tunka depression [4, 10, 11] was made. The mapping of the air temperature field of the Tunka bolson was made on a landscape basis [13].

The method of landscape-interpretation mapping [14] of the characteristics of the temperature field is applied in this paper. The method is based on the fact that a certain type of geosystem (for example, a class of facies) has homogeneous natural characteristics throughout its range. This allows extrapolating characteristics (including air temperature) measured at one or several points of the range of a certain class of facies to the entire range of this class of facies.

Mapping of the air temperature field is made based on a landscape-typological map of the territory at the level of facies classes M 1: 100000 [15]. The study area is represented by nine classes of facies, belonging to four groups of geomes (figure 1).

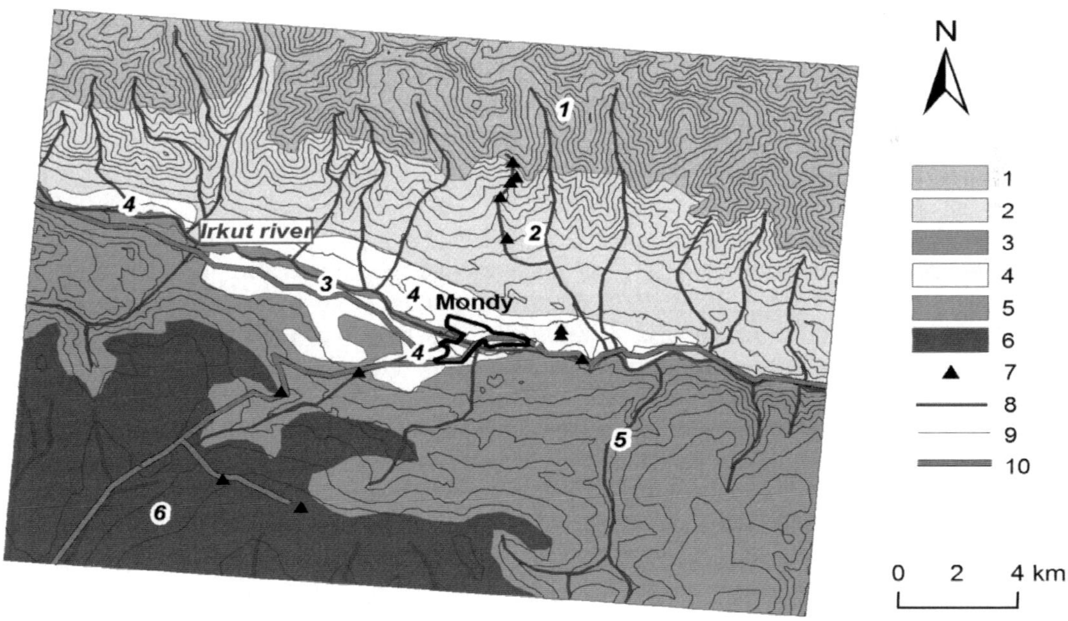


**Figure 1.** Landscapes of the Mondy depression: 1-6 – classes of facies; Tunka Goletz ridge: 1- Class of goletz facies; 2 - Class of mountain taiga (larch fwith Siberian pine) facies on steep slopes; 3 - Class of meadow-steppe facies of the Irkut valley; 4 - Class of steppe facies on slopes; Khamar-Daban ridge: 5- Class of mountain taiga (larch with Siberian pine) facies on smooth slopes; 6 - Class of mountain taiga (larch with Siberian pine and dwarf birch) facies on flat interfluve, 7 – thermographs locations; 8 – rivers. 9 – absolute isohyps (in 100m), 10 – roads.

For micrometeorological measurements of the surface air temperature, autonomous thermographs DS-1922 are used, which allow solving the problems of climate research at a local level. The thermograph is an electronic recorder (logger), protected from external influences, accumulating in its own memory the values of the temperature of its surrounding environment, with reference to real time.

At present, 12 observation sites are located on the territory of the Mondy bolson in the altitude range from 1100 to 2100 m above sea level (table 1), which form a transect from the right to the left side of the basin and situated into the range of each class of facies. The devices are installed at a height of 2 m from the underlying surface and programmed for measurements at a frequency of 3 hours, synchronously with measurements at weather stations. The measurement data are read from the thermal registers once a year. For the analysis of microclimatic features of the temperature regime of air in different landscapes, year-round data for 2013 were used.

## 3. Results and discussion

For each class of facies, the average monthly air temperature values for each month of 2013 are calculated (table 1) and are presented in the form of maps (figure 2).

**Table 1.** Average monthly air temperature within different classes of facies of the Mondy depression

| Classes of facies | Range of altitudes | N[a] | Average monthly air temperature ° C, 1 – 12 - months of the year | | | | | | | | | | | |
|---|---|---|---|---|---|---|---|---|---|---|---|---|---|---|
| | | | 1 | 2 | 3 | 4 | 5 | 6 | 7 | 8 | 9 | 10 | 11 | 12 |
| 1 | 1915-3081 | 1 | -15.8 | -18.6 | -9.7 | -4.3 | 2.1 | 7.1 | 9.9 | 9.2 | 1.6 | -3.9 | -8.8 | -12.2 |
| 2 | 1199-2550 | 4 | -15.5 | -17.4 | -8.3 | -2.3 | 4.2 | 8.8 | 11.0 | 10.6 | 3.3 | -2.0 | -6.5 | -11.2 |
| 3 | 1182-1429 | 1 | -22.1 | -19.3 | -8.0 | 0.1 | 7.5 | 11,8 | 14.4 | 13.6 | 4.9 | -1.9 | -6.7 | -14.1 |
| 4 | 1235-1534 | 2 | -18.0 | -17.0 | -7.1 | 0.8 | 8.0 | 12.1 | 14.4 | 13.6 | 5.3 | -1.0 | -5.7 | -12.4 |
| 5 | 1182-2072 | 2 | -18.5 | -18.4 | -8.2 | -0.5 | 6.6 | 10.8 | 13.0 | 12.3 | 4.6 | -1.5 | -6.0 | -12.3 |
| 6 | 1576-2017 | 2 | -18.5 | -19.0 | -9.5 | -2.8 | 3.5 | 8.8 | 11.1 | 10.3 | 2.3 | -3.5 | -8.3 | -13.3 |

[a]Note: number of sensors

The average annual air temperature (according to the data of the installed sensors) in the territory of the Mondy depression varies from -3.6 °C on slopes to -0.6 °C in the central part. These values are average according to the data from the sensors from all over the basin and do not reflect the microclimate peculiarities in certain types of landscapes. The coldest and warmest months are January (on slopes - February) and July, the average monthly temperature of the basin air in these months is -18.1 (-18.3) and 12.3 °C, respectively.

Microclimatic features of the distribution of air temperature in various landscapes of the Mondy depression are most visible in January, the difference between the warmest and the coldest class of facies is 6.6 °C. The lowest mean monthly air temperature in January is fixed in the landscapes of the central part of the basin: this is the class of meadow-steppe facies of the Irkut valley (-22.1 °C). Slopes of the Tunka goletz ridge are 3 °C warmer than the slopes of the Khamar-Daban ridge. This is due to the presence of winter temperature inversions, when the temperature rises with altitude. Therefore, the landscapes of the central part of the basin are the most cold, and the most elevated slopes of the Tunka goletz ridge are the warmest. Vegetation conditions on the slopes do not have a significant effect on the average monthly temperature values in the winter, the differentiation of the temperature along the slope is due to the presence of powerful temperature inversions.

The distribution of the average monthly temperature in July according to the classes of facies is more homogenous than in January. The difference between the warmest and the coldest class of facies is 4.5 °C. The most warm in the summer are the facies classes located in the central part of the basin, the average monthly temperature here is 14.4 °C.

IOP Conf. Series: Earth and Environmental Science **190** (2018) 012038 doi:10.1088/1755-1315/190/1/012038

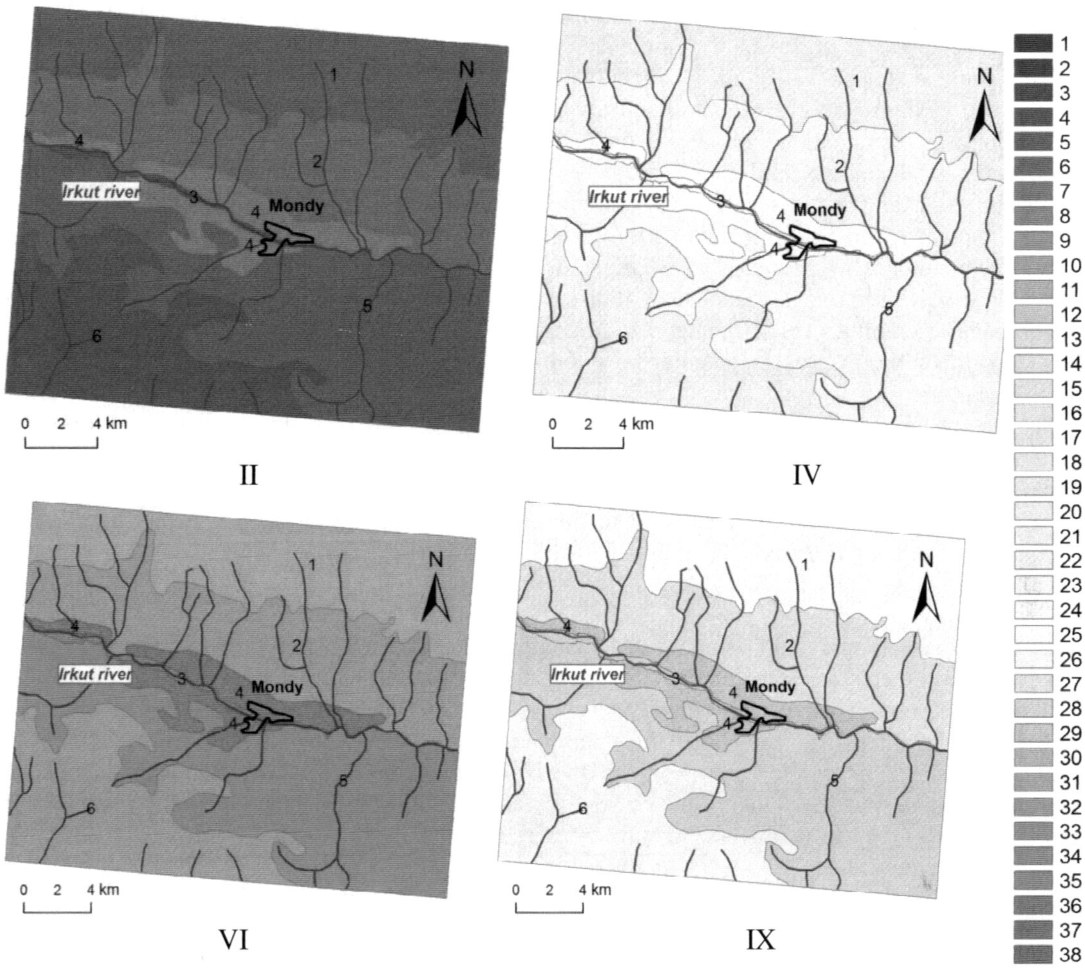


**Figure 2.** The spatial distribution of the mean monthly air temperature (єC) within the various classes of facies of the Mondy depression: II - IX - months of the year; 1– -23 - -22; 2– -22 - -21; 3– -21 - -20; 4– -20 - -19; 5– -19 - -18; 6– -18 - -17; 7– -17 - -16; 8– -16 - -15; 9– -15 - -14; 10– -14 - -13; 11– -13 - -12; 12– -12 - -11; 13– -11 - -10; 14– -10 - -9; 15– -9 - -8; 16– -8 - -7; 17– -7 - -6; 18– -6 - -5; 19– -5 - -4; 20– -4 - -3; 21– -3 - -2; 22– -2 - -1; 23– -1 - 0; 24– 0 - 1; 25– 1 - 2; 26– 2 - 3; 27– 3 - 4; 28– 4 - 5; 29– 5 - 6; 30– 6 - 7; 31– 7 - 8; 32– 8 - 9; 33– 9 - 10; 34– 10 - 11; 35– 11 - 12; 36– 12 - 13; 37– 13 - 14; 38– 14 - 15

On the Khamar-Daban ridge, the average monthly July temperature in the lower part of the slope is 13.0 °C (the class of mountain taiga (larch fwith Siberian pine) facies of smooth slopes) and decreases to 11.1 °C in the upper part of the slope (the class of mountain taiga (larch with Siberian pine and dwarf birch) facies of flat interfluve). On the Tunka Goletz ridge, the average monthly temperature in July falls from 14.4 °C in the foothills (the class of mountain taiga facies) to 9.9 °C in the goletz belt.

Thus, relative to January, the picture of the change in mean temperature values from the center of the basin to its sides is transformed: the central part of the basin is heated in summer, the values of the temperature decrease upon transition to the mountain taiga landscapes, the higher is altitude the lower is temperature.

In February, March, and also in October-December, the difference in mean monthly temperature between different classes of facies is insignificant. It peaks in January and May. From December to

February, there are insignificant temperature inversions, the maximum in January, in the remaining months the temperature decreases with altitude.

To extrapolate the data of point observations of air temperature, various models are used, taking into account predominantly the relief characteristics, as well as the circulation of the atmosphere, the distribution of solar radiation and other parameters [16, 17]. Such models are often based on the data of meteorological stations and they contain general patterns of temperature changes with other parameters, for example, a decrease in temperature with altitude. On the territory of the Mondy bolson due to temperature inversion processes, the general laws of temperature decrease with altitude are violated, which leads to erroneous calculations of such models. Also in recent decades remote data have been widely used to analyze the spatial distribution of the temperature field [18]. However, these data contain uncertainties [11, 19], and also reflect only a certain time sample (time of survey). Using these data, it is difficult to assess the diurnal and seasonal temperature changes, especially the average monthly values.

The advantage of this work is the use of a dense network of ground-based observations with the use of thermographs DS-1922, as well as the method of landscape-interpretation mapping. The landscape approach in the mapping of meteorological parameters allows to take into account the influencing factors that have already been considered in the compilation of the landscape map, which greatly simplifies the mapping process

## 4. Conclusions

Thus, the conducted study and mapping of the seasonal air temperature of various landscapes of the Mondy basin in the facies class section showed that, by average annual temperature, the landscapes located in the central part of the bolson are cooler than the landscapes of the sides of the basin. However, the temperature regimes of the landscapes in summer and winter are very different. In winter, the landscapes of the central part of the basin are characterized by the lower temperature values than the landscapes of the slopes of ridges, which is due to the presence of winter temperature inversions; in the summer the reverse picture is observed.

## References

[1] Isaev A A 1988 *Statistics in Meteorology and Climatology* (Moscow: Moscow State University Press) p 248

[2] Vyrkin V B, Kuzmin VA and Snytko VA 1991 The generality and differences of some features of the nature of the Tunka branch of the hollows *Geography and Natural Resources* **4** 61-8

[3] Lamakin V V 1935 Past relief formation in the Tunkinsky Baikal region *Earth science* vol 38 (Moscow — Leningrad: ONTI) pp 1-26

[4] Vasilenko O V and Voropay N N 2015 Features of formation of the climate of the depressions of the southwestern Baikal region *Izvestiya RAS, Geographic Series* **2** 104-11

[5] Belousov V M, Budhe I Yu and Radziminovich Y B 2000 *Physico-Geographical Characteristics and Problems of Ecology of the Southwestern Branch of the Baikal Rift Zone* (Irkutsk: Irkutsk State University Press) p 160

[6] Kartushin V A 1969 *Agroclimatic Resources of the South of Eastern Siberia* (Irkutsk: East-Sib) p 100

[7] Larin S I 1991 The main stages of development of landscapes of the Tunka hollows *Historical and Geographical Studies of Southern Siberia* (Irkutsk: Publishing House of the IG SB RAS) pp 70-85

[8] Holboyeva S A and Namzalov B B 2000 *Steppes of the Tunka Depression South-Western Pribaikalye* (Ulan-Ude: Publishing house of the Belarusian State University) p 116

[9] Istomina E A 2012 Geoinformation mapping of landscapes of the Tunka depression basing on the method of factorial-dynamic classification *Geodesy and Cartography* **4** 32-9

IOP Conf. Series: Earth and Environmental Science **190** (2018) 012038 doi:10.1088/1755-1315/190/1/012038

[10] Voropay N N, Istomina E A and Vasilenko O V 2011 Investigation of the temperature field of the terrestrial surface of the Tunka depression with the use of Landsat satellite images *The Atmosphere and Ocean Optics* **1** 67-73

[11] Istomina E A and Vasilenko O V 2015 Geoinformation analysis of the temperature field of the landscapes of the Tunka depression with the use of Landsat satellite imagery and ground data *Geography and Natural Resources* **4** 162-70

[12] Scientific and Technical Laboratory «Electronic Instruments» 2007 (in Russian) Income accessed online on 11th May 2017 via http://www.thermochron.ru

[13] Vasilenko O V, Istomina E A and Voropay N N 2017 Mapping of the air temperature field of the Tunka depression on a landscape basis *Geography and Natural Resources* **2** 182-9

[14] Cherkashin A K 2005 *Landscape–Interpretation Mapping* (Novosibirsk: Nauka) p 424

[15] Istomina E A and Ovchinnikova E V 2018 Geoinformation mapping of landscapes of the Mondy basin *Geodesy and Cartography* **4** 23-30

[16] Stahl K, Moore R D, Floyer J A, Asplin M G and McKendry I G 2006 Comparison of approaches for spatial interpolation of daily air temperature in a large region with complex topography and highly variable station density *Agricultural and Forest Meteorology* **139** 224-36

[17] Global Climate Data 2016 (in Russian) Income accessed online on 13th June 2018 via http://worldclim.org/

[18] Li Z, Tang B, Wu H, Ren H, Yan G, Wan Z, Trigo I F and Sobrino J A 2013 Satellite-derived land surface temperature: Current status and perspectives *Remote Sensing of Environment* **131** 14-37

[19] Kochendorfer K P J, Dumas E J, Guillevic P C, Baker C B, Meyers T P and Martos B 2015 Comparison of in-situ, aircraft, and satellite land surface temperature measurements over a NOAA Climate Reference Network site *Remote Sensing of Environment* **165** 249-64

# Regional tendencies in air temperature at the southwestern Pribaikalie

**O V Vasilenko**[1], **N N Voropay**[1,2]

[1]Sochava Institute of Geography SB RAS, Russia, Irkutsk
[2]Institute of Monitoring of Climatic and Ecological Systems SB RAS, Russia, Tomsk

E-mail:oksa_na85@mail.ru, voropay_nn@mail.ru

**Abstract.** The regional features of long-term changes in air temperature at the south-western Baikal region on the background of global climate changes were studied. To estimate changes in air temperature in the basins of the south-western Pribaikalie, the long-term series of air temperature were used for the Tunka weather station. The correlation coefficients between the series of air temperature at the Tunka weather station and the temperature averaged for the temperate latitudes, the Northern Hemisphere, and the Globe were analyzed. Analysis of the variations in the anomalies of the mean annual temperature averaged for different zones, have shown a pronounced increase in the growth rate of the annual air temperature since the beginning of the 1970s for all the series of data. To estimate the relationship between the change in the surface air temperature in south-western Pribaikalie and large-scale atmospheric circulation mechanisms, correlation coefficients between the series of temperature and the characteristics of the atmospheric circulation have been calculated. The results of the analysis showed that the closest relationship between air temperature in the basins of the south-western Pribaikalie exists with the Scandinavian index (SCAND), the western type of atmospheric circulation of Wangengeym-Girs (W) in winter, the pressure in the center of the Siberian High in December.

## 1. Introduction

The current climate change is characterized by continuing warming, the main indicator of which is the near-surface air temperature, calculated as the average surface air temperature (at 2 m above the surface) over the continents and the water surface temperature of the ocean surface [1, 2].

According to the results of the Intergovernmental Panel on Climate Change (IPCC) from 1880 to 2012, the increase in global near-surface air temperature on the continents and oceans was 0.85°C (0.65 to 1.06°C). Since 1951, the rate of growth of surface air temperature was 0.12°C per 10 years (from 0.08 to 0.14), and for 1998-2012 it was only 0.05°C / 10 years (from 0.05 to 0.15). The slowdown in temperature growth was due to natural fluctuations in the climate system and can not serve as evidence of the cessation of the global warming [3].

It is known, that the change in the near-surface air temperature for the Northern Hemisphere in the XX and early XXI centuries was also not homogeneous. The anomaly of the mean annual near-surface temperature averaged for the Northern Hemisphere varied unevenly during the period 1850-2014 [4]. Until 1945, there was a temperature increase, then until 1970 the average annual air temperature of the

Northern Hemisphere was lowered. Since 1970, the phase of active growth of near-surface air temperature has been observed, with a maximum (0.719°C) in 2005 [5]. The most intensive is considered warming, which began after 1976. At the same time, interest in research on the problem of climate change has grown [6].

A large number of works have been devoted to the analysis of the ongoing climate changes on the territory of Russia [1, 4, 5, 7-11]. According to the IPCC Fifth Assessment Report, climate change in Russia as a whole (on average for the year and for the territory) should be characterized as continuing warming, noting that the trend towards slowing down of warming is traced only in winter [3]. At the same time, many authors emphasize the very important role of the natural variability of the climate system over the decades, especially for individual regions

## 2. Objects, data and methods

The territory of the southwestern Pribaikalie is represented by mountain-hollow landscapes. Local conditions are imposed on the zonal features of the climate of this region due to a combination of high-mountainous terrain and relatively low intermountain depressions, latitudinal orientation of the main orographic elements, regional peculiarities of the atmospheric circulation. In addition, the complex terrain and individual geomorphological conditions contribute to the formation of a unique microclimate of different landscapes, due to the interaction of the circulation and radiation factors and the properties of the underlying surface. The description of the main meteorological parameters and their changes is based on the data of meteorological stations located on the territory of the southwestern Pribaikalie. As it was shown, the data of the Tunka weather station are the most representative for the study area [12]. To assess the changes in air temperature in the basins of the south-western Baikal region, against the backdrop of current climate changes, long-term series of temperatures at the weather station Tunka [13]. In addition, the anomalies of the mean annual near-surface air temperature averaged for the globe, the Northern Hemisphere and the temperate latitude zone from 1888 to 2014 [14] were studied.

One of the leading climate-forming factors that affect the fluctuations of the regional climate is the large-scale mechanisms of atmospheric circulation. The connection between the circulation systems and the temperature regime of the Northern Hemisphere has been confirmed by many authors [1, 4, 5, 15, 16]. When the relationship between large-scale atmospheric circulation mechanisms and components of the climate system is identified, scientists use long-term series of atmospheric circulation indices that are linked to specific geographic sectors. According to the results of modern studies for the territory of Siberia, the most informative are the indices SCAND (Scandinavian Index), NAO (North Atlantic Oscillation Index), as well as the frequency of Wangenheim-Girs atmospheric circulation types (W - western, E - east, C - meridional) and elementary circulation mechanisms according to the classification of Dzerdzievsky. In addition, the Siberian anticyclone (Siberian High) has a significant influence on the climate of Eastern Siberia [17-19]. In the present paper, to determine the relationship of regional changes in the surface temperature in the south-western Pribaikalie with large-scale atmospheric circulation mechanisms, the monthly and annual values of the atmospheric circulation indices, the frequency of atmospheric circulation types according to the Wangenheim-Girs and Dzerdzievsky classifications, and the position of the center of the Siberian High were used. To estimate the relationship between the change in the surface air temperature of the basins of the southwestern Baikal region (according to the data of the Tunka weather station) and large-scale atmospheric circulation mechanisms, the coefficients of correlation between the series of temperature and the characteristics of the atmospheric circulation were studied.

## 3. Results and discussion

To estimate changes in air temperature in the basins of the south-western Pribaikalya, against the background of modern climate changes, long-term series of temperatures were used for the Tunka meteorological station. As it was shown, that the data of the Tunka weather station are the most representative for the investigated territory. Long-term changes in the temperature regime in the basin area correspond to global trends in air temperature. The coefficient of correlation between the series of air temperature at the Tunka weather station and the temperature averaged for the temperate latitude zone (44-64 N) is 0.72, for the Northern Hemisphere is 0.65, for the globe is 0.49. Correlation coefficients are statistically significant (at p <0.05). When analyzing the graph of the variation in the anomalies of the mean annual temperature averaged for different zones, there is a pronounced increase in the growth rate of the average annual air temperature since the beginning of the 1970s for all the series of data (figure 1).

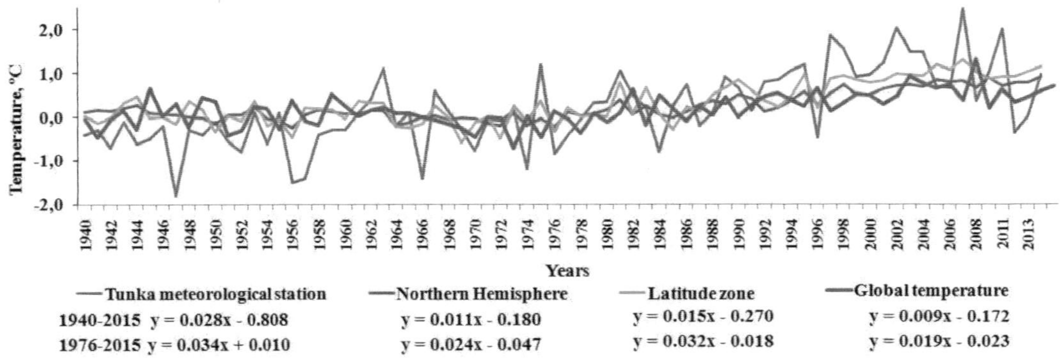


**Figure 1.** Change in anomalies in the average annual air temperature.

When analyzing trends in air temperature, special attention has been paid to the period since 1976, because during this interval the most intensive warming is observed. During the period of instrumental observations at the Tunka weather station, positive trends in temperature were noted during the whole period of observations (1939-2015 – 0.28°C/10 yr). During the period 1976-2015 trend of air temperature was higher (0.34°C/10 yr), than for the base period (1961-1990 – 0.08°C/10 yr) and for the whole period of instrumental observations (1939-2015), which is generally consistent with the global trends of long-term changes in air temperature.

In order to analyze the features of regional climate change on the territory of the Russian Federation we investigate the standard physiographic regions [20]. The territory of the southwestern Pribaikalie is a part of the Pribaikalie and Transbaikalia regions, the main area of the region is represented by the Altai-Sayan and Baikal Mountain countries. Comparison of the linear trend coefficients for 1976-2006 and 1976-2012 showed that the annual trend values on the territory of Russia have not changed. In the Baikal region and Transbaikalia and at Tunka station, the rate of increase in the average annual temperature became less (table 1).

**Table 1.** The values of the linear trend coefficient (°C/10 years) of air temperature.

| | Winter | Spring | Summer | Autumn | Year |
|---|---|---|---|---|---|
| | | 1976 – 2006 | | | |
| Russia | 0.35 | 0.52 [a] | 0.41 [a] | 0.43 [a] | 0.43 [a] |
| Pribaikalie and Transbaikalie | 0.39 | 0.66 [a] | 0.58 [a] | 0.21 | 0.46 [a] |
| Weather station Tunka | 0.74 [a] | 0.59 [a] | 0.29 | 0.46 [a] | 0.52 [a] |
| | | 1976 – 2012 | | | |
| Russia | 0.18 | 0.56 [a] | 0.44 [a] | 0.54 [a] | 0.43 [a] |
| Pribaikalie and Transbaikalie | 0.06 | 0.53 [a] | 0.52 [a] | 0.26 | 0.32 [a] |
| Weather station Tunka | 0.50 [a] | 0.41 [a] | 0.24 | 0.46 [a] | 0.43 [a] |

[a]Note: significant coefficients of the linear trend are marked in bold (p <0.05).

Comparison of estimates of the linear trend coefficient over two periods (1976-2006 and 1976-2012) in the territory of the Russian Federation, Pribaikalie and Transbaikalie and the weather station Tunka showed that the annual trend values in the territory of the Russian Federation did not change. In the region Pribaikalie and Transbaikalie and at the weather station Tunka, the rate of increase in the average annual temperature, taking into account recent years, become less. However, with the addition of the series for 6 years, the magnitude of the trend of air temperature in Russia in all seasons, except in winter, has increased. In the region Pribaikalie and Transbaikalie, the average annual temperature trend, with the addition of a series of observations for 6 years, became smaller (0.32°C/10 years). In addition, the trend for the winter period has significantly changed - the trend values have decreased to 0.06°C/10 years. In the other seasons of the year, the change is insignificant, mainly indicators have become lower, but still remained positive. The maximum rate of warming in the territory of the Russian Federation and in the region is observed in the spring. The values of the linear trend are 0.56°C/10 years in Russia and 0.53°C/10 years in the Pribaikalie and Transbaikalie region.

In the territory of the southwestern Baikal region, the values of the linear trend, with the addition of a time series of air temperatures for 6 years, as well as in the region, have changed in the smaller values. The rate of change in the average annual temperature from 1976 to 2006 were 0.52°C/10 years, for the period 1976 to 2012, the trend value is lower (0.43°C/10 years). According to the weather station Tunka data, the maximum values of the long-term trend are recorded in the winter period, and the minimum values are recorded in the summer. In the Pribaikalie and Transbaikalie region, as well as on the territory of the Russian Federation, the maximum trend values fall on the spring months.

Correlation coefficients between the series of temperature and the characteristics of atmospheric circulation have been calculated to estimate the relationship between the change in the surface air temperature and the southwestern Pribaikalie (according to the data of the Tunka weather station) and large-scale atmospheric circulation mechanisms (figure 2).

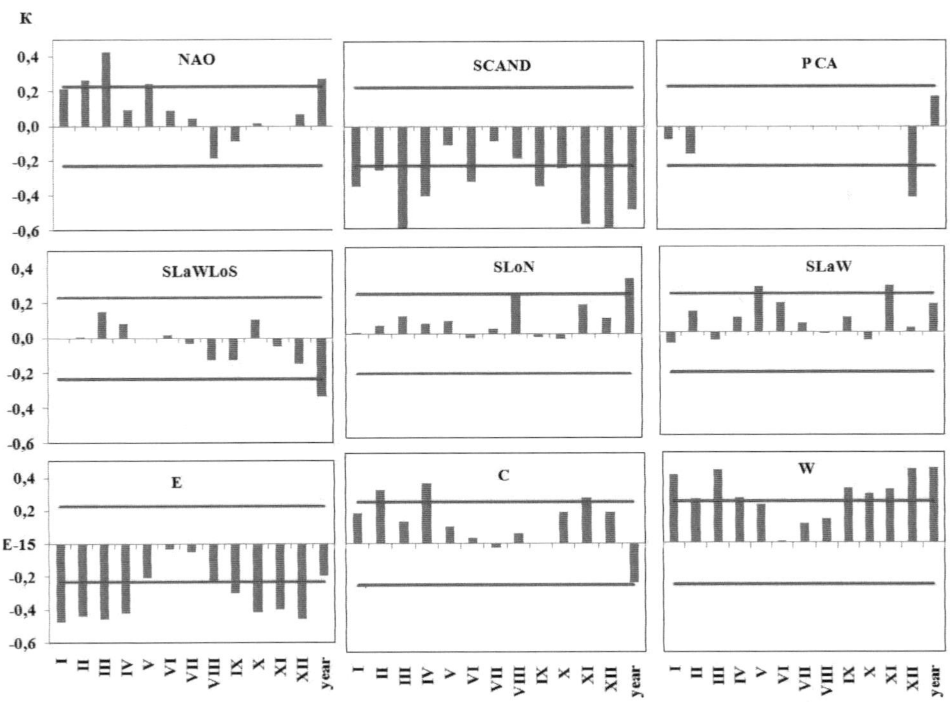


**Figure 2**. Correlation coefficients of mean monthly and mean annual air temperature values at the Tunka weather station and atmospheric circulation indices (NAO is the index of the North Atlantic Oscillation, SCAND is the Scandinavian index, PCA is the pressure in the center of the Siberian High, SLaWLoS, SLoN, SLaW are the elementary circulation mechanisms according to the Dzerziyevsky classification, W, E, C are the types of atmospheric circulation of Wangengeym-Girs, red line - a significance level for $p < 0.05$).

The results of the analysis showed that the closest relationship between air temperatures in the basins of the southwestern Pribaikalya exists with the Scandinavian index (SCAND), the western type of atmospheric circulation of Wangengeym-Girs (W) in winter (table 2). The pressure in the center of the Siberian High in December (-0.41).

**Table 2.** Correlation coefficients between air temperature series at Tunka weather station and atmospheric circulation characteristics (1940-2014).

| Months | I | II | III | IV | V | VI | VII | VIII | IX | X | XI | XII | Год |
|---|---|---|---|---|---|---|---|---|---|---|---|---|---|
| SCAND | **-0.35**[a] | **-0.25**[a] | **-0.60**[a] | **-0.41**[a] | -0.11 | **-0.32**[a] | -0.09 | -0.19 | **-0.35**[a] | **-0.24**[a] | **-0.57**[a] | **-0.66**[a] | **-0.49**[a] |
| W | **0.37**[a] | **0.24**[a] | **0.40**[a] | **0.25**[a] | 0.21 | 0.01 | 0.11 | 0.13 | **0.30**[a] | **0.27**[a] | **0.30**[a] | **0.41**[a] | **0.41**[a] |

[a]Note: Significant coefficients are marked in bold ($p < 0.05$).

## 4. Conclusions

Thus, long-term changes in the air temperature of the basins of the southwestern Baikal region occur synchronously with the global changes. However, a high variety of types of underlying surface and a significant difference in the heights of the mountain-hollow landscape affect regional climate change. The winter months make the main contribution to the increase in the average annual air temperature in the basins of the south-western Pribaikalie during the period from 1976 to 2012, while in the territory of the Russian Federation and the Pribaikalie and Transbaikalie region - spring and summer.

Circulation processes play a significant role in the formation of the climate of the basins of the southwestern Pribaikalie, especially in the winter season, when the solar radiation incoming the territory is minimal. In the cold period of the year, the influence of Western transport on the territory of Eastern Siberia is weakened as a result of the formation of the Siberian anticyclone, which leads to a violation of zoning in the temperature distribution. The effect of global circulation, as a climate-forming factor, is manifested in air temperature relationships with SCAND indiex and the frequency of western circulation in the winter months. In addition, an increase in the linear trend coefficient of the average annual air temperature at the Tunka weather station from 1976 to 2012 is due to the temperature increase during the winter months, during the formation and development of the maximum capacity of the Siberian High. This may indicate the influence of the power of the Siberian anticyclone on the formation and trends in the thermal regime of the basins of the southwestern Pribaikalie.

## References

[1] Perevedentsev Y P, Vereshchagin M A, Naumov E P, Shantalinsky K M and Nikolaev A A 2005 Modern climate change in the Northern Hemisphere of the Earth *Scientific Notes of Kazan State University* **1** 91-106

[2] *Evaluation Report on Climate Change and its Consequences on the Territory of the Russian Federation: Climate change* 2008 vol I (Moscow: Roshydromet) p 277

[3] Stocker T F, Qin D, Plattner G-K, Tignor M, Allen S K, Boschung J, Nauels A, Xia Y, Bex V and Midgley P M 2013 *Climate Change 2013: The Physical Science Basis. Contribution of Working Group I to the Fifth Assessment Report of the Intergovernmental Panel on Climate Change* (Cambridge: Cambridge University Press and NY, USA) p 1535

[4] Perevedentsev Yu P and Shantalinsky K M 2014 Estimation of current changes in air temperature and wind speed in the troposphere of the Northern Hemisphere *Meteorology and hydrology* **10** 19-31

[5] Perevedentsev Y P, Wilfand R M and Shantalinsky K M 2016 Low-frequency changes in atmospheric pressure and near-surface air temperature in the extratropical latitudes of the Northern Hemisphere *Proceedings of the Hydrometeorological Research Center of the Russian Federation* **360** 5-25

[6] Gruza G V, Rankova E Ya and Rocheva E V 1989 Analysis of global data on changes in surface air temperature during the period of instrumental observations *Meteorology and Hydrology* **1** 22-31
[7] Ippolitov I I, Kabanov M V, Komarov A I and Kuskov A I 2004 Modern natural and climatic changes in Siberia: the course of average annual surface temperatures and pressure *Geography and Natural Resources* **3** 90-6
[8] Izrael Y A, Pavlov A V, Anokhin Y A, Ball L T and Sherstyukov B G 2006 Statistical estimates of the dynamics of changes in climate elements in permafrost areas on the territory of the Russian Federation *Russian Meteorology and Hydrology* **5** 27-38
[9] Voropay N N and Kichigina N V 2018 Long-term changes in the hydroclimatic characteristics in the Baikal region *IOP Conference Series: Earth and Environmental Science* **107**(1) 012042
[10] Voropay N N, Maksyutova E V and Riazanova A A 2016 Hydrothermal conditions at the south of East Siberia during the ongoing warming *IOP Conference Series: Earth and Environmental Science* **48** doi:10.1088/1755-1315/48/1/012003
[11] Ryazanova A A and Voropay N N 2017 Droughts and Excessive Moisture Events in Southern Siberia in the Late XXth - Early XXIst Centuries *IOP Conference Series: Earth and Environmental Science* **96**(1)
[12] Vasilenko O V and Voropay N N 2015 Features of formation of the climate of the depressions of the southwestern Baikal region *Izvestiya RAS*, *Geographic Series* **2** 104-11
[13] Bulygina O N, Veselov V M, Razuvaev V N and Aleksandrova T M 2014 Description of the array of urgent data on the main meteorological parameters at Russian stations *Certificate of State Registration of the Database 2014620549* (in Russian) Income accessed online on 18th July 2018 via http://meteo.ru/data/163-basic-parameters#description of the mass-data
[14] Climatic Research Unit University of East Anglia (UK) 2017 Income accessed online on 10th May 2018 via https://crudata.uea.ac.uk/cru/data/temperature/HadCRUT4-nh.dat
[15] Groysman P Y 1983 About changing of some characteristics of atmospheric circulation in global warming and cooling processes *Russian Meteorology and Hydrology* **11** 26-9
[16] Gorbatenko V P, Ippolitov I I, Kabanov M V, Loginov S V, Podnebesnykh N V and Kharyutkina E V 2011 Influence of atmospheric circulation on the temperature regime of Siberia *Atmospheric and Ocean Optics* **1** 15-21
[17] Ippolitov I I, Kabanov M V, Loginov S V, Podnebesnykh N V and Kharyutkina E V 2012 Variability of the temperature regime in the Asian territory of Russia during the period of global warming *Atmospheric and Ocean Optics* **2** 122-31
[18] Kabanov M V 2015 Regional climate-forming factors in western Siberia *Geography and Natural Resources* **3** 107-13
[19] Borzenkova A V and Shmakin A B 2013 Modern changes of climate characteristics of heating period in Russia and their communication with the atmospheric circulation *Izvestiya RAS*, *Geographic Series* **4** 59-69
[20] *Second Assessment Report of Roshydromet on Climate Change and its Consequences on the Territory of the Russian Federation* 2014 vol I (Moscow: Roshydromet) pp 18-235

# Assessment of current state of conserved areas of mountain rocks in strip mining conditions

**N V Vlasova, I B Vorobyeva**

Sochava Institute of Geography SB RAS, Irkutsk, Russia

E-mail: vlasova@irigs.irk.ru

**Abstract.** In the conditions of the development of today´s mining industries, many coal mining enterprises (even though the production volumes are currently increasing) are looking for the ways to conserve previously mined areas and monitor their ecological state. The extraction of fossils from the depths of the earth produces natural and manmade complexes that have a considerable impact on vast territories. The largest impact lies in turning the lands for general use into commercial mining areas and land allocations. Open cut mining leads to the emergence of quarries, stopes and embankments of different sizes, changes in relevant and absolute elevation marks compared to the initial surface, which is connected with the construction of quarry dumping complexes. Direct dumping methods contribute to the formation of the system of narrow crests that are combined with leveled plateau-like parts of a spoil pile. The slopes of spoil piles have high angles of gradient (14-40°) and are subject to sheet and gully erosions which form large alluvial cones. Technogenic sediments formed at their bottom reach the height of several meters and spread up to a few hundred meters from the dumps. During coal mining, the rocks are stored in external and internal dumps of coal strip mines without any preliminary sorting, which also leads to soil and water pollution with heavy metals.

## 1. Introduction

Environmental protection in coal mining enterprises is a priority task at all levels of mining operations, especially during temporary conservation. The most damaging impact on natural complexes is produced by open cut mining. In this connection, the Russian legislation obliges enterprises to implement the systems and programs of environmental observation and monitoring. One of such systems is the monitoring of conservation areas [1]. All environmental events require special attention and careful planning.

Work objective: assessment of the current state of the areas of temporary conservation in the framework of ecological monitoring and their impact on natural territories.

## 2. Methodical approach and research findings

Target of research: area of the temporary conservation of coal strip mines, including the research of natural environments which have been changed and/or lost their productivity in the result of a negative impact. The research was carried out in accordance with all-Union State Standards (GOST) methods and requirements of regulatory documents. Selection, storing and preparation of samples were held in accordance with the norms set by state standards.

Open cut mining is considered ecologically hazardous and harmful to the environment [2]. Its impact is complex and diverse. It is not limited to mining alone but includes gas and dust pollution, contamination of surface and subsurface waters, hydrobalance disturbance, technogenic soil disturbance, withdrawal of productive agricultural and forest lands, change of the qualitative composition and integrity of subsurface resources, multiple industrial waste, electromagnetic and sound pollution, activation of destructive natural physiogeographic processes (gullying, flooding, aridization, etc.) [3]

The researchers of the V.B. Sochava Institute of Geography of the Siberian Branch of the Russian Academy of Sciences were assigned with the task of assessing the current state of conserved areas and creating an ecological monitoring program during their collaboration with the mining companies *Tulunugol Ltd* and *Vostsibugol* at the base of Azeiskii mines. As there was no program of ecological monitoring of conserved mining areas available at that moment, the creation of such program became a necessity. At the first stage of the scientific research, special attention was given to the initial analysis of the current state of conserved mining areas, examination of the sources of exposure, and monitoring of the soil thickness based on qualitative and quantitative indicators in the monitored areas.

The ecological monitoring program is based on the assessment of the state of the environment in conserved areas and its natural counterpart, i.e. an analogue area before the commencement of mining operations. The identity of the analogue area is expressed in the slope exposure, identical wind patterns, uniform geological environment, soil characteristics and vegetation layer. Besides that, the analogue territory must not be affected by any coal mining activity, yet at the same time, must be relatively close to the conserved area (figure 1).

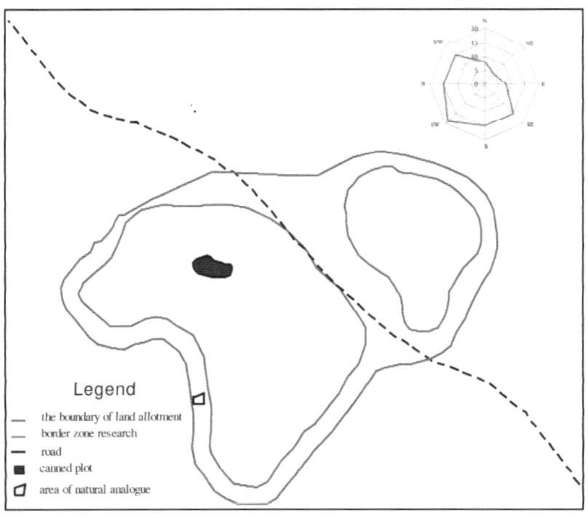


**Figure 1.** Arrangement of conserved site and natural counterpart.

By the time of the commencement of the current research, the areas withdrawn for the conservation were in different environmental states and had a different time of stoppage of mining operations. In this regard and in accordance with the developed monitoring program, the research sites were located in both "moonscape" and vegetated areas. The vegetation in the latter area is diverse and represented in various successive species from the pioneer stage on crests and slopes of the dumps to the tallgrass aspen, birch tree and pine forest in the perimeter. Most of the vegetation species of the community are represented by grassland vegetation [4]. To obtain a more complete picture of the transformation of the territory and its pollution, we examined the soil in the conserved areas and that in the analogue area to make subsequent comparisons [5]. The research showed that the chemical qualities of man-made soils have common features, notably weak-acid and or neutral reaction of the environment (from 6.4 to 7.3) in the upper profile horizons, pH is often higher due to the biogenic base accumulation. At the same time, the pH indicators of the dump crests in the conserved areas, with pioneer vegetation or without it, vary from 4.7 to 4.9, which is the consequence of a heavy mixing of all layers of the soil sample and bedding rocks. The technogenic soil is low in humus (0.2 – 0.8), nitrogen and phosphorus. Exchangeable cations include $Ca^{2+}$, $Mg^{2+}$, occasionally aluminum and hydrogen ion [6]. The sum of absorbed cations is not high. One of the main indicators of soil fertility is the composition of water extracts. Their analysis showed low content of water-soluble salts in the ground and technogenic soils. Their content is higher in technogenic soils due to the technogenic input. $HCO_3^-$, a $Cl^-$ и $SO_4^{2-}$ are the dominant anions, their content is much lower than in the natural soils [7]. Dominant cations are $Ca^{2+}$ and $Mg^{2+}$. The analogue territory located in the pine and shrub forest on the meadow wood podzolized soil has a clear distribution of pH in the section and varies from 4.4 to 6.0. The natural soil is characterized by a highly sufficient accumulation of humus. Its content conforms to the norms adopted for these particular soils, and has a good depthwise distribution and decreases with at a depth from 13.5 cm in the sod soil horizon with the depth of 4 cm to 0.25% at a depth of 61 cm [8]. The soil is rich in nitrogen whose content also decreases as the depth increases. The analysis of water extracts showed that the content of water-soluble salts in the meadow wood podzolized soil is not very high. Their maximum concentration can be seen in the upper part of the section. As for the salt composition, $HCO_3^-$ is a dominant anion, and $Ca^{2+}$ is a dominant cation.

The main pollutants of the technogenic soils in the industrial areas are heavy metals. They penetrate the environment of the coil mining complex mostly through atmosphere. Dust pollution dominates in the area, because we did not detect a high concentration of polluting elements in the snow cover. The comparison of the concentrations of the total forms of Mn, Co, Pb, V in the natural and technogenic soils was carried out considering the maximum permissible concentrations (MPC) and established hygienic standards. The comparison of the concentrations of the total forms of Cr, Ni and Cu was carried out considering their approximate permissible concentrations (APC).

To increase the reliability and informativeness of the monitoring data in assessing and identifying the trends of the rehabilitation of natural environment, we took into account the regional background (RB): the content of an element in bedding rock, bottom sediments, surface waters, snow cover, soil, vegetation, fauna which have been identified in an experimental way and are available in scientific press [9]. Regional background helps to increase the accuracy of impact assessment of the objects of ecological monitoring on the components of natural environments. The comparison of the MPC values of heavy metals with the received indicators in the analogue territory showed that their content is within the norms of the permissible values. Copper and nickel have the highest indicators exceeding maximum permissible concentrations by a factor of five. However, they are much lower than the regional background which means that the natural area is not polluted.

One of the monitored zones located almost in the center of a coal strip mine and experiencing most of its impact was picked for a more detailed analysis of the technogenic soil of the conserved areas. Two small catenas with the inclusion of the highest point at the top of the dump and flattened surface at the bottom of the dump were set in the coal strip mine. The first one includes points B-1, B-2, and B-3, the second one – points B-4 and B-5. The five-year observations showed fluctuations in the content of metals in the upper layer of the technogenic soil. For example, maximum accumulation of copper and nickel was detected at the top of the dump in 2010 and 2011 respectively (figure 2).

The charts of the concentrations of these elements in catenas are identical, although the level of content in the adjacent points is considerably lower. At the same time, there is a sharp increase in the concentration of copper at the bottom of the dump in point B-5, where the concentration of the element takes place with a following sharp decline. Chrome and cobalt have a high migration capacity, as evidenced by their high content in the lower part of the dump. Taking into account the migration capacity of the combinations of these elements and content fluctuations in the soil, it can be concluded that the pollution occurs through dust emissions of the enterprise itself during overburden removal. The main impact is produced by removed soils and rocks, or more precisely, their chemical composition. At the same time, the charts indicating their content in the technogenic soil show sharp increases and declines, which are aided by the soil compositions. It is mostly large coarse-grained material with low content of fine earth, predominantly in the upper part of a coal strip mine mixed with a high quantity of coal waste.

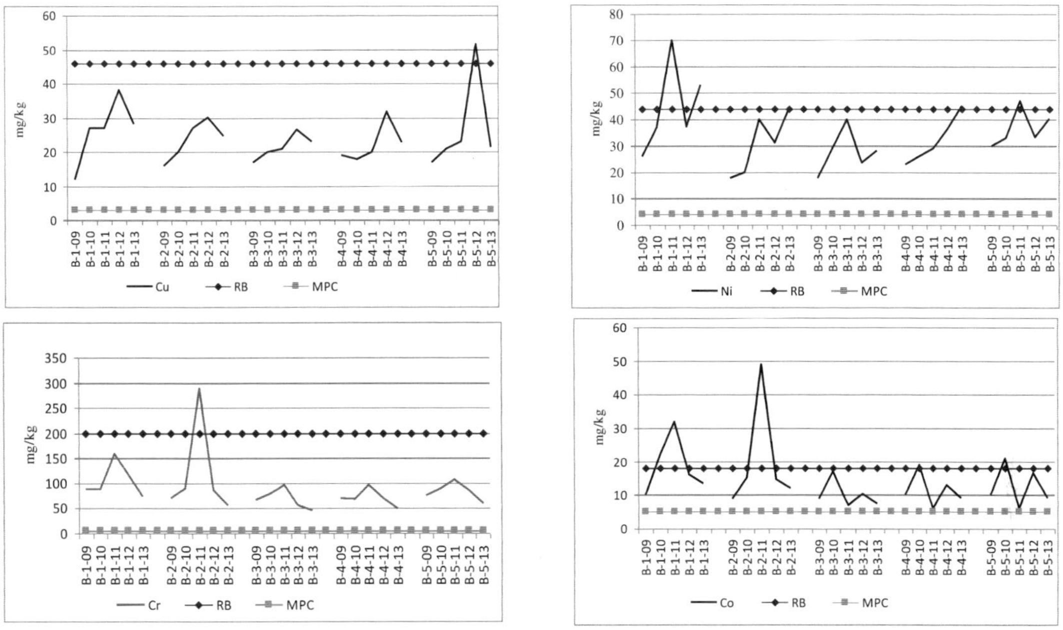


**Figure 2.** Heavy metal contents in soils of techno-sealed plots (rooms are located on the x axis points selection with an indication of the year)

The dust precipitating on the tops of the dumps is actively washed off by rain and snow waters which contributes to radial and lateral migration of the elements and their combinations. Despite the precise distribution of the elements along the catenary section, their content exceeds MPC 10. This pollution level in conserved areas is considered critical, thus the areas need to be detoxicated, even if they are brought back into production. The content of toxic elements in the analogue area is twice as low as in the conserved areas. The comparison of the content of pollutants with regional background was carried out to identify the technogenic load and pollution level of the dumps. The analysis showed that their concentration in the monitored area is lower than the average values in the region. Excess peaks do occur, but they are quite rare and show a sharp decline in the next observation period.

Besides the elements mentioned above, there has also been identified an elevated content of manganese in the upper part of the soil section covered with vegetation communities. This characteristic highlights the capacity of technogenically disturbed geosystems to return to their natural state, because Mn is a biophile element which is actively absorbed from soil by vegetation. Mn is re-emitted to soil during vegetation die-off and accumulated in the upper horizon [10].

## 3. Conclusion

Ecological monitoring in the conserved areas of mining activities allows to assess the current state of technogenic soil and its capacity for self-regeneration. The conducted research showed that technogenic soils have a good accumulation and migration capacity. The assessment of the ecological state of the conserved areas based on monitoring activities which follow the criteria of MPC alone leads to erroneous conclusions. The implementation of regional background criteria in the program of ecological monitoring increases the accuracy of determining the pollution level of the natural environments of conserved areas by more than 50 %.

IOP Conf. Series: Earth and Environmental Science **190** (2018) 012040 doi:10.1088/1755-1315/190/1/012040

**References**

[1] Ashihmina T Y 2008 *Ecological Monitoring* (Moscow: Nauka Press) p 416

[2] Kharyonovsky A A, Litvinov A R, Danilova M Y and Mahmud T 2016 Evaluation of the environmental impact of open and underground coal mining methods *J. Vestnik of the Scientific Center for the Safety of Work in the Coal Industry* **4** 113-8

[3] Litvinov A R and Kharyonovsky A A 2012 State of the environment in the coal industry *Coal* **1039**(10) 74-8

[4] Dubynina S S 2017 Temporal and spatial changes of above-ground mass in the dump ecosystems after mining operations *Achievements in Today's Natural Science* **9** 57–62

[5] Motuzova G V and Bezuglova O S 2007 *Ecological Monitoring of Soils* (Moscow: MSU Press) p 237

[6] Vorobyeva I B and Vlasova N B 2012 Geochemical assessment of lands disturbed by strip mining of brown-coal strip mines in eastern Siberia *Journal of Orenbourg State Agrarian University* **38**(6) 10-3

[7] Vorobyeva I B and Vlasova N B 2013 Research of technogenically altered seasonally-frozen soils of Southern Taiga *International Journal of Applied and Fundamental Research* **8**(1) 34-7

[8] Kuzmin V A 1988 *Soils of the Baikal and Northern Transbaikal* (Novosibirsk: NSU Press) p 175

[9] Grebenshchikova V I, Lustenberg E E, Kitaev N A and Lomonosov I S 2008 *Geochemistry of the Environment of Baikal Region* (Baikal geoecological polygon) (Novosibirsk: NSU Press) p 234

[10] Perelman A I and Kasimov N S 1999 *Geochemistry of the Landscape* (Moscow: MSU Press) p 768

# A comparative assessment of the aridity indices for analysis of the hydrothermal conditions

**N N Voropay**[1,2], **A A Ryazanova**[1]

[1]Institute of Monitoring of Climatic and Ecological Systems SB RAS, Tomsk, Russia
[2]Sochava Institute of Geography SB RAS, Irkutsk, Russia

E-mail: voropay_nn@mail.ru, raa@scert.ru

**Abstract.** Comparison between hydrothermal coefficient of Selyaninov (HTC), Ped` drought index ($S_i$) and indices SPI, SPEI are presented. Indices were calculated for the territory of Southern Siberia (50-65 N, 60-120 E). Functionality of the web-GIS "CLIMATE" was extended by a software module based on the integrated package to calculate indices. The ECMWF ERA-Interim reanalysis data on air temperature and precipitation were used as input data for the study. Analysis were carried out for the warm season from 1979 to 2017. The highest coefficients of linear correlation were obtained by comparing SPI and SPEI (0.8-0.9), Si and HTC (0.7-0.9) from June to August. Distribution of the number of strong and extreme droughts over the study period according to SPI and SPEI indices is the same, 2-3 strong droughts was observed per period. According to the analysis of Si, the highest frequency of strong droughts (up to 4) was observed also in July and August. But it was recorded only at the Sayan Mountains region. According to the results of the anomaly of HTC analysis, June and August are most droughty (2-3 strong droughts per period). The identified intensity and duration of droughts, based on these indices do not always coincide.

## 1. Introduction

When studying the geographical features and dynamics of changes in the natural environment, in addition to studying the vibrations of individual climatic elements, it is of particular interest to study the integral characteristics of the territory, in particular hydrothermal conditions. The latter are described by different coefficients, which in most cases represent a combination of air temperature and precipitation. Great importance is given to the climatic features of the emergence and spatial distribution of droughts. Drought is a complex natural phenomenon with the strongest regional anomalies of temperature and humidity (precipitation, soil moisture content) [1]. In general, a drought means a temporary decrease in the humidity of the environment in relation to its average state. Due to the complex nature of the drought, it is usually limited to studying one of its sides, manifested either in the atmosphere or in the soil. A qualitative analysis of the climate humidification conditions of previous decades makes it possible to give a more accurate prediction of arid / wet periods in which specialists from different branches of science and economics are interested.

Numerous works devoted to the study of droughts consider various methods for determining it, the characteristic types and conditions for the formation of this phenomenon, the most representative aridity indices are being developed, the catastrophic droughts are analyzed and the drought cataloged [2, 3 etc.]. Meanwhile, as indicated by D A Ped', there is no generally recognized quantitative indicator of drought. In recent years a number of works have been devoted to comparison of aridity indices [4].

## 2. Objects, data and methods

The comparative analysis of the most known Russian and international quantitative indicators of aridity, such as the hydrothermal Selyaninov coefficient (HTC), the Ped' aridity index ($S_i$), the standardized precipitation index (SPI) and the standardized precipitation and evaporation index (SPEI) for the period 1979-2017 years was performed.

The hydrothermal coefficient of Selyaninov [5] is now widely used in the practice of Roshydromet as the main quantitative indicator of the ratio of heat and moisture. The values of the coefficient are determined by the following formula:

$$HTC = \frac{\sum r}{\sum t_{>10}} \quad (1)$$

where $\sum r$ is the sum of precipitation during the vegetation period at air temperature above 10°C, $\sum t_{>10}$ is the sum of air temperature over the vegetation period with a temperature above 10°C. Based on this definition, the HTC is the ratio of precipitation amounts over a certain period to sums of temperatures above 10°C, reduced by a factor of 10, over the same period. Thus, the hydrothermal coefficient as an indicator of the assessment of the territory according to humidification conditions has an advantage over precipitations that characterizes not only the incoming part of the water balance, but also the unproductive expenditure of moisture (evaporation from the soil surface, vegetation, etc.), closely related to the air temperature . Therefore, the HTC criteria for determining the drought and its intensity depend on the moisture content and are variable throughout the territory. To reduce this dependence, it is advisable to estimate the negative anomalies of the HTC with respect to its mean value, normalized to its standard deviation [6].

Ped' index [7] is a simple algebraic difference of the normalized air temperature and precipitation anomalies for a given time interval (often in a month):

$$S_i = \frac{\Delta t_i}{\sigma_t} - \frac{\Delta r_i}{\sigma_r}, \quad (2)$$

where ***i*** is the number of a certain period, $\Delta t_i = t_i - t_{norm}$ is the temperature anomaly in the ***i***-th period, $T_{norm}$ is the long-term average air temperature, $\sigma_t$ is the standard deviation of the temperature. For precipitation (***P***) notations are similar.

Ped' index does not depend on the moisture content of the territory and therefore can be applied for a vast area, which includes several zones with different humidity regimes. Also, it clearly does not depend on the average values of temperature and precipitation, since the drought is determined in relation to them.

The standardized SPI precipitation index [8] was recommended by WMO in 2012 to all National Meteorological and Hydrological Services around the world to characterize meteorological droughts in addition to other drought indices they use in their work [9]. To calculate this index, only precipitation data are actually used, and it is desirable to have data series at least 20-30 years (optimally 50-60 years or more). SPI can be calculated for different time scales (1, 3, 6, 12, 24 and 48 months), which allows to identify different types of droughts. This indicator is comparable in any points of its calculation (the location of meteorological stations or grid nodes) and does not depend on zonal wetness. A detailed description of the algorithm for calculating this index is given in [10].

In order to take into account such an important factor as evaporation, in 2010 Vicente-Serrano S.M. et al. [11] upgraded the SPI into a standardized precipitation and evapotranspiration index (SPEI). To calculate SPEI, the difference between the sum of precipitation and potential evapotranspiration (PET) is used, which is a simple indicator of the water balance. The simplest version of the calculation of PET is possible through the equation of Tronthwaite [12], the parameters of which are the average temperature over the period and the latitude of the point for which the calculation is performed. There is a more complicated version of PET calculation through the Penman-Monteith equation [13], which takes into account a larger number of parameters (radiation balance, wind speed and water vapor deficit at a height of 2 m). The algorithm for calculating the SPEI index is similar to the algorithm for calculating SPI, a detailed description of it can be found in [14]. This index can also be calculated for different time scales (1, 3, 6, 12, 24 and 48 months), it is also comparable at any points of its calculation (meteorological station locations or grid nodes) and does not depend on zonal wetness.

The classification of droughts for each index is shown in table 1.

**Table 1.** Classification of droughts based on the aridity indices of the HTC, Ped' Index, SPI and SPEI.

| Drought intensity | Normalized anomaly of HTC | Ped` Index | SPI/SPEI |
|---|---|---|---|
| **Weak** | -1.25< $\Delta HTC_{norm} \leq$-1 | $1 \leq S_i <2$ | -2 < SPI/SPEI $\leq$-1 |
| **Medium** | -1.5< $\Delta HTC_{norm} \leq$-1.25 | $2 \leq S_i <3$ | -3 < SPI/SPEI $\leq$-2 |
| **Severe** | -1.75< $\Delta HTC_{norm} \leq$-1.5 | $3 \leq S_i <4$ | -4 < SPI/SPEI $\leq$-3 |
| **Extreme** | $\Delta HTC_{norm} \leq$-1.75 | $S_i \geq 4$ | SPI/SPEI $\leq$-4 |

The main tool in our work for the calculation of these indices is the previously developed web-based GIS "CLIMATE" [15]. The "CLIMATE" system, built on the basis of web and GIS technologies, is part of the hardware and software complex for "cloud" analysis of climate data, including various sets of climate and meteorological data, as well as special interactive tools for their search, sampling, processing and visualization. Using this system greatly facilitates and accelerates the work with large volumes of geospatial climatic data, allowing a user who is not an expert in information technologies to remotely perform their statistical analysis using any modern desktop PC connected to the Internet. To calculate the hydrothermal Selyaninov coefficient and the Ped' index, software modules have been previously developed and integrated into the system [16, 17]. To calculate the normalized anomaly of the HTC on the basis of the old module, a new one has been developed. To calculate SPI / SPEI, it was decided to use the finished statistical package (https://cran.r-project.org/web/packages/SPEI/index.htm) [11] of the programming language R (https://www.r-project.org/). A detailed description of the procedure for integrating R language packets into the web-GIS "CLIMATE" is presented in [18]. Based on this statistical package, a software module for the SPI / SPEI calculation system has been developed.

Many researchers use the observation data obtained on the meteorological station network to calculate the hydrothermal conditions of the territory. In this paper, 6-hour temperature and precipitation data of the European ERA Interim reanalysis [19] with a grid spacing of 0.75 × 0.75 ° were used as the initial data for calculating the above-mentioned drought quantitative data. Previously, the reanalysis precipitation data was corrected according to weather station data [20], which allowed to improve their quality on the average. The procedure of comparison with weather station data and the subsequent correction of sediment reanalysis data was proposed and described earlier in [16]. Calculations were performed for each month of the warm season (May-September) for the period from 1979 to 2017 for the territory of Southern Siberia (50-65 N, 60-120 E). Using a monthly scale allows us to treat the identified droughts only as atmospheric.

To perform a comparative statistical analysis of quantitative indicators of droughts in the "CLIMATE" system, additional software modules were developed to calculate the Pearson correlation coefficient, the number of severe droughts and the percentage of coincidences in the occurrence of severe droughts. All characteristics are calculated in each node of the reanalysis grid for each month from May to September for the entire study period.

## 3. Results and discussion

At the first stage of the work, a comparison of the aridity index series for individual months within the growing season (May-September) was carried out. The method of Pearson's linear correlation was used. In the mountain regions and in the north of the territory in May and September, the Selyaninov HTC was not calculated due to the absence of average daily air temperatures above 10°C in the series of observations (which is a prerequisite for the calculation). Consequently, the relationship between the HTC and other indices for May and September in the mountains is not defined. Statistically significant correlation coefficients were obtained by comparing the SPI and SPEI, $\Delta HTC_{norm}$ and $S_i$ series. The closest connections are observed during the active vegetation period (June-August). The correlation coefficient between SPI and SPEI is maximal in June-July (r = 0.9-1.0) throughout the territory. The relation is weakening in August in the territory of Western Siberia to r = 0.8 and almost disappears in September throughout the territory (r = 0.2-0.4), remaining statistically significant only in the mountain areas in the south of the territory (r = 0.6-0.7). In May, the correlation coefficients are negative (r = -0.6 ... -0.7), but statistically significant. If there is insufficient amount of precipitation in May, the change in the sign of the coefficient is affected by the inclusion in the SPEI calculation formula of potential evapotranspiration. The linear relationship between the series of the Ped' index and the normalized anomalies of the HTC was traced throughout the warm period. The maximum connection ($|r|$ = 0.8-0.9), as in the case of comparison of SPI and SPEI indices, is observed in June-July. In August, the average in the territory of the value $|r|$ = 0.6-0.7 with a maximum ($|r|$ = 0.8) in the Ob-Irtysh interfluve and the middle values of the Yenisei. In August, the correlation coefficient of $\Delta HTC_{norm}$ and $S_i$ at the greater part of the territory is not statistically significant, except for the foothills of Altai and Sayan Mountains, where $|r|$ = 0.6-0.7. In May, the correlation coefficients are significant ($|r|$ = 0.6-0.8) in the south of the territory, the north and the mountains areas are characterized by statistically insignificant correlation coefficients or no data (average daily air temperatures above 10°C are not observed).

When assessing the distribution of the number of serve and extreme droughts over the period under study, similar results were obtained using the SPI and SPEI indices. Repeatability of severe droughts in the coverage of the entire territory is maximal in July and August (2-3 droughts per period). However, hotspots with maximum repeatability (up to 5 droughts per period) were detected in May and August. According to the results of analysis S, the highest frequency of severe droughts (up to 4) is also observed in July and August. But, unlike SPI and SPEI data, it is recorded only in the Sayan Mountains region. According to the results of the analysis of the normalized HTC anomalies, the most droughty summer months (June-August), with a uniform distribution of the frequency of droughts throughout the territory (2-3 droughts per period).

Of particular interest in the study of contemporary climate change is the assessment of extreme events, which include strong and extreme droughts. According to research [3], the general drought is better identified by the spread of the negative anomaly of the Selyaninov HTC. Therefore, we compared the cases of observation of severe and extreme droughts, identified by means of different indices with the presence of the indicated phenomena, determined by the anomalies of the HTC. As a result of the calculations, it was found that the maximum number of coincidences for $\Delta HTC_{norm}$ and $S_i$ (7.5% of the territory) is observed in determining severe droughts in June and August. That is, less than 10% of the territory for the period under consideration, the proposed aridity indices show identical results. In most cases, both in space and in time, the consistency of the conclusions obtained in the analysis of various indices is absent. Since all indices considered give a numerical representation of the intensity of droughts, it is necessary to use the characteristics of vegetation or the state of the soil cover as a criterion for the intensity of drought in the future.

## 4. Conclusions

The highest coefficients of linear correlation were obtained by comparing the indices SPI and SPEI (0.8-0.9), $S_i$ and $\Delta HTC_{norm}$ (0.7-0.9) during the active vegetation period (June-August). In the foothills correlation coefficients are high (about ± 1.0), which is explained by the large contrasts of the sums of atmospheric precipitation on the windward and leeward slopes, in comparison with the flat territory. In the remaining cases, the relationship between the aridity indices is statistically insignificant (± 0.2) over most of the territory under consideration. Synchronicity of changes in the aridity indices in question is confirmed by high correlation coefficients between them. However, the identified extreme events in some years (intensity, duration of drought) with the help of these indices do not always coincide. In order to determine the most representative index of aridity in the territory of southern Siberia, it is proposed to use long-term series of vegetation characteristics.

**Acknowledgments**
The reported study was funded by Russian Foundation for Basic Researches and Tomsk Region Administration, according to the research project grant № 18-45-700015

**References**
[1] Golicin G S and Vasilev A A 2001 *Hydrometeorological Disasters* (Moscow: KRUK Publisher) p 296
[2] Meschcerskaya A V, Blazhevich V G and Zhitorchuk U V 1978 Hydrothermal coefficient and connection with yields of agricultural crops *Proc. of the Voeikov Main Geophysical Observatory* **400** 134-49
Meshcherskaya A V, Blazhevich V G, Golod M P and Belyankina I G 1986 Long-term fluctuations of aridity indices in the warm period of the year in the main agricultural zone of the USSR *Proc. GGO* **505** 120-9
Rauner Y L 1981 *Climate and Productivity of Cereals* (Moscow: Nauka) p 163
Drozdov O A 1980 *Droughts and Dynamic of Wetting* (Leningrad: Hydrometeo Pub) p 96
Pokrovskaya T V 1979 The relationship between grain crop yields in Northern Kazakhstan and meteorological conditions *Proc. GGO* **403** 22-8
Zolotokrylin A N and Cherenkova E A 2013 Moistening tendencies in grain-growth zone of Russia in beginning of 21st century *Problems of Ecological Monitoring and Ecosystem Modelling* **25** 251–64
[3] Zolotokrylin A N, Vinogradova V V and Cherenkova E A 2007 Dynamics of droughts in European Russia in the situation of global warming *Problems of Ecological Monitoring and Modeling of Ecosystems* **21** 160-81
[4] Cherenkova E A and Zolotokrylin A N 2016 About comparability of some quantitative indices of drought *Fundamental and Applied Climatology* **2** 79-94
Strashnaya A I, Tischenko V A, Bereza O V and Bogomolova N A 2015 Using standardized precipitation index (SPI) for detection droughts and for forecast of quantitative assessment of yields of grain and grain-bean crops *Proceedings of the Hydrometcentre of Russia* **357** 81-97
Zadornova O I 2015 Comparative characteristics of the foreign indices for drought assessment main grain subjects of European Russia *Proceedings of Voeikov Main Geophysical Observatory* **578** 126-39
*Droughts of the East European Plain According to Hydrometeorological and Tree-Ring Data* 2017 (Moscow, SPb.: Nestor-Historia) p 360
Vinit K J, Rajendra P P, Manoj K J and Hi-Ryong B 2015 Comparison of drought indices for appraisal of drought characteristics in the Ken River Basin *Weather and Climate Extremes* **8** 1-11
Olivia K, Dev N and Frank D 2016 Marks Contribution of land falling tropical system rainfall to the hydroclimate of thee astern U.S. CornBelt 1981–2012 *Weather and Climate Extremes* **13** 54-67
Ganeshchandra M, Vimal M, Dev N, Shivam T and Rao S G 2016 Trends and variability of droughts over the Indian monsoon region *Weather and Climate Extremes* **12** 43-68
Pena-Gallardo M, Gamiz-Fortis S R, Castro-Diez Y and Esteban-Parra M J 2016 Analisis comparativo de indices de sequia en Andalucia Para el periodo 1901-2012 *Cuadernos de*

*Investigacion Geografica* **42**(1) 67-88
Coll J R, Aguilar E, Prohom M and Sigro J 2016 Long-term drought variability and trends in Barselona (1787-2014) *Cuadernos de Investigacion Geografica* **42**(1) 29-48
Yuan S, Quiring S M and Patil S 2016 Spatial and temporal variations in the accuracy of meteorological drought maps *Cuadernos de Investigacion Geografica* **42**(1) 167-83
[5] Selyanoinov G T 1928 About agricultural estimate of climate *Proc. of Agricultural Meteorology* **20** 165-77
[6] Zolotokrylin A N and Cherenkova E A 2006 Changes in the indicators of the ratio of heat and moisture to productivity in the zonal plain landscapes of Russia in the second half of the 20th century *Izvestiya RAS. Ser. Geogr.* **3** 19-28
Cherenkova E A 2013 Quantitative Evaluation of Atmospheric Drought in Federal Districts of the European Russia *Izvestiya RAS. Ser. Geogr.* **6** 76-85
[7] Ped D A 1975 About index of drought and excessive wetting *Proc. of the Hydrometcentre of USSR* **156** 19-38
[8] McKee T B, Doesken N J and Kleist J 1993 The relationship of drought frequency and duration of time scales *Eighth Conference on Applied Climatology American Meteorological Society* **8** 179-86
McKee T B, Doesken N J and Kleist J 1995 Drought monitoring with multiple time scales *Ninth Conference on Applied Climatology American Meteorological Society* **9** 233-6
[9] *World Meteorological Organization: Guidance for Users of the Standardized Precipitation Index* 2012 ed M Svoboda et al (Geneva WMO) p 1090
[10] Edwards D C and McKee T B 1997 Characteristics of 20th century drought in the United States at multiple time scales (Colorado State Univ., Ft. Collins, CO) *Climatology Report* **97**(2).
[11] Vicente-Serrano S M, Beguería S and López-Moreno J I 2010 A Multi-scalar drought index sensitive to global warming: The Standardized Precipitation Evapotranspiration Index - SPEI *Journal of Climate* **23** 1696-718
[12] Thornthwaite C W 1948 An approach toward a rational classification of climate *Geographical Review* **38** 55–94 doi:10.2307/2107309.
[13] Allen R G, Smith M, Pereira L S and Perrier A 1994 An update for the calculation of referenceevapotranspiration ICID *Bulletin of the International Commission on Irrigation and Drainage* 35–92
[14] Beguería S, Vicente-Serrano S M, Fergus R and Borja L 2014 Standardized Precipitation Evapotranspiration Index (SPEI) revisited: parameter fitting, evapotranspiration models, kernel weighting, tools, datasets and drought monitoring *International Journal of Climatology* **34** 3001-23
[15] Gordov E P, Shiklomanov A, Okladnikov I G, Prusevich A and Titov A G 2016 Development of Distributed Research Center for analysis of regional climatic and environmental changes *IOP Conf. Series: Earth and Environmental Science* **48**
[16] Riazanova A A, Voropay N N, Okladnikov I G and Gordov E P 2016 Development of computational module of regional aridity for web-GIS "Climate" *IOP Conf. Series: Earth and Environmental Science* **48**
[17] Voropay N N, Maksyutova E V and Riazanova A A 2016 Hydrothermal conditions at the south of East Siberia during the ongoing warming *IOP Conference Series: Earth and Environmental Science* **48** doi:10.1088/1755-1315/48/1/012003
Ryazanova A A and Voropay N N 2017 Droughts and moistening periods in South Siberia in the late XX - early XXI centuries *IOP Conference Series: Earth and Environmental Science* **96** doi:10.1088/1755-1315/96/1/012015
[18] Ryazanova A A, Okladnikov I G and Gordov E P 2017 Integration of modern statistical tools for the analysis of climate extremes into the web-GIS "CLIMATE" *IOP Conference Series: Earth and Environmental Science* **96** doi:10.1088/1755-1315/96/1/012014
[19] Dee D P et al 2011 The ERA-Interim reanalysis: configuration and performance of the data

assimilation system *Quarterly Journal of the Royal Meteorological Society* **137** 553–97
[20] Bulygina O N, Veselov V M, Razuvaev V N and Aleksandrova T M Description of the array of urgent data on the main meteorological parameters at Russian stations *Certificate of State Registration of the Database 2014620549* (in Russian) Income accessed online on 28th July 2018 via http://meteo.ru/data/163-basic-parameters#description of the mass-data

# Exogenous processes and problems of rational nature management in the Selenga river transboundary basin

**V B Vyrkin, D V Kobylkin, V A Golubtsov**

Sochava Institute of Geography SB RAS, Irkutsk, Russia

Email: vyrkin@irigs.irk.ru

**Abstract.** The problem of soil degradation as a result of deflation remains particularly relevant for the territory of the Selenga midland. Based on historical data, it led to huge losses in agriculture, the dust storms, sand filling of settlements and roads. The problem of desertification is closely connected with the processes of aeolian relief formation. Therefore, the development of effective policies and practices to prevent desertification requires investigations of geomorphological systems response to climate change and anthropogenic impact. The necessary data can only be obtained in the course of interdisciplinary research involving geography and geology methods and historical materials. This paper presents the data on the dynamics of aeolian relief formation during the Holocene and historical era in the central part of the Selenga midland. Based on the study of soil-sedimentary sections and its radiocarbon dating the stages of soil formation corresponds with an relatively favorable climatic conditions were distinguished. Arid phases correspond to activation of aeolian processes and manifest themselves as horizons of loess and sand deposits overlying the buried soils. The most intense, they were proceeded at the Late Glacial – Holocene border (12.9 – 11.7 kyr BP) and Early Holocene (10.5 – 9.4 kyr BP), in the Atlantic (8.6 – 6.9 kyr BP), beginning (~ 5.4 – 4.8 kyr BP) and the end (~ 3.4 – 2.9 kyr BP) of Subboreal period of Holocene. Analysis of historical and literature data revealed the process of farmland degradation into desert landscape, unsuitable for further agricultural use. For this territory results of investigation of the morphological structure, the rate of aeolian landforms formation and particle size distribution of the constituent of sand are presented.

## 1. Introduction

The development of issues of rational nature management requires knowledge of modern exogenous processes taking place on the background of the growing influence of human activity on nature. The most important element in studying the relationship between relief formation and features of land use is the assessment of various exogenous processes in connection with the problems of soil fertility conservation and protection of land resources.

The restructuring of the entire system of socio-political and economic relations in Mongolia, which has not yet been completed, has largely devalued the previously accumulated statistical-economic, land-use and demographic materials, has led to the aggravation of a number of traditional economic and economic problems for the region (pasture degradation and breaking traditions their optimal use, the abandonment of foci of agricultural development and irrigated meadow-growing, the intensification of land erosion, desertification etc.). On the background of global warming and intensified anthropogenic impact on landscapes in the modern era tendencies of desertification of some territories have been outlined, the traces of which have clearly become apparent in Northern Mongolia. Therefore in the new situation increased the need to assess the current state of steppe and forest-steppe landscapes of transboundary territories, careful consideration and study of environmental management processes in Russia, as well as Northern Mongolia, the most well-endowed with water and pasture resources, and which can become the backbone of the country's economy for the coming years [1].

This problem can be most fruitfully solved on the example of the comprehensive analysis of exogenous processes and issues of rational nature management of the Selenga river transboundary basin.

## 2. Objects, data and methods

In the Selenga river basin covering much of northern Mongolia and Western Transbaikalia, the most common among the exogenous processes are aeolian, slope water-erosion, slow hydrothermal movements of soil masses on the slopes (creep), as well as fluvial [2]. Creep processes despite widespread distribution do not have a significant impact on landscapes due to the slow (centuries) displacement of the loose sediments cover. Fluvial processes although they can actively influence landscapes but their impact is limited mainly by riverbeds and less flood plains (during flood periods). Among the exogenous processes in the region, slope water-erosion and aeolian processes are the most dangerous from the point of view of protecting land resources and preserving soil fertility under the conditions of a sharply continental climate the flat and mountain-hollow relief of Northern Mongolia and Western Transbaikalia. Their negative impact is most pronounced in the forest-steppe and steppe landscapes in the areas of distribution of sand, sandy and sandy loamy soils [3, 4]. The increasing human impact on the environment often leads to their activation.

Historical chronicles, literary sources, archival materials, various cartographic material, and remote sensing data were used to study exogenous processes. Reconstruction of the intensity of various exogenous processes during the Holocene was carried out as a result of an analysis of the analytical data of subaerial and lake deposits in various regions of the Selenga middle mountains. For representative sites maps reflecting geomorphological elements were made with the separation of the main relief-forming process. An example of such a map for aeolian processes is the section of the Khudan basin where aeolian processes have a high activity (figure 1).

## 3. Results and Discussion

During the late Pleistocene and Holocene the exogenous processes experienced repeated activation and decline, followed by other leading processes, as evidenced by the presence of aeolian, deluvial and polygenetic loess-like deposits on the slopes, watersheds, in the valleys of temporary streams separated by buried soils. For this period of geological time there are several phases of the shift of the leading processes in the central part of the Selenga middle mountains, depending on global climatic changes.

The earlier investigations of the soil-sedimentary sequences (see figure 1) have allowed us to reconstruct the dynamics of sediment accumulation and pedogenesis in this territory during the last 15 kyr BP [5]. Multiple changes of sedimentation regimes has been distinguished. The deluvial and deluvial-proluvial sands are replaced by carbonate-rich sandy loams, the particle size composition of which is dominated by fractions of fine sand and coarse silt. Similar composition changes of the deposits reflect a change in the leading exogenous process. There is a decrease in the intensity of erosion-accumulative processes and the main role in sedimentation begins to play the aeolian processes. The longest and most intense periods of activation of aeolian sedimentation occurred on the study territory at the boundary of the Late Glacial and Holocene (12.9 – 11.7 kyr BP) and in the early Holocene (10.5 – 9.4 kyr BP). Less prolonged phases of an increase in aeolian activity were manifested in the Atlantic (8.6 – 6.9 kyr BP), in the beginning (~ 5.4 – 4.8 kyr BP) and the end (~ 3.4 – 2.9 kyr BP) of Subboreal periods [5].

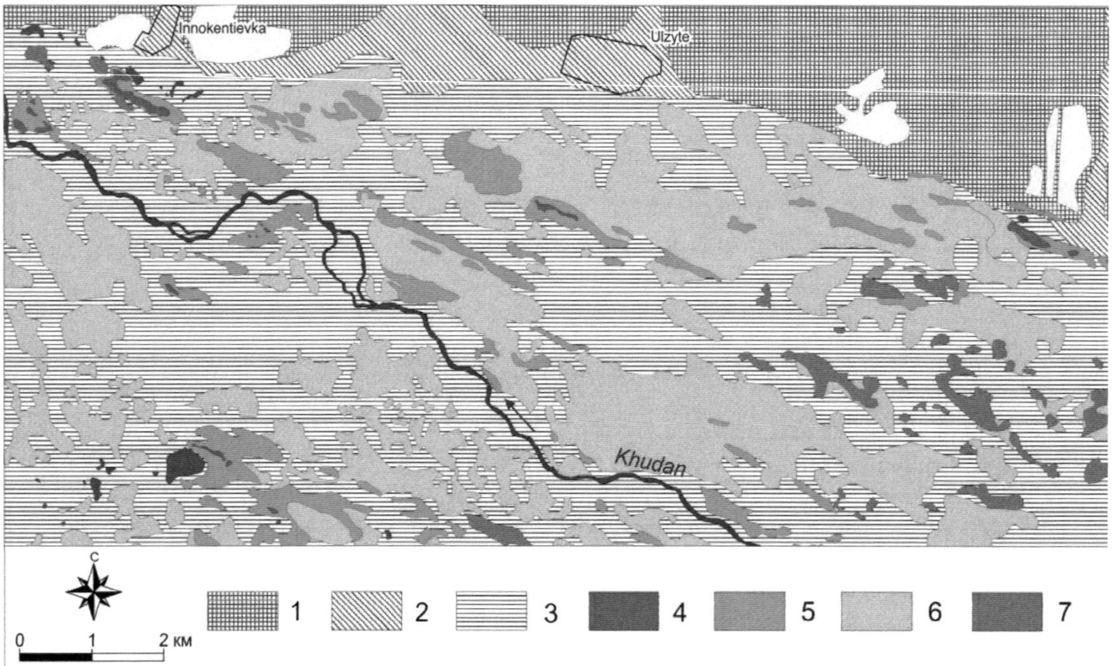


**Figure 1.** Map of the distribution of aeolian sands on the key section of Khudan depression. Geomorphological elements: 1 – ridge slopes, 2 – alluvial fans and valleys of temporary streams, 3 – floodplain and river terraces. Sands: 4 – mobile, 5 – semi-fixed, 6 – fixed by grassland-steppe vegetation, 6 – fixed by forests.

The prevalence of slope water-erosion and aeolian processes in the structure of the modern exogenous relief formation of the Selenga river basin is observed in some depressions and low mountains of the Khangaye-Khantei and Transbaikalian climatic and geomorphological regions.

By comparison of lake sedimentation corresponding to the humid period with adjacent regions it is most correlated with Northern Mongolia. High lake levels are recorded here after the Early Holocene low level. This time corresponds to the basal deposit layers of bottom sediments of the Rybnoe Lake with lake sedimentation regime. After boundary of 7000 years gradual and asynchronous decrease in levels begins.

The 5000-year boundary from which the gradual decline in the levels of Mongolia's lakes begins is reflected by interlayers of aeolian sandy deposits, indicating a pulsating character of level fluctuations. The temporary increase in the levels at the Middle and Late Holocene boundary (3000-2000 years) corresponds layer of 40-70 cm in the study section. The further regime of existence of the lake (later 2000 years) was apparently close to the modern one.

Modern aeolian relief formation in Northern Mongolia and Western Transbaikalia is widely distributed in intermontane basins, as well as within individual plains and lowland plots with arid and semiarid climate. The main centers of aeolian relief formation are sandy massifs in intermontane basins and windward slopes of low mountains facing them. The slopes of the ridges of the Selenga middle mountains (Malkhansky, Zagansky, Tsagan-Daban, Burgutui etc.) are among the latter. Aeolian processes occuring here together with slope water-erosion, providing dominance in the morphogenesis of aeolian processes in dry, and water-erosive in wet years.

The predominance of aeolian processes in the structure of the modern exogenous relief formation of the study territory noted in individual valleys, basins and low mountains is due to the combined influence of relief morphology, features of the lithological structure of the Cenozoic deposits and the climate. In the Selenga middle mountains, the genetic and spatial connection of zones of deflation and aeolian accumulation, dune-hollow and hummocky relief, aeolian sands and loess-like deposits are most distinctly traced.

Aeolian relief formation is actively continuing mainly within the mobile sand massifs. Deflation of many areas of agricultural land in the 21st century weakened due to the decrease in anthropogenic pressure and carried out in the second half of the 20th century phytomelioration [6]. The relationships between the nature of land use and exogenous processes in the studied region can be characterized by the example of key sections located in the Kuitunka river basin and the southern part of the Gusinoozersk depression.

In the Kuitunka river basin aeolian-deluvial loess-like sandy loam is widely distributed with a capacity of 2–40 m. For this territory topographic maps of 1911 and 1972-1973 were analyzed [7]. It turned out that in the beginning of twentieth century the arable land was located on gentle slopes, low watersheds and wide valleys of the upper and middle reaches of this river. In the 60-70s of the last century, the areas of cultivated land have grown substantially due to the plowing of the lower part of the valley, which is most dissected by permanent and temporary watercourses.

The growth of arable areas, the use of heavy mechanisms, plowing bottoms and slopes of wide valleys caused activation of erosion-accumulation processes. The potential erosion of soils from arable land was estimated at 1080-19090 km / km2, the density of the basin is 0.89 km / km2, the vertex density is 3.9 units / km2 [8]. Annually from the gullies an average of 190 tons / km2 of loose sediments is carried out. If their movement during the washing away of soils on the slopes occurs over a distance of the first tens of meters the watercourses in the gullies are able to carry suspended and entrained particles many hundreds of meters to floodplains and river beds. In the bottoms of the valleys, at the foot of the slopes and on the floodplain there is an accumulation of flushing and washing products. The active development of erosion processes led to the predominance of the basin component in the sediment and the formation of mudflows in strong showers. For example, such a phenomenon on the Kuitunka River was recorded on July 13, 1968.

The plowing of soils of light mechanical composition (sandy, sandy loamy) in combination with strong winds and arid conditions of the spring period led to the appearance of eroded (deflated) soils and areas of waving sand in the lower reaches of the Kuitunka river.

One of the most pronounced examples of human intervention in natural processes is the degradation of the soil cover and the formation of a foci of moving sand that is not suitable for economic use. It is located on the southern sides of the Gisinoozersk depression on the right side of the Selenga River and has a name for the former village of Old Nomokhonovo. By the beginning of the 20th century, there was a thriving settlement with pastures and plots of arable land [9]. Gradually, all large territories were involved in economic circulation, which ultimately led to the complete destruction of the soil cover and the appearance of dune ridges. Despite the fact that the Nomokhonovo place has not been used for a long time, the movement of the dunes continues actively. In the winter of 2013, to measure the speed of motion, we set the benchmarks for the largest and active dunes. The movement of sand in their front area was 1.6 m during the spring period of the most active winds, and 2 m in the whole year [10].

## 4. Conclusion

In the steppe and forest-steppe landscapes Selenga river basin activity of aeolian and slope water-erosion processes during the study period varied significantly depending on climatic parameters. Several time intervals synchronous to the global climatic cycles of the Holocene are distinguished in terms of the structure of the leading exogenous landforming processes.

During the period of active economic development of the territory, their activation arise with an increase in anthropogenic pressure on landscapes (felling, fires, expansion of arable land, overgrazing). Huge areas were deprived of natural vegetation and the plowing of the granulometrically light deposits led to a disturbance of the soil cover that promoted the broad development of the processes of water and wind erosion.

**Acknowledgements**
The reported study was funded by RFBR according to the research project № 17-29-05089.

**References**

[1] Vyrkin V B, Belozertzeva I A and Anchtaivan D 2013 *Landscapes Prichubsugulja: State of the Art and Rational Use* (Irkutsk: Publishing house of the Institute of Geography of the SB RAS)
[2] *The Ecokological Atlas of the Baikal Basin* 2015 ed A R Batuev et al (Irkutsk: Publishing house of the Institute of Geography of the SB RAS) p 145
[3] Ivanov A D 1966 *Eolian Sands in Western Transbaikalia and the Baikal Region* (Ulan-Ude: BKNII SO AN SSSR) p 232
[4] Vyrkin V B 1995 *Modern Exogenous Relation Between Hollows of the Baikal Type* (Irkutsk: Publishing house of the Institute of Geography of the SB RAS) p 175
[5] Ryzhov Y V, Golubtsov V A and Kobylkin D V 2016 Development of Exogenous processes and soil formation in steppe landscapes of the Selenga Middle Mountains (Western Transbaikalia) in the late Glacial and Holocene *Doklady Earth Sciences* **6** 82-93
[6] Vyrkin V B 2010 Aeolian relief formation in the Prebaikalia and Transbaikalia *Geography and Natural Resources* **3** 25-32
[7] Vyrkin V B, Kobylkin D V and Ryzhov Y V 2009 *Land Use and Exogenous Processes in the South of East Siberia* Land cover and land use change in North East Asia: Problems of sustainable nature management (Vladivostok: Dalnauka) p 172-5
[8] Reimche V V 1986 *Erosion Processes in Forest-steppe Landscapes of Transbaikalia (by the Example of the Kuitunka River Basin)* (Novosibirsk: Nauka) p 120
[9] Shipek T, Vika S, Snytko V A and Buyantuev A B *Facies Blown Sands Chikoy-Selenga Interfluve in the Western Transbaikalia* (Irkutsk: Publishing house of the Institute of Geography of the SB RAS) p 71
[10] Kobylkin D V, Golubtsov V A and Batozyrenov E A The dynamics of aeolian morphogenesis in the Central Part of the Selenga Middle Mountains *The bulletin of Irkutsk State University Series. Earth Science* **20** 43-52

# Conjugations of transport networks between southern regions of the Russian Far East and neighbouring countries

**P Y Baklanov**

Pacific Geographical Institute, FEB RAS, Vladivostok, Russia

E-mail: pbaklanov@tigdvo.ru

**Abstract**. All modes of transport are developed in the Far East region of Russia, namely: the railway, road, river, maritime, aviation and pipeline ones. The following groups of functions performed by regional transport are distinguished like provision of the intra-district, inter-district, foreign economic and transit links as well as the joints of various modes of transport, including those between land and maritime transports. The various types and spatial structures of conjugations of railway transport in the southern regions of the Russian Far East with neighbouring countries, PRC and DPRK are determined. The main connecting role in such conjugations is played by the trans-boundary links of the transport network and the pairs of cross-border transport nodes. Generalized assessments of conjugations between the different modes of transport in the Far Eastern region of Russia as well as the presence of conjugations of different modes of transport in the region with transport networks of neighbouring foreign countries, PRC, DPRK, USA and Japan, are given.

## 1. Introduction

The Russian Far East as a whole is a frontier region, which borders on foreign countries in the south and east. In-land in the south, the region borders on China for 4,209.3 km and DPRK for about 17 km. In the east, by a huge coastline of about 26 thousand km, the region faces the seas and the Pacific Ocean, including bordering on the US over the Bering Strait and Japan (Hokkaido Island) over the La Perouse Strait and the Kuril Strait. Thus, the Far Eastern region has about 4,200 km of the total length of terrestrial state borders, and over 20 thousand km of the maritime borders.

Taking into account the geographic location of the region, both terrestrial and maritime modes of transportation were created here. By now, nearly all modern modes of transportation are represented in this vast space: railway, road, river, sea, aviation, and pipeline. All of them are used to implement various foreign economic relations with neighbouring and more distant foreign countries. Availability of various connections of the Russian transport links with foreign ones is a required condition in this process.

## 2. The statement of the problem

Many studies have been devoted to the development of the transport network in the Far Eastern region, including those by V.N. Bugromenko, R.G. Leont'ev, Ya.N. Semenikhin, L.B. Vinokur, R.V. Vakhnenko, and others. They analysed the geographical features of the transport network both in the region as a whole and in some of its southern portions, where the transport network is more developed

[1, 2]. In a row of works, the problems of development of seaports and formation of their specializations were considered [3].

Since the 90s of the last century, foreign economic ties have increased significantly in Russia and in its eastern regions, especially in the Far East. That led to the growth of scientific interests to the development of international transportation and the formation of the international transport corridors in the Far East [1, 2, 4]. However, geographic analysis of the conjugation between the transport networks of the Russian Far East, both terrestrial and maritime, with foreign transport networks has received insufficient attention so far. Recently, international conjugations started to form in pipeline transport and in the transmission of electricity, which also requires its relevant consideration.

In addition, in the long term, a significant increase in export orientation of the region is expected due to forming the territories of advanced development in the Far East, giving the status of a free port to several seaports in the region. It will also require new studies.

## 3. Key findings and discussion

Taking into account the most important features of the economic and geographical situation of the region - its considerable remoteness from the central most developed areas of the country and its direct access to the state border of a number of the countries like PRC, DPRK, Japan and the USA as well as its broad exit to the seas of the Pacific, its regional transport system performs four groups of functions: 1) ensuring internal transport socio-economic links; 2) ensuring a connection of land transport modes with maritime ones; 3) realization of many more effective foreign economic relations of the region through exits to foreign countries and their markets; 4) providing of transit export-import shipments of cargoes and goods from other regions of Russia. As it follows from the content of the functions, the second, third and fourth groups are directly tied with the implementation of various foreign economic relations.

Spatial features of the basic infrastructure, which ensures the performance of all these functions, namely - transportation of goods, cargoes and passengers in general, have a linear-node network expression. Transportation takes place in the linear links, and processing of cargoes and conjugation of various modes of transports occurs in the nodal links. The conjugation of land transport modes - railway, road, pipeline transports with maritime transport is of particular importance for the Far Eastern region (table 1). Such connections and conjugations are realized in specially engineered sections of the coast - seaports, which are linked with ground-based modes of transport and where cargoes handling takes place. At the same time, practically all modes of transport have connections and conjugations with the corresponding links of transport systems of foreign countries (table 2). Land transport modes like railway, road and pipelines have the most pronounced connections and conjugations.

**Table 1.**The presence of conjugations between the different modes of transport in the Far Eastern region of Russia.

| Transport Modes | The presence (+) or absence (-) of links between modes of transport | | | | | |
|---|---|---|---|---|---|---|
| | Railway | Highway | River | Sea | Aviation | Pipelines |
| **Railway** | + | + | + | + | + | - |
| **Highway** | + | + | + | + | + | - |
| **River** | + | + | + | + | + | - |
| **Sea** | + | + | + | + | + | - |
| **Aviation** | + | + | + | + | + | - |
| **Pipeline** | - | - | - | + | - | + |

The conjugation with transport networks of neighbouring countries is understood here as availability of one or several transport links joining national transport arteries (or sections of the transport network), passing at some distance from the state border.

**Table 2.**The presence of conjugations of various modes of transport in the Far Eastern region of Russia with foreign countries.

| Transport Modes | Presence (+) or absence (-) of links with countries | | | | |
|---|---|---|---|---|---|
| | P.R.C. | DPRK | Japan | USA | Other countries |
| Railway | + | + | - | - | + |
| Highway | + | - | - | - | - |
| River | + | - | - | - | - |
| Sea | + | + | + | + | + |
| Aviation | + | + | + | + | + |
| Pipeline | + | - | - | - | - |

The links connecting the national transport networks (sections of railways or highways, etc.) and crossing the state border, are essentially trans-boundary ones. They consist of two national sections of the transport network and direct transport crossing the border. The latter may have certain equipment, for example, for a change of wheel sets to another track, for changing locomotives, electric locomotives, buses, etc. Within the border crossing sites, transfer of cargoes from one vehicle to another, compilation of new trains and so on take place. Therefore, the cross-border transport link is generally represented by a section of the transport network (of one or another type of transport), which crosses the state border, and by two national frontier transport nodes, where cargoes handling, logistics, technical maintenance, etc. are performed (figure 1).

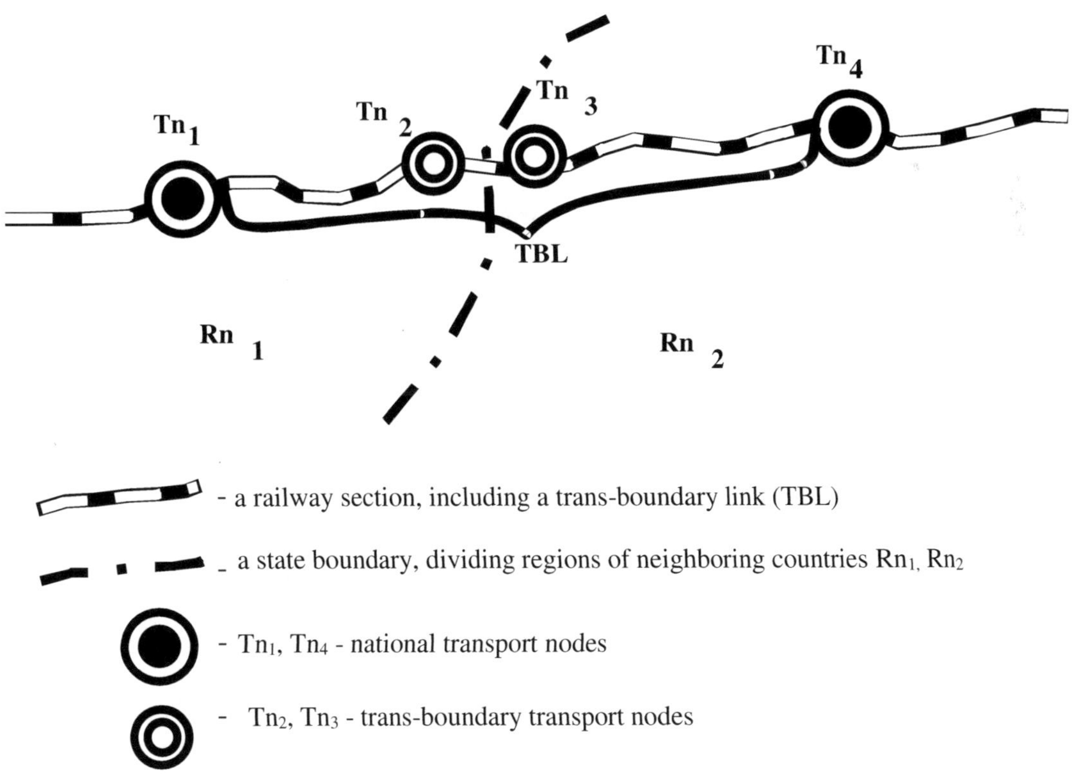


**Figure 1.** A cross-border transport link in the conjugation of the railways of two neighbouring countries.

In terms of spatial features, trans-boundary transport links may have different types of conjugations with national transport networks or its parts (figures 2a, b, c, d).

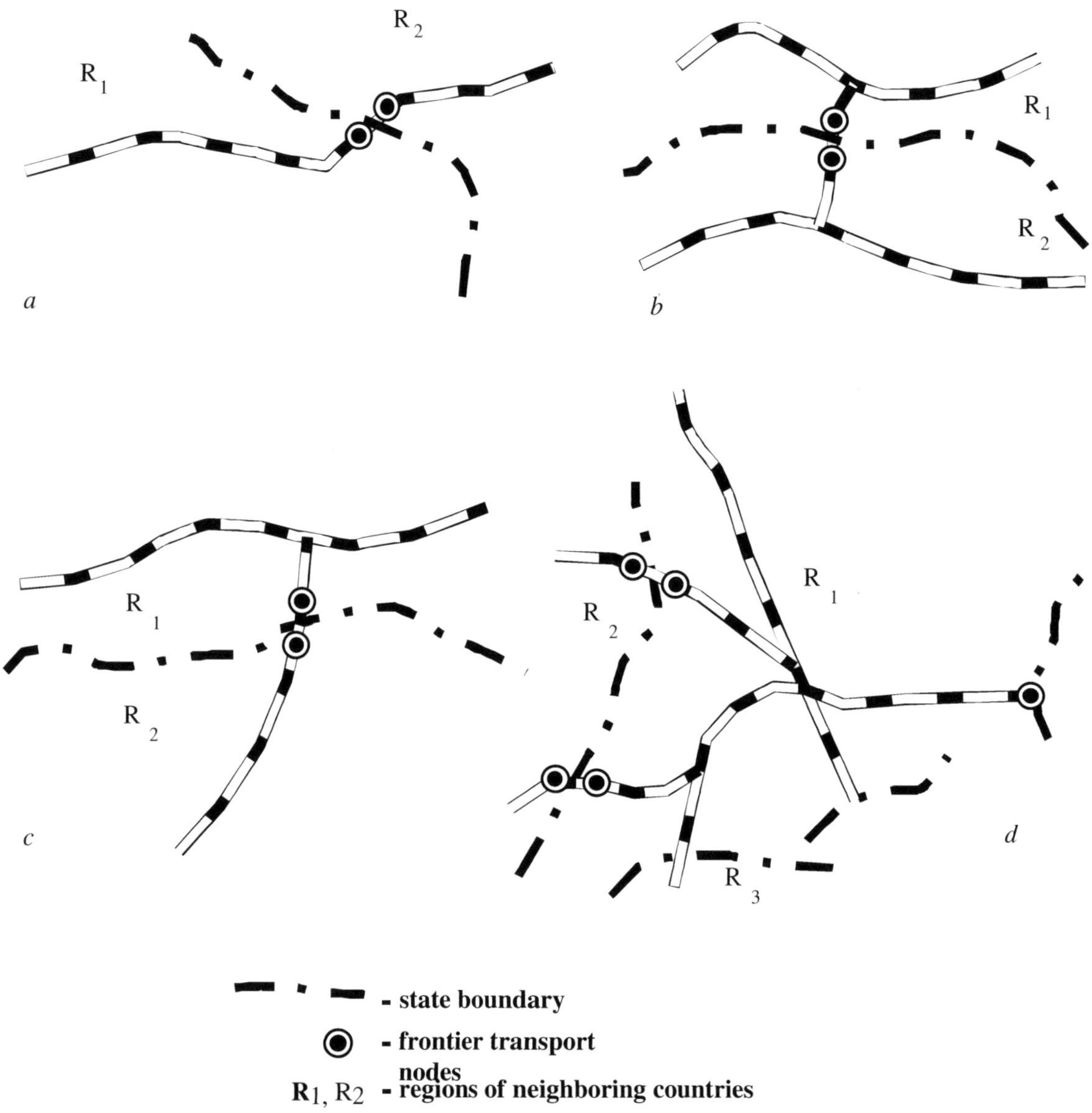


**Figure 2.**The types of conjugations of transport links of two neighbouring countries (*a-d*).

Type a is represented by an essentially unified transport artery that crosses the state border. At the same time, the national section of the road goes to a trans-boundary link, and then to another national road section. Prior to the trans-boundary link, domestic and international transportations are carried out generally along such national areas, and only international transportations along the trans-boundary link.

Type ***b*** is typical of the regions of neighbouring countries with sufficiently developed internal transport networks. At the same time, at some stage of its development, the trans-boundary transport links connecting them are being built.

Type ***c*** is formed at various levels of development of a transport network in neighbouring regions of two countries. For example, in R1 the transport network is more developed, and domestic transportations are more developed. A branch to a neighbouring country (region R2) was built from one of the transport arteries. The section of this branch is a trans-boundary transport link, and then it goes into the national transport line.

Type ***d*** reflects the option where a large transport artery terminates, approaching, for example, the seashore. At the same time, several branches are developed from it: firstly to seaports, where maritime transport is matched with foreign countries, and, secondly, for transport transitions to other neighbouring countries, to its ground transportation networks. This type is the most complex, it is typical, e.g., for the final portion of Trans-Sib, which approaches Vladivostok and has the branches to the ports of Nakhodka and the cross-border transitions to PRC and DPRK.

**Acknowledgments**
The study was carried out with the financial support of the RGO-RFBR grant under Project No. 17-05-41044.

**References**

[1] Bugromenko V, Fil S and Shubin A 2009 *Long-term Strategy of Primorsky Krai Road Network Development (A White Book of Primorsky Krai Roads)* (Moscow: MixPrint) p 128
Vinokur L 1994 *The Perspective Directions of Transport Development in the System of Foreign Economic Relations of Primorsky Krai* (Vladivostok: the Publishing House of the Far Eastern Research Center of the Academy of Transport of Russia) p 30
[2] Leontyev R 1998 *Economics, Transport and Nature Management of the Far East of Russia* (Khabarovsk: Publishing House of The Far Eastern State University of Communications) p 214
[3] Semenikhin Y and Averchenkov P 1992 Maritime transport of the Far East *Bulletin of the Far Eastern Branch of the Russian Academy of Sciences* **2** 69-75
Semenikhin Y 2011 Current state and development perspectives of the seaports in Primorsky Krai and in the Far East *Proc.8th Int.Conf. on the Problems of Transport in the Far East* (Vladivostok: Publishing House of The Far Eastern State University of Communications) pp 7-9
Vakhnenko R 1990 *Transport and Mobility of the Population of the South of the Far East* (Vladivostok: Publishing House of the Far Eastern Branch of the Academy of Sciences of the USSR) p 137
Vakhnenko R 1998 *The seaports of the Sea of Japan Region (Economic and Geographical Characteristics)* (Vladivostok: Dal'nauka) p 132
[4] Vinokur L 1994 *The Perspective Directions of Transport Development in the System of Foreign Economic Relations of Primorsky Krai* (Vladivostok: the Publishing House of the Far Eastern Research Center of the Academy of Transport of Russia) p 30
Baklanov P, Romanov M and Ignatov A 2004 The strategy for the development of Primorsky Krai road network up to 2025 *Vestnik DVGAEU* **3** 5-17
Baklanov P and Romanov M 2008 The program for the Development of Primorsky Krai Road Network until 2025 *Russian Customs Policy in the Far East* **3** 99-101

# The "Silk Road of China" and economic priorities of the Pacific Russia

**P Y Baklanov[1], V N Bocharnikov[1], E G Egidarev[1,2]**

[1]Pacific Geographical Institute, Far East Branch, Russian Academy of Sciences, Vladivostok, Russia
[2]World Wildlife Fund (WWF), Amur Branch, Vladivostok, Russia

E-mail: vbocharnikov@mail.ru

**Abstract.** The article discusses the strategic initiative of China, "One Belt, One Road" (OBOR), aimed at creating a new model of international economic cooperation. This megaproject, except establishment of new and modernization of centuries-old trade routes for the direct supply of goods from China through Eurasia to Europe and Africa, is aimed at the cardinal improvement of transport, logistics, energy and telecommunications infrastructure, which requires a fundamentally different level of coordination in the field of macroeconomic policies of states, readiness to reduce trade barriers and build more open contacts between people. "The new Silk Road of China" provides a new vector to a single Eurasian economic space, in which China expects to receive a system-forming role. In essence, this multi-purpose project leads to the practical testing of a new model of globalization, and at the same time inevitably generates a completely different context of anthropogenic impact on the environment, including environmental impact, and providing huge social, cultural, economic, institutional multi-level impacts. An example of this is the situation in the transboundary basin of the Amur River. The conservation value of the Amur ecoregion is selectively represented here; the differences in the transport accessibility of the border administrative facilities of the three countries are shown in comparative terms, the most likely options for the formation and development of infrastructure projects are discussed, the necessary restrictions as the conditions for ensuring the nature conservation values of the territory are discussed.

## 1. Introduction

The project of revival of the Great Silk Road, which is an ancient transcontinental trade route, which was opened before our era and united the Chinese and Roman empires, gave the name to the new strategic initiative of China "One Belt, One Road" (OBOR). In the framework of this megaproject, the two largest Chinese strategic projects are actually united: the "Economic belt of the Silk Road" and the "Silk Road of the XXI century", implemented in Asia, Africa and Europe, respectively, in order to create a new model of international economic cooperation. In total, this grandiose project, announced by the Chairman of the People's Republic of China Xi Jinping in September 2013, covers more than 60 countries, including 40% of global GDP and 4.4 billion of the world's population.

In the future, the new Silk Road, besides Africa and Europe, can cover Latin America. Within the framework of this initiative, six major existing and planned transport corridors are provided: Bangladesh-China-India-Myanmar, China-Mongolia-Russia, China-Central Asia-West Asia, China-

Indochina, China – Pakistan, and the Eurasian Continental Bridge through the territory of Kazakhstan [1].

It seems that with such a large-scale economic and social implementation of territorial transformations of raw materials, real environmental threats must be assessed. The most important threat is illegal logging and forest fires in Russia. Overgrazing by livestock sporadically occurs in Mongolia and Russia, and is rapidly growing throughout northeast China. Expansion of agriculture and infrastructure is largely an issue in China where rapid economic growth has enabled large government investments in irrigation, flood control, and road-rail infrastructure. Transboundary transport infrastructure is rapidly growing to connect Russia and Mongolia with China and other Asian markets. Excessive harvest of biological resources: timber, terrestrial wildlife is also triggered by international market demand. Fish resources are overexploited and total catch decreased manifold in recent decades, which necessitated Russia and China to agree on common fishing rules. Transboundary pollution and degradation of spawning rivers by logging and mining adds pressure on remaining fish populations [2].

Against this background, it is obvious that the extensive resource development model existing in Russia, dominating in the forestry and agrarian sector, will inevitably be complicated by the often poorly conceived options for a hasty creation of a new environmental management infrastructure that provides only immediate interests of investors but does not ensure the country's environmental and economic security. Therefore, in this report we focus on the fact that the alleged intensive development of a vast infrastructure and production, providing fundamentally different levels of transport access to the natural resources of the Russian territory (construction of new roads, bridges, pipelines, power lines, etc.), and construction of export-oriented large infrastructure facilities (hydroelectric power stations, ore mining and processing plants, coal mines, mines, etc.) will lead to large-scale changes in Siberia and the Far East.

We believe that large scale local-level programs and projects, in particular, Russian projects for the creation of infrastructure for Advanced Development Territories (ADT, or TOR) and the Free Port of Vladivostok (FPV), should also be included in the assessment of large-scale impacts. The state policy in the sphere of land use is also in need of revision; the ongoing process of practically uncontrolled implementation of the state program "Far Eastern Hectare", fundamentally changing the ownership structure in Russia due to the free distribution of land for the purpose of settling the Far Eastern region of Russia, needs reconsideration.

## 2. Materials and Methods

Within the framework of the Chinese megaproject, it is planned to establish seven "belts": transport, energy, trade, information, scientific and technical, agrarian, and tourist. The result may indeed be the emergence of a large-scale free trade zone from the north-western provinces of China, Central Asia, to Central and Eastern Europe. About 3 billion people live in the project area. Domestic geographers understand the surface of the Earth as a complex that arose and develops under the influence of mutually interconnected and interpenetrating processes, which unfold on land, in the atmosphere, in the hydrosphere, and in the organic world, which jointly make up the geosphere. In March 2015, Russian First Deputy Prime Minister Igor Shuvalov said about Russia's decision to take part in the strategy of the Economic belt of the Silk Road: "We are pleased with the opportunity to increase cooperation in the format of China and the Eurasian Union... Free movement of goods and capital within the framework of the EAEC brings together the economies of Europe and Asia, which echoes the initiative of the economic belt of the Silk Road put forward under the Chinese leadership. We are sure in the Russian Federation that the joint work on the development of the Eurasian partnership and economic belt of the Silk Road will create additional opportunities for the development of the countries of the Eurasian Union and China "[3].

The geographical approach to the study of human habitat is based on the identification and consideration of various natural and social factors, where the natural environment constitutes the basic part [4]. The work considers methodological issues that are important for the ecological economy, necessary for taking into account when optimizing the location of production, agriculture, practical implementation of the sustainable development paradigm and sustainable nature management, biodiversity conservation and the study of ecosystem services in Russian territorial policy. Territorial resources represent a special group of geographic resources, since they must represent in an adequate form a very complex combination of the space-time diversity of the dynamically changing world.

Geosystem concept, landscape concept, and concept of territorial socio-economic systems can be noted as the most applicable in domestic geography, and the basic theory of the geosystems by Acad. V.B. Sochava provided a principle basis for applying system principles for assessing the human impacts on the environment. Geoinformation approach and GIS allow to provide generalization of phenomena on general geographic, natural-ecological and socio-economic multiscale maps, provides hierarchical decomposition of phenomena, situations and processes, allows modeling of internal structure and variety of geosystems [5]. A special condition is the choice of the basic spatial units: the primary component in geography is landscape (geosystem); the population is characterized by indicators of population, location, and many social and demographic indicators; the economy can be comprehensively defined through the establishment of volumes of goods and location of production facilities, and identification and mapping of infrastructure and communications providing certain geographic "linkage" [6].

We draw attention to environmental priorities, which are mainly expressed by the issues of biodiversity conservation, ecosystem services, maintenance of biosphere functions and preservation of ecological balance in the largest region of the world and Eurasia. Previously, we formulated and implemented the scientific tasks of analytical comparison and generalization of various schemes of physical-geographical, ecological-geographical and natural-economic zoning, with the purpose of visualizing the actual degree of anthropogenic disturbance in Russia [7]. When different geo-informational layers of landscape landscapes, population density maps, land use and various environmental impacts are compared, certain connections and patterns are revealed to judge the current state of Russian regions that could potentially be involved in the Chinese Silk Road project.

We believe that elaboration of an eco-regional interactive public "no go zones" schematic map, which will reflect the degree of risk to the integrity of the landscape when implementing large investment projects, can be an important tool in carrying out an environmental assessment. In our assessment of the degree of preservation of natural areas, the minimum size of an integral natural site of ≥500 square kilometers was adopted. The working scale of all comparisons, as well as our spatial analysis, respectively, was based on 1: 1 000 000 topographic maps of the Digital Chart of the World Data (DCW), which in 2012 were updated by the Russian company 'DATA+'. In addition to topographic maps, various electronic resources [8] of information on infrastructure changes in the territory of the Russian Federation have been used recently. A standard GIS software package was used in our work (ESRI software product, version ARCGIS 10.1).

We have revealed that a special role should be given to the anthropogenic changes in the landscape, which are progressing for the recent centuries. This is the reason for a significant transformation of local ecosystems, causing significant losses of biodiversity, reducing the total biological productivity and quality of ecosystem services, markedly amplified by negative manifestations of global climate change. Monitoring of these processes requires the identification and maintenance of unique "reference points", the extreme poles of which can be recognized as constructs - "wilderness" and its opposite, named as cultural, rural and urbanized areas or all together recognized as "cultural landscapes". Based on such assumptions, and taking ecoregional approach as a basis for the vast territory of Asian Russia, a comprehensive assessment of the degree of anthropogenic disturbance and its differentiation within the boundaries of the previously proposed Natural Economic Areas (NEA) were implemented (figure 1) [9].

## 3. Results and Discussion

In our theoretical work [10], etc. we underline that a fundamentally important aspect of the development of socioeconomic geography is its constructiveness or focus on solving certain practical problems. Identification and evaluation of all features of territorial socio-economic systems and their structures existing in the country or region is an important and interesting scientific task. It is important that these systems and their structures already exist, function, and operate. But how effective they are, how they can be changed to become more effective, what are the possible variants and trends of their changes and dynamics, should be explained for the need of identification of territorial socio-economic systems (structures) that provide a more definite picture of the natural and social complementarity and interdependence in the context of understanding the processes of interaction through the system of actions expressed by certain economic activities and use of natural resources.

The economic belt of the Silk Road as a whole opens new opportunities for Russia. However, these opportunities are available only if Russia accepts reasonable, strategically weighed, favorable conditions of participation in this project, which is based on China's serious growing economic opportunities.

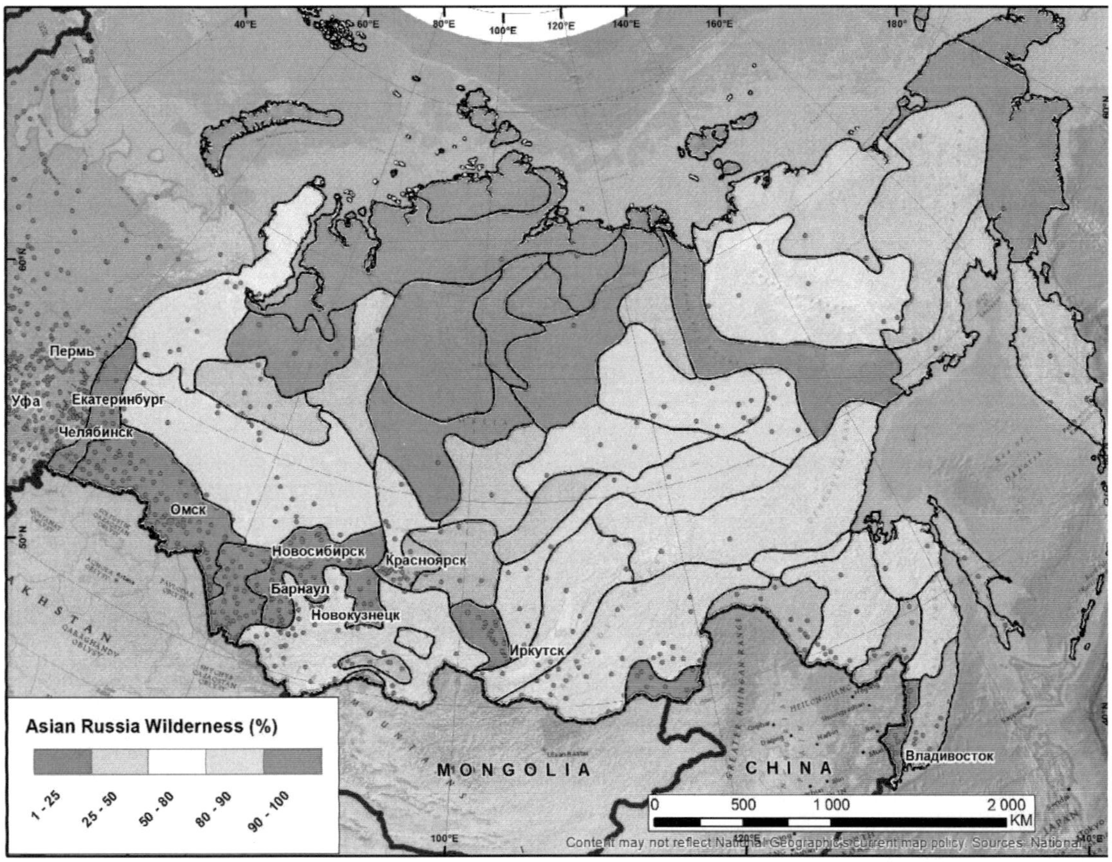


**Figure 1.** Wilderness in the Natural-Economic Areas (NEA) of Asian Russia.

The "New Silk Road" is regarded as the "Second Eurasian Land Bridge" between the economic poles of this continent. According to Chinese experts, it is the longest land bridge and it can become the most potentially significant on the Earth. Given the very large scale of the economy and foreign trade turnover of China and the European Union, providing high economic density throughout the entire considered geo-space, the high potential significance of this transport corridor is unquestionable. The length of the route from Lianyungang to Rotterdam is 10,900 km and, according to China estimates, the New Silk Road will form an economic zone that covers up to 23 percent of world exports, involving more than 30 Eurasian countries with large market, huge potential and population of about 3 billion people.

Their formation in the context of globalization and activation of foreign economic relations and, at the same time, increased competition, geopolitical tensions in the modern world, is viewed as a real mechanism for the strategic unification of the Eurasian continental space, creating conditions for the sustainable development of its economy and improving the geopolitical relations between the poles. To date, it is expected that a group of "eastern" transport projects planned within the "PRC-Mongolia-RF" corridor will be limited in the east to the corridors "Primorye-1" and "Primorye-2". If we consider that the capacity of the existing port facilities confined by Primorye-1/-2 is limited even with regard to modernization, by the time of "Mongolian" projects, the issue of finding new ports for these projects is very likely to arise.

At the strategic level, in the medium term, the development of the PRC-Mongolia-RF corridor will lead to high probabilities of: a) reduction of presence and activities of the PRC in the form of large national projects in the south of the RFE. The majority of Chinese investors in the south of the RFE will be regional JVC companies interested in expanding their markets or purchasing raw materials, which at large do not meet the standards of the "green economy"; b) formation of export-oriented resource-producing clusters in the eastern Mongolia and adjacent areas of Inner Mongolia on the basis of existing and new transport corridors and power lines from Transbaikalia. If this scenario of environmental management is being implemented, this will exacerbate existing environmental, primarily water, problems of the Upper Amur, which is very significant for the southern RFE.

We believe that the Transsiberian railway (or 'Transsib') was considered for a long time as the first Eurasian transcontinental transport corridor [11]. Within its borders, the whole previous period was a significant opportunity to successfully implement integration functions in the relations of Russia and other countries (Japan, Republic Korea, DPRK, northeast provinces of China) of the Far East with Europe through development of the economic space along it, but for many objective and subjective reasons this route cannot yet provide this sufficiently.

The change in the vector of the strategic development of the PRC from the eastern (Northeast Asia) to the western one, which is stated in the package of documents "The Belt and the Way", is significant for the entire space of Northern Eurasia, including its eastern margin (Eastern Mongolia, RFE, NEA); however, it provokes a problem of modernization of transcontinental transport corridors and the formation of economic "West-East" axes along them, as well as foundation of a high-speed transport corridor and economic axis "North-South" along the Pacific coast of China and Russia. Significant potential opportunities are open for possible inclusion of the Arctic Ocean Route in the Maritime Silk Road of the 21st century. Still, the issues of ecological safety and sufficient level of conservation of the natural environment in Russia are sharply emerging.

Quite a lot has been done in our country earlier in the context of anthropogenic impact, but in the face of new global challenges and megaprojects, new geographical assessments and studies are required. In modern conditions, based on the calculation and application of the wilderness index, a specific work was carried out to preserve the natural geosystems within a "grid" of the natural and economic regions of Russia [7]. Previously, the existence of 106 natural and economic regions was substantiated for the territory of Russia [9]. Therefore, it can be argued that considering the potential impact of the implementation of the "Chinese Silk Road" project in the European part of Russia, more concern should be paid to the development of innovative green economy programs than preserving "impact-free" natural landscapes. Such landscapes most likely have the status of protected natural areas or areas with a special regime for nature management.

For centuries, the vast expanses of Russia from the Baltic to the Pacific Ocean have been a symbol of the greatness of the country. But if we map out and allocate very large areas of wildlife, huge foci of the economic development consequences are particularly noticeable in the Southern Siberia, South of the Far East, Central Siberia, and the Northeast. It is well known that Russia has strong disparities in regional development, and it is very likely that the development of the Chinese megaproject of the Chinese silk road will lead to new changes in the economic and social life of our country. According to our calculations, it should be noted that wilderness was preserved only within 79 out of 106 total allocated NEA. In addition, five NEAs in the European part of Russia contain less than 1% of the wilderness, and in other 10 the share of wildlife does not exceed 10%; in general, 28% of NEAs in Russia contain less than a quarter of the wilderness.

A completely different picture is emerging in the Asian part of Russia, where more than half of the NEA is presented. The zoning of the territory according to the degree of conservation of wilderness displays that the maximum level of transformation of geosystems is registered in the Southern Siberia and in the south of the Russian Far East (Figure 1), noting that all these areas have a well-developed infrastructure within their boundaries, with main transport corridors and economic belts. More than half of the territory of Siberia and the Far East still contain half to four fifths of the remaining large areas of wilderness. The maximum level of preservation of the natural environment is recorded for the Arctic and mountainous regions of the Asian part of Russia.

## 4. Conclusion

Russia is not like the other, even the largest, states: it is characterized by vast territory, different models and economic structures, significant differences in the economic-geographic, geo-economic situation and the availability of mineral resources-all these and other reasons determined the dichotomy of the Russian space, but they also allow to find our way in the 21st century. It is obvious that the completed zoning is only the most general and preliminary assessment, more detailed surveys and calculations are needed. The search for solutions to major environmental problems requires new ways of thinking, and most importantly, the historically evolved heterogeneity of Russia's economic space reflects an exceptional variety of natural, geopolitical, socio-economic, and national cultural conditions in various parts of the country. Nevertheless, it can be noted that, in an ecological respect, the territory of Russia is still a unique phenomenon that provides environmental balance and considerably underestimated natural capital [12].

Many Russian regions actively support the Chinese initiative to include Russia's economic and social space in the megaproject of the Chinese Silk Road. This is facilitated by growing uncertainty in the development of the world economy and the global energy sector, unfavorable demographic and migration trends, along with the development of new technologies, reducing the number of working employment, while increasing the requirements for the educational level, and the unknown consequences of large-scale application of bio-, nano-, and cognitive technologies. We believe that at the state level it is necessary to introduce new standards for assessing large infrastructure objects such as strategic environmental assessment (SEA) in government programs and documents, and most importantly, joint active international work on the topic under consideration.

## References

[1] Gladky Y N and Bocharnikov V N 2017 New "Silk Road" and the place of Russia in globalization "in Chinese" *Society. Environment. Development* **2** 22-8
[2] Chukhlebova L M 2010 Ecological estimation of the Amur river condition and Amur fish quality *Vestnik KrasGAU* **1** 80-4
[3] Novozhenova M F and Panteleyeva V N 2017 Participation of the Russian Federation in integration processes in the Asia-pacific region (review of the latest publications) *Russia in the Global World* **11** 113-26
[4] Baklanov P Y 2014 On the object, subject and tasks of modern socio-economic geography *Vestnik ARGO* **3** 9
[5] Samsonov T E, Podolsky A S and Yurova N D 2013 Transformations of image methods and their relationship to the hierarchy of mapping units. *Vestn. Moscow State University. Ser. 5. Geography* **4** 15-29
[6] Cherkashin A K 2006 Geography and Geoinformatics *Geography and Natural Resources* **4** 19-28
Fotheringham A S, Brunsdon C and Charlton M 2000 *Quantitative Geography: Perspectives on*

IOP Conf. Series: Earth and Environmental Science **190** (2018) 012044 doi:10.1088/1755-1315/190/1/012044

*Spatial Data Analysis* (London: Sage Publications) p 272
[7] Bocharnikov V N and Egidarev E G 2017 Wildlife in the landscapes and ecoregions of Russia *Geography and natural resources* **4** 38-49 DOI: 10.21782/GIPR0206-1619-2017-4(38-49)
[8] EY 2013 *Map of Russia's infrastructure facilities* (in Russian) Income accessed online on 31th August 2018 via https://www.ey.com/ru/ru/issues/business-environment/ey-road-to-2030-russian-infrastructure-in-global-context
Federal Geological Fund of Russia (ROSGEOLFOND) and Federal Agency Rosnedra 2018 *Interactive electronic map of subsoil use of the Russian Federation* Income accessed online on 31th August 2018 via http://map.mineral.ru/Auth/logon.aspx?returnurl=%2FStandard%2F%3Fl%3Dru&l=en
[9] Antipova A V 2001 *Geography of Russia. Ecological and Geographic Analysis of the Territory. Uch. Allowance* (Moscow: MNEPU) p 208
[10] Baklanov P and Romanov M 2013 On the uniqueness of the geopolitical position of Pacific Russia *Problems of the Far East* **6** 29-38
[11] Romanov M T 2009 *Territorial Organization of the Economy of Poorly Developed Regions of Russia* (Vladivostok: Dalnauka) p 318
[12] Zakharova Z A and Matrehina D E 2017 The role of Russian Federation in the deepening of trade and economic cooperation of the EEU and China *Federalizm* **2** 87-96

# Roadside service as factor of regional development

**V N Bogdanov, G B Dugarova**

Sochava Institute of Geography SB RAS, Irkutsk, Russia

E-mail: victvss@gmail.com, geldugarova@gmail.com

**Abstract.** The article presents results of investigation of roadside service in Irkutsk oblast carried out by the authors in order to improve the current network and for rational placement of new facilities. The major factors and problems of roadside service development were revealed and mathematical calculations and thematic maps were made. A geoinformation system based on applied GIS MapInfo with original database was created by the authors. The study is of great theoretical, methodological and practical value for future development of roadside service in all Russian regions. Particular practical significance of the study is that roadside service facilities scheme layouts for each municipality of Irkutsk oblast were created. Many of the calculations were made for municipality of the first level which is most useful and valuable for authorities and departments controlling roadside service market development, as well as for private entrepreneurs. The basic ideas and findings in the form of concrete recommendations for Irkutsk oblast as a whole and individual municipalities are currently used by the federal and regional authorities to address many problems of roadside service system development.

## 1. Introduction

Many developed countries have extensive experience in development of roadside service. The roadside service in Russia began to emerge recently and has a lot of various problems which are particularly typical for such underpopulated regions as Irkutsk oblast.

Relevance of the study is obvious; it is determined by the problem of improving the current network and efficient deployment of new roadside service facilities in Irkutsk oblast. It is also determined by the lack of theoretical and practical researches in this field in Russia, lack of roadside service facilities scheme layouts and development concepts, as well as lack of consistent information support for management optimization.

Roadside service is a sphere which provides for the needs of people and vehicles on the road. The structure of roadside service includes the following services [1]: Sale and hire of transport vehicles and spare parts, diagnostics, repair and maintenance services, supporting technical operation of vehicles (sale of fuels, supplies and accessories, car wash, etc.), roadside infrastructure of human everyday life (catering services, hotel, health and information services, trade, communication services, sanitary-hygienic services), etc. However, at present many services, including personal services, leisure, car hire, information, health care, communication services aimed at providing proper rest for drivers and passengers are not sufficiently developed in Russia [2].

The roadside service system is currently accumulated around the most significant service hubs: filling stations and complexes, service stations. The representatives of these segments are now most organized and competition rate among them is the highest. Therefore we will focus in greater detail on this segment of roadside service as it precedes the creation and development of the whole roadside service complex and indirectly affects the demand for other customer services.

Other roadside services such as automobile gas-filling station, NGV filling station and multi-fuel filling station are less developed.

## 2. Models and Methods

To analyze the current state of the roadside service in Irkutsk oblast we carried out a number of activities: a large amount of methodological and statistical material was collected and analyzed; inventory of existing network of filling complexes and roadside service stations was performed; roadside service facilities scheme layouts for municipalities of different levels were created; traffic capacity and motorization level were calculated; the demand for filling stations and complexes was defined; the thematic maps (the length of road network, the volume of consumed fuel, traffic intensity, the road length per one filling station, the amount of automobiles per one km and one filling station, etc.) were made.

We also created a geoinformation system based on applied GIS MapInfo with original database for Irkutsk oblast, plans of individual locations, network of all motorways and the option of enabling additional sources of information such as large-scale satellite survey and online cartographic services of third-party developers [3].

All activities mentioned above allowed establishing the location of 566 facilities in Irkutsk oblast which sell petroleum products of all types and 106 facilities of auxiliary roadside service (diagnostic services, minor repairs, washing, food trade and catering services, etc.). The shares of facilities according to the types of fuel are as follows: liquid motor fuel (filling stations and complexes) – 501, natural gas motor fuel (gas-filling stations) – 65.

Gas-filling stations are distributed unevenly in Irkutsk oblast, both northern cities and towns and most of rural municipalities lack them. Retail sale of liquefied natural gas in Irkutsk oblast is implemented mainly by special-purpose companies (OAO Irkutskoblgas, OOO Gasomarket, OOO Conterra+Kraisneft'). It is related to existing monopolization of trade of liquefied gas.

There is substantial lack of NGV filling stations in Irkutsk oblast, as in most of regions of Russia, due to the absence of main gas pipeline and distribution network to final customers, as well as due to stricter security requirements. At present there is only one NGV filling station in the city of Bratsk.

The developed and extensive filling stations and complexes network is concentrated around large cities of Irkutsk oblast. This is due to the fact that the demand for the services is significantly higher in a large location.

On the basis of comprehensive analysis and mapping of roadside facilities, we proved the territorial dependence: the farther it is from the big city, the less roadside facilities there are and the less diverse range of services there is.

Irkutsk oblast is one of the largest producers of petroleum products in the country. There is Angarsk Petrochemical Company (ANKhK) which is part of the structure of the largest petroleum company "Rosneft". Delivery of petroleum products to Irkutsk oblast is not limited to AnKhK, provision for the retail network is also made by other enterprises of "Rosneft'" from other regions of the country (e.g., Achinsk and Komsomolsk oil refineries). The average annual production for 2012-2013 at ANKhK made approximately 1300 thousand tons of diesel fuel and 960 thousand tons of gasolines [4].

The crucial factors for roadside service development are the car fleet and its growth trends. According to the data of the State Traffic Safety Inspectorate, there are 752 800 cars registered in Irkutsk oblast as of 2013. Their distribution across the territory is extremely uneven; most of them are concentrated in the regional center. We estimated the level of motorization (per 1000 citizens) in Irkutsk oblast which comprises 300-320 cars on average being slightly above nationwide rate. In municipalities with developed industries the level of motorization is higher (Zalarinsk, Bodaibinsk municipalities). The lowest rates are in rural municipalities and a number of northern municipalities with poorly developed road network. The indicator of the number of cars per 1000 citizens also emphasizes the level of welfare of the population and is a source of data in the calculation of appropriate density of filling stations and complexes. Visual representation of some statistic data gives a clear picture of location of the resource, agricultural and industrial regions in Irkutsk oblast.

In general, the number of vehicles in the country tends to grow at average 10% a year. Therefore in the future, according to different sources, motorization level in Irkutsk oblast may reach maximum values of all regions of Russia (500-600 cars). The global motorization experience indicates marginal growth of 700-800 cars in developed countries; further growth is limited by the road infrastructure overload, well-developed public transport system and many other factors, including contrived limitations on the part of authorities [5].

Irkutsk oblast is lowly urbanized, therefore a significant growth of motorization level in larger part of its territory is possible under favorable economic condition. There is only one territory (urban agglomeration Angarsk Irkutsk-Shelekhov) where road overload and public transport development can limit growth of motorization level.

Location of roadside service facilities depends directly on car traffic intensity.

Maximum traffic intensity of more than 10 000 cars per day (c/d) was registered on highways of the south of Irkutsk oblast which links the regional center with suburbs and large locations. Most of highways in the region have low traffic intensity of not more than 1000 c/d and only some parts of roads have significantly more intense traffic, like in that of Bratsk and Ust'-Ilimsk.

## 3. Results and Discussion

By analyzing the estimated results, we can distinguish three groups of municipalities experiencing lack of filling stations and complexes: northern industrial regions, southern regions and cities (Irkutsk, Bratsk, Sayansk).

Southern regions and big cities, despite the general redundancy of filling stations and complexes, still have a deficit per transport unit due to the faster motorization progress and low quality of service at filling stations and complexes.

General redundancy of filling stations and complexes in the south of Irkutsk oblast is associated with the period of rapid development of petroleum products retail business and conditioned by a number of facts: rapidly increasing motorization in the country, low starting level of Russian retail prices for petroleum products, lack of state regulation of prices, etc. All of them allowed for high profitability and quick return of investment due to poor technical equipment and underdeveloped auxiliary services [6].

According to some estimates [7], for return of investment to a standard filling station / complex of more than 50 mln Ruble in five years and to profitability at profitability rate of 15% for petroleum products and 25% for related products, at least 4.5 thousand tons of petroleum products per year must be sold.

According to our estimation, there are an average of 3.5 thousand tons of petroleum products sold per year at one filling station / complex in Irkutsk oblast (figure 1).

This number is presumably somewhat overstated as there is no possibility to calculate fuel sales directly to customers, not counting intermediaries. Generally speaking, turnover of large network companies with filling complexes equipped with modern fuel dispensers which have favorable location exceeds 4-5 thousand tons a year. However, majority of filling stations and complexes sell much smaller volumes of petroleum products. Consequently, in order to maintain minimum profitability, the owners have to purchase and sell low-quality petroleum products, turning to illegal suppliers, or focus on auxiliary roadside services. This situation leads to gradual replacement of small businesses from retail market with large vertically integrated oil enterprises which have their own facilities for production, storage and transportation of petroleum products and can control retail prices in a wide range.

Filling stations and complexes of large oil enterprises can retain a lower price compensating for loss of profit at the expense of other items of income, such as oil extraction. Therefore independent filling stations and complexes can not compete and are forced to close down or move under control of large oil enterprises. It should be noted that small filling stations and complexes in many developed countries earn mainly though auxiliary businesses (adjoining cafe, grocery shop, etc.).

Auxiliary services and roadside service facilities in Irkutsk oblast are well-developed only along federal highways and in large towns. There are the following types of facilities: service stations, tire centers, car washes, catering and recreational stations (cafes, hotels, shops) or complex of these facilities. Though there are very few such complexes. This is a characteristic feature of roadside service not only in Irkutsk oblast but in all Russian regions.

Having analyzed various factors, after having performed field surveys and some calculations, we concluded that the lack of comprehensive roadside service of high level along federal and regional highways is characteristic for Irkutsk oblast in general. At the same time, construction of such facilities on local road network is unprofitable due to low traffic intensity. As for countryside, it is logical to locate them in populated localities and target at local population. The majority of roadside service facilities in Irkutsk oblast do not meet international requirements.

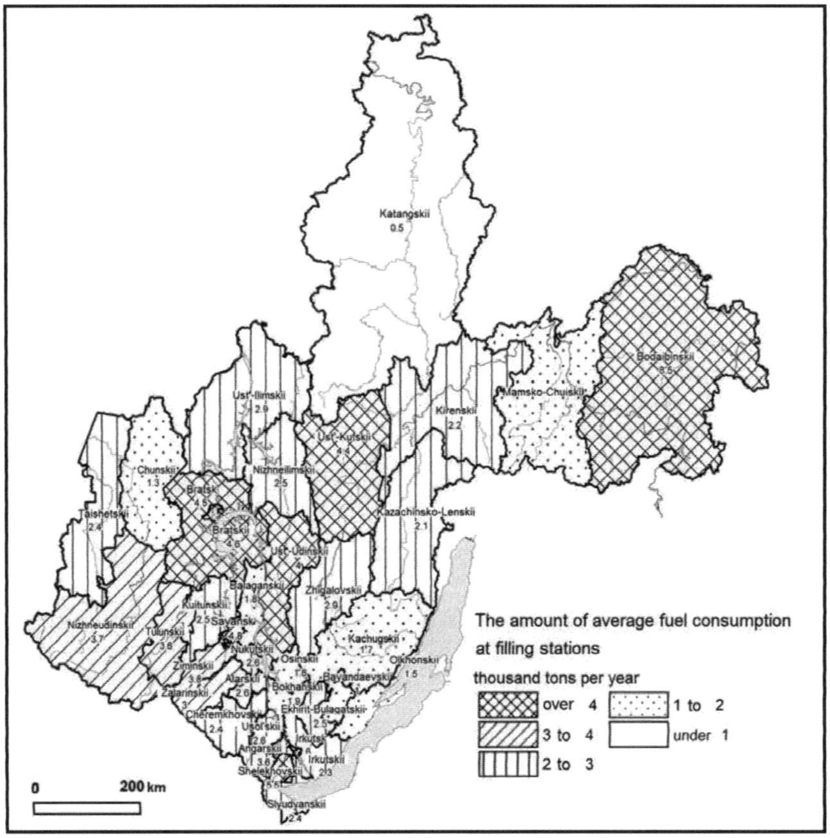


**Figure 1.** The amount of average fuel consumption at filling stations.

The major concerns of roadside service development are:

1) Uneven placement of roadside service facilities on both federal and regional highways; pronounced deficit of facilities offering a complex of services. Personal, health care, communication services, recreational stations and sanitary-hygienic services stations almost are not presented.

2) Low level of service and lack of appropriate qualifications of roadside service workers. Physical infrastructure of a number of facilities often does not meet sanitary standards and rules; moreover, the assortment list provided in the menu is often not observed;

3) There is no integrated database of roadside service facilities in Irkutsk oblast, facilities location approval procedure is often not performed causing erratic development of roadside service and neglecting traffic safety requirements.

Thus, roadside service system remains unorganized. In these conditions an acute need appears for modernizing and reconstructing existing roadside service facilities or creating new stations with high level of service and a wide range of services provided. Development of roadside service in Russia falls significantly behind motorization of population. The possibility appears of replacement of small and medium-size businesses on major highways with large companies. As a result, regional and local motorways are left for small business owners. So, at the present stage of roadside service development in Irkutsk oblast, government regulation is of great importance as this development is associated with increasing significance of automobile transport for socio-economic life of society and growth of tourist-recreational industry. International experience shows especially important role of government participation in approval of planning and adequate provision of necessary services on roads with low traffic intensity [8].

Possessing rich natural and socio-economic potential and being a transit zone between the western and east parts of the country, Irkutsk oblast can become an important factor in the development of the economic corridor "China-Mongolia-Russia".

## 4. Conclusion

Summing up the results of our study, the following findings can be formulated:

1) Roadside service market in Irkutsk oblast is underdeveloped. Fuel supply and maintenance services are currently the most important part of roadside service in the region. Other auxiliary services are poorly developed being located only at big cities and federal highways.

2) Placement of roadside service facilities in Irkutsk oblast is uneven; they are developed mostly in the south of the region (in big cities and along federal highways). It is conditioned by the fact that roadside service has been formed spontaneously, without consideration of the degree of satisfying needs of car owners and absence of monitoring.

3) Lack of information on roadside service facilities, their location and types of services has a negative effect on development and management. Hence, our paper is one of the first studies in Irkutsk oblast and the material collected can be the basis for creating a complex scheme of effective placement of roadside service facilities. We developed schemes of roadside service facilities placement for 36 municipalities of Irkutsk oblast using territorial planning schemes as municipality's basic planning document.

4) In general, the region needs gradual modernization of filling stations and complexes network as well as of roadside service facilities, founding multifunctional complexes on transit highways that would offer a wide range of high quality roadside services and sell different types of motor fuel.

In our view, in countryside with low level of motorization low profit facilities combined with auxiliary roadside service should be built and their location should be regulated according to the amount of transport and distances between facilities. In many cases such facilities are socially important and their absence worsens quality of life of rural population.

Thus, in this article we presented main results of our study as one of the most important stages of integrated scientific and practical research.

## References

[1] Khokhlova V V 2012 Domestic and foreign experience of service industry development at traffic infrastructure objects *Theory and Practice of Social Development* (Krasnodar: Publishing House "HORS") 7 224-227

Vinogradova M V 2003 *Integrated Approach to Formation of Roadside Service in Russia* (Moscow: MSFU) p 212

[2] Vuymin V V 2003 *Territorial Placing of Objects of Roadside Service in Region (On an Example of Smolensk Area)* (M: Moscow State University of Service) p 179

[3] Batuev A R, Bogdanov V N and Dugarova G B 2014 Cartographic support for the development of roadside service *Izvestia IGU "Earth Sciences"* **10** 22-31
[4] Dugarova G B and Bogdanov V N 2015 Analysis of the current state of the roadside service in Irkutsk oblast *ECO* **2** 117-26
[5] *Russian Statistical Yearbook* 2016 (M: Rosstat) p 725
Gorin V, Perepakov V 2009 Transport policy of Russia: gloomy experience of reforms and bright prospects *Economic Strategies* **4** 70-7
[6] Vinokurov M A 2002 *The Economy of the Irkutsk Region* (Irkutsk: BGUEP) Vol 3 p 432
[7] Dmitriev A V 2010 The system of transport service performance indicators *Strategy of Development of tools of Commerce* (St. Petersburg: Publishing House SPbGuEf) p 348
[8] Dunets A N 2002 Prospects for the development of roadside service *The Economy. Service. Tourism, Culture. Materials of the 3rd Int. Conf.* (Barnaul: Publishing house AltSTU) pp 27-32

# Agricultural trade pattern and potential in China and Russia

**X Bu[1,2], S Dong[1,], F Li[1], Y Li[1]**

[1]Institute of Geographic Sciences and Natural Resources Research, Chinese Academy of Sciences, Beijing, China
[2]Ningxia Polytechnic, Yinchuan, China

E-mail: dongsc3@163.com

**Abstract**. It makes use of the trade complementarity index, trade integration index and export concentration index to conduct an in-depth analysis of trade development process and the import and export structure of agricultural products between China with Russia from 2006 to 2017. The results show that degree of integration of China's agricultural trade with Russia continues to increase, and its trade structure is highly complementary. At the same time, the types of agricultural trade between them have increased significantly. The trade pattern of agricultural products is dominated by medium- and low-end primary agricultural products in China and Russia. The share of deep-processed agricultural products is relatively small, and the trade structure of agricultural products is monotonous, the bilateral trade potential has not yet been fully realized between China with Russia. Under the background of the "One Belt and One Road" initiative and the construction of China- Mongolia-Russia Economic Corridor, the trade potential between China and Russia will increase rapidly and the trade space will further expand.

## 1. Introduction

Many scholars have conducted some research on the strategies of the China-Mongolia-Russia Economic Corridor, mainly in three areas:

- Research on trade patterns and potential. Scholars mainly use trade integration index [1], trade specialization index [2], RCA index [3] and trade complementarity index [4] for statistics and calculations. It was found that China's fruit and vegetable products and vegetables have relatively comparative advantages. Russia and Mongolia have comparatively weak comparative advantages in fruit and vegetable products. Live animals, animal and vegetable oils and other products are weak in complementarity. At the same time, the two sides still lack the in-depth industrial cooperation, trade structure is relatively single, it is necessary to expand and deepen practical cooperation.
- Research on the factors affecting trade [5]. Some scholars have found that economic aggregates, border trade, and distance have a significant impact on China-Russia export trade [6].
- Geo-strategic significance and political structure. Domestic scholars from the strategic characteristics of the China-Mongolia-Russia Economic Corridor, the relevant theoretical basis of regional economic integration [7], the role and importance of China-Russian, China-Mongolian cooperation, the bilateral trade development process of China, Russia, and Mongolia, and the potential in regional economic and trade cooperation Study [8].

The results of the study helped promote the implementation of the "Belt and Road Initiative" and promoted the great potential for trade development between China, Russia, and Mongolia. Overall, the above studies have analyzed the economic and trade cooperation between China, Russia, and Mongolia from different perspectives [9]. However, most of the research focuses on China-Russian, China-Mongolian cooperative research [10], and detailed analysis of agricultural cooperation is relatively fragmentary. There are many researches on the development potential of agricultural products market [11], but most of the studies have no data analysis. At present, China, Mongolia, and Russia are considered as regions, and there is relatively little research on the analysis of China's market demand and Russian-Mongolian exports.

Therefore, this study will use time-series data for the import and export of product types between regions. The trade integration index, trade complementarity index, and export concentration index were used to analyze the trade development process and evolution trends between China and Russia. Based on the spatial pattern and types of agricultural products, the international spatial pattern of trade between China and Russia, and the spatial pattern of import and export of key products are further analyzed, revealing the trade pattern between China and Russia. Finally, using the regression analysis method to predict the trade development trends of China, Russia in the next 30 years, so as to provide reference for the formulation of relevant Chinese plans and strategies.

## 2. Models and Methods

### 2.1. *Data Sources and Processing*

The data includes the import and export total volume of agricultural products, and the import and export volume of agricultural products by category in China and Russia in 2006-2017 and the import and export data of sub-categories for Russia in various provinces and regions in China in 2017. The total import and export volume of agricultural products from China and Russia is from UNCOMTRADE DATABASE, FAO data, and China Customs Statistical Yearbook for calculating trade complementarity index and export concentration index in 2006-2017. The import and export data of agricultural products of various provinces, regions and cities came from the China Customs Information Network and the General Administration of Customs of China in China and Russia in 2006-2017, which are used to analyze the status quo of trade patterns between China and Russia.

The agricultural products in this study mainly refer to primary products and processed products from planting, animal husbandry, and fishery. That is, plants, animals and their products obtained in agricultural activities include fruits and vegetables, flower seedlings, food crops, and aquatic products. The classification method adopted in the 1992 "Coordination of Trade Names and Coding System" (HSI 1992) includes agricultural products in Chapters 1-24, 41, 43, 44, 50, 51, 52 (table 1).

### 2.2. *Methods*

*2.2.1. TII of China's Agricultural Trade with Russia.* Trade intensity index (TII) indicates the degree of trade dependence between two countries, the proportion of a country's exports to a trading partner country to its total export volume and the total exports of that trading partner country to world total imports. The greater the value, the closer the trade links between the two countries and the stronger the dependence. The specific formula is:

$$TII_{ij}=(E_{ij}/E_i)/(\ I_j/(\ W))$$

In the formula, $E_{ij}$ represents the export value of i to country j, $E_i$ indicates the total export of i to the world, $I_j$ indicates the total import of country j, and W indicates the total import of the world. The greater the TII, the closer the trade links between the two countries; the smaller the TII, the looser the trade links between the two countries.

*2.2.2. TCI of China's Agricultural Trade with Russia.* The trade complementarity index (TCI) measures the degree of complementarity of the trade structure between the two trading countries and analyzes the export expansion potential of the two countries. This paper is widely used internationally to formulate a calculation method for international trade policies, as follows:

$$TCI=1-\sum\left\{\left|\left(M_i^k/M_i\right)-\left(X_j^k/X_j\right)\right|\right\}/2 \quad (1)$$

$$C_{ij}^k = RCA_{xi}^k * RCA_{mj}^k \quad (2)$$

$$RCA^k_{mi}=(M_i^k|M_i)/(X_W^k|X_W) \quad (3)$$

$$RCA^k_{xi}=(X_i^k|X_i)/(X_W^k|X_W) \quad (4)$$

In this formula, $M_i^k$ denotes the import value of agricultural products k in i countries, Mi denotes the import volume of agricultural products in i countries, $MX_j^k$ denotes the export value of agricultural products k in country j, and Xj denotes the export value of agricultural products in country j. This formula is used to indicate that the agricultural products exported in country j are exported to the world. Another country i imports k agricultural products from the world.

**Table 1.** Agricultural products in "Coordination of Trade Names and Coding System" (HSI 1992).

| | |
|---|---|
| **HS01** | Live animals |
| **HS02** | Meat and edible meat offal |
| **HS03** | Fish, crustaceans, mollusks, aquatic invertebrates ne |
| **HS04** | Dairy products, eggs, honey, edible animal product nes |
| **HS05** | Products of animal origin, nes |
| **HS06** | Live trees, plants, bulbs, roots, cut flowers etc. |
| **HS07** | Edible vegetables and certain roots and tubers |
| **HS08** | Edible fruit, nuts, peel of citrus fruit, melons |
| **HS09** | Coffee, tea, mate and spices |
| **HS10** | cereals |
| **HS11** | Milling products, malt, starches, inulin, wheat glute |
| **HS12** | Oil seed, oleagic fruits, grain, seed, fruit, etc, ne |
| **HS13** | Lac, gums, resins, vegetable saps and extracts nes |
| **HS14** | Vegetable plaiting materials, vegetable products nes |
| **HS15** | Animal, vegetable fats and oils, cleavage products, et |
| **HS16** | Meat, fish and seafood food preparations nes |
| **HS17** | Sugars and sugar confectionery |
| **HS18** | Cocoa and cocoa preparations |
| **HS19** | Cereal, flour, starch, milk preparations and products |
| **HS20** | Vegetable, fruit, nut, etc food preparations |
| **HS21** | Miscellaneous edible preparations |
| **HS22** | Beverages, spirits and vinegar |
| **HS23** | Residues, wastes of food industry, animal fodder |
| **HS24** | Tobacco and manufactured tobacco substitutes |
| **HS41** | Raw hides and skins (other than furskins) and leather |
| **HS43** | Furskins and artificial fur, manufactures thereof |
| **HS44** | Wood and articles of wood, wood charcoal |
| **HS51** | Wool, animal hair, horsehair yarn and fabric thereof |
| **HS52** | Cotton |

*2.2.3 ECI of China's Agricultural Trade with Russia.* The Export Concentration Index (ECI) measures the concentration of a country's export product categories. Calculated as follows:

$$\text{ECI}=\sqrt[2]{\sum\left(X_i^k|X_i\right)^2} \tag{5}$$

In the formula: $X_i^k$ is the export amount of k product in i country; $X_i$ is the total export value of i country. ECI usually takes value between $\frac{1}{\sqrt{n}}$~1. The smaller the ECI, the more scattered the structure of the agricultural export products of country i; the bigger the ECI, the more concentrated the export agricultural product structure of country.

*2.3. Models*

Based on agricultural trade data during 2006-2017, regression analysis was used to curve the growth of China's agricultural trade, and forecasted the trade volume of agricultural products in 2025, 2035, and 2050. By comparing several common fitting functions, a function with a good fitting effect (R2 maximum) is selected as a prediction model, table 2.

**Table 2.** Agricultural products trade predicted value of China with Russia based on regression model.

| Export—Import | Forecast model | $R^2$ | F | Forecast Value (Billion US) | | |
|---|---|---|---|---|---|---|
| | | | | In 2025 | In 2035 | In 2050 |
| China-Russia | $Y=0.1634t^2+3.078t+5.6729$ | 0.8309 | $125.0^{**}$ | 40.75 | 114.02 | 285.1 |
| Russia-China | $Y=0.0554t^2-0.1175t+13.438$ | 0.7674 | $71.9^{**}$ | 16.04 | 29.27 | 69.89 |

## 3. Results and Discussion

*3.1. The China's agricultural trade with Russia.*

China is the world's most important agricultural trade country. It has close bilateral trade with Russia in the agricultural sector. The total trade volume of agricultural products between China and Russia increased rapidly. The growth of import trade was significantly higher than that of export trade (figure 1). The proportion of agricultural products imported by China from Russia has increased slightly in China's agricultural trade, indicating that China has a greater demand for agricultural products in Russia. It can be seen that the trade cooperation between China and Russia has increased significantly, and its share in China's agricultural trade has also increased.

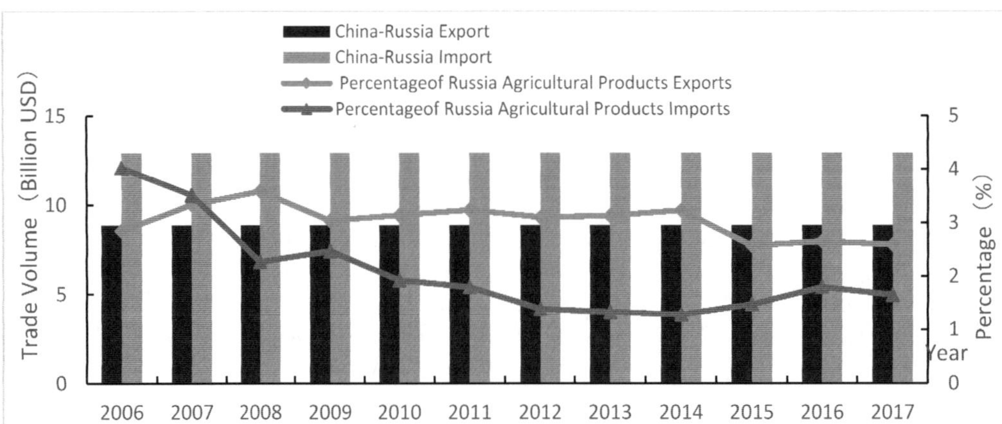


**Figure 1.** Agricultural product trade volume of China with Russia during 2006-2017.

*3.2. Agricultural products trade structure for China and Russia*

There are major differences in China's export agricultural products trade with the Russia. China exports more labor-intensive primary agricultural products, and imports are mainly land-intensive. Russia are important trading partners of China. With the "One Belt and One Road" initiative, China and Russia will have a huge potential for the release of their agricultural product trade potential.

From the perspective of the exports of the China-Russia, the structure of China's export products is relatively concentrated, with labor-intensive agricultural products as the main factor, accounting for 42.37%; China's exports to Russia are relatively concentrated, and export categories have not changed much. It mainly exports grain (grain), fruits and horticultural products represented by vegetables and flowers (figure 2).

China's imports of agricultural products from the Russia are more concentrated, with land-intensive agricultural products accounting for 49.10% of the total. China's imports of agricultural products from Russia are still concentrated, and they mainly import resource-intensive products such as aquatic products, oilseeds, and nuts. After 2010, they added bulk agricultural products such as feed products, oil crops, soybeans, and corn. At present, Russia is the largest importer of aquatic products in China, and its imports account for about 30% of China's total imports of aquatic products. In the future, China will have a stable demand for Russian dairy products.

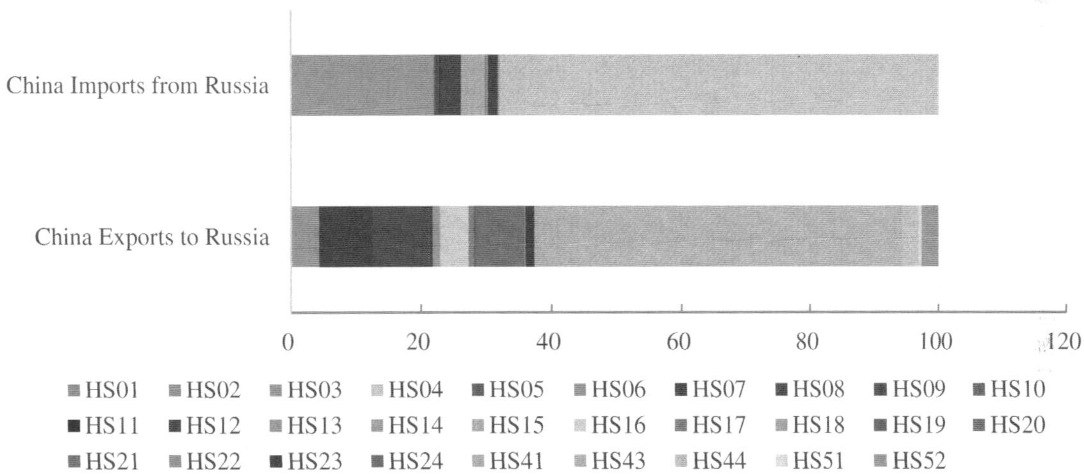


**Figure 2.** Structure of trade agricultural products for China and Russia.

*3.3. International spatial pattern of agricultural trade for China and Russia*

In 2017, the import market for Chinese agricultural products was dominated by Japan, Hong Kong, and the United States. Exports were mainly dominated by the United States, Brazil, and Australia, accounting for 37.61% of China's total exports of agricultural products and accounting for 45.34% of total imports. The relatively concentrated agricultural product trade market has increased the risk of China's agricultural product trade market, and has adversely affected China's effective use of international agricultural resources. The top three countries in Russian agricultural product exports are Turkey, South Korea and China, which account for a large amount of agricultural products for export; Brazil, Germany and Lithuania have the top three countries in import (figure 3). From the perspective of the international spatial pattern of Russian agricultural products imports, China's agricultural exports to Russia are facing fierce competition from other exporting countries. China's agricultural production costs have increased with the increase in labor costs and land rents, which shows that China's export of labor-intensive agricultural products to Russia needs to be adjusted, the types of exports should be diversified, and they will face more fierce competition at the same time.

*3.4. Agricultural trade Provincial Spatial Pattern for China- Russia economic corridor*

There are significant spatial differences in the pattern of import and export trade between China's provinces and Russia. Judging from the inter-provincial spatial distribution, China's Inner Mongolia, Xinjiang, Shanxi, Heilongjiang, Jilin, Shandong, Jiangxi, Chongqing, Gansu, Shaanxi, Henan, Ningxia provinces and regions and Russia's Far East region to carry out agricultural trade (figure 4), with border trade as the mainstay. China and Russia import and export trade is mainly concentrated in Heilongjiang and eastern coastal provinces. There are major differences in the participants and spatial patterns of China-Russia import and export trade. In 2017, China's total imports from Russia accounted for 35.1% of Heilongjiang's total, accounting for one-third of its total trade volume; of Russia's export trade, Heilongjiang's share was only 9.1%, while Zhejiang, Jiangsu, and Guangdong accounted for approximately 50% of the total. It can be seen that in China's import trade of Russian agricultural products, border trade plays an important role; in the export trade of Russian agricultural products, the eastern coastal areas with a relatively large economic aggregate play an important role.

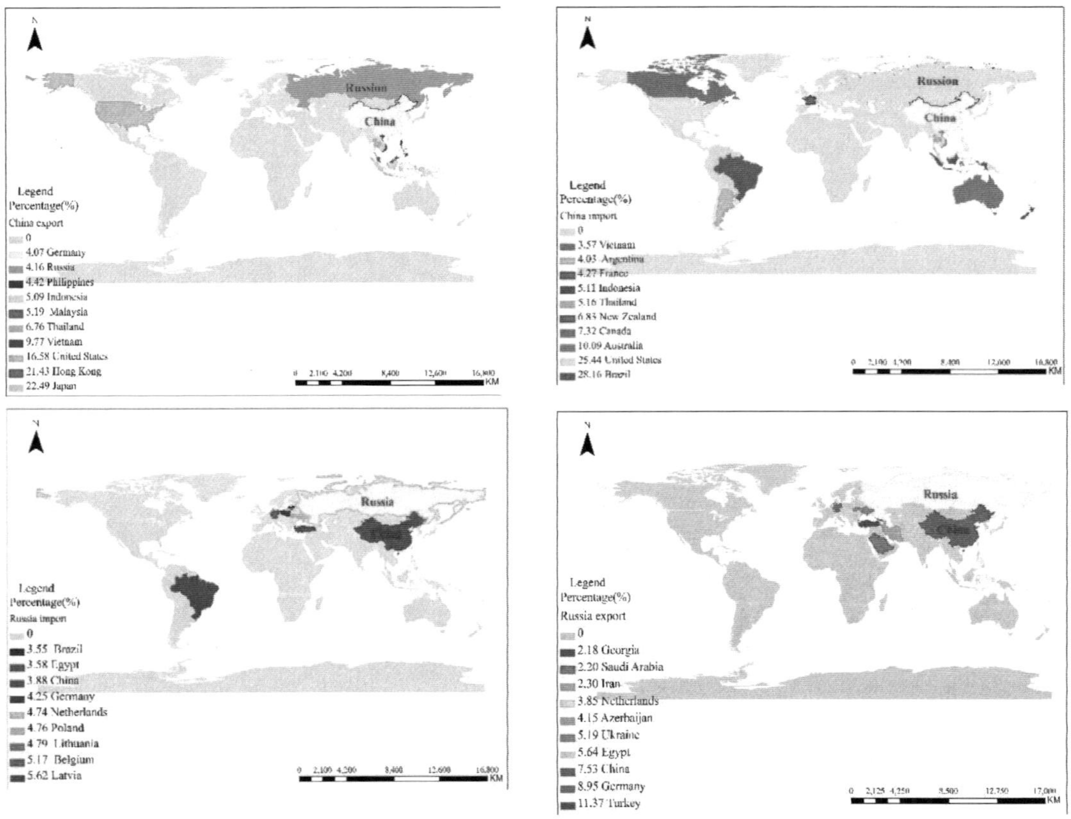


**Figure 3.** Top ten national of agricultural products trade in China and Russia.

The above results confirm to some extent the existing research results that the economic aggregate and border trade have a significant impact on the trade volume between China and Russia. China–Russian trade accounts for a large share, Russia's total economic volume and total trade volume are relatively large, and the economic impact has a relatively wide geographical range.

*3.5. Agricultural trade potential for China and Russia*

The trade volume will show an upward trend in the next 30 years between China and Russia (table 2). In 2025, China's trade volume with Russia's agricultural products will reach $56.79 billion; in 2035, it is expected to reach $143.29 billion. The strength of agricultural economic and trade ties between

IOP Conf. Series: Earth and Environmental Science **190** (2018) 012046 doi:10.1088/1755-1315/190/1/012046

China and Russia will continue to deepen. The trade space will further expand between China and Russia with the advancement of the "Silk Road Economic Belt" and the "Chinese-Mongolia-Russia Economic Corridor".

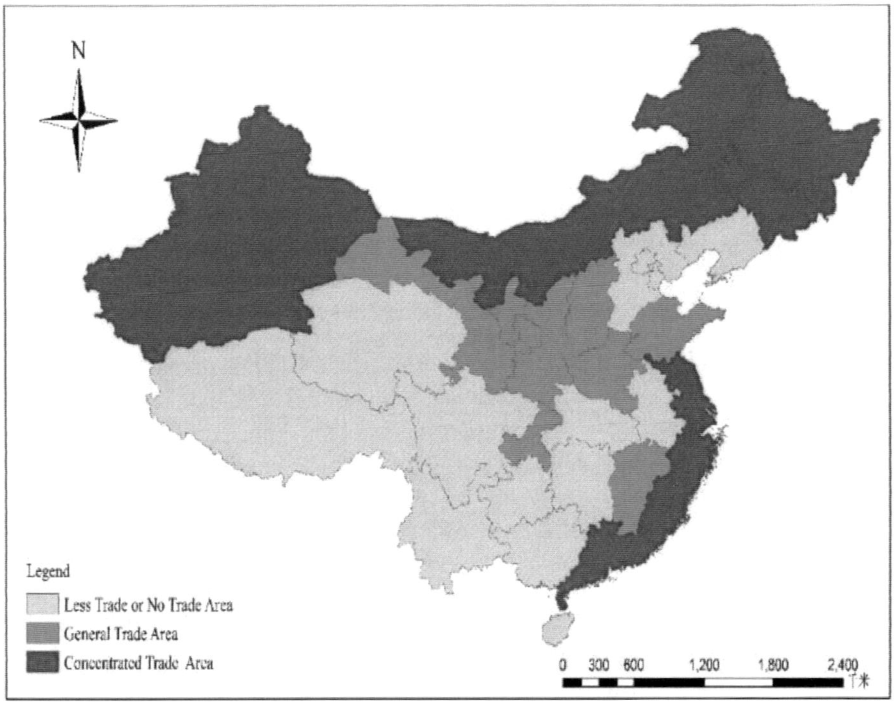


**Figure 4.** Provincial distribution of China's trade with Russia in 2017.

In the specific agricultural product structure, the types of agricultural products imported by China from Russia will increase in the future, and vegetables and fruits will continue to be used. The structure of china's export agricultural products will show a trend of advanced. Seeing from the inter-provincial spatial distribution, China's Inner Mongolia, Xinjiang, Shanxi, Heilongjiang, Jilin, Shandong, Jiangxi, Chongqing, Gansu, Shaanxi, Henan, Ningxia and other provinces and regions have more trade with Russia.

## 4. Conclusion

Chinese trade links are getting closer with Russia. Russian agricultural products have become more and more important in Chinese agricultural products.

China's agricultural exports to Russia are facing fierce competition from other exporting countries. China's export of labor-intensive agricultural products to Russia needs to be adjusted, the types of exports should be diversified. China's Inner Mongolia, Xinjiang, Shanxi, Heilongjiang, Jilin, Shandong, Jiangxi, Chongqing, Gansu, Henan, Ningxia and other provinces and regions have more trade with Russia.

The types of agricultural products imported will increase by China from Russia in the future, and vegetables and fruits will continue to be used. The structure of china's export agricultural products will show a trend of advanced.

Three countries have different resources and endowments with different advantages. Agricultural products have strong mutual complementarity, high degree of integration, and great potential for trade cooperation. With the comprehensive advancement of the "Silk Road Economic Belt", three countries trade cooperation will continue to expand, and the space and potential will remain enormous for future.

Optimize trade structure and innovate industrial cooperation. Agricultural industrial parks will be actively build along the economic corridors, promote industrial clusters and platforms through investments, and promote three countries agricultural trade. It will optimize structure of agricultural products and expand the types of agricultural trade.

With the "One Belt and One Road" initiative and the advancement of the China-Mongolia-Russia Economic Corridor, China and Russia will have a huge potential for the release of their agricultural product trade potential. Tap the potential of China's import trade cooperation and promote the balanced development of bilateral trade cooperation.

**Acknowledgements**
This work was supported in China by the Multidisciplinary Joint Scientific Expedition of China-Mongolia-Russia Economic Corridor [grant number 2017FY101304]; and the Ningxia college first-class discipline construction projects (grass) [grant number NYYLXK2017A01].

**References**
[1] Tiiu P 2000 The Gravity Approach for Modeling International Trade Patterns for Economies in Transition *International Advancesin Economic Research* **6** 633-48
[2] Liu W and Ding Y B 2016 An Analysis of Trade Pattern and Trade Potential of Sino—Mongolia-Russia Economic Cooperation Corridor *Commercial Research* **474** 24-31
[3] Seung J K 2012 South Korea's trade intensity with ASEAN countries and its changes over time *International Review of Business Research Papers* **8**(4) 63-79
[4] Yang X Y and Wang D 2005 Analysis on the trade complementarity between China and Russia *World Economy Study* **7** 71-7
[5] Ai Sai T J and Guo Y D 2011 A comparative analysis of China's trade with Russia and Central Asia *Social Sciences in Xinjiang* **4** 83-7
[6] Gong P P, Song Z Y and Liu W D 2015 A study on trade pattern of China with Russia and Central Asia *Geographical Research* **34**(5) 812-24
[7] Yu J P 2003 Comparative advantage and trade complementary between China and other Asian economies *The Journal of World Economy* **26**(5) 33-40
[8] Seung J K 2012 South Korea's trade intensity with ASEAN countries and its changes over time *International Review of Business Research Papers* **8**(4) 63-79
[9] Raballand G and Andrésy A 2007 Why should trade between Central Asia and China continue to expand *Asia Eur J.* **5**(2) 235-52
[10] Xie W X 2017 An Analysis of Economic and Trade Corporation between China-Mongolia under Building The belt and Road *Economic Issues* **2** 14-8
[11] Li F J, Dong S C, Yuan L N, Cheng H, Chen F, Li Y, Li Zehong and Gu Y Y 2016 Study on Agriculture Patterns and Strategy of the Belt and Road *Chinese Academy of Sciences* **31**(6) 12-4

# Formation of effective cooperation instruments for development projects in the Lake Baikal basin

**I V Bychkov, I I Orlova**

Irkutsk Scientific Center of the SB RAS, Irkutsk, Russia

E-mail: ivbychkov@mail.ru; maksimova.irina.il@gmail.com

**Abstract**. Key factor of Russia-Mongolia conciliation procedures in development projects on the Lake Baikal basin territory is the status of Baikal as a World Heritage site. Both countries are the parties to the World heritage Convention. Hydropower construction projects in the Lake Baikal basin, which currently are under consideration, are indicative examples of elaboration of procedures in achievement of harmonized decisions of Russia and Mongolia with participation of international institutions. This paper gives analysis of problem solving instruments, as well as consolidated actions of authorities, science and the public in international talks on threat prevention in Baikal region. The paper emphasizes not only necessity of scientific support as one of the mandatory conditions of any cooperation, but also highlights formation of working groups and their mandates; different levels of talks for harmonization of views; interaction of science, regional and federal authorities; role of NGOs and citizens, public consultations. The following preliminary conclusions are justified: accomplished scientific and public actors can be a key element in harmonization of interests; there is a subjective factor that has a crucial significance – availability of specialists, who provide substantive work and organizational management of actions; it is reasonable to regulate main stages of harmonized actions with identification of existing legal and institutional instruments.

## 1. Introduction

The Lake Baikal's status of World Heritage site [1] sets international requirements to member-countries of the world cultural and natural heritage Convention (Paris, 23 November 1972), inclusive obligations "not to take any deliberate actions that can be directly or indirectly damaging to cultural and natural heritage, located on territories of other member-countries of the Convention" (par. 3, Article 6), as well as "by joining the Convention, the member-country acknowledges the universal status of the World heritage sites and all countries must work together for their protection" (par.1, Article 6 of the Convention).

Mongolian hydrotechnical construction projects in the Selenga river transboundary basin, main tributary of Baikal, are object matter of active cooperation between Russia and Mongolia, as well as scientific organizations, the public and international society. During searching for mutually acceptable solutions there is a process of formation and practicing of effective cooperation instruments for development projects in the Lake Baikal basin.

## 2. Analysis and assessment of institutions influencing interstate decision making

The sense of analysis of responsibilities and actions of stakeholders in decision making in transboundary water resources development projects is: assessment of effectiveness of their influence instruments; determination of maximum effective ways to consolidate national actors; formation and provision of technologies of interstate interest harmonization; elaboration of suggestions on legitimation of effective technologies for harmonization of the Lake Baikal basin development projects; formulation of suggestions on introduction of amendments and additions to regulatory framework of the Russian Federation and regions in order to realize potential of different actors for preservation of Baikal and prevention of threats to its unique ecosystem.

Participation of international institutions is the fundamental factor of problem solution in the Lake Baikal basin [2]. Russia and Mongolia are parties to the Convention on the Wetlands of international importance especially as wildlife habitat (Ramsar, 2 February 1971) [3] and the Convention for the protection of the world cultural and natural heritage (Paris, 23 November 1972) [4].Besides, Russia is a party to the Convention on the protection and use of transboundary watercourses and international lakes(Helsinki, 17 March 1992) [5].

Effectiveness of the World heritage institute [6] in solving of energy development problems in Mongolia enables us to use it as one of key instruments for prevention of potential threats to the Lake Baikal [7]: due to hydrotechnical construction plans in Mongolia potential impact of the Selenga river flow regulation on the object "Lake Baikal" is the key factor of Russia-Mongolia conciliation procedures. Since 2012 the World heritage Committee has been adopting resolutions for preservation of the World heritage site "Lake Baikal" with the focus on problems of potential impact on Baikal from planned hydro constructions in the Selenga river basin in Mongolia.

Besides the list of concrete actions, which have to be taken by member-countries of the Convention, the made decisions also include directives on preliminary harmonization of projects for their approval. The 2016 Decision [8] for Mongolia includes the following point: "Not approve any of the projects until the above – mentioned EIAs and assessment of cumulative impacts have been reviewed by the World Heritage Centre and IUCN", and also message to Russia and Mongolia about collaborative development of Strategic ecological assessment for any future hydropower and water projects, which potentially can impact the object considering any existing and planned projects on territories of both countries. The World heritage committee (WHC) again confirmed its demands in Decision 41 at the 2017 session [9].WHC decision impact is done through member-countries of the Convention, and the member-countries must follow the decisions made. Moreover, the countries must annually present the WHC decision compliance reports and reports on safe state of the World heritage sites and fulfillment of recommendations. In addition to interactions with official countries' bodies, the WHC and the International Union for Conservation of Nature and Natural Resources (IUCN) cooperate with different experts and NGOs.

Due to development projects with use of water resources, bilateral relationships between Russia and Mongolia are directly regulated by the Protection and use of transboundary waters Agreement between government of the Russian Federation and government of Mongolia(Ulaanbaatar, 11 February 1995).Unfortunately, legal and institutional norms of the Agreement do not include effective and reliable instruments for prevention of ecological threats in the Russian Federation [10].The main conceptual problem of the existing Agreement is the term "transboundary waters" [11], which includes only State border crossing rivers, streams, lakes and other surface water objects, as well as groundwater deposits (Article 1 of the Agreement). In this regard there are different definitions of planned projects qualifying them to the regulation sphere of this Agreement. Mongolia [12] and Russia [13] have different interpretations of achieved agreements. Correcting the norms in order to set the basin approach to scope of the Agreement will enable us to comprehensively consider all development projects. Another serious problem here is a mandate limited institutional mechanism of cooperation – institute of Plenipotentiaries [14], as well as that the Agreement lacks liabilities of the parties for consultations in elaboration of the Selenga basin water use and protection measures. Moreover, the former 1974 Agreement on rational use and protection of the Selenga river waters

(Moscow, 3 July 1974), which was terminated by adoption of the 1995 Agreement, stated that "Negotiating Parties without mutual understanding are not going to take water related actions, which could negatively impact the change of water regime and cleanliness of the Selenga river basin waters" (article 9). Recovery of this requirement in the existing bilateral Agreement will be a reliable guaranty of harmonized policy and measures of the Lake Baikal basin water use.

The existing bilateral agreement lacks mandatory provisions in project planning in transboundary basin of the Selenga River, as well as defined procedure of mutually acceptable decision takings. Therefore, it stipulates complicated and time consuming harmonization process, involving external and internal actors for achieving compromise for all interested social, economic and political groups in both countries.

Interests of the Russian Federation are expressed by competent federal executive authority in this field – Ministry of natural resources and environment of the Russian Federation, as well as Russian-Mongolian commission for economic-trade and scientific-technical cooperation, Meeting of government Plenipotentiaries of the Russian Federation and Mongolia on fulfillment of the Russian-Mongolian governmental Agreement on protection and use of transboundary waters, and Joint Russian-Mongolian working group on fulfillment of the Agreement. Due to above mentioned limitations discussion of impact of the Selenga river flow regulation, planned by Mongolia, at the intergovernmental meetings in 2012-2015 did not go beyond intents of elaboration of the possibilities to collaboratively study the projects considering possible ecological risks, and expression of concern from the Russian side [15].

In 2014-2015 scientific institutes of the Siberian branch of the Russian Academy of Sciences (primarily, Energy systems Institute of the SB RAS and Irkutsk Scientific center of the SB RAS) prepared formulation of the problem of assessment of impact on the Lake Baikal and Russian part of the Selenga river basin, done by water flow regulation in Mongolian part of the Selenga river basin. It was presented in 2015 in the Ministry of natural resources and environment of the Russian Federation at the sitting of expert group on assessment of the Lake Baikal ecosystem impact, done by Shuren hydropower construction project and other hydrotechnical construction projects, which are planned to be implemented on the Selenga river and its tributaries in Mongolia. Joint work of the Ministry of natural resources and environment of the Russian Federation and scientific institutions of the SB RAS, done in the framework of Interagency commission on the Lake Baikal protection, as well as in process of preparation and holding of special meetings and expert consultations of representatives of Russia and Mongolia, led to the ministerial decision on necessity of carrying out a targeted scientific research of this problem, which was included into 2016-2017 R&D scope of implementation of the Federal targeted program "Development of water industry of the Russian Federation in 2012-2020" [16]. The goal of the research was elaboration of science based assessments of how planned hydro constructions in Mongolia impact the transboundary basin of the Selenga river within the Russian territory [17]. The science based assessments and recommendations, resulted from research works, became the rationale for Russia's position in negotiations with Mongolia, international organizations, international financial institutions.

Analysis of mandates and practical actions of authorized institutions enables us to make conclusions about productivity of negotiation positions under condition of joint efforts of the authorities and scientific organizations: political mandates and scientific justifications are complementary necessary conditions of formation of consolidated state position, what was the essential factor for beginning of productive Russian-Mongolian talks on problems of transboundary water projects in 2016-2017.

Considerable part of hydropower projects in Mongolia (for example, Shuren Hydropower project, Orkhon-Gobi drainage project) is financed by the World bank, which has to adhere to international Conventions as mandatory and endorsed norms of international law. Such are the Convention on environment impact assessment in transboundary context (Espoo, 25 February 1991), and UNECE Protocol on strategic environmental assessment(Kiev, 21 May 2003),for example, in cases when realization of a certain strategic plan or program may lead to substantial ecological consequences outside the country of the plan or program.

Consideration of public opinion in decision making is an integral and principle requirement of above mentioned international legal acts, therefore those documents is an important instrument of influence on decisions of the World Bank. NGOs not only give an opinion of local community position, which can be concerned in a result of project realization, but also form and implement active interaction with international institutions, including WHC, UNESCO, IUCN. An opportunity of use of public impact on international financial institutions has a crucial significance: Greenpeace Russia and international coalition "Rivers without boundaries" managed to put the Hydropower construction projects in Mongolia for additional consideration in the World Bank. As a result, the projects were put for public discussion, and in spring 2017 there were public consultations in the Russian Federation (Buryatia and Irkutskaya oblast) [18]. Joint actions of the federal agencies, firstly, Ministry of natural resources and environment and Ministry of Energy, authorities of territorial entities of the Russian Federation, scientific community, NGOs, municipalities allowed to achieve a very important result – decision of the World Bank on procedural segregation of Regional ecological assessment (REA) of the planned projects considering existing hydropower projects in Mongolia and development of Environment impact assessment (EIA) and feasibility study, which will be elaborated only after preparation and harmonization of EIA results with Russian side.

## 3. Effective institutions of interstate interest alignment

Elaborated technologies of interaction of authorities, science and the public, based on existing mandates, qualifications, cooperation instruments, as well as full mutual trust, allowed for productive institute of Russian-Mongolian expert consultations with development of common ground for the main part of principal problems, what was introduced in Irkutsk in April 2018 at the Russian-Mongolian Working group meeting on comprehensive consideration of issues, coming from the hydrotechnical construction plans in Mongolian part of the Selenga river catchment area. Using exclusive instruments of each organization, participated in the alignment procedure, areas of responsibilities were identified and procedures of agreed solution development were carried out. The exclusive instruments are:

- Ministry of natural resources and environment of the Russian Federation – political functions, including interaction with diplomatic structures, international organizations and partner-country;
- Scientific organizations – development of science based assessments and justification of the country's position; preparation, explanation and submission of content materials to authorities and the public;
- NGOs –interaction with international organizations, world financial institutions; collaboration with local population.

Other functions and joint actions were also distributed during preparations to the meeting (table 1). The main condition was full alignment of all actions of each party with other actors, as well as agreements on presentation of a common position from the Russian expert side, and management control of those agreements.

Besides Russian participants, mentioned in the table, who were involved in preparation and holding of the meeting, other designated federal and regional authorities took part in the event, as well coastal municipalities, scientific institutions, primarily located in the region and working with Mongolian scientific organizations, experts and employees of protected areas, etc. All mentioned experts and organizations worked in contact with key actors.

As a result of the meeting, the following principal positions were aligned by Russian and Mongolian experts:

- Execution of UNESCO World heritage committee decision on putting all planned hydrotechnical construction projects under consideration; as well as inclusion of a separate section about preservation of the Lake Baikal as the World heritage site;

- Involvement of Russian experts in elaboration and consideration of technical specifications for all stages of impact assessments;
- Formulation of main goal of the REA – assessment of projects' realization tolerance and possibility of their implementation resignation;
- Deep study of alternatives to presented hydrotechnical construction projects; and others.

**Table 1.** Distribution of responsibility areas in preparation and holding of talks of Russian and Mongolian experts for consideration of REA project technical specification of the Selenga river hydrotechnical constructions

| Areas of responsibility | Ministry of nature resources of Russia | Authorities of Irkutskaya oblast | Irkutsk Scientific center of the SB RAS | NGOs |
|---|---|---|---|---|
| **Political and official support** | Key functions | Support functions | Support functions | Support functions |
| **Scientific researches** | Coordination | | Key functions | |
| **Analytical support** | Coordination | | Key functions | Key functions |
| **Consultation management** | Coordination | | Key functions | |
| **Organizational support** | Coordination | Key functions | Support functions | |
| **Funding** | Coordination | Key functions | | |
| **Agreement performance monitoring** | | | | Key functions |

In fact, considerable part of disputes was arranged. Thus, format of the meeting provided results that were maximum possible in current conditions of interstate negotiation positions. Basing on the achieved results one can make the following assessments:

- Productivity of interstate decision making in development projects in transboundary water basin is possible only under conditions of active official support from international bodies; eligible science based position of the state; political will of designated agency management; consolidation of authorities, science and NGOs; support of local population, municipalities and regions.
- Maximum effectiveness of national actors is possible only under conditions of use of exclusive mandates. Other functions and responsibilities should be fulfilled in interaction and alignment with other actors – stakeholders.
- High degree of trust among authorities, science and public leaders, from the view of professionalism and reliability of compliance with decisions.
- Format of expert meeting is productive under subjective conditions, such as: common position of national experts; professional conduct of discussion; qualified preparation of content materials.

Experience of interest harmonization among authorized experts from the countries of the Lake Baikal basin allows justifying suggestions on legal coverage of involvement of scientific organizations and NGOs in preparation of the state's position in international talks, as well as their inclusion into coordination governmental bodies responsible for decision making on federal and regional levels. Realization of this suggestion is possible under condition of legal coverage for liabilities on implementation of scientific research results, done according to government order, and appropriate funding.

## 4. Conclusion

The Lake Baikal is reckoned by world community as universal global merit; and this status sets high requirements to development projects in the Lake Baikal catchment basin, which includes territories of Russia and Mongolia.

The goal of current hydrotechnical construction plans in the Selenga river basin is energy security of Mongolia and water supply for population and industry. Elaboration of harmonized decision system with safety guarantees to the Lake Baikal will allow finding a solution to economic problems in Mongolia together with guaranties of prevention of the Lake Baikal ecosystem threats.

The project work is currently in progress, including processes within the scope of bilateral working and expert groups. A number of essential issues, which go together with achievement and realization of mutually acceptable decision, is worth paying attention to: media representation of projects and processes of solution searching, monitoring of decision realization by all stakeholders and financing of harmonized decision development.

Monitoring of harmonized decision realization – is the most important function, which is actually done by NGOs. Inclusion of obligation for submission of full and clear information into the normative requirements to preparation procedures of applied project can allow for timely and objective consideration of all stakeholders' suggestions.

One of the principal issues of preparation of joint Russian-Mongolian mutually satisfactory development plans in the Lake Baikal basin is the financing of elaboration of such projects. Interests of the countries can be realized only under condition of equal or independent funding. Research works or projects carried out using the funds of one of the countries (for example, scientific researches on planned hydrotechnical construction impact done by the SB RAS using the Federal Targeted Program funds in Russia, research works on Egiin-Gol Hydropower plant impact on hydrological regime of the Selenga river and the Lake Baikal done by French-Belgian company «TRACTEBELEngineering» using the funding from Mongolian government)or using investments from one of the countries, always arouse mistrust.

Russia and Mongolia have positive experience of joint development of projects: Strategic Action Program [19] for the transboundary basin of the Lake Baikal as well as Transboundary Diagnostic Analysis [20] was elaborated in 2011-2015 within the scope of the Lake Baikal basin Project, implemented by United Nations Development Program. The project was endorsed by competent authorities in Russia and Mongolia, and was highly appraised by United Nations. Problems of use of transboundary water basins were not considered in this work due to political reasons. Resumption of joint work within the scope of UN program can currently be regareded as maximum independent formation instrument for mutually accepted decisions on securing interests of both countries in the sphere of use of the Lake Baikal basin water resources.

## Acknowledgments

This work has been done within the scope of scientific research projects: "Activation of inner development potential in resource specialized regions (through the example of the Baikal region)" 0341-2016-0003 and "Fundamental researches and breakthrough technologies as basis for advanced development of the Baikal region and its interregional ties"0341-2016-0001

## References

[1] Grachev M A 2002 *Concerning Modern Condition of the Lake Baikal Ecosystem* (Novosibirsk: SB RAS) p 156

[2] Laverov N P et al 2008 Suggestions on ecological and economic justification of Eastern Siberia - Pacific Ocean oil pipeline (options beyond the Lake Baikal catchment area) *Problems of National Security: Expert Conclusions, Analytical Materials, Suggestions* 9–49

[3] Russi D, Brink P, Farmer A, Badura T, Coates D, Förster J, Kumar R and Davidson N 2013 *The Economics of Ecosystems and Biodiversity for Water and Wetlands* (Institute for European Environmental Policy (IEEP) & Ramsar Secretariat) p 84
[4] Alvarez J E 2006 International organizations: then and now *American Journal of International Law* **2** 324-47
[5] Ryazanova M A 2015 European cooperation in the sphere of transboundary watercourse use and protection from pollution. International public and private law *Yurist* **2** 25–9
[6] Maksimova I I 2013 Use of the world heritage status for preservation of the Lake Baikal *Geography and Natural Resources* **2** 30-5
[7] Kuzmin M et al 2016 *Lake Baikal Special Issue of RFBR Journal* **1** 104
[8] UNESCO 2016 Decision of the World Heritage committee on the Lake Baikal. Income accessed online on 15th July 2018 via http://whc.unesco.org/archive/2016/whc16-40com-7BAdd-en.pdf
[9] UNESCO 2017 Decision of the World Heritage committee on the Lake Baikal. Income accessed online on 15$^{th}$ July 2018 via http://whc.unesco.org/archive/2017/whc17-41com-7B-en.pdf
[10] Ubugunov L L, Borisova N G, Anenkhonov O A, Baldanov D R, Bazova N V and Ubugunov V L 2015 Possible ecological and biological consequences of hydropower construction on the Selenga river and its tributaries in Mongolia *Izvestiya of Irkutsk State University. Biology, Ecology Series* **12** 58-72
[11] Government of Mongolia. Ministry of environment and green development 2013 *Integrated Water Management Plan of Mongolia* (Ulaanbaatar) p 340
[12] Mongolia Now 2016 Russia agrees with scientists' conclusion on hydropower construction on the EgiinGol river. Income accessed online on 17th July 2018 via http://www.mongolnow.com/26_05_001.html
[13] Russian embassy in Mongolia. Income accessed online on 17th July 2018 via http://www.mongolia.mid.ru/press_573.html
[14] River basin commissions and other institutional mechanisms in the sphere of transboundary water cooperation 2009 *Strengthening of Potential of Water Cooperation in Eastern Europe, Caucasus and Central Asia* (New-York, Geneva: UN) p 67
[15] Russian-Mongolian commission on trade-economic and scientific-technical cooperation 2012, 2013, 2014, 2015, 2016 Protocols of XVI, XVII, XVIII, XIX and XX sessions. Income accessed online on 30th March 2018 via https://www.mnr.gov.ru/
[16] Bychkov I V, Nikitin V M and Maksimova I I 2017 Hydropower projects in Mongolian part of the Selenga river transboundary basin: possible risks for the Russian Federation *Region: economics and sociology* **2** 269-86
[17] Bychkov I V, Nikitin V M, Abasov N V, Berzhnyh T V, Maksimova I I and Osipchuk E N 2017 Possible changes in hydrological characteristics due to flow regulation in the Selenga river basin *Geography and Natural Resources* **3** 75-86
[18] World Bank project on mining infrastructure investment support (MINIS) 2017 Consultations on pre-feasibility study of Regional ecological assessment in the Selenga river basin and Lake Baikal in the context of the projects of Shuren hydropower plant and Orkhon river drainage (Russia, Irkutskaya oblast, Buryatia) Income accessed online on 15th July 2018 via http://www.minis.mn/n/s/58-76
[19] *Strategic Action Programme for the Lake Baikal Basin* 2015 (Ulan-Ude: UNOPS) p 76
[20] *Transboundary Diagnostic Analysis of the Lake Baikal Basin* 2013 (Ulan-Ude: UNOPS) p 232

# Infrastructural approach and geospatial data processing services in the tasks of territorial development management

**I V Bychkov[1], G M Ruzhnikov[1], V V Paramonov[1], A S Shumilov[2], R K Fedorov[2], K G Levi[3], S Demberel[4]**

[1]Siberian Branch of Russian Academy of Sciences Irkutsk Scientific Center, Irkutsk, Russia
[2]Matrosov Institute for System Dynamics and Control Theory of Siberian Branch of Russian Academy of Sciences, Irkutsk, Russia
[3]Institute of the Earth Crust of Siberian Branch of Russian Academy of Sciences, Irkutsk, Russia
[4]Institute of Astronomy and Geophysics of the Mongolian Academy of Sciences, Ulaanbaatar, Mongolia

Email: bychkov@icc.ru

**Abstract**. The technology for creating an information and analytical environment (IAE) for spatial data processing based on the use of spatial data infrastructure (SDI) components and a service-oriented paradigm was proposed for the management of territorial development. The environment is characterized by its spatial nature and distribution of territorial information resources. At the same time, emphasis is placed on the transition from a standard SDI, as an access medium for the exchange of spatial data, to an SDI of an integration type, as an environment for access, exchange, creation, storage of new sets of spatial and thematic data and processing services. The use of Service-oriented Architecture (SOA), open standards by OGC (Open Geospatial Consortium), Web technologies, standardization of the browser's interface and Web services, allowed to move from local to distributed and “cloud” computing in which information and computing resources are provided as Web services. The technology for creating an information and analytical environment involves the use of sets (combinations) of thematic Web services with standardized interfaces integrated with standard protocols (SOAP, WSDL, etc.).

## 1. Introduction

The management of territory development is a multidimensional process, viewed as the combination of social, ecological, economic goals, which is managed through a wide range of strategies, programs, concrete actions and administrative decisions to ensure sustainable and balanced reproduction of the economic, social and natural potentials of the territory aimed at improving living standards of the population. The effectiveness of the territorial development directly depends on the level of informatization and the availability of objective data.

At present, there is a gap between the formation of territorial information resources and their use in the preparation and support of management decision-making. This is due to:

- increasing volumes and distribution of territorial information resources of state authorities of local government (SALG), institutions of the scientific and educational complex (SEC),

business, etc., implemented on different platforms and located in distributed and heterogeneous networks;
- localization (private use) of information systems, models and data in the SALG, SEC, as well as the complexity of their inclusion in the solution of complex problems of territorial development management;
- limitations associated with the use of legacy programming;
- weak use of the service-oriented paradigm in SALG;
- inadequate efficiency of existing information and analytical systems (platforms) providing information services for SALG, SEC, business and population;
- limited use and creation of distributed geoportal information and analytical systems (platforms) for processing spatial and thematic data for SALG, SEC, business and population, etc.

Territorial informatization is formed on spatially embedded levels, but large volumes of distributed, multi-format data limit their use in systemic problem-oriented analysis of the features of territory development.

Modern decision-making software for the support of the territory development management should include service-oriented information and analysis environment for processing distributed spatial and thematic data, using open OGC standards for interoperability, Web technologies for standardizing the browser interface and Web services. This makes it possible to move from local to distributed and "cloud" computing, in which information and computing resources are provided to the user as WPS (Web Processing Service) services. The transfer of processing and storing of SALG data using remote servers allows to simplify their updates, maintenance and implements such functionality as openness, scalability, provision of common classifiers, availability of data and processing services, etc. [1].

Thematic WPS-services and their combinations provide the solution of specific tasks and carry out processing of distributed spatial and thematic data, integrated use of analytical and forecast models, means of displaying and visualizing the results. Spatial information functionality is used for the formation, processing and storage of spatial data, implementation of territorial sampling and imaginative visualization of the results in the IAS.

The article considers the creation of geoportal service-oriented information and analytical environment for managing territorial development, as well as its thematic WPS-services.

## 2. Models and Methods

An infrastructural approach to the formation of IAS [2] was used in order to make decisions on the management of territorial development. The approach is based on:
- considering the requirements of standardization and unification of software and technical and regulatory support;
- using modern principles of organizing and creating distributed information systems based on SOA [3];
- free access and use of spatial data and metadata through services using OGC specifications (WMS, WCS, WFS, WPS, CSW, KML, etc.) [4];
- using not only Web, SaaS, search and retrieval services for spatial data, but also services for correct data creation, analysis and processing (including complex ones) (cartographic Web applications, as well as Web applications that have GIS functionality and etc.);
- introduction of orchestration (service management) to combine small Web services into more complex, higher-level ones that can be included in the management processes [5];
- creation of geoportals, data storage and processing datacenters;
- continuity of existing spatial and thematic information resources of the territories.

Using SOA, OGC standards, Web technologies, WPS-services and geoportals at the creation of IAS allows to obtain not only infrastructural, but concrete results, taking into account the mode of operation of already existing information management systems for territorial development.

The main component of the service-oriented IAS is the geoportal, which is a common point of access to information resources and processing technologies and supports the exchange of spatial information between users, as well as transmitting in a special form data on the design and methods of information presentation (figure 1) [6]. The basic components of a typical geoportal are: data and service catalogs, subsystems for starting services and displaying data; a system of input and editing of relational data with spatial attributes, providing access to data services within the framework of the WPS standard [7].

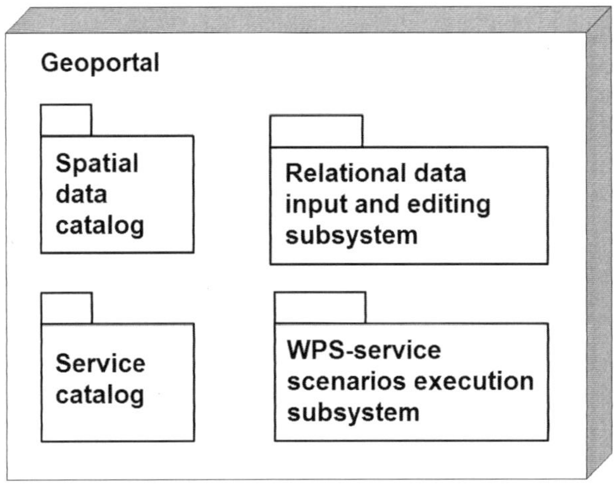


**Figure 1.** The structure of a typical geoportal

The multifunctionality of territorial development tasks justifies the need for creation of a cluster of thematic geoportals that support the exchange of spatial and thematic data and Web services for their processing based on OGC standards, which increases load balancing of the data network and simplifies the usability of the system [8]. The set of templates for new geoportal instances in a form of virtual machines based on different platforms for deploying WPS services was developed. The PostgreSQL DBMS is used to store spatial data. The use of a common DBMS for all geoportals in cluster makes it possible to simplify data exchange and use various user tables in queries. Mapserver and SMD Server are required to display spatial data in accordance with the WMS standard. Mapserver displays the constantly changing data in the PostgreSQL database. Compared to other systems, SMD Server displays static data at high speed without the use of caching systems.

For a comprehensive analysis of data or the application of processing services, unification is carried out by additional information (structural specifications) on the tables. To automate the search for services and data, a specification is developed in JSON format that defines service input data, describes the structures of relational tables. This specification can be converted into one of commonly used standards (XSD, WFS, WSDL-S). In addition to the standard fields of data set names and field descriptions (name, data type), structural specifications contain: metadata in accordance with ISO 19115; units for numeric data; possible values for enumerable data types; controls. The controls implement the methods needed to form the user interface for adding, editing, and displaying attribute data.

Structural specifications are used to create tables, search for relevant data services and are stored in a specialized catalog and arranged in hierarchies. A mechanism has been developed for converting relational data to DBF or SHAPE (if there are spatial attributes) for applying them as input to WPS services. The IAS implements the tasks of forming, integrating and transmitting data, launching services, organizing the computing process in the cloud infrastructure, presenting results, etc.

The catalog of thematic WPS-services was created, it includes following services:

- monitoring and analysis of the socio-ecological and economic situation of the territories;

- ranking of territories by local and integral indicative indicators;
- spatial analysis and assessment of trends in the formation of the transport system of the territory;
- monitoring of changes in the floristic composition of the territory depending on its socio-economic development;
- vegetation index;
- SRTM radar survey data processing;
- calculation of the density of point and linear objects in cells of a regular grid;
- decryption of the GRID data set using the classifier and the method of reference vectors;
- converting spatial data of various formats;
- primary statistical data processing;
- ETL integration of unstructured tabular data, etc.

Combinations of WPS services are implemented as programs in the JavaScript language, which allows reducing the task of setting the interaction between services to the task of writing a standard program in one of the most common functional languages [9]. When the service combinations are executed, automatic parallelization and scheduling of the executed services occurs in order to reduce the overall execution time of the combination [10]. The process of executing combinations is resistant to changes in the distributed environment (switching nodes on and off, varying times of execution of services, etc.).

## 3. Results and Discussion

Due to the backbone nature of spatial data and the digital model of the territory, a geoportal information and analytical environment that uses distributed spatial and thematic data is an effective tool for collecting, processing, analysing information, and modelling possible ways to improve the territorial development process. Structurally, the geoportal IAS is invariant to the organizational structure of the territorial management system.

IAS uses distributed and "cloud computing" in which information resources of processing, data storage are provided as Web services. WPS standard allows specifying any calculation (algorithms, models, spatial data processing functions, etc.), setting the data format and how to transfer it, gives access to the geoprocessing functions, defines how to start and end the process and how to obtain results. The data required for WPS service can be obtained over the data network or be available on the server. To conduct analytical calculations on the territorial development, as a rule, combinations or orchestrations of small services are used and implemented as applications or portals at higher levels. Transferring the processing and storage of territorial administration data to remote servers allows to simplify its update, maintenance and implements such functionality as openness, scalability, provision of common classifiers, availability of data and services for their processing, etc.

## 4. Conclusion

The effectiveness of the use of the infrastructure approach for the creation of a geoportal information and analytical environment for the territorial development was presented. The characteristics of the environment are: large volumes of spatial and thematic data and its distribution, application of a service-oriented paradigm, OGC standards, Web technologies and compositions of distributed thematic WPS-services.

## References

[1] Sevugan P, Purushotham S and Chandran A 2017 Expert System through GIS-Based Cloud *Advances in Computational Intelligence and Robotics (ACIR) Book Series* (Hershey: IGI Global) pp 387-98

[2] Bychkov I V et al 2016 *Infastructure of Infomational Resources and Technologies of Informational-Analytical Systems of Regional Management* (Novosibirsk: SB RAS Press) 241 pp (in Russian)

IOP Conf. Series: Earth and Environmental Science **190** (2018) 012048 doi:10.1088/1755-1315/190/1/012048

[3] Alawadain A, Fielt E, Korthaus A and Rosemann M 2016 Empirical insights into the development of a service-oriented enterprise architecture *Data & Knowledge Engineering* **105** 39-52

[4] *Open Geospatial Consortium OGC* Income accessed online on 19th July via 2018 via http://www.opengeospatial.org/

[5] Chu V W, Wong R K, Chi C H and Hung P C K 2014 Web Service Orchestration Topic Mining *21st IEEE International Conference on Web Services* pp 225-32

[6] Zelentsov V A and Potrjasaev S A 2017 Architecture and example of implementation of informational platform for development and distribution of thematical servces using remote sensing data *Proceedings of St. Petersburg Institute for Informatics and Automation of the Russian Academy of Sciences* **6**(55) 86-113 (in Russian)

[7] *OpenGIS Web Processing Service (WPS) Implementation Specification, v1.0.0.* Income accessed online on 19th July 2018 via http://www.opengeospatial.org/standards/wps

[8] Yue P, Zhang M, Taylor T, Xie J, Zhang H, Tong X, Yu J and Huang J 2016 Open Standards in Practice: An OGC China Forum Initiative *IOP Conf. Ser.: Earth Environ. Sci.* **46** 012038

[9] Sukhoroslov O, Volkov S and Afanasiev A 2015 A Web-Based Platform for Publication and Distributed Execution of Computing Applications *14th International Symposium on Parallel and Distributed Computing IEEE* pp 175-84

[10] Fedorov R K, Bychkov I V, Shumilov A S and Rugnikov G M 2016 Planning and executing web-service compositions in heterogeneous dynamic environment *Computational technologies* **21**(6) 18–35 (in Russian)

IOP Conf. Series: Earth and Environmental Science **190** (2018) 012049 doi:10.1088/1755-1315/190/1/012049

# Transportation industry patterns and strategy for the countries along "The Belt and Road"

**H Cheng, S Dong, F Li**

Institute of Geographic Sciences and Natural Resources Research, Chinese Academy of Sciences, Beijing, China

E-mail: lifujia2003@163.com

**Abstract**. "Silk Road Economic Belt" and "The 21st Century Marine Silk Road" are the major strategic concepts of China to build a new sustainable international political and economic order, to face the complex and changeable international geopolitical economic patterns and combat the global economic crisis. Transportation is the dominant force to shape and change economic spatial patterns. In this research, through the study of existing literature and data, the author built the transportation patterns of the countries along "The Belt and Road", revealed that among the countries along "The Belt and Road", the infrastructure constructions were perfect, the trade development of transportation industry was not balanced, the developed countries had obvious advantages, and the developing countries were developing rapidly. Then, the author established the transportation measures of the countries along "The Belt and Road", which were that China should innovate cooperation mechanism, establish cooperation platforms, play a vanguard role of the civil aviation, set high-speed rail construction as the breakthrough point, establish international logistics system, set up International Logistics Association, and so on. This research took the international perspective and scientific analysis to support the national strategy of "The Belt and Road", and the conclusions will provide a strong theoretical basis and strategic support for the construction of "The Belt and Road".

## 1. Introduction

As a historical concept, the silk road has been a channel and bridge for exchanges and economic cooperation between eastern and western civilizations since ancient times. "Silk Road Economic Belt" and "The 21st Century Marine Silk Road" are the major strategic concepts of China to build a new sustainable international political and economic order, to face the complex and changeable international geopolitical economic patterns and combat the global economic crisis [1]. "The Belt and Road Initiative" is a major strategic concept for comprehensively strengthening China's international economic and trade cooperation with central Asia, the Middle East, Europe, north Africa, South Asia and southeast Asia, benefiting countries along "The Belt and Road, combating the international economic crisis and building a new international political and economic order of sustainable development [2]. "The Belt and Road" includes 65 countries in Asia, Europe and Africa. In addition, there are 29 other countries within the radiation range. The total area accounts for 53.6% of the world [3].

Transportation is an important force to create economic spatial pattern and dominate economic spatial pattern evolution [4]. In May, 2017, in the opening ceremony of Belt and Road Forum for International Cooperation, Chinese President Xi Jinping declared that: the interconnection of infrastructure is the foundation for the cooperation of "The Belt and Road", we will promote the interconnection of land, sea, airspace and internet, and connect the highways, railways, shipping lines and air lines network in future [5]. Construction of transportation system within the countries along "The Belt and Road" is the precondition and important basis for implementation and development of "The Belt and Road Initiative", the essential bond and bridge for the interconnection of infrastructure, trade and people, and the priority task that must be broken through [6].

## 2. Models and Methods

This study set the 65 countries along "The Belt and Road" as the study area. According to the analysis on the transportation infrastructure index and the difference between import and export in transportation trade, we obtained the transportation industry patterns for the countries along "The Belt and Road", and put forward the transportation industry development strategy for the countries along "The Belt and Road".

## 3. Results and Discussion

### *3.1. Transportation Industry Patterns for the Countries along "The Belt and Road"*

*3.1.1. Transportation facilities are complete, and the condition of coastal areas are better than inland areas.* The transportation infrastructure of countries along "The Belt and Road" is relatively perfect, such as railways, roads, air and sea transport. The index of transportation facilities in coastal areas such as Western Europe, southern Europe, the Middle East, southeast Asia and east Asia is significantly better than that in inland areas (Figure 1). Among them, the railway and road transport networks in Western Europe, southern Europe, east Asia and South Asia have taken on a large scale. The maritime transport networks on the east Asia-southeast Asia - South Asia - west Asia - Middle East - south Europe are extremely developed, and the density of civil aviation infrastructure in Western Europe, southern Europe, east Asia and southeast Asia is significantly higher than that in other regions.

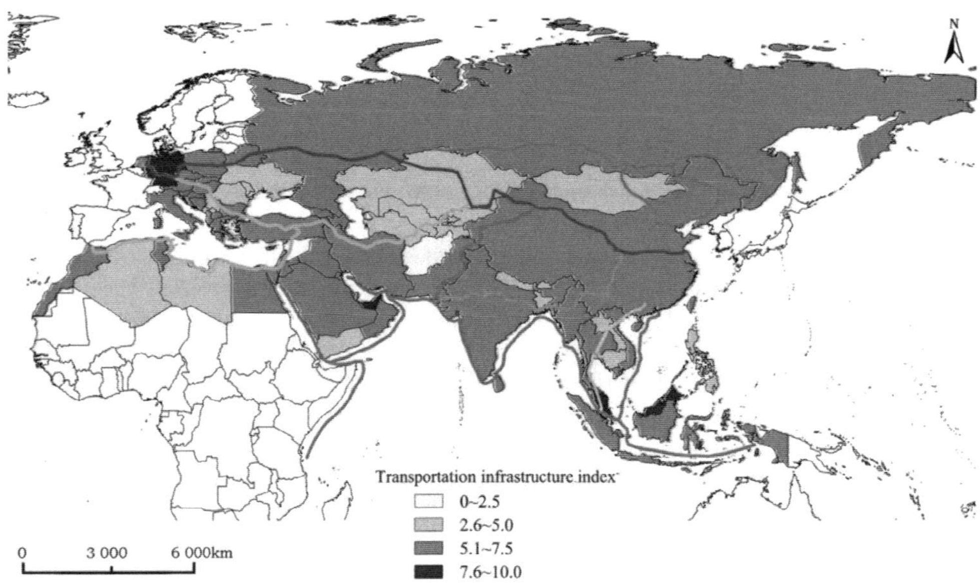


**Figure 1.** The transportation infrastructure index for the countries along "The Belt and Road" in 2013 (Data source: BBS global competitiveness report 2014-2015).

*3.1.2. The development pattern of the transportation trade is unbalanced, the developed countries dominated it.* According to 2013 data from the world bank database, among countries along "The Belt and Road", western European countries have the highest share of transportation exports, 29.15%. Eastern Europe was next at 17.45 per cent. East Asia and central Asia accounted for 10.14% and 0.82% respectively. Compared with the import volume, it is found that the import share of the transportation industry in western European countries is 21.14%, and the import status is lower than the export status. The import shares of transportation industry in southeast Asia, South Asia and east Asia are 19.20%, 15.00% and 12.26% respectively. It can be seen that in the transportation trade of countries along the "The Belt and Road", the development between regions is extremely unbalanced. Europe is the main export region of transportation industry, and it is also the source of trade surplus of transportation industry (figure 2).

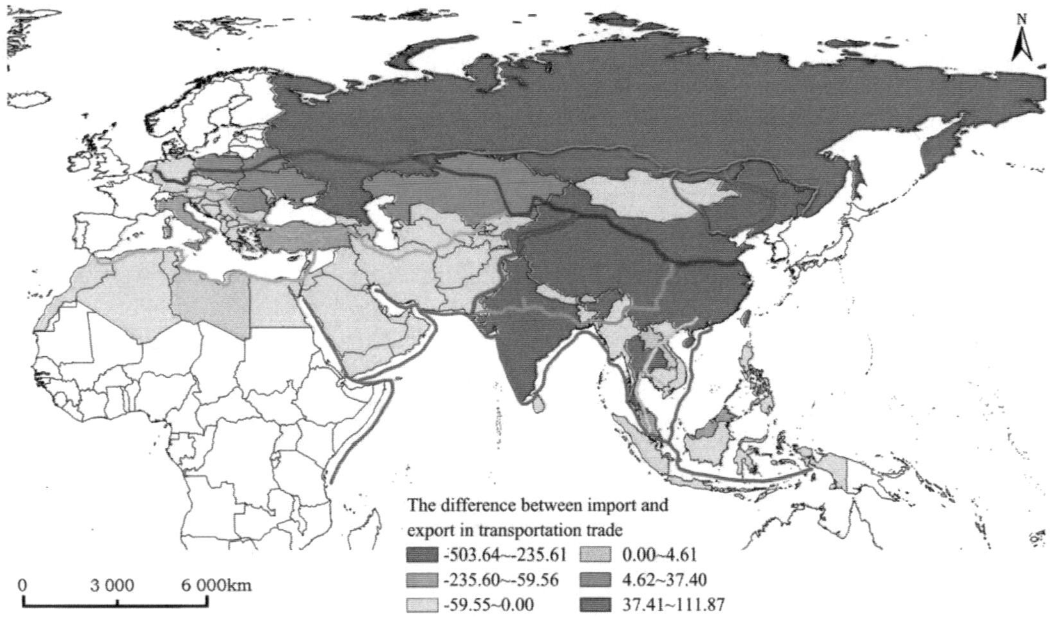


**Figure 2.** The difference between import and export in transportation trade for the countries along "The Belt and Road" in 2013 (Data source: The World Bank Database).

## *3.2. Transportation Industry Strategy for the Countries along "The Belt and Road"*

*3.2.1. We will innovate cooperation mechanisms and establish a transportation cooperation platform between China and central Asian countries.* It is of great significance to innovate the cooperation mechanism for coordinating and developing the development of transportation industry between China and central Asian countries along "The Belt and Road". The establishment of the china-central Asia transport cooperation mechanism should be divided into two aspects: first, the establishment of the transport system, namely the adjustment and allocation of the functions and responsibilities of the transport cooperation organization; Second, the establishment of transportation system, including traffic laws and regulations and various rules and regulations [7].

*3.2.2. Give play to the vanguard role of civil aviation transportation, high-speed rail construction as an important breakthrough.* For central Asia, west Asia and other regions along "The Belt and Road", it is easier to open air routes than road, railway and other land routes in terms of technology and existing civil aviation infrastructure. According to the strategic layout of "The Belt and Road

Initiative", the leading role of civil air transport in the delivery of goods is to give priority to east Asia - central Asia - west Asia - Middle East and east Asia - central Asia - eastern Europe, to open air routes and flights between important cities along the route as soon as possible, and to build a new Eurasian land bridge in the air [8].

According to the theory of late-development advantage, emerging economies and low-income countries can directly use the latest technologies when they catch up in infrastructure construction. Therefore, high-speed rail has become the infrastructure choice of many emerging economies and low-income countries. China can take high-speed railway construction as an important breakthrough for the external development of the transportation industry [9].

*3.2.3. We will open seven international logistics channels and build a modern international multimodal transport logistics system.* Cooperate with all the way "area" development and action guide, strengthen the "area" all the way along the regional logistics infrastructure construction, regional logistics hub city as the core, to the coastal port, station, port of customs clearance for the portal, actively to strengthen the construction of highway, railway, port, logistics, comprehensive through 7 international logistics channels, docking of the countries along interstate and international transportation corridors.

## 4. Conclusions

Among the countries along "The Belt and Road", the infrastructure constructions were perfect, the trade development of transportation industry was not balanced, the developed countries had obvious advantages, and the developing countries were developing rapidly. We established the transportation measures of the countries along "The Belt and Road", which were that China should innovate cooperation mechanism, establish cooperation platforms, play a vanguard role of the civil aviation, set high-speed rail construction as the breakthrough point, get through international logistics corridor, establish international logistics system, set up International Logistics Association, confirm international standards of logistics, construct international logistics information network, and so on. This research took the international perspective and scientific analysis to support the national strategy of "The Belt and Road", and the conclusions will provide a strong theoretical basis and strategic support for the construction of "The Belt and Road".

**Acknowledgements**

Foundation item: National Science and Technology Basic Resources Survey Specialized "Multi-disciplinary Joint Investigation of China-Mongolia-Russia International Economic Corridor" (No.2017FY101300) "Investigation on Economic and Social Development and Investment Environment of China-Mongolia-Russia International Economic Corridor" Project (No.2017FY101304), and National Natural Science Foundation of China (No.41701639). First author: CHENG Hao, Post-doctoral, E-mail: chenghao0011@163.com

**References**

[1] China news 2013 *President Xi Delivered an Important Speech Calling for the Establishment of the Silk Road Economic Belt* (in Chinese) Income accessed online on 11th May 2018 via http//www.chinanews.com/gn/2013/09-07/5257748.shtml

[2] Dong S, Huang Y and Li Z 2014 The Economic Development Pattern and Regional Economic Integration Mode of Silk Road Economic Belt *Resources Science* **36**(12) 2451-8 (in Chinese)

[3] Research group of Institute of Geographical Science and Resources, Chinese Academy of Sciences (RG IGSNRR CAS) 2014 Analysis on the Sustainable Development Mode of the Silk Road Economic Belt *China National Conditions and Strength* **10** 24-6 (in Chinese)

[4] Dong S 1994 *Economic Geography* (Beijing: Science Press) (in Chinese)

Xie X 2015 The Development Trend of Transportation under One Belt and One Road Strategy *Maritime China* **9** 34-5 (in Chinese)

[5] Belt and Road Forum for International Cooperation *2017 President Xi Jinping delivered the opening speech for the round table meeting of the "Belt and Road Forum for International Cooperation"* (in Chinese) Income accessed online on 18th May 2018 via http://www.beltandroadforum.org/n100/2017/0515/c24-415.html
[6] Dong S, Cheng H and Guo P 2016 Transportation Industry Patterns and Strategy of the Belt and *Road Bulletin of Chinese Academy of Sciences* 31(6) 663-70 (in Chinese)
[7] Zhang Q, Cai C 2014 "No Difference Space" and Regional Cooperation under The Concept of "Silk Road Economic Belt" – on The Transportation Cooperation between China and Central Asia *Russian Central Asian & East European Market* **6** 66-124 (in Chinese)
[8] Huang Y 2015 The development opportunity of civil aviation in One Belt and One Road strategy *Air Transport & Business* **3** 13-5 (in Chinese)
[9] Jin X 2015 High-speed Railway Will Be A "One Belt and One Road" Strategic Booster *The Earth* **2** 50-1 (in Chinese)

# Trans-Siberian railway and interrelation of economic development of regions

**A A Cherkashin**

Sochava Institute of Geography SB RAS, Irkutsk, Russia

E-mail: akcherk@irnok.net

**Abstract**. The extending of Trans-Siberian railway is considered as homotopic parameter, i.e. a factor of a geographical location and connectivity of economic space, forming a line of functional order along which the economy of Siberia and the Russian Far East is transformed. It is possible to pass judgment on a similarity of economies by correlating the regional investments on industry and agriculture VRP. For this purpose the economic connectivity (similarity) of regional economies along the railway is calculated with using a Jacobi determinant for vectors of coefficients of investment acceleration. The comparisons are carried out concerning parameters of Irkutsk region economy. The indexes reflecting advantages of the regional environment of economic activity or regional investment climate are calculated. The main spatial and temporary trends of formation of geoeconomic relations of regions are revealed. It is possible to consider Baykal region with intracontinental position and unique natural features as peculiar "reference point" for the comparative quantitative geographical analysis of economic activity conditions.

## 1. Introduction

Presence of the alternate transport paths of movement of people and goods from the East on the West and in the opposite direction was important geohistorical advantage of the Euroasian continent at all times. The movement was on sea and river water ways and numerous roads in the course of migration of Asian people and Russian development of the territory. Construction and exploitation of the Trans-Siberian railway (TSR) had allowed to establish uninterrupted year-round transport connection of regions of Siberia and the Far East with the European part of Russia since beginning of the 20th century. It affected both positive and negative on traditional economic activity of the population and further was important in the history of the whole country and for economic development of its east lands. As a result the economic belt was created as an economic complex of open resource economies of regions arranged in geographical order along TSR and developed in the conditions of the market relations with special place of any region in all-Russian and international division of labor.

Due to the occurrence of fundamental problems of scheduling new transcontinental railways and the related economic belts [1], it makes sense to reveal the created regularities of interregional relations along TSR and feature of the geographical environment of economic activity with use of the modern models of the analysis of territorial development opportunities.

## 2. Models and Methods

The quantitative methods are widely and efficiently used at the solution of transport and economic problems [2]. As a first approximation, the nature of interregional connections can be estimated by means of formulas of intensity of movement on different sites of the railway, taking in account of various factors of influence [3]. Integrated approach in the analysis of space data from various sources [4] is used at a research of regional economy penetrated with TSR on the example of development of the cities.

The Trans-Siberian railway (TSR) is a network of the railways connecting Moscow during more 100 years with east regions of Russia and the countries of the Far East [5]. It is an infrastructure factor of a connectivity of economic space, a homotopic path of the functional order along which the economy of Siberia and the Russian Far East sequentially is transformed.

Homotopy is called the continuous map $F(t,x):[0,1]\times X\to Y$ which is continuously depended on parameter $t\in[0,1]$. Time or the space position of an object can be accepted as parameter $t$. This parameter has initial value (location) $t$=0 and terminal maximal value corresponding to $t$=1. For example, it is possible to take as $t$=0 a starting point of TSR, i.e. the station Moscow-Yaroslavskaya, and as $t$=1 a terminal point Vladivostok which railway station is located apart $L_m$ = 9289 km from Moscow. Parameter $t$ for any location $L$ on this way is calculated by a simple formula $t=L/L_m\in[0,1]$, where $L$ is distance from Moscow by railway.

The space $X$ is interpreted as a set of values of various ($i$-th) economic and ecological reasons, i.e. factors and potentials $x=\{x_i\}\in X$ of territorial development. The space $Y$ corresponds to a set of $k$-th results $f(x)=\{f_k(x)\}\in Y$ of economic activity. The reasons $x(t)$ and results $f(t,x)$ are parametrized with time moments or geographical coordinates of a location $t$. There are $x(0)$ and $f(0,x)$ for Moscow location and $x(1)$ and $f(1,x)$ for Vladivostok position. We will coordinate regions $j$ with values $t_j$, that correspond to their administrative center locations $L_j$ on railway track. In this case the correlation of economies is reflected in comparison of economic assessment functions $z_j=f(t_j,x)$ for different regions. For this purpose the Jacobi determinant [6] is used. For two variables $x_1$ and $x_2$ and two estimations $z_1=f(t_1,x)$ and $z_2=f(t_2,x)$ of different regions $j$=1 and $j$=2 the determinant is equal to

$$J=\begin{vmatrix}\dfrac{\partial z_1}{\partial x_1} & \dfrac{\partial z_1}{\partial x_2}\\ \dfrac{\partial z_2}{\partial x_1} & \dfrac{\partial z_2}{\partial x_2}\end{vmatrix}=\begin{vmatrix}a_{11} & a_{21}\\ a_{12} & a_{22}\end{vmatrix}=a_{11}a_{22}-a_{12}a_{21} \qquad (1)$$

Here $a_{ij}=\partial z_j/\partial x_i$ is the partial derivative expressing sensitivity of change of the function $z_j$ on factor $x_i$; $a_j=\{a_{ij}\}$ is dual vector (covector) of a vector of factors $x=\{x_i\}$.

If the module is $J_m=|J|=0$ in some range of variables $x_1$ and $x_2$, then functions $z_1$ and $z_2$ in this range are interdependent. These functions are independent at case $|J|\neq 0$ [7]. It is supposed, the more deviation the value $J_m$ from zero, the connection of $z_1$ and $z_2$ is worse. The last statement theoretically is proved by linking $J$ with a gramian $\Gamma$ which value is a degree of independence of vectors and functions in indexes of a deviation of positive value $\Gamma$ from zero [8].

It is possible to pass of judgment on a similarity of economies by correlation of investment processes in which the dependence of the investment $z_j$ on sectorial VRP $x_i$ is considered. In this case coefficient $a_{ij}$ has economic meaning of investment process acceleration. The bilinear connection of $z_j$ with variables $x_i$ is proved by theoretical and statistical methods [6]:

$$z_j(x_j)-z_{0j}=\sum_{i=1}^{n}a_{ij}(x_{ij}-x_{0ij}),\ H(a_i)=z_{0j}-\sum_{i=1}^{n}a_{ij}x_{0ij} \qquad (2)$$

where values $z_{0j}$, $x_{0i}$ characterize geographical conditions of economic activity (environmental shift) or investment climate of regions. The value $H(a_i)$ characterizes this environment in a complex. Coefficients $a_{ij}$ are calculated by method of the rolling linear regression on time-series data. In this case the value $H(a_i)$ has meaning of the intercept term in the linear equation. In according to equation (2), we find values $z_{0j}$, $x_{0i}$ with comparing $H(a_i)$ and $a_j = \{a_{ij}\}$.

## 3. Results and Discussion

The explained algorithm is implemented on the example of the research of dependence (2) of gross domestic investments $z_j(x)$ (million rubles/year) on volumes of production in the industrial $x_{1j}$ and agricultural $x_{2j}$ enterprises (million rubles/year) in regions. Calculations are based on indexes (in actual prices) of a social and economic condition and development of territories from the available database of Goskomstat (State committee of statistics) of Russia for period 1999-2013. Calculation of regional coefficients $a_j = \{a_{ij}\}$ is carried out with rolling regression on three points of the corresponding time-series data. On this basis with the equation (1) the indicator $J_j$ of a connectivity of any region economy (along TSR) with Irkutsk region economy by years and root-mean-square value $J_{0j}$ for the entire period are estimated. Hereafter these values are compared with the homotopic parameter $t_j$ of each region for identification of the hidden regularities.

The Irkutsk region is situated almost in the center of the Asian continent on the main railway thoroughfares connecting Europe to the Russian Far East districts and the countries of the Southeast Asia. The Irkutsk city is situated on 5042 km of TSR from Moscow in 126 km from the coast of Lake Baikal (Slyudyanka railway station). Possessing a large volume of unique natural resources the Irkutsk region is industrially developed territory and the industrial center of the Baikal region where also there are the Republic of Buryatia and Zabaykalsky Krai (Trans-Baikal territory, Chita region). The leading industries of Irkutsk region are power and fuel industries, ferrous and nonferrous metallurgy, chemical and petrochemical production, mechanical engineering and metal working, forest, woodworking, pulp and paper industries.

The index of economic similarity $J_{mj}$ in time and regions changes in a different way that indicates an originality of the regional investment process (figure 1). However there are historical periods of high similarity of dependence of the investment volumes on sectorial VRP. There is the period before world financial and economic crisis of 2008. The crisis in the investment sphere in 2008 similarly influenced on economy of Russian regions ($J_{mj}$<2).

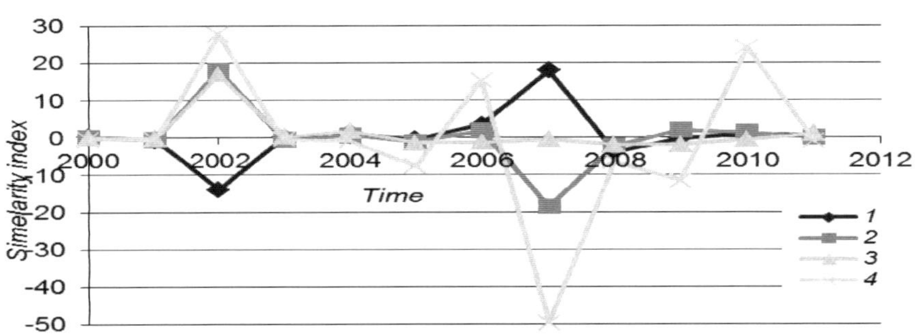


**Figure 1.** Comparison by years of function of dependence of regional investment on the industry and agriculture VRP with function for the Irkutsk region by indicator $J_j$ calculated with using Jacobi determinant and acceleration coefficients of investment process. Points are the calculated indexes. Regions: 1 – Vladimir region; 2 – Omsk region; 3 – Altai Krai (Barnaul); 4 – Primorsky Krai (Vladivostok).

The dependence of mean values $J_{0j}$ on a difference of homotopic indexes $\Delta t=|t_j-t_0|$ (figure 2) is ambiguous. Value $\Delta t_j$ corresponds to the conditional distance by railway between each region and the region of comparison with the index $t_0$ (Irkutsk). Within boundaries of the Baikal region the connectivity index of investments $J_{0j}$ decreases at increasing a distance from the Irkutsk region in the direction to the Republic of Buryatia and Zabaykalsky Krai. The connections $J_{0j}$ < 20, which is weakly dependent on distance for industrial developed regions, are widespread. However the main trend expected is expressed in a positive linear dependence of the connectivity on distance and accordingly the homotopic index $\Delta t_j$. This trend is unidirectional, but link lines are shifted for different regions. The dependence is described by the equation

$$J_{mj} = a(\Delta t_j - n_j \Delta t_0) \quad (3)$$

where $a$, $\Delta t_0$ are constants; $n_j$ is the quantitative characteristic of the geographical investment environment. There are $a$=205.7, $\Delta t_0$ = 0.164 per calculation. The characteristic of the geographical environment is evaluated by an equation $n_j = (a\Delta t_j - J_{mj})/a\Delta t_0$. The value $n_j$ increases both to the East and the West and depends on distances to the Baikal region, especially the Republics of Buryatia (figure 3).

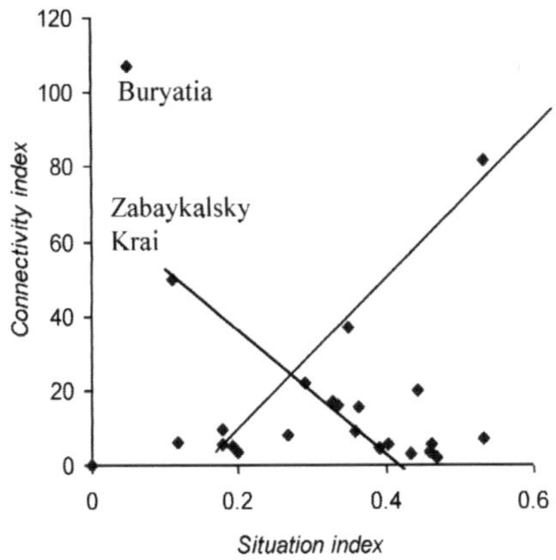


**Figure 2.** Dependence of mean values of connectivity index $J_{0j}$ of economies on the relative geographical location $\Delta t_j$. Straight lines are the main trend of interconnection of indexes.

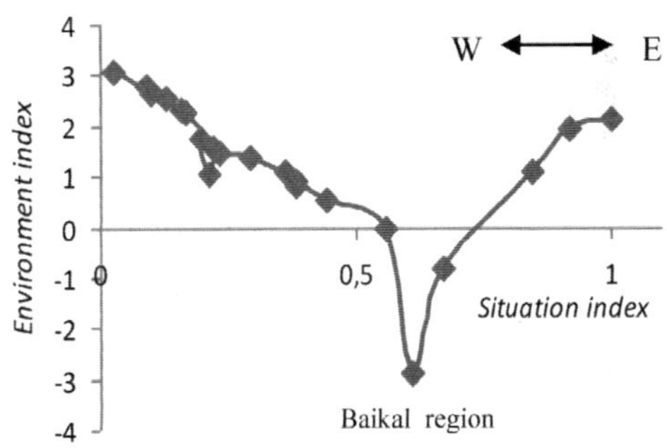


**Figure 3.** Dependence of the calculated characteristics of the geographical environment $n_j$ in regions on index of space situation $t_j$. Curve line is the space trend of index $n_j$ changing.

This regularity is appeared in compliance with the equation (3) as result of low space variability of the connectivity index of economies. Developing economy of Transbaikalia performs in Russia important transit function, connecting east and western regions and neighboring states, first of all with Mongolia and China. High freight density of traffic flows by the railroads in comparison with other regions of Siberia and the Far East is characteristic of this territory. Both midland and peripheral position and a lack of energy resources constrain growth rate of the economy. It is expressed in low negative values of the investment environment. The factor of a proximity to the political, administrative and economic center of Russia (to Moscow) and to seaports defines higher rates $n_j$ of the geographical environment of regions.

Using the equation (3) the linear dependence of connectivity indexes of economies of two any regions $s$ and $j$ is established:

$$J_{ms} - J_{0ms} = \frac{n_s}{n_j}(J_{mj} - J_{0mj}), J_{0mj} = a\Delta t_j \quad (4)$$

where $J_{0mj} = a\Delta t_j$ is the spatial and environmental shift of an connectivity index. The ratio $K_{js} = n_s/n_j$ is a refractivity of geographical environments at transition from one region to another one. This is simple analogy of refraction of a light ray on borders of physical mediums in the geometrical optics.

Refraction coefficient $K_{js}$ between the neighboring regions of Siberia and the Far East does not increase with value $\Delta t_j$, i.e. the spatially close regions are uniform in characteristics of the economical and geographical environment.

## 4. Conclusion

For the identification of the hidden regularities of the space organization of territories of Siberia and the Russian Far East the economic connectivity (similarity) of regional economies in the line of TSR is estimated using the mathematical models and methods. The extending of TSR is considered as an infrastructure factor of a connectivity of economic space, a homotopic pathway of the functional regularity along which the economy of east regions of Russia sequentially is transformed and the international transport corridor between the countries of the southerneast Asia and Europe is created. In geohistorical aspect from the TSR line in the directions to the North and the South were constructed and now is formed the parallels and meridians of the railroads for economic development of the countries of the Asian continent.

The analysis of similarity is based on comparison of economic functions of dependence of internal investments in the region on industry and agriculture VRP. Statistically coefficients of acceleration of investments in each group of economic branches are evaluated. On the basis of these coefficients using Jacobi determinant the indicator of interregional connectivity of investment functions is calculated.

The economy of regions with these functional indexes was compared to economy of the Irkutsk region (part of the Baikal region) possessing unique natural resources and the developed industry. In the temporary and spatial relation the connectivity indicator is very changeable. In order to identify the quantitative regularities the indicator of the geographical environment is evaluated. The indicator increases to the West and the East from the Baikal region. It makes sense to consider this region possessing intracontinental position and unique natural features as peculiar "reference point" for the comparative quantitative geographical analysis of economic activity conditions.

**Acknowledgements**

The research is supported by the project "Innovative development, territorial organization, and quality of life improvements for the population of the Siberian and Arctic regions of Russia" funded by the Russian Foundation for Basic Research N 16-02-00570(a).

## References

[1] Bazarov B V 2017 "A new silk road": to statement of problems of strategic interaction of Russia and China *Power* **25**(11) 7–12 (in Russian)

Nobis A 2017 The New Silk Road, old concepts of globalization, and new questions *Open Cultural Studies* **1** 203–13

[2] Golts G A and Kartavenko G G 2008 Macroeconomy and transport: intercountry comparisons and regularities *Bulletin of Transport Information* **11** 25–35 (in Russian)

Golts G A 2009 Turnpike freight transportation and gross internal product: an historiometric research for the purposes of prediction *Problems of Prediction* **113**(2) 151–7 (in Russian)

[3] Mekhedov M I and Muginstein L A 2015 On problems of the organization of driving and effectiveness of use of traffic capacities of the station *Railway Transport* **7** 20–7 (in Russian)

[4] Uchida S, Takeuchi W, Hatoyama K and Mazurov Yu 2016 Socio economic impact of Trans Siberian railway after the collapse of Soviet Union by integrated spatial data analysis *IOP Conference Series: Earth and Environmental Science* **37** 1–6

[5] Mikhailoff M 1900 The Great Siberian railway *The North American Review* **522**(170) Income accessed online on 18th July 2018 via https://enwikisourceorg/wiki/The_North_American_Review/Volume_170/Issue_522/The_Great_Siberian_Railway

Westwood J N 1964 *History of Russian Railways* (London: George Allen and Unwin Ltd) 326 p

Westwood J N 2002 Soviet Railways to Russian Railways *Studies in Russian and East European History and Society* (London: Palgrave Macmillan) p 262

Trans Siberian Railway 2015 *Lonely Planet Publications* pp 19-38. Income accessed online on 15th July 2018 via http://medialonelyplanetcom/shop/pdfs/3105

Liliopoulou A M, Roe M and Pasukeviciute I 2005 Trans Siberian Railway: from inception to transition *European Transport* **29** 46–56

[6] Myadzelets A V and Cherkashin A K 2016 Spatial and temporary indicators for comparison of conditions of economy development in regions of Russia *Regional Researches* **3** 22-31

[7] Korn G A and Korn T M 1968 *Mathematical Handbook for Scientists and Engineers: Definitions, Theorems, and Formulas for Reference and Review* (New York, San Francisco, Toronto, London, Sydney: McGraw-Hill Book Company) p 1130

[8] Courant R and Hilbert D 1953 *Methods of Mathematical Physics* Vol 1 (New York: Interscience Publishers Inc) p 560

# Assessment of connectivity of Russian cities with passenger air transport links with Mongolia and China

**I A Dets**

Sochava Institute of Geography, Siberian Branch, Russian Academy of Sciences, Irkutsk, Russia

E-mail: igordets@ya.ru

**Abstract**. Aviation passenger traffic is becoming increasingly important in international economic relations. The development of the China-Mongolia-Russia economic corridor is no exception. Despite the fact that the media focuses on projects for the construction of the new high — speed Railways and oil and gas pipelines, a significant part of economic cooperation is implemented through air transport links. The study examines all the air routes between Russia and China, as well as Russia and Mongolia, available in the peak summer schedule of 2018. All routes connect 13 cities in Russia and 30 cities of China and the capital of Mongolia, forming a separate air corridor between the three countries studied. The population of cities is certainly one of the essential factors determining the level of development of aviation ties. Nevertheless, other factors play an important role: the high centrality of the Russian economy, the dominance of the Moscow air hub as a transfer point, the weak development of interregional air transport and others. Among the identified trends, it is a high proportion of flights to Irkutsk from Mongolia and China — with not so significant economic potential of the city, the main factor determining the leading position of the city was the tourist attraction of the region for travelers from the countries in the study.

## 1. Introduction

The intensive development of economic and social relations in the economic corridor "China-Mongolia-Russia" is reflected in the increase of transport links between the states. In recent years, the construction of the first oil trunk pipeline connecting Russia and China ("Eastern Siberia — Pacific Ocean" with a branch to China) was completed, the construction of the "Sila Sibiri" ("Power of Siberia") gas pipeline and bridges across the Amur River was started. Other projects linking China and Russia, including the development of freight and passenger transport by high-speed rail, are also being discussed. The discussion of the perspective construction of gas transport infrastructure on the route Russia-Mongolia-China has also started. A significant part of the transport projects is supported by the Chinese initiative "One Belt and One Road".

Nevertheless, passenger traffic, currently mainly aviation, remains a vivid reflection of the existing economic ties between the territories. Passenger aviation flows illustrate the existing network of connections, its saturation, and highlight the cities important for economic relations within the framework of the economic corridor being formed. Undoubtedly, it is not an easy task to divide passenger traffic into tourist and business traffic, which have different economic significance. Nevertheless, it should be taken into account that tourist flows not only do create the potential for economic growth of territories, but also form a common cultural image of countries, contributing to the development of further international cooperation.

## 2. Data and Methods

For the purposes of the study, all Russian airports with regular flights to the airports of Mongolia and China were selected. The largest number of regular flights in 2018, as in previous years, fell on the period from May to October, due to the more intense tourist flows in the summertime (the most frequent flights are in June, July and August). In the schedule of each airport, the weekly frequency of flights in different directions was analyzed, and adjacent flights were carefully verified. A fragment of the generated database is presented in table 1, which shows the number of flights (one way) for the 4 most intensive destinations between Russia and China / Mongolia (more than 20 flights).

**Table 1.** Number of weekly flights (one way) between Russian cities and the most connected cities of China and Mongolia.

| | People's Republic of China | | | Mongolia | All flights |
|---|---|---|---|---|---|
| | Beijing | Shanghai | Urumqi | Ulaanbaatar | to these cities |
| **Moscow** | 41 | 28 | 9 | 10 | 88 |
| **Saint-Petersburg** | 7 | 10 | 4 | - | 21 |
| **Yekaterinburg** | 4 | - | - | - | 4 |
| **Novosibirsk** | 10 | 1 | 6 | - | 17 |
| **Krasnoyarsk** | 2 | - | - | - | 2 |
| **Irkutsk** | 11 | 2 | 2 | 8 | 23 |
| **Ulan-Ude** | 2 | - | - | 2 | 4 |
| **Chita** | 2 | - | - | - | 2 |
| **Khabarovsk** | 2 | - | - | - | 2 |
| **Vladivostok** | 6 | 2 | - | - | 8 |
| All flights to Russian cities | 81 | 41 | 21 | 20 | - |

Among Russian cities, only 13 have scheduled flights with China and Mongolia in the summer flight schedule in 2018. Flights connect Russia with 30 cities in the PRC and the capital of Mongolia, i.e. on average, each Russian city accounts for 2.38 cities in China and / or Mongolia. In fact, the number of connections varies from 1 to 23 cities in the presence of only three cities with more than five connections (Moscow — 23, Irkutsk — 12, Novosibirsk — 6). St. Petersburg, Krasnoyarsk, Vladivostok and Khabarovsk had links with four cities; the other half of Russian cities had fewer connected points. Chinese cities have significantly lower connectivity rates to Russian cities, as the average number of connected cities is 2.3. The distribution of the number of routes between Chinese cities is less centralized: most Chinese cities are connected with one city in Russia. Only a few cities have more connections: Hong Kong — 4, Urumqi and Shanghai — 5, Harbin — 6, Sanya and Beijing — 9. Ulaanbaatar is connected with only 3 cities in Russia.

The apparent imbalance in the connectivity of cities is even greater when taking into consideration the weekly number of flights. The average number of flights per week for Russian city is 22.9, but of the total number of flights is about 300, almost half is in Moscow (49.4%). Irkutsk, which has half the number of connected cities, provides almost 3 times fewer flights per week (18.2%), and St. Petersburg and Novosibirsk have almost equal number of flights per week (7.7 and 7.4%, respectively). Further in the list of the number of flights are located Vladivostok (4.7%), Khabarovsk (3.0%), Krasnoyarsk and Yekaterinburg (2.7% each). For Chinese cities, the average number of flights per week is lower — 9.6. Beijing accounts for less than 1/3 of Chinese flights to Russia (31.4%), Shanghai — 15.5%, Urumqi — 7.6%, Guangzhou — 6.5%.

Based on the obtained database, graphs and maps with diagrams were created to analyze the existing economic and geographical factors and trends. Special attention was paid to the impact of economic development of cities and tourist flows. The study also took into account the number of unique connections and the share of the city in flights in all directions.

## 3. Results and Discussion

The processed data allow us to present a map of the main regular air passenger links formed between Russia, Mongolia and China to date. The map shown in figure 1 represents all cities that have regular passenger flights with the rest of the countries in the study. The size of the symbol of the city allows to estimate the total number of flights to the countries of the study per week, the most intensive links between the cities are also displayed.

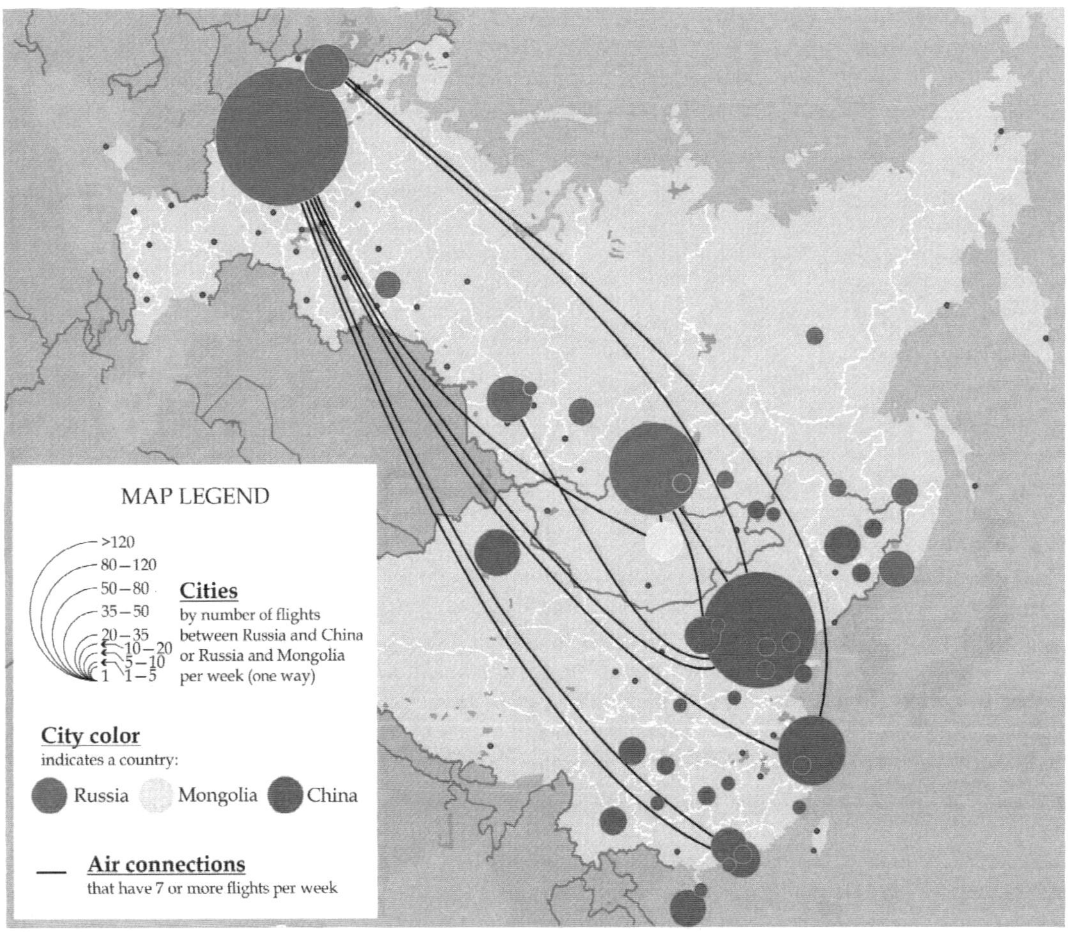


**Figure 1.** Cities of Russia, Mongolia and China with aviation links with the countries of study and the most intensive air routes with daily or more frequent flights.

Among the general trends, it is worth noticing first the over-concentration of communication with China and Mongolia in the European part of Russia: Moscow as the country's largest agglomeration draws on almost all aviation contacts with the countries of the study. Only St. Petersburg from the cities of the European part of Russia has its own communication with China, the next city in Russia with direct flights (Yekaterinburg) is already on the border with the Asian continent. In China, the concentration of air communications is not so noticeable-in several cities located close enough to Beijing there is a direct air communication with Russia.

In total, only 11 destinations between the countries of the study have more than seven departures per week (i.e., an average of 1 per day). In Russia, five of them are carried out in Moscow, three in Irkutsk, two in St. Petersburg and only one in Novosibirsk. Two of these flights go to Mongolian Ulaanbaatar, four to Beijing, two to Shanghai, one each to Hong Kong, Guangzhou and Ordos.

Among other things, the relationship between the number of flights per week from the Russian city and the size of its gross city product was analyzed. The linear correlation coefficient for these two indicators is 0.92, which is attributed to a very strong relationship. However, the urban gross product indicators have not been calculated for all the cities presented in the study in the existing statistical sources, which limits the possibility of using this indicator. The correlation between the number of flights and the population of Russian cities was almost as strong — 0.9. The use of this indicator is more preferable, even though the Moscow population indicator makes the graphical way of displaying the relationship less obvious — most cities form a very dense group of points due to the huge difference in population with Moscow (see figure 2).

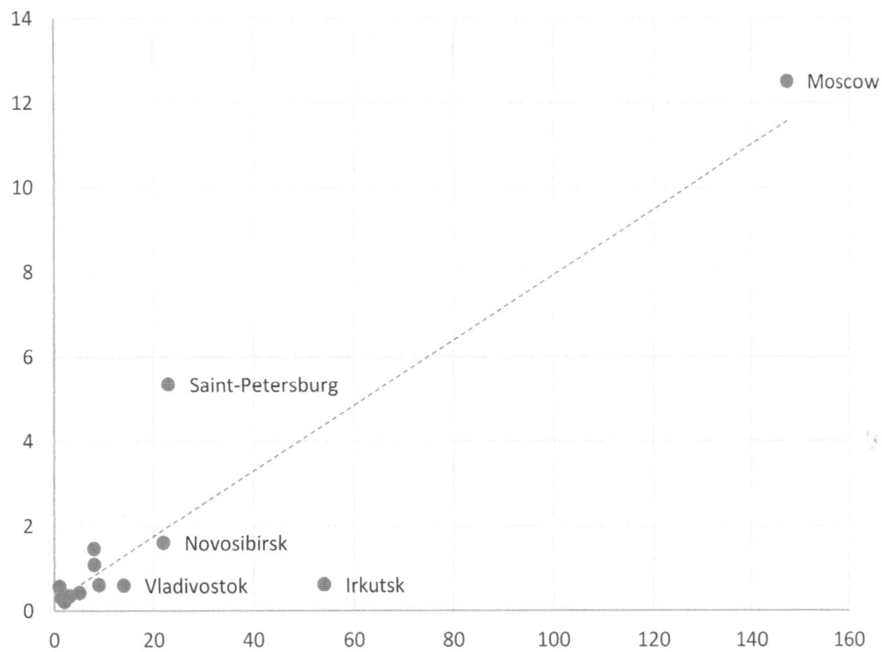


**Figure 2.** The relationship between the urban population of Russian cities (vertical axis, million) and the weekly number of flights to China and Mongolia (one way).

Especially noticeable deviation from the average in Irkutsk — with an available population of about 620 thousand people, the number of departures per week is 54, which is 9–11 times greater than that of similar Russian cities. Such a great importance of the city in the structure of the Russian-Chinese air traffic is primarily due to the significant tourist attractiveness of Lake Baikal in China. Tourists come to the city from different parts of China (see figure 3), thus Irkutsk forms a dense network of air links with this country. Irkutsk is also connected with two directions unique for Russian cities — Shenyang and Ordos do not have other regular air routes to Russia. Nevertheless, the frequency of air traffic shows that the strongest ties in Irkutsk are with the Northern territories, and such directions as Shanghai and Hong Kong, which are the largest financial centers of China, do not send a large number of tourists to the region. Irkutsk's special role in relations with Mongolia is also notable - the number of flights to Ulaanbaatar is only slightly less than between this city and Moscow.

Irkutsk has a long tradition of close economic and cultural ties with Mongolia, which continues to affect its place in the aviation communication with this country.

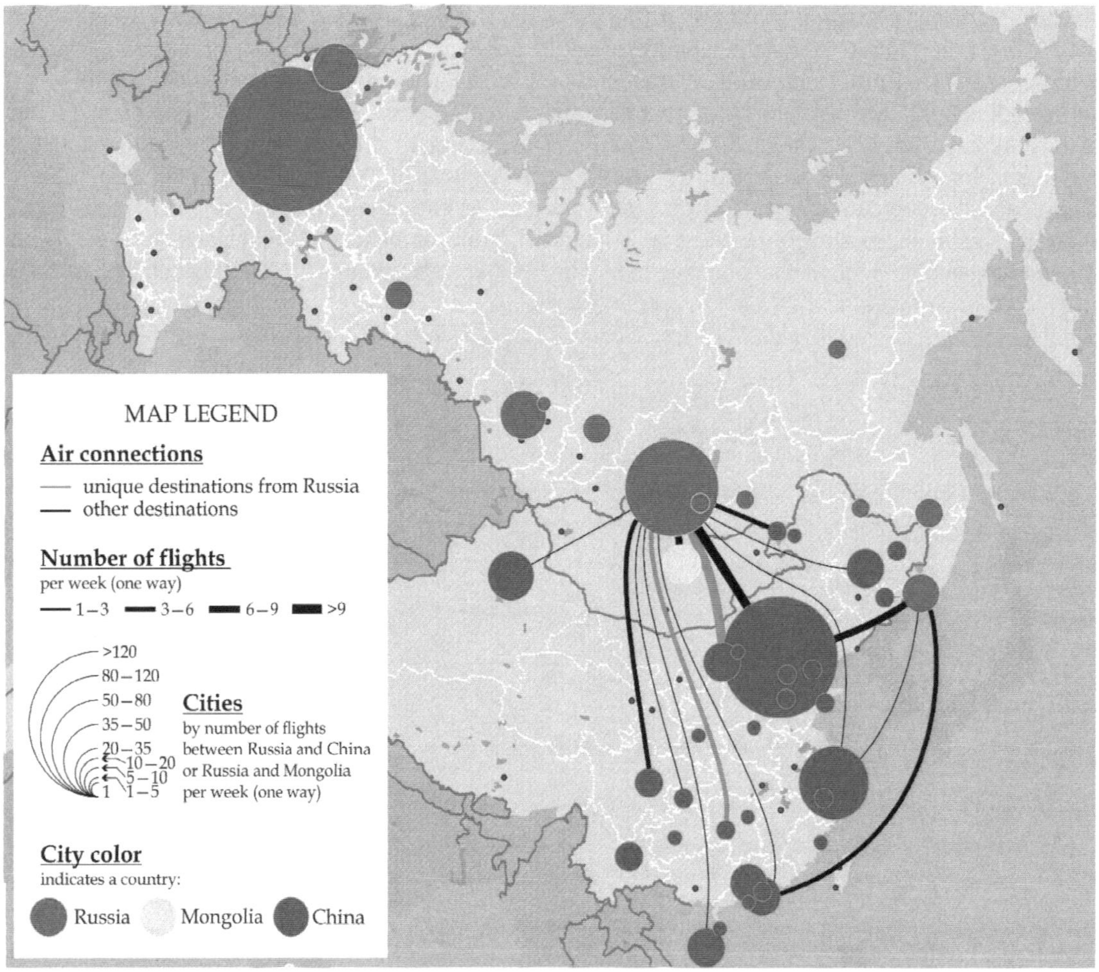


**Figure 3.** Regular passenger air routes from Irkutsk and Vladivostok to the cities of China and Mongolia

Another important regional centre in the East of Russia — Vladivostok. This city has a much smaller number of regular flights to China, and has no links with Mongolia at all. Nevertheless, the city is connected not only with the most important cities of China: Beijing, Shanghai and Hong Kong, but also has a unique destination in Mudanjiang.

The relations of the two remaining Russian cities with the most intensive air traffic with China are shown in figure 4. Novosibirsk has 3.5 times fewer residents than St. Petersburg, but the number of regular routes to China is greater. The main economic center of Siberia is connected with the most important cities of China, as well as with Harbin, Urumqi and the resort city of Sanya. Such a good connection with the Chinese cities is due not only to the relative territorial proximity to this country, but also a kind of central factor of the Novosibirsk airport for neighboring major cities. This effect leads to the fact that of all the surrounding cities (among which several have a population of more than 0.5 million people) only Tomsk has regular flights with the only city in China — Sanya.

St. Petersburg is the second largest city in Russia and one of the most important tourist centers of the country. However, direct regular flights to China are only available from Beijing, Shanghai,

Urumqi and Chengdu. An important feature that limits the development of air communication with China, in this case is also the predominance of the aviation hub of Moscow. At the same time, Chinese travelers not only go to Moscow as a transit center, but also often visit both cities in one trip to Russia, which also reduces the need for direct flights to St. Petersburg.

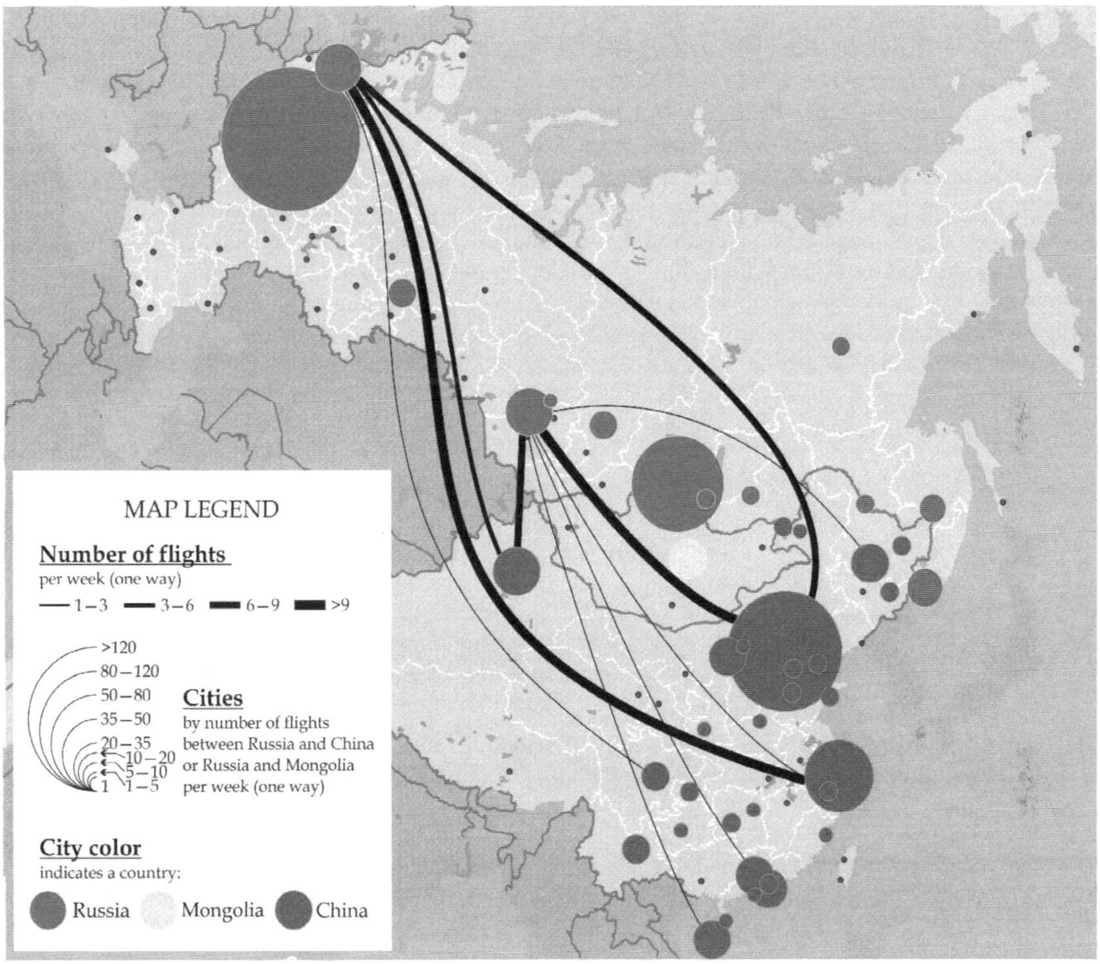


**Figure 4.** Regular passenger air routes from St. Petersburg and Novosibirsk to the cities of China and Mongolia

The importance of the development of air communication between Russia, Mongolia and China has not yet found its place in the discussions on the project "one Belt and one Road". Currently, the main focus is on the construction of a high-speed railway from China to Europe via Russia [1]. While the development of domestic and international air traffic can be expected to have a positive effect on the growth of employment and gross national product [2] especially on the example of low-cost airlines [3] the development of high-speed rail services may not always lead to an unambiguous positive economic effect [4].

Another important area for further research may be the impact of air traffic between Russia, Mongolia and China on the development of tourism. As we can see on the example of Irkutsk, tourism is now one of the most important factors in the development of air traffic. However, further trends may be subject to significant changes under the influence of many factors [5]. Pop culture and destination images can also become such factors [6]. Nevertheless, in reality, only after a certain time it will be possible to determine the real effect of the established relationship: whether it will be a joint co-evolution, whether the air communication will facilitate the development of trade or no significant effect for the connected cities and economies will be achieved [7].

## 4. Conclusion

In general, the results of the study show the presence of especially intensive independent (not passing through Moscow) ties with China and Mongolia in the Asian part of Russia with a prominent role of Irkutsk. At the same time, the important place of Irkutsk in the relations between Russia and Mongolia has its roots from the Soviet time, while the intensification of the city's relations with China occurred in recent years and is largely due to the growth of the tourist flow to Baikal.

The role of the Moscow aviation hub in the established regular air service is also evaluated as highly concentrated. This effect suppresses the development of almost all regional airports, which leads to the fact that significant air links with China appear only in the Asian part of Russia. The development of economic relations between Russia and Mongolia at the level of regular air passenger traffic is limited to three cities with the complete domination of Moscow and Irkutsk.

**Acknowledgements**

The study was funded within the framework of the Scientific Project No. 0347-2016-0006 "Socio-geographical factors of ecological and socio-economic transformations in the Siberian regions from the perspective of sustainable development".

## References

[1] Dong S, Yang Y, Li F, Cheng H, Li J, Bilgaev A, Li Z and Li Y 2018 An evaluation of the economic, social, and ecological risks of China-Mongolia-Russia high-speed railway construction and policy suggestions *J. Geogr. Sci.* **28**(7) 900–18 doi:10.1007/s11442-018-1512-y

Ostrovskii A V 2017 Prospects to interface the Silk Road economic belt and the Eurasian Economic Union projects *Her. Russ. Acad. Sci.* **87**(6) 480–90 doi:10.1134/S1019331617060041

[2] Irwin M D and Kasarda J D 1991 Air Passenger Linkages and Employment Growth in U.S. Metropolitan Areas *Am. Sociol. Rev.* **56**(4) 524 doi:10.2307/2096272

Fernandes E, Pacheco R R 2010 The causal relationship between GDP and domestic air passenger traffic in Brazil *Transp. Plan. Technol.* **33**(7) 569–81 doi:10.1080/03081060.2010.512217

[3] Donzelli M 2010 The effect of low-cost air transportation on the local economy: Evidence from Southern Italy *J. Air Transp. Manag.* **16**(3) 121–6 doi:10.1016/j.jairtraman.2009.07.005

[4] Zhang Y and Zhang A 2016 Determinants of Air Passenger Flows in China and Gravity Model: Deregulation, LCCs, and High-speed Rail *J. Transp. Econ. Policy* **50**(3) 287–303 http://dx.doi.org/10.2139/ssrn.2775501

[5] Spasojevic B, Lohmann G and Scott N 2018 Air transport and tourism – a systematic literature review (2000–2014) *Curr. Issues Tour* **21**(9) 975–97 doi:10.1080/13683500.2017.1334762

Andrades L and Dimanche F 2017 Destination competitiveness and tourism development in Russia: Issues and challenges *Tour Manag.* **62** 360–76 doi:10.1016/j.tourman.2017.05.008

Sheresheva M and Kopiski J 2016 The main trends, challenges and success factors in the Russian hospitality and tourism market *Worldw. Hosp. Tour. Themes* **8**(3) 260–72 doi:10.1108/WHATT-02-2016-0004

[6] Whang H, Yong S and Ko E 2016 Pop culture, destination images, and visit intentions: Theory and research on travel motivations of Chinese and Russian tourists *J. Bus. Res.* **69**(2) 631–41 doi:10.1016/j.jbusres.2015.06.020

[7] Van De Vijver E, Derudder B and Witlox F 2014 Exploring causality in trade and air passenger travel relationships: the case of Asia-Pacific, 1980–2010 J. Transp. Geogr. **34** 142–50 doi:10.1016/j.jtrangeo.2013.12.001

IOP Conf. Series: Earth and Environmental Science **190** (2018) 012052 doi:10.1088/1755-1315/190/1/012052

# The economic corridor "China-Mongolia-Russia": problems and development prospects

**Z Dondokov**

Buryat Scientific Centre, Siberian Branch of the Russian Academy of Sciences, Ulan-Ude, Russia

E-mail: dzorikto@mail.ru

**Abstract.** The article describes the problems of creating the economic corridor China-Mongolia-Russia. The issues of interaction between the three countries on joining the Chinese initiative of the economic belt of the Silk Road, the Mongolian concept of the "Steppe Way" and the Russian idea of building the Eurasian Economic Union are studied. The problems of implementation of the "Program of creation of economic corridor China – Mongolia – Russia" are revealed. The necessity of specification of the list of investment projects and determination of the membership of the participants is substantiated. The expediency of the creation of a permanent tripartite intergovernmental Commission for the implementation of the Program of creation of the China – Mongolia – Russia economic corridor, headed by high officials of each country, is justified. The commission is proposed to allocate several subcommissions in various areas, including transport infrastructure, trade, energy, industry, humanitarian cooperation and other areas of activity. The necessity of the establishment in each of the three countries of the state commissions for the implementation of the Program, the activities of which should be aimed at the preparation of specific projects, the appointment of stakeholders, the involvement of specialists and experts is substantiated. The expediency of placing the state commissions of Russia and China in the administrative centers of the regions, most connected with the economic corridor is shown in Ulan-Ude (the Russian Federation) and Hohhot (China) is pointed out.

## 1. Introduction

In the context of globalization, the creation of economic corridors as instruments of international cooperation in transport, logistics, trade and other spheres of activity is an important area of development of neighboring countries [1].

Several corridors are being promoted through regional cooperation initiatives in sub-Saharan Africa [2]. In the Asia-Pacific region, regional corridors are being developed with the support of the Asian development Bank [3]. The creation of economic corridors with neighboring countries in the framework of the development of Eurasian cooperation is an urgent task for Russia [4]. In this context, the initiative to create an international economic corridor China-Mongolia-Russia, approved by the leaders of the three countries, is very important. It fits in with the plans of trilateral cooperation, allows to "dovetailing" regional development strategy: the Chinese initiative of the economic belt of the Silk Road, the Mongolian concept of "Steppe way" and the Russian idea of building the Eurasian economic Union [5]. The creation of a trilateral economic corridor can become a strong impetus for the development of the adjacent Siberian regions [6]. The increase in freight traffic will increase the efficiency of the use of the Trans-Siberian Railway, highways and other infrastructure.

## 2. Models and Methods

The author analyzes the factors influencing the efficiency of the economic corridor China-Mongolia-Russia.

The terrestrial nature of the corridor reduces its competitiveness in comparison with sea communications: the delivery of cargo for long distances by water transport is much cheaper. Thus, the opening of the Suez Canal made the transportation of tea to Central Russia by sea more profitable than by the Great tea route [7]. In general, relatively high transport tariffs on the land route reduce its attractiveness for interested parties (trade companies, buyers, etc.).

The competitive advantages of the China-Mongolia-Russia economic corridor are the security of the transport route running through conflict-free areas, as well as the high speed of cargo delivery [8]. It is necessary to study the market for possible transport by type of transport, distance, volumes and types of cargo, as well as comparative analysis with alternative transport routes taking into account safety, speed of delivery and insurance costs [9].

A significant part of the economic corridor passes through Mongolia and Siberian regions of Russia, characterized by low population density and low economic potential. Therefore, the activities carried out within the framework of the tripartite corridor have little impact on the development of the respective territories. On the contrary, in the densely populated regions, the creation of transport communications (railroads and highways, bridges, ports, etc.), aimed at increasing transit traffic, is a powerful impetus for the development of the local economy. The relatively low potential for economic activity in the territories adjacent to the economic corridor China-Mongolia-Russia reduces the attractiveness of relevant investment projects for potential investors, including the Asian Bank for Infrastructure Investments, the Asian Development Bank.

Due to the high cost of investment projects to create new transport infrastructure, including railway and road sections, it is advisable to concentrate on the main transport corridors. By rail, this route is Erlian-Ulan-Bator-Naushki-Ulan-Ude [10].

By road, the main route is Erlian-Ulan-Bator-Kyakhta-Ulan-Ude. The Eastern and Western corridors, including the Trans-Baikal territory and, accordingly, the Republic of Tuva, are complementary and can be developed in the implementation of relevant investment projects.

The effectiveness of the economic corridor depends to a large extent on institutional factors, including the existence of a stakeholder engagement mechanism.

In 2016, the heads of the three States signed the "Program for the creation of the China – Mongolia – Russia economic corridor" (hereinafter – the Program), which aims to provide conditions for the development and expansion of trilateral cooperation [11].

The Program provides for the implementation of joint projects aimed at increasing trade, ensuring the competitiveness of products, facilitating cross-border transport and infrastructure development.

The projects cover the construction of a cross-border highway and railway, the development of transit transport, the expansion of cooperation in the industrial, energy, tourism, educational and scientific and technical sectors, in the field of agriculture, environment, health and humanitarian exchanges.

In accordance with the Program, from each country, the authorized executive bodies responsible for its implementation are established - the Ministry of Economic Development of the Russian Federation, the Ministry of Foreign Affairs of Mongolia, the State Committee for Development and Reform of the People's Republic of China. The Program contains a general provision on the mechanism of its implementation: the authorized bodies should meet at least once a year to monitor the implementation of the Program and projects, as well as to agree on the necessary measures for their further implementation.

At the same time, at present the Program is a framework document specifying the goals, directions and principles of interaction. Most of the declared projects are not, in fact, projects. It is necessary to specify them, to allocate investment projects and to define the list of participants.

The author identifies the main organizational problems of the implementation of the China-Mongolia-Russia Economic Corridor Program. First, there are significant difficulties in interaction organizing between stakeholders. It is not always possible to establish cooperation between the state, business structures, public and expert organizations within one country. It is much more difficult to do this at the international level. Moreover, all three participating countries - Russia, Mongolia and China, vary greatly in economic potential, state structure, development goals and priorities. Difficulties in interstate communications are significant, as the state languages belong to different language groups. Secondly, the Program is multidisciplinary and covers issues of interaction in various fields of activity - from transport infrastructure to humanitarian cooperation. All this complicates the implementation of the Program and determines the need to create appropriate mechanisms.

It should be noted that these organizational problems, with all the features in general, are of a general nature. They were typical for the creation of other economic corridors, including the Great Mekong, the Central Asian Regional Economic Cooperation, and the Economic Corridor of Bangladesh - China - India - Myanmar.

The unique features of the studied economic corridor include:

- participation in the corridor of the two largest countries in the world-Russia and China;
- the transit character of Mongolia;
- a huge length of the territory of one country - Russia.

The large scale of the economic corridor determines the existence of a specific problem of coordination of activities between the central and regional governments of Russia and China.

## 3. Results and Discussion

First, it is reasonable to establish a permanent trilateral intergovernmental Commission on the Implementation of the China – Mongolia – Russia economic corridor Program, headed by high officials of each country. It should be composed of several sub-commissions (working groups) in the following areas: transport infrastructure, trade, energy, industry, humanitarian cooperation, etc. This will allow solving issues of trilateral cooperation between the countries quickly and effectively.

The main task of the tripartite commission will be selection of projects and executors, identification of funding sources. The projects are expected to be financed from public and private investments, as well as from the resources of international financial institutions that were established relatively recently: the Asian Bank for Infrastructure Investments, the New Development Bank of BRICS, the Interbank Association of the Shanghai Cooperation Organization (SCO), the Silk Road Fund. In order to provide scientific and expert support to the projects, the Association of Expert Centers of Russia, Mongolia and China (hereinafter - the Association) should be involved in the work of the intergovernmental commission, whose activities should be aimed at collecting, processing and analyzing information on the problems of the China-Mongolia-Russia economic corridor.

Secondly, for the qualitative and timely preparation of documents related to the implementation of the Program, in each country of the three countries it is necessary to create state commissions for the implementation of the Program. In our opinion, these commissions could be headed by representatives of the authorized executive bodies responsible for the implementation of the Program - the Ministry of Economic Development of the Russian Federation, the Ministry of Foreign Affairs of Mongolia, and the State Committee for Development and Reform of the People's Republic of China. The main task of the commissions could be the preparation of documents for the meetings of the tripartite intergovernmental commission. Their activities should be aimed at the preparation of specific projects, the identification of stakeholders, the involvement of specialists and experts, etc.

In order to promptly resolve the issues, the state commissions of Russia and China with the corresponding state should be located in the regions, most connected with the economic corridor in

Ulan-Ude (Russian Federation) and Hohhot (China). The Russian commission should include a plenipotentiary representative of the President of the Russian Federation in the Siberian Federal Region, governors and heads of interested subjects of the Russian Federation, heads of territorial departments of federal agencies related to the development of the economic corridor, representatives of the expert center of Russia on the economic corridor. One of the tasks of the Commission should be coordination of activities for the preparation and implementation of projects between the participants of the Program-Federal and regional governments, business structures, foreign partners, investors, etc.

Thirdly, it is necessary to build in the system some activities for the implementation of the Program. The author suggests a sequence of stages in the reverse order in time.

1. The most important event for the implementation of the Program is the meeting of the heads of Russia, Mongolia and China. In 2014-2017, such meetings were held within the framework of the summit of the SCO member states.

2. In order to prepare for the summit meetings, in our opinion, it is necessary to create and hold a tripartite intergovernmental commission for the implementation of the Program with the participation of representatives of the Association of Expert Centers of China, Mongolia and Russia. It seems expedient to hold preliminary meetings of think tanks within the framework of the Association Forum.

3. In each participating country the meetings of state commissions should be held for the implementation of the Program with the participation of representatives of the Association.

4. Preparation of documents for the meeting of the state commission. In each participating country the collection and preliminary selection of projects and proposals for the implementation of the Program are to be organized and conducted.

For the successful implementation of the Program, in addition to the institutional mechanism, it is necessary to determine the financial mechanism, including the involvement of public, private and international financial institutions. Especially, it is necessary to highlight the legal and regulatory framework of the Program, paying particular attention to interstate interaction in various fields.

## 4. Conclusion

The Program for the creation of the economic corridor China-Mongolia-Russia is a large-scale and at the same time a framework document. The need to fill the Program with concrete projects determines the urgency of the task of developing mechanisms for its implementation. For coordinated interaction between the three countries, it is advisable to create a tripartite intergovernmental Commission with the allocation of sub-commissions in certain areas. It is proposed to organize and streamline the interaction of the representatives of the Association of Expert Centers of Russia, Mongolia and China with the authorities of the respective countries.

Creation of clear mechanisms for the implementation of the Program, including the sequence of organizational arrangements, is a prerequisite for its success.

**Acknowledgements**

The work was supported by the comprehensive program of fundamental scientific research of the SB RAS No. II.1

## References

[1] Asian Development Bank 2014 *Economic Corridor Development for Inclusive Asian Regional Integration: Modeling Approach to Economic Corridors* Income accessed online on 22th May 2018 via https://think-asia.org/bitstream/handle/11540/63/economic-corridor-development. pdf? sequence=1

[2] Srivastava P 2011 Regional Corridors Development in Regional Cooperation *ADB Economics Working Paper Series* **258** Income accessed online on 10th June 2018 via https://www.adb.org/sites/default/files/publication/28889/ewp-258.pdf

[3] Parshin P 2014 Projects for the development of the Mekong River subregion *Analytical Reports of the Institute of International Studies* (Moscow: Moscow State Institute of International

Relations Press) Income accessed online on 25th May 2018 via http://imi-mgimo.ru/images/pdf/analiticheskie_doklady/AD443.pdf
[4] Atanov N 2017 Eurasian doctrine in practical implementation of development prospects *Problems of Modern Economy* **1** 6-12
[5] Uyanaev S 2016 Russia - Mongolia - China: activation of tripartite rapprochement (the outcome of the Second Forum of the Association of Experts Centers of Russia, Mongolia and China) *Problems of the Far East* **6** 30-5
[6] Borisov G, Dondokov Z and Namzhilova V 2017 The economic corridor China – Mongolia – Russia: the standby mode *ECO* **5** 98-106
[7] Epifanova G 2016 The Great Tea Way: History and Modernity *Civilization-Society-Man* **1** 36-42
[8] Bezrukov L 2016 Transsib and the Silk Road: global infrastructure and regional development *ECO* **7** 21-36
[9] Malov V 2018 Mongolian transit: tactical steps to solve strategic goals *Region: Economics&Sociology* **1** 217-30 https://doi.org/10.15372/REG20180110
[10] Namzhilova V 2018 Economic corridors in the Eurasian space: China's initiatives and world experience *Russia and the APR* **1** 65-81
[11] Frolova I 2017 Perspectives of trilateral cooperation of Russia, China and Mongolia *Problems of National Strategy* **6** 84-102

IOP Conf. Series: Earth and Environmental Science **190** (2018) 012053 doi:10.1088/1755-1315/190/1/012053

# Ecological environment risks and green development modes of China-Mongolia-Russia economic corridor

**S Dong[1,2], Y Li[1,2], Z Li[1,2], F Li[1*], H Cheng[1], Y Yang[1,2], A Bilgaev[1,2], J Zheng[1,2], T Bazarzhapov[1,2]**

[1]Institute of Geographic Sciences and Natural Resources Research, Chinese Academy of Sciences, Beijin, China
[2]University of Chinese Academy of Sciences, Beijing, China

E-mail: lifujia2003@163.com

**Abstract**. China-Mongolia-Russia Economic Corridor is a key strategic area that China, Mongolia, and Russia are all dedicated to developing. However, China-Mongolia-Russia Economic Corridor cover three countries, and passes through multiple geographical units, and there are huge differences in economic, social, and ecological development within the regions. The construction of China-Mongolia-Russia Economic Corridor has many potential risks. Therefore, this paper studied the current socio-economic and environmental situation of main regions along the China-Mongolia-Russia Economic Corridor, including Primorski Krai, Khabarovsk Krai, Amur oblast of Fareast Region, Irkutsk oblast, Zabaykalsky Krai and Republic of Buryatia of Baikal region and Novosibirsk region, Mongolia, and Liaoning, Jilin and Heilongjiang of Northeast China. Then we revealed the main ecological environment risks, and put forward green development mode to achieve sustainable socio-ecological-economic development of the corridor, which includes the ecological civilization mode, four hierarchies of Chinese circular economy mode, green service industry and low carbon tourism, establishing green and low carbon eco-tourism mode and establishing eco-city. Then prior action to achieve sustainable development of the China-Mongolia-Russia Economic Corridor was proposed. Including establish green, low-carbon, circular economic system, establish low carbon eco-tourism pilot area and cultivate green development and ecological civilization.

## 1. Introduction

As one of the six core corridors planned in "the Belt and Road (B&R) Initiative", China-Mongolia-Russia Economic Corridor fits within the "Eurasian Economic Union" proposed by Russia and the "Prairie Road Program" proposed by Mongolia. This corridor is a key strategic area that China, Mongolia, and Russia are all dedicated to developing. However, China-Mongolia-Russia Economic Corridor cover three countries, and passes through multiple geographical units, including the North China Plain, Northeast China Plain, Mongolian Plateau, and Eastern Siberian Plateau. The overall eco-logical environment is characterized by transitivity, diversity, complexity, vulnerability, and sensitivity [1]. At the same time, there are large differences in economic, social, and ecological development within these regions [2]. Therefore, the construction of China-Mongolia-Russia Economic Corridor has many potential risks, which needs to be studied carefully. So, this paper studied the current socio-economic and environmental situation of main regions along the China-

Mongolia-Russia Economic Corridor, revealed the main ecological environment risks, and put forward further steps towards a sustainable socio-ecological-economic development of the corridor.

## 2. Socio-economic situation of the China-Mongolia-Russia Economic Corridor

### *2.1. Socio-economic situation for Russia*

*2.1.1. Social-economic of Lake Baikal Region and Far Eastern regions.* The permanent population of Burytia has an increasing trend from 2005 to 2016, and the entire population in 2016 is 983 200. Also, we observe a 4.4% increase in the urban population during the period under review from 2008 to 2016 [3]. This is the largest growth among the regions under consideration.

The occupation level showed an increasing trend from 2005 to 2016 in regarding to whole Russia, in some regions it slightly decreased in 2016 (table 1). The population with incomes below the subsistence minimum is decreasing during recent years (table 2). Despite the increase of occupation level, we see an increase in the number of people with incomes below the subsistence minimum.

**Table 1.** Occupation level (%) [3]

| Regions | 2005 | 2010 | 2011 | 2012 | 2013 | 2014 | 2015 | 2016 |
|---|---|---|---|---|---|---|---|---|
| **Russian Federation** | 61.3 | 62.7 | 63.9 | 64.9 | 64.8 | 65.3 | 65.3 | 65.7 |
| **Siberian Federal District** | 59.8 | 61 | 61.5 | 61.9 | 62 | 62.6 | 62.5 | 62.5 |
| **Republic of Buryatia** | 53.2 | 58.2 | 58 | 58.9 | 58.6 | 58.4 | 58.6 | 57.7 |
| **Irkutsk oblast** | 60.2 | 60.4 | 61.8 | 62.4 | 62.9 | 62.1 | 64.2 | 63.7 |
| **Zabaykalsky Krai** | 60 | 55.8 | 57.1 | 57.6 | 58 | 59 | 59.4 | 59.6 |
| **Far Eastern Federal District** | 61.3 | 62.6 | 64.4 | 65 | 65.2 | 65.6 | 65.8 | 66.3 |
| **Primorsky Krai** | 60.2 | 61.2 | 62.4 | 63.6 | 63.9 | 64.2 | 64.9 | 65.9 |
| **Khabarovsk Krai** | 62.7 | 62.4 | 65.2 | 65.4 | 66 | 66.3 | 65.9 | 67.4 |
| **Amur oblast** | 54.5 | 60.9 | 65.4 | 63.7 | 61.9 | 63.6 | 62.9 | 63.6 |

**Table 2.** Population with income below the subsistence minimum (% of total population of the region) [3]

| Regions | 2005 | 2010 | 2011 | 2012 | 2013 | 2014 | 2015 | 2016 |
|---|---|---|---|---|---|---|---|---|
| **Russian Federation** | 17.8 | 12.5 | 12.7 | 10.7 | 10.8 | 11.2 | 13.3 | 13.4 |
| **Republic of Buryatia** | 32.6 | 19.2 | 20.1 | 17.7 | 15.9 | 16.9 | 17.7 | 18.3 |
| **Irkutsk oblast** | 21.3 | 18.1 | 19.2 | 16.8 | 17 | 18.6 | 20.1 | 20.6 |
| **Zabaykalsky Krai** | 26.1 | 19 | 18.9 | 17.6 | 16.2 | 18 | 20.4 | 21.4 |
| **Primorsky Krai** | 27.4 | 16.3 | 15.7 | 14.2 | 15.9 | 14.7 | 15.3 | 15.7 |
| **Khabarovsk Krai** | 20.7 | 15.9 | 15.8 | 14.3 | 12.5 | 13.2 | 13.7 | 12.7 |
| **Amur oblast** | 30.1 | 23.7 | 20.4 | 16 | 15.1 | 14 | 14.2 | 16.1 |

*2.1.2. Socio-economic situation of the Novosibirsk region.* Novosibirsk region locates in the central part of the West Siberian Plain, it is one of the largest regions, the most industrially developed, the largest scientific center and the largest municipal unit in Russia. The permanent population has an increasing trend from 2012 to 2018, and the entire population in 2018 is 2 788 849. The proportion of urban population increased from 77.5% to 79.1% in 2012–2018, while the proportion of the rural population decreased from 22.5% to 20.9%.

IOP Conf. Series: Earth and Environmental Science **190** (2018) 012053 doi:10.1088/1755-1315/190/1/012053

*2.2. Social-economic situation for Mongolia*

Population gathered in Ulaanbaatar contributed 46.2% and that along the China-Mongolia-Russia railway is 58.45% of Mongolia total population in 2016 [4]. The population of Ulaanbaatar exceeded the carrying capacity of the city, and had severe impacts on the environment. GDP is 49.9% of Mongolian total in the areas along China-Mongolia-Russia railway

The level of employment fluctuated acutely during 2005-2014 in the Mongolian part, except Ulaanbaatar, because the resource development labor demand fluctuated with the resource market price.

*2.3. Social-economic situation of Northeast China*

In 2015, population in Liaoning, Jilin and Heilongjiang were 43.82 million, 27.53 million and 38.12 million respectively, the total population of the northeast China was 109.47 million [5]. In 2005-2015, the population of the northeast China change little.

Urbanization rate grew rapidly in the past decade. Urbanization rate of Liaoning, Jilin and Heilongjiang increased from 58.7%, 52.5%, 53.1% to 67.4%, 55.3% and 58.8% respectively, which grew rapidly.

GDP is huge and increased rapidly. In 2005-2015, GDP of Liaoning, Jilin and Heilongjiang grew rapidly, which increased from 800.9, 362.03 and 551.15 billion RMB to 2866.9, 1406.31 and 1508.37 billion RMB, the annual growth rate in 2005-2015 were all above 9%.

## 3. Environmental situation of the China-Mongolia-Russia Economic Corridor

*3.1. Environmental situation for Russia*

*3.1.1. Environmental situation of Lake Baikal Region and Far Eastern regions.* Emissions (mln. t):Irkutsk oblast (642) > Primorskii Krai (186) > Amur oblast (135) > Zabaykalskii Krai (122) > Khabarovsk Krai (114) > Republic of Buryatia (94) ; Waste Water (mln. m3): Irkutsk oblast (514) > Primorskii Krai (276) > Khabarovsk Krai (172) > Amur oblast (74) > Republic of Buryatia (38) > Zabaykalskii Krai (34) .

For Buryatia, the air pollution showed increasing trend from 2010 to 2014, in 2016 we can see its decrease in Republic of Buryatia. The sector of the production and distribution of electricity, gas and water accounts for the core factor in air pollution of the Republic of Buryatia. It is obvious that the increase of the production rate inevitably leads to increase of the load on the environment. However, a sufficient volume of investments aimed at its protection, proportional to their distribution among all groups of activities, and timely conduct of environmental work, will allow to prevent critical deterioration of the environmental situation in the region.

*3.1.2. Environmental situation of Novosibirsk region.* Atmospheric air. The growth of cities, the development of industry, the increase in the number of road transport, combined with the lag in the development of appropriate modern road transport infrastructure cause an intensive increase in the mass of emissions of pollutants into the air.

*Groundwater.* New sources of ground water pollution are detected every year. The main reason for the receipt of pollutants in groundwater is the failure to comply with sanitary standards in the operation of wells (figure 1).

*Floods.* Negative of channel siltation of river beds, riverbed over growing trees and shrubs.

*Waste material.* Waste disposal of territories, including water protection zones, forests; loss of secondary material raw materials; pollution of the environment with particularly hazardous waste generated by the population (figure 2).

*Preservation of flora and fauna.* Violation by citizens and legal entities of the rules of hunting; non-compliance with the regime of specially protected natural areas of regional significance; insufficient destruction of animals harmful to hunting.

IOP Conf. Series: Earth and Environmental Science **190** (2018) 012053 doi:10.1088/1755-1315/190/1/012053

*3.2. Environmental situation of Mongolia*

The China-Russia-Mongolia railway goes through the Central Plateau and Southern Gobi in Mongolia, and hilly area in Zabaykalskii Krai in Russia. The climate of the study area is multiplicity. The climate type of Mongolian part area is continental temperate grassland climate, and the Russian part is extreme continental climate. The Mongolian part area along the China-Russia-Mongolia railway belongs to arid and semi-arid regions, with average annual precipitation below 400 mm or 200 mm [4].

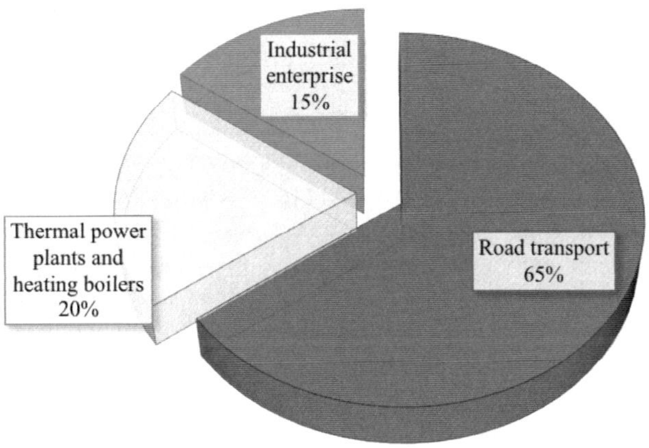


**Figure 1.** The structure of air pollution

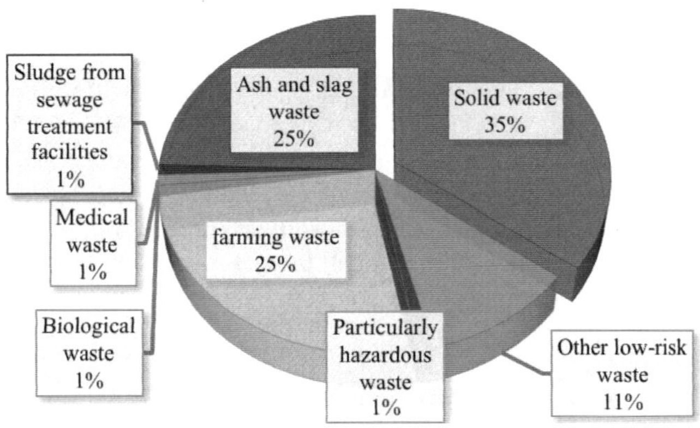


**Figure 2.** The structure of waste

*3.3. Environmental situation of Northeast China*

*3.3.1. Shortage of resource.* The rapid growth of urbanization and GDP in northeast China has consumed large quantity of resources and leads to shortage of resource. The total amount of water resource reduced from 249 billion m3 to 132 billion $m^3$. Water resource is short and shows a declining trend.

The continuous expansion of construction land. While the urbanization rate in Liaoning, Jilin and Heilongjiang has increased rapidly, their land for construction has increased from 1.40, 1.07 and 1.49 million hectares in 2008 to 1.62, 1.08 and 1.62 million hectares [6]. The continuous expansion of urban space and construction land has increased resources consumption and pollutant emissions, and increased the ecological environment load.

Black soil degradation and loss. The black soil layer of cultivated land in the Northeast Plain has fallen, while the forming 1 cm will take about three hundred or four hundred years. The rate of loss is hundreds of times more than that of forming.

*3.3.2. The emissions of pollutants mainly show increasing trend.* Environmental Pollution is huge in Northeast China. Discharge of waste water in Liaoning and Jilin shows an increasing trend, the industrial waste gas emission of Liaoning and Jilin shows an increasing but that of Heilongjiang started to decline in 2014. In 2005-2015, industrial solid waste generated of Liaoning, Jilin and Heilongjiang shows an increasing trend, that of Liaoning increased especially rapid [6]

## 4. Green Development Modes

To implement UN SDGs along the China-Russia-Mongolia economic corridor, the best way is to explore typical modes which are suitable for the basic regional conditions, for their circumstances based on distribution patterns of its resources, environment, ecology, society and economy.

### *4.1. Ecological civilization mode*

Human-beings created agricultural civilization and industrial civilization, in which man and nature were conflict and contradict. Human is marching towards Eco-Civilization, which establishes harmonious relationship between human and nature. The core ideology is to respect, obey, and protect nature, which is priority to Green "One Road One Belt" and its sustainable development [7]. China is implementing Eco-Civilization Building Strategy.

Passing through agricultural civilization and industrial civilization, human society is marching towards Ecological civilization which advocates coexistence harmonious between human and nature. And the core ideology is to respect nature, to follow nature, and to protect nature, and to drive human sustainable development.

The general mode of Ecological civilization is: the basic premise is protecting the spatial ecology, the crucial support is good ecological environment, the drive force is flourishing ecological economy, the leading ideology is advanced ecological culture, the important guarantee is thorough ecological regime, the direct behavior is high quality living environment. Finally, Silk road ecological civilization system of six in one will be set Ecological civilization is sublimation of traditional industrial civilization, the core ideology is respecting, obeying, and protecting nature, which is priority for Green Lake Baikal and its sustainable development (figure 3).

### *4.2. Four hierarchies of Chinese circular economy Mode*

It is vital for the Countries, like China, to develop four sorts of circular economy, which are enterprise circular economy, industries circular economy, regional circular economy and social circular economy, formulating grade circular economic system through whole process of production, circulation and consumption, promoting green, circular and low-carbon development, at principle of Reduce, Reuse, Re-cycling.

Following is the four-layer circular system which integrated the four-layer circular economy system, it's consists of enterprise circular economy, industries circular economy, regional circular economy and social circular economy [8], formulating grade circular economic system through whole process of production, circulation and consumption with the principle of 3R [9] (figure 4).

*4.2.1. Enterprise circulation.* From 2005 until now, carrying out more than 300 cleaner production projects. Main industry: metallurgy industry, chemical industry, building new material industry. Main continent: implementing compulsory clean production audit for Enterprises that discharge pollutants in hierarchies exceeding standards. Encouraging enterprises to employ submerged arc furnace for ferroalloy industry and calcium carbide industry in order to recycle CO and waste heat in the burner gas. Promoting comprehensive utilization of dry coke quenching and technology using gas to generate electricity.

IOP Conf. Series: Earth and Environmental Science **190** (2018) 012053 doi:10.1088/1755-1315/190/1/012053

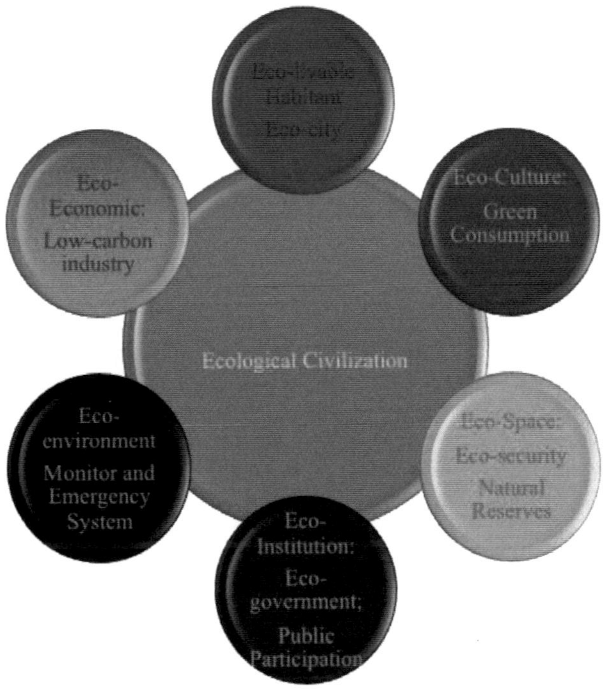


**Figure 3.** Ecological civilization mode

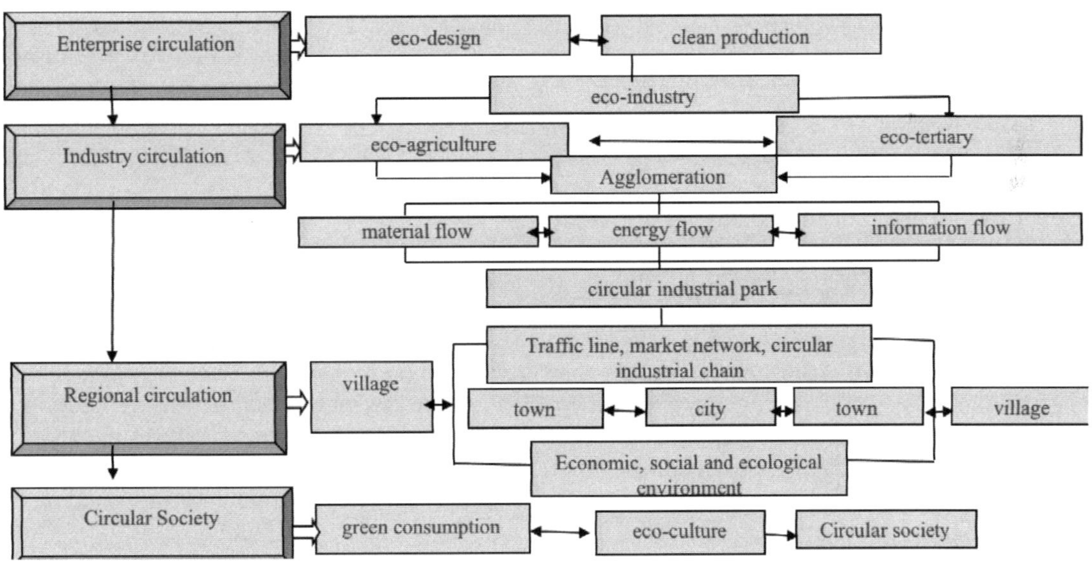


**Figure 4.** Four hierarchies of Chinese circular economy Mode

*4.2.2. Industry circulation.* Since 2005, combined with present situation of Shizuishan City's industry, government identified some circular economy projects in fields of city's comprehensive utilization of resources, energy-saving and cost-reducing, cleaner production etc.

*4.2.3. Region circulation.* In a certain region such as a county, a prefecture, a province, a country, a natural geographical area which includes parts of several administrative region, an ecological region or economical region, and even region which includes parts of several countries and so on, circular economy is carried out by connecting raw materials, production and wastes of the enterprises together within this region. On the basis of benefit sharing, virtuous cycles of cash flow, information flow, people and goods flow are promoted through connection of industrial chains, market network, financial network, information network, resources and ecological environment according to labor division and cooperation in various regions.

*4.2.4. Society circulation.* Forming macroscopic recycling of resources consumption-environment-resources in whole society from the level of social consumption through establishing a sense of resource saving, mobilizing whole people to participate green consumption and garbage recycling use.

### *4.3. Green service industry including low carbon tourism*

Green service industry refers to supplement or replace traditional services by using green, ecological ideas, to reduce resource consumption and pollution emissions, to improve the comprehensive utilization of resources and services for the ecological value of the service industry. It includes green financial system, green logistical industry, green environmental protection services and green eco-tourism.

To develop of green service industry, Lake Baikal region should comprehensively utilize laws and regulations and other coercive means and economic incentive measures to stimulate producer behavior and consumer behavior of services enterprises, to improve the comprehensive utilization level of services industries, to reduce production of living and producing waste and pollution emission levels, strengthen green management and management services, building green service industry[10].

### *4.4. Establishing green and low carbon eco-tourism mode*

Eco-tourism characterized by its unique ecological environment, refers to the tourism mode of regarding sustainable development as goal, protecting environment as premise, realizing harmonious development of human and nature as principle, based on the beautiful natural eco-environment and unique cultural ecological system, adopting eco-friendly way to carry out the ecological experience, ecological education, cognitive and psychosomatic pleasant.

Lake Baikal as the world's natural heritage, its superior ecological environment is its most valuable asset. In recent years, with the rise of Lake Baikal tourism, tourism economy has become an important component of Lake Baikal economy. In this case, Lake Baikal tourism must take ecological tourism, so as to protect the ecological environment of Lake Baikal, scientifically assess tourism carrying capacity of Lake Baikal, exploring tourism resources properly and moderately, so as to protect the world natural heritage.

*4.4.1. Green consumption.* Green consumption contains green consumption and Green lifestyle. Green consumption refers to the consumption behavior which is characterized by conserving resources and protecting the environment as a goal, mainly for advocating thrift, reducing loss and waste, choosing efficient, environmentally friendly products and services, and reducing the consumption of resources in the process of consumption and pollution emissions. Green lifestyle refers to by advocating the use of green products, advocating people to establish green development values, sharing ideas, to promote green travel, green living, green life, people nature, environmental protection, frugal and healthy life style.

*4.4.2. Green culture.* As a kind of cultural phenomenon, green culture is associated with the awareness of environmental protection, ecological consciousness, life consciousness and other green ideology, it's the sum reflection of living mode, code of practice, thinking ways and values of harmony between mankind and nature, co-prosperity and common development, Green culture is the soul of green development.

*4.5. Establishing eco-city*
Eco-city contains six layers: safe city, convenient city, circular city, green city, innovation city and harmony city (figure 5). The safe city is the basic requirements, the core of the city's social order is good, perfect infrastructure, comprehensive disaster prevention level. The convenient city is the core. It is the three-dimensional traffic oriented urban layout led by public transport. The circular city is the eco-industrial chain as the core, modern ecological agriculture and ecological service industry as the auxiliary system of circular economy. The green city is the core of city - ecological landscape and beautiful environment of the living environment, suitable for living and business. The innovation city is high level of eco city requirements, science and technology and high concentration of talent, the development of high-tech industries, and the construction of innovative cities. The harmony city is the highest level of requirements, the core is a high level of ecological and cultural and social civilization [9, 11].

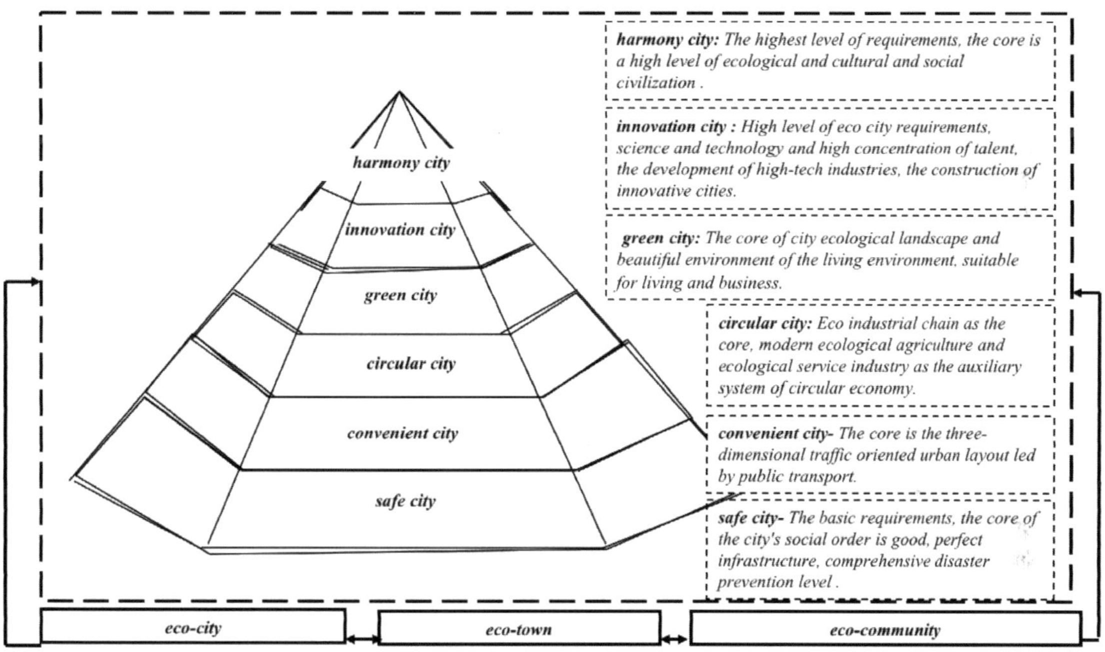


**Figure 5.** Six-layer eco-city

## 5. Prior Action Suggestions to achieve sustainable development of Lake Baikal

*5.1. Establish green, low-carbon, circular economic system demonstration area*
Establish Circular Economy Park, introduce in advanced technology and equipment, speed up clean and green renovation of equipment, set strict emission standards, to achieve the low resource consumption of production, less environmental pollution, good economic benefits.

*5.2. Establish low carbon eco-tourism pilot area*
Develop low carbon ecological tourism mode, protecting the natural resources of Lake Baikal as the premise, reasonable planning and developing tourism resources of Lake Baikal. Strengthen the construction infrastructure, in particularly the construction of garbage recycling stations, sewage treatment plants, public toilet, etc., to build a international pilot area of low carbon ecological tourism in Lake Baikal.

### *5.3. Cultivating green development and ecological civilization*

Strengthen propaganda and education of concept of green development and ecological civilization to promote formation of value of green production, green consumption and green culture, promoting merging of green ideology into production, life, ecological construction.

**Acknowledgements**

The research was funded by National Social Science Fund of China (No.17VDL016); Key international cooperation projects of the Chinese academy of sciences (No. 131A11KYSB20160091)

**References**

[1] Dong S C, Li Z H, Li F J and Li Y 2017 The green development mode of the "B&R" *China Econ. Times* 05-11

[2] Dong S C, Yang Y, Li F J, Cheng H, Li J N, Bilgaev A, Li Z H and Li Y 2018 An evaluation of the economic, social, and ecological risks of China-Mongolia-Russia high-speed railway construction and policy suggestions. *J. Geogr. Sci* **28**(7) 900-18

[3] Russian Statistical Yearbook http://www.gks.ru

[4] Mongolia Statistic Yearbook http://www.en.nso.mn/index.php

[5] Liaoning Statistical Yearbook http://www.ln.stats.gov.cn/tjsj/sjcx/ndsj/otherpages/2016/indexch.Htm
Jilin Statistical Yearbook http://tjj.jl.gov.cn
Heilongjiang Statistical Yearbook. http://www.hlj.stats.gov.cn

[6] China Environmental Statistic Yearbook. http://www.stats.gov.cn/tjsj./tjcbw/201706/t20170621_ 1505 831.html

[7] Li Z H, Wang J L, Zhao Z P, Dong S C, Li Y, Zhu Y Q and Cheng H 2014 Eco-environment patterns and ecological civilization modes in the Silk Road Economic Zone *Resour. Sci.* **12** 2476-82

[8] Li Z H, Dong S C and Tang S Y 2008 Constructing a circular economy with Chinese characteristics *Theor. Explor.* **1** 88-90

[9] Li Y, Dong S C, Wang F, Wang Z and Zhang L J 2012 Study on development pattern and path of eco-city based on circular economy *Urban Stud.* **11** 146-8

[10] Dong S C, Zhao M Y, Guo P, Shi G Y, Li Y, Li Z H, Wang J N and Zhu S Q 2016 Development mode and countermeasures for international ecotourism zone along the Belt and Road *Bull. Chin. Acad. Sci.* **6** 647-55

[11] Zhang X J, Dong S C, Yin W H and Wang L L 2009 The transform of the resource city in china: A case study of Shizuishan eco city in Ningxia Hui Autonomous Region. *J. Arid Land Resour. Environ.* **4** 6-11

# Classification of educational tourism objects according to their location

**Z M Khasheva, V P Kazachinsky, I I Filipovich**

Southern Institute of Management, Krasnodar, Russia

E-mail: zarema_muratovna@mail.ru

**Abstract.** Tourism as a subject of research is a complex and not yet fully studied phenomenon, despite the fact that in practice of economic activity it occupies an increasingly prominent place and tends to steady growth. One of the most important tasks of understanding this phenomenon is building an adequate and comprehensive classification system that could be used by both theorists and practitioners of tourism business. The article proposes to classify historical, cultural and natural objects of educational tourism according to the location with regard to each other and in recreational and tourist centers of the water areas of educational tourism of the Krasnodar region. There are three basic models: compact, annular and linear. Compact-focal, compact-linear and semi-circular models are referred to mixed variants. The received data of the preliminary assessment of recreational tourist zones models, which have received classification characteristics, allow us take into account the specifics of historical, cultural and natural objects in the structure of water areas of educational tourism differentially and systematically. Accordingly, the proposed models are evaluated on the consolidated indicators, which is the basis for further research of socio-economic, infrastructure, environmental and other aspects in the planning of development prospects of the recreational and tourism industry.

## 1. Introduction

Tourism is an independent and growing complex in the world economy, the main task of which is to meet the diverse and increasing demand for various types of tourism services [1]. In 2010, the average growth rate of tourism amounted to 4.6% per year, the tourism sector accounted for: about 10% of world investment, each 11th workplace in the total world employment; 7.8% of the world GDP. The most important direction of the industry development is educational tourism, the share of which according to the estimates of the world tourism organization of the United Nations is about 10% of the world's tourist flow. In Europe, the share of educational tourism is much higher - up to 25% of tourist arrivals in the European Union. In such circumstances, the problem of effective use of the Russian regions tourism potential, in particular the Krasnodar region becomes particularly relevant [2]. The cluster approach has become the main tool of effective strategy of tourism business development and management in countries with a high level of competitiveness. In this regard, the issue of tourist and recreational zoning of the Russian Federation regions for the development of educational tourism becomes relevant [3].

IOP Conf. Series: Earth and Environmental Science **190** (2018) 012054 doi:10.1088/1755-1315/190/1/012054

## 2. Models and Methods

The construction of an adequate classification in tourism sphere is one of the most relevant, but methodologically scantily explored problems. In the theory and practice of management, the most common and used are classification features based on the allocation of goals and directions of tourist arrivals, features of destinations, the differentiation of accommodation, describing the diversity of tourist services, the length of tourists stay in a particular region, types of tourism, depending on the characteristics of tourist activities, etc. Such a variety of approaches raise the question of methodological principles that can be the basis of typology and classification, taking into account the features of the tourism industry [4].

In this article comparative multivariate system analysis is taken as a scientific method, it allows to consider the objects under study and their features in the relationship and in relative independence; study of historical, cultural and natural objects of the region, water areas of educational tourism, recreational and tourist areas and centers as tourist and educational areas; the transition from accumulation of private knowledge about certain objects of educational tourism, collection and systematization of disparate information about them to their generalization, scientific, historical, cultural and natural synthesis. For development of separate issues, such methods of analysis as theoretical, cartographic, typological, modeling and field surveys are applied.

## 3. Results and Discussion

The authors of the paper "Systematization of tourist resources: evidence from Krasnodar region" [5], have initiated a debate on the feasibility of cognitive-territorial zoning of recreational and tourist space, which will allow a systematic approach to the evaluation of relationship and interdependence of historical, cultural and natural objects, required amount of infrastructure construction and its development. In this regard, the task was set to produce zoning of the Krasnodar region territory in the water areas of educational tourism, reflecting the distribution and location of historical, cultural and natural sites.

The water area of educational tourism is understood as a recreational and tourist zone, where historical, cultural, natural and infrastructure facilities are located on land, underground and on water.

The following grouping of cultural tourism areas is proposed: Gelendzhik, Mostovskaya, Primorsko-Akhtarsky. Gelendzhik area includes the following recreation and tourism zones: Gelendzhik (G), Novorossiysk (G1), Abinsk (G2), Goryacheklyucheskaya (G3) and Tuapse (G4). Mostovskayaarea includes:Mostovskaya (M), Belorechenskaya (M1),Absheron (M2), Labinskaya (M3) and Sochi (M4). Primorsko-Akhtarsky area includes:Primorsko-Akhtarsky(P), Yeisk (P1), Temryuk (P2) and Anapa (P3) (figure 1).

It should be noted that at present the structure of Gelendzhik, Mostovskaya and Primorsko-Akhtarsky water areas of educational tourism (WAET) is traditionally determined on the basis of such factors as climate, social, economic, etc. Inclusion of educational tourism objects (ETO) in the strategy of tourism industry development occurs spontaneously, without appropriate scientific support. To establish the level of modern development of tourism industry and specific requirements of the inclusion of ETO in the structure of water areas of educational tourism, regulations and recommendations, it is necessary to classify the objects of educational tourism. The classification is based on the location of recreational and tourist centers, historical and cultural and natural objects with respect to each other. At all variety the existing models can be combined into three main classes: compact, annular and linear (figure 2). Compact-focal, compact-linear and semi-circular models can be referred to mixed variants (figure 3).

Compact model is determined by the location of ETO within the boundaries of recreational and tourist centers. Here, pedestrian and transport links conveniently connect historical, cultural and natural sites. The following recreational and tourist areas of cultural tourism have the compact model: Novorossisky (G1) with recreational-tourist center of Abrau-Dyurso village; Temryuksky (P2) with recreational-tourist center station Tamansky; Anapa (L3) with recreational-tourist center Sukko village; Sochi (M4) with recreational-tourist center Sochi resort town and Krasnodar with recreational-tourist center of Krasnodar town.

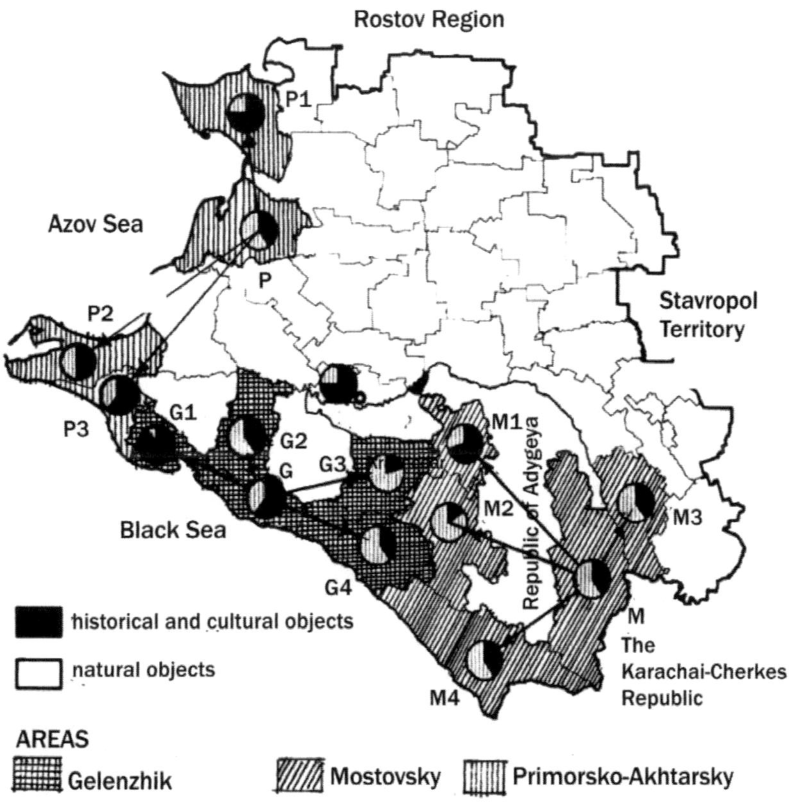


**Figure 1.** Zoning of Kuban territory in the water area of educational tourism and distribution diagrams of historical, cultural and natural sites in the water areas and recreational and tourist areas: G - Gelendzhik area (G1 – Novorossisky, G2 – Abinsky, G3 – Goryacheklyucheskaya, G4 – Tuapsinsky); M – Mostovsky area (M1 – Belorechensky, M2 – Apsheronsky, M3 – Labinsky, M4 – Sochi); P – Primorsko-Akhtarsky area (P1 – Yeisky, P2 – Temryuksky, P3 – Anapa).

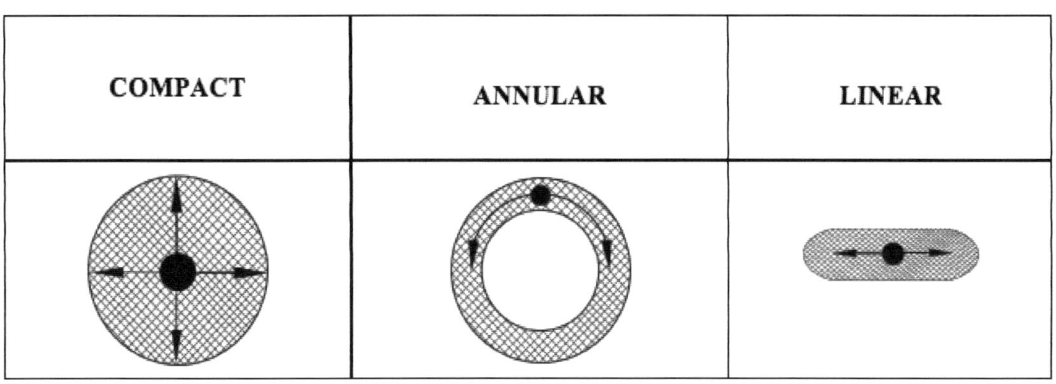


**Figure 2.** Basic models of educational tourism objects and recreation and tourist centers location with respect to each other.

IOP Conf. Series: Earth and Environmental Science **190** (2018) 012054 doi:10.1088/1755-1315/190/1/012054

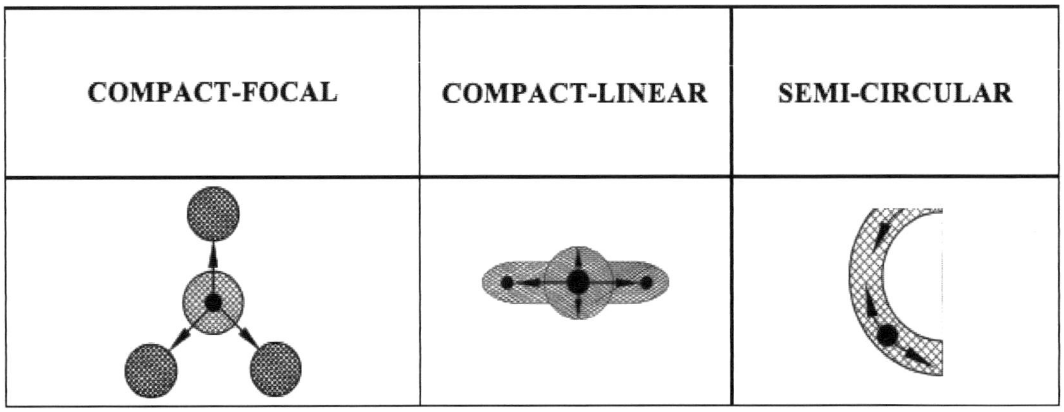


**Figure 3.** Mixed models of educational tourism objects and recreation and tourist centers location with respect to each other.

Annular model characterizes location of recreational and tourist centers, historical, cultural and natural sites in the form of an interconnected circular formation. At the same time, the RTC is located in the geometric center of the circular models, and historical, cultural and natural objects are beyond its borders. With this model, it is possible to organize an additional recreational and tourist center.

Annular model has three recreational and tourist areas. The total number of historical, cultural and natural sites is 29, that is 25% of the total number of ETO main classification models. Annular model is characterized by a large number of natural objects – 23, that makes 79% of the total number of ETO of the annular model.

Recreational and tourist area with annular model is located in Gelendzhik water area of educational tourism – Abinsky (G2) with recreational-tourist center of Abinsk town and station Shapsug; Goryacheklyucheskaya (G3) with recreational-tourist centers of Goryachiy Kluych and Fanagoriysky village; Primorsko-Akhtarsky area (P) with recreational-tourist centre of Primorsko-Ahtarsk town. In Mostovsky water area of educational tourism there is no annular model.

Linear model of location of historical, cultural and natural objects with respect to each other and recreation and tourist centers includes the case of placing ETO consistently one after another, linearly and sequentially. As examples are the following recreational and tourist areas: Gelendzhik (G) with recreational-tourist center of Arkhipo-Osipovka village; Belorechenskaya (M1) with recreational-tourist center of Belorechensk town; and Labinsky (M3) with recreational-tourist center of Labinsk town. Total number of educational tourism objects of the linear model is 33, which is 30% of the total number of ETO of this model. It is necessary to mention almost equal number of historical and cultural (17) and natural (16) objects. Recreational-tourist zones with linear model are located mainly in Mostovsky water area of educational tourism. This model is absent in Primorsko-Akhtarsky water area.

Linear model is characterized by location of educational tourism objects within the borders or near the borders of recreational-tourist centers. In this case, pedestrian and transport links are arranged linearly, and the RTCs are located in the geometric center. With this model, it is possible to organize additional recreational-tourist sub-centers.

Compact-focal model of location of RTC, historical, cultural and natural objects is the most common among the main models, and among the mixed variants. The largest number of ETO – 80 is located here, which is 36% of the total number of historical, cultural and natural sites of the region. In this model, the main RTC is located in the geometric center. The objects of educational tourism are dispersed with respect to the centers and to each other, pedestrian and transport links are developing curvilinear.

The following recreational-tourist areas of educational tourism have compact-focal model: Novorossisky (G1) with recreational-tourist center of Novorossiysk town; Mostovsky area (M) with recreational-tourist center of Psebay village; Apsheronsky (M2) with recreational-tourist center of Mezmay village; Sochi (M4) with recreational-tourist center of Krasnaya Polyana village; Yeisky (P1) with recreational-tourist center of Yeisk town; Temryuksky (P2) with recreational-tourist center of Temryuk town; Anapa (P3) with recreational-tourist center of Anapa town. All the water areas of educational tourism in the region have this model.

Compact-linear model is determined by the location of historical, cultural and natural objects compactly, within the boundaries of the RTC and sequentially one after another along pedestrian and transport links. This model has only one water area of educational tourism – Gelendzhik (G) with main recreational and tourist center of Gelendzhik town and recreational-tourist sub-centers of Kabardinka and Dzhankhot village.

Semi-circular model of location of recreation-tourist centers, historical and cultural objects with respect to each other includes a variant in the form of an incomplete ring. In this case, the RTCs are located in the center and on the periphery of semicircular formation. An example is the Tuapsinsky (G4) recreational-tourist area, with main recreational- tourist centers of Tuapse town and Dzhubga village, and also the sub-center of Gorniy village.

As an illustration of the possible direction of classification models development, the following variants of their spatial and planning organization are proposed:

- parallel-sector, where objects of educational tourism of RTC are placed in parallel, and separate historical, cultural and natural objects are organized in the form of sectors. When natural objects are located outside the boundaries of RTC, it is advisable to use them for infrastructure facilities arrangement;
- in parallel-annular, where objects of educational tourism are parallel, and some historical, cultural and natural objects are presented in the form of an interconnected circular formation. Allocation of additional recreational-tourist center is preferable in the geometric center of the model at the junction of existing and projected infrastructure facilities;
- autonomous, where historical, cultural and natural objects are located outside the boundaries of recreational-tourist centers at different distances from them; transport and pedestrian links between objects of educational tourism, as a rule, are combined with the existing infrastructure; natural objects are used to arrange infrastructure;
- circular, where single objects of educational tourism are located at the junction of transport and pedestrian links of infrastructure objects that unite them with each other. In this case, roads can be designed as a circular network linking historical, cultural and natural objects.

## Conclusion

The role of established models of the location of recreational-tourist centers, historical, cultural and natural objects with respect to each other in practice and theory of tourism development is necessary to establish the principles of approach to the assessment of educative-territorial opportunities of water areas of educational tourism. In future, within the framework of proposed classification models, it is necessary to develop a methodology for infrastructure facilities arrangement in water areas of educational tourism.

## References

[1] Bieger T 2008 *Destinations Management* (München und Wien: Oldenbourg Wissenschaftsverlag) p 132

Cohen E 1998 Authenticity and Commoditization in Tourism *Annals of Tourism Research* **153** 371-86

WTO 2001 *Cultural Heritage and Tourism Development* (Madrid: World Tourism Organization) p 3

Lee T-H and Crompton J 1992 Measuring Novelty Seeking in Tourism *Annals of Tourism*

*Research* **19** 732-51
Nash D and Smith V L 1991 Anthropology and Tourism *Annals of Tourism Research* **18** 12-25
Smith V L 1980 Anthropology and Tourism: A Science-Industry Evaluation *Annals of Tourism Research* **71** 13-33

[2] Morozov M A 2014 Peculiarities of and opportunities for the evaluation of the potential of the Russian destination of health tourism *Economics, Statistics and Informatics* **4** 133-9
Orekhov S A and Kruglov V N 2014 Features of regional innovation policy while putting aside models of sustainable development of the territory *Economics, Statistics and Informatics Vest MO* **6**(2) 304-8
Sevastyanova S A 2010 *Regional Planning of Tourism Development and Such Economy* (Moscow: KnoRus) p 255
Voskolovych N A 2010 *Tourism in a Changing World* (Moscow: Moscow State University Press) p 142

[3] Kazachinsky V P and Khasheva Z M 2017 Classification of recreational-tourist areas of Krasnodar region *Scientific Bulletin of the Southern Institute of Management* **4** 64-7
Kazachinsky V P and Khasheva Z M 2017 The question of educational tourism influence on zoning of recreational-tourist areas and zones Economic and legal aspects of Russian modernization strategy: the search for model of effective social development *Collection of Articles of the Int. Scientific and Practical Conf.* (Moscow: Research Institute of history, Economics and Law) pp 132-6

[4] Kharitonova T V Vapnyarskaya O I and Golikova O M 2014 Evaluation of the competitiveness of Russia as a single destination tourists (based on the RB methodology) *Service in Russia and Abroad* **53**(6) 119-39
Tenova Z Y Anichenko E A Kastueva A O and Gelyaeva I A 2016 Tourism as an object of development management Integra *Economic and Legal Issues* **1** 121-4

[5] Kazachinsky V P, Khasheva Z M and Shulgatyi O L 2017 Systematization of tourist resources: evidence from Krasnodar region *International Journal of Economic Perspectives* **11**(4) 83

# China strategic initiative "One Belt, One Road" (OBOR) as tool for development and algorithm for effect on Central Asia countries

**B A Krasnoyarova[1], A V Krotov[2]**

[1]Institute for Water and Environmental Problems SB RAS, Barnaul, Russia
[2]Altai State University, Barnaul, Russia

E-mail: bella@iwep.ru, krotov_av@mail.ru

**Abstract.** The concept of Central Eurasian Region is introduced and grounded; the role of the Region in geopolitical and geo-economic entities of world leading actors is considered. Multiple interests of remote and adjacent to Central Asia countries involved in integration/disintegration processes are identified. Probable effects of the initiative "One Belt, One Road" (OBOR) on the territory (through which transnational transport and logistics routes will pass) are discussed. The emphasis is made on mountainous inland landscapes and their ecological value for national communities and states. Authors describe major risks associated with the proposed mega-project, which accounts for China interests. The danger of receiving most dividends by few stakeholders and possible negative consequences for local population are considered. We propose the methods for optimal development of this ultra-continental region. Here, a coordinated development of numerous spheres is required that will contribute to progressive movement forward and prevent the extreme interregional and cross-country competition under limited investments and resources. Transboundary projects ("Tashkent-Shymkent and Rubtsovsk-Semey industrial hubs") can be realized along with OBOR in comply with own interests. The need for Russia and adjacent countries cooperation with further establishment of a specific economic-political union aimed to improve their position on the world stage is specified.

## 1. Introduction

Central Asian countries located in the heart of Eurasia have a huge demographic, territorial, natural resource and environmental potential, which is certainly of great importance to the world leading actors, especially in the context of the proposed initiative "One Belt, One Road" (OBOR). This Region is of a strategic interest to Russia due to its historical, geopolitical and geo-economic aspects.

## 2. Study object

In the USSR, the Central Asian countries were called Central Asia [1]. More recently, Mongolia and Kazakhstan were referred to the Central Asian countries. According to the classical country studies, Mongolia and China belong to East Asia [2]. Sometimes Afghanistan is also referred to Central Asia though the distance from its southern borders up to ocean makes up only 542 km. In our opinion, some territories of Russia and Xinjiang Uygur Autonomous Okrug of China should be a part of Central Asia (figure 1), but is it legitimate or not? It is rather the Central Euroasian Region, the boundaries of which quite logically fit into mathematical-statistical proportions of a central position with extreme points marked by longitude-latitude coordinates of Eurasia.

**Figure 1.** Schematic map of Central Eurasian territories

Combining these territories into a single macroregion is based on both common nature and territory and concurrent joint development of cultural, political, economic, linguistic, ethno-religious, transport and logistics spheres in the past. All these territories are extremely remote from the sea and the largest world political and economic centers. Being ultra-continental, they are distinguished by rather low competitiveness in the economy.

## 3. Models and Methods

The study is based on the system-dialectical approach to any regional system development at different stages (i.e. formation, functioning and development), the application of which is particularly topical under transformation of the world economic system induced by global and macro-regional factors. At the global level, it is climate change, progressive desertification of plains and glaciers melting in the mountains as well as turnover of technology and leading world actors, redistribution of spheres of influence and markets. At the macro-regional level, a widespread effect on geopolitical, geo-economic and environmental features of the territory generated by the construction of the transnational transport and logistics corridor was revealed. The study of the implemented transcontinental projects, characteristics and models of countries' development, including their economic and geographical characteristics enabled to propose some ways for their development under financial, economic and geopolitical pressure from China and third countries [3]. The use of statistical data and mathematical-statistical tools proves the fact of central intracontinental position of the Region with all the ensuing difficulties and opportunities. Such an economic –geographic concept differs our study from the previously made investigations [4].

## 4. Results and Discussion

Geopolitical and geo-economic interests of the world leading actors come together in this middle part of Eurasia, which is of paramount importance for China, Russia and Iran in case of full exemption from sanctions of the latter. As their major opponent, the USA is interested in controlling main world

IOP Conf. Series: Earth and Environmental Science **190** (2018) 012055 doi:10.1088/1755-1315/190/1/012055

competitors. These countries, with their own models of social-cultural development, have common borders in this region and, accordingly, can participate in joint transboundary mega-projects. In case of their effective implementation, each of the mentioned countries may receive significant benefits. No wonder, the uniting of Central Eurasia countries into the Association, similar to the EU or NAFTA (especially NATO) is absolutely unacceptable for the USA and Western countries which consider them as three main "troublemakers". Moreover, in case of a success the fourth world economy– India, and its main opponents Pakistan and Turkey may join this super union. Together, they will be able to settle the situation in Afghanistan and get tremendous opportunities for the creation of a common economic (and maybe even political or geopolitical) space, which is home to about 3.5 billion people.

The studied territory has a great potential for its development; ethnic, resource, political and cultural conflicts are also highly probable here. Currently, such conflicts do exist already; many of them are latent and controlled by external, more powerful countries. The construction of transit transcontinental highways (the Economic Belt of the New Silk Road (EBNSR) proposed by China President Xi Jinping in 2013) is one of the ways for the development of independent states in the Region. Though the discussed initiative is economically beneficial, this mega-investment project is extremely risky. Let us consider some risks.

*Economic risks.* The territories, through which transcontinental roads will pass, immediately become open to international competition. The EBNSR promotes the liberalization of international trade. India can join the Chinese transit corridor, being interested in developing its own "silk road" called "North-South" [5]. China will be definitely a winner in this situation because it clearly pursues a policy of observing its own interests, consistently implements the concept of establishing the East-Central Asian economic zone and the idea of building the Eurasian mainland bridge for connecting the SUAR - Kashgar - Osh - Fergana valley-Iran with the Trans-Siberian and other recent projects. Local producers will lose in case of modern infrastructure creation to deliver Chinese goods to Europe [6]. The mentioned projects do not envisage any protection against cheap imports and involvement of formal and public institutions in control and regulation of social-environmental aspects related to foreign investors. Xavier Riche states that in this situation you can get "one winner and many losers" [7].

Obviously, rental income is a significant share in the economies of Central Eurasian countries. Instead of national economies stabilization, in case of "a transit economy" establishment, a narrow circle of elite representatives may appropriate the income from transport and transit activities; it may look as a charity in the form of direct assistance to the rest of the population (or a part of it) and indirect redistribution of incomes. Therefore, incentives for high-tech industries development may not appear at all.

*Geopolitical risks.* Market liberalization contributes to the open society formation, however, most Islamic countries of the studied region are not ready for it yet. Further stratification of the society, increased xenophobia and sinophobia, conflicts between radical groups and so-called "westerners" oriented to the European way of life and values are probable. In addition, the heterogeneity of Islam itself, insoluble contradictions between its many currents and the strengthening of terrorism, often having Islamic roots, complicate the issue. All this contributes little to the development of these countries within the framework of a single strategy.

*Environmental risks* of the OBOR project are also very high. Such risks are associated not only with the highways construction and operation but with further industrialization of the adjacent territories, their environmental colonization. One can hardly assume that Chinese investors with their "brown economy" are ready to introduce resource-saving and environmentally safe technologies of the "green" economy in the transit countries. Investors are focused rather on getting a quick return on their investments, not on high-tech and eventually competitive jobs creation. All this may lead to an extra anthropogenic load on the overcrowded plains and give impetus to population and production growth in the ecologically valuable endangered mountain ecosystems.

The real problem of water shortage in the region is complicated by unsolved issues of water allocation, when most water resources are formed in the mountain regions of Tajikistan and

IOP Conf. Series: Earth and Environmental Science **190** (2018) 012055 doi:10.1088/1755-1315/190/1/012055

Kyrgyzstan, but consumed in Kazakhstan, Turkmenistan and Uzbekistan. Political basis that strictly determines the volume and mode of water withdrawal is not available. In Kyrgyzstan, for example, 75.2% of Syr Darya river runoff is formed; in Tajikistan- 2.7 % of Syr Darya and 74% of Amu Darya; in Uzbekistan - 15.2 and 8.5 %, respectively. Currently, most countries of the region face water deficit; legal regulation of water use in these transboundary river basins is absent, and water-related international agreements have not been signed yet [8].

Mountain areas along the southern borders of the Asian Russia, Kazakhstan, Kyrgyzstan, Tajikistan and Afghanistan, which due to transport and infrastructure development can form an advanced development zone of currently inaccessible natural resources (i.e. biological, water, mineral and recreational) deserve a particular attention. Special ecosystem status of the mountain areas for the development of exclusive specializations (e.g. tourism) should be taken into consideration. Switzerland, offshore mini-states of Liechtenstein, Andorra and San Marino are positive examples of the mountain areas development. In Central Asia countries, high per capita indicators of production or living standards are associated either with relatively large investments from the centers, or the increased environmental costs. For example, relative economic success made East Kazakhstan Oblast a world leader in dangerous pollution. Tourism development in Altai and Tyva were realized due to the state mega-infrastructure projects.

In fact, local mountain ultra-continental regions can have a success only through the involvement into the projects of external territories with a much higher human, industrial and technological potential. Table 1 presents distances to the world largest seaports from capitals of regions and Central Eurasia countries.

**Table 1.** Distance to seaports from capitals of regions and Central Eurasia countries.

| Distance to hubs (km) | Urumchi | Barnaul | Tashkent | Ust-Kamenogorsk | Kyzyl | Ulaanbaatar |
|---|---|---|---|---|---|---|
| **Arkhangelsk** | 3699 | 2626 | 3168 | 2917 | 3344 | 4257 |
| **Novorossiysk** | 3912 | 3427 | 2574 | 3378 | 4167 | 5148 |
| **Vladivostok** | 3534 | 3666 | 5046 | 3772 | 2940 | 2010 |
| **Shanghai** | 3271 | 3890 | 4747 | 3803 | 3171 | 2228 |
| **Mumbai** | 3079 | 3931 | 2498 | 3546 | 4093 | 4436 |
| **St. Petersburg** | 4161 | 3255 | 3364 | 3407 | 3944 | 4892 |
| Average distance | 3609 | 3465 | 3566 | 3470 | 3609 | 3828 |

High efficiency of such projects will be provided only if the strategies meet the interests of politically and economically more powerful actors. Russia has its own transport and transit interests in the Region. According to L. B. Vardomsky, the country needs to build meridional connections through the Central Eurasian territories to link Novosibirsk and China, Tyumen and India, Krasnoyarsk and China, Saratov and Iran [9]. The realization of the Trans-Eurasian belt project along the Trans-Siberian railway proposed by JSC "Russian Railways" will ensure transport infrastructure modernization and give impetus to the economy change in the adjacent territories.

The Central Eurasian areas that are not a part of the RF is of great interest because, first, export of Russia depends on raw materials. Secondly, these territories are rich in almost identical resources and compete with each other at international markets, and their production costs are often lower than that in Russia. Thirdly, there is still a market for the national high-tech goods. Fourthly, Russia just needs "quiet" frontiers and the countries with friendly attitude to our people, language and culture [10].

## 5. Conclusion

The analysis of the Central Eurasian countries development suggests that despite risks the development of transit transcontinental routes can provide extra investments and further stable income. In addition, it may unite and stabilize the studied territories as well as to obtain a synergetic effect from the integrated and coordinated development of individual countries and regions. To do it, a common socio-economic strategy for their development with regard to local features of the heterogeneous territory is required. One important point to remember is that excessive competition among the Central Eurasian countries deprives them of maximum benefits from real and potential advantages of the territories. In addition to the institutional standards of cooperation, this strategy should include a number of major investment projects of macro-regional importance, for example, the creation of industrial and innovative clusters for the development and processing of minerals, the construction of hydroelectric power and water facilities, the creation of hubs and other infrastructure facilities in the border areas An example is the establishment of the Tashkent-Shymkent hub and the construction of the Rubtsovsk-Semey transboundary industrial hub.

**References**

[1] Gvozdetsky N A and Mikhailov N I 1978 *Physical Geography of the USSR. Asian Part.* (Moscow: Mysl Press)
[2] Volsky 2003 *Economic Geography of Foreign Countries* (Moscow: Drofa Press)
Badenkov Y P 2017 *Life in the Mountains. Natural and Cultural Diversity: a Variety of Development Models* (Moscow: GEOS Press)
Tsvetkov V A, Zoidov K Kh and Medkov A A 2014 *Formation of Evolutionary Model of Transport and Transit System of Russia in the Conditions of Integration and Globalization* (Saint Petersburg: Nestor-Istoriya Press)
[3] *Eurasian Integration: Background, Problems, Prospects* 2017 (SPb: University at the IGA of EAEC)
[4] Zoidov K K, Medkov A A and Zoidov Z K 2017 *Public-Private Partnership as the Basis for Innovative Development and Transition Economy Safety* (Moscow: Economic education) p 528 https://doi.org:10.26653/2017/978-5-7425-0186-2
[5] Shepard U 2018 India creates its own Silk Road. *War and Peace* Income accessed online on 18th June's 2018 via http://www.warandpeace.ru/ru/news/view/122402
[6] Glinkina S P *2014 Chinese Strategy for Post-Soviet Space Development and Fate of Eurasian Economic Union (Russia in the XXI Century: Global Challenges and Prospects for Development: Proc. of III Intern. Forum*) (Moscow: INR RAS Press) p 59
[7] Richer X 2017 *One Winner and Many Losers* (Moscow: Gudok Press)
[8] Vinokurov Y I and Krasnoyarova B A 2012 *Water-Ecological Gradient of International Cooperation in Central Asia* (Novocherkassk: SFU Press) pp 387–94
[9] Vardomsky L B 2015 Transit potential of Kazakhstan in the context of Eurasian integration *ECO* **8** 59–78
[10] Krotov A V 2018 Geopolitical and geo-economic challenges to the Central Asian territories in the context of their ultra-continental position *Society and Security Insights* **2**(1) 139–58

# Territorial environmental management in conditions of strict environmental limitation (on example of territories with special status)

**S I Lesnykh**

Institute of Geography SB RAS, Irkutsk, Russia

E-mail: tyara@irigs.irk.ru

**Abstract**. The mechanism of territorial environmental management in the possessing unique natural and geographical characteristics is proposed. Their uniqueness is an objective criterion for giving them a special status that does not depend on the government interests. The main task of managing such territories is the competent combination of various instruments for regulating the use of natural resources and protecting the unique parameters. For this purpose, in addition to general environmental legislation, the standards of a special system of government regulation are applied. Therefore the same land plot falls within the scope of the several legal regimes, which causes difficulties in definition the priority regime. On example of Lake Baikal a mechanism of legal environmental zoning has been developed that allows making decisions in the condition of a lot environmental limitations oriented to protecting unique territorial properties. Legal zoning is informational mechanism of territorial management that regulates relations in the field of land use, reflecting contradictions in real and permitted use and allowing a long-term forecast of territorial development. With use of geographic information systems operation of laws on the territory in the form of the map of legal zoning is visualized that allows to be easy guided in legislative space.

## 1. Introduction

Any activity in the territory is related to the attitude of people to the land, to its various areas, types of use, possession, and lease, protection from negative consequences and restoration of potential, i.e. to everything that, one way or another, is perceived in the system of relations of ownership of land. To determine the nature and intensity of the use of land, the natural possibilities of the available resources and the limits of these possibilities must be taken into account. The use of these resources without taking into account the limits of ecological opportunities leads to the destruction of ecosystems [1] (environmental damage) and, as a result, to economic damage.

Any use of nature is impossible without the implementation of general land use. Therefore, in addition to the activities shown on the ground, the system of relations to resources is fixed in normative and legal acts of different levels, imposing various limitations on the use of the territory and showing preferable standards of action on them. Often, the same land is subject to several legal regimes: a specific category of land, a territorial zone and a zone with special conditions for the use of territories [2], causing difficulty in determining its priority legal regime and as a consequence, priority of use and protection.

IOP Conf. Series: Earth and Environmental Science **190** (2018) 012056 doi:10.1088/1755-1315/190/1/012056

In addition, the existing environmental protection mechanism in the Russian Federation is oriented to the consideration of the situation as a whole and does not allow for taking into account regional and local (municipal) features of the territory where the complex of unique spatially interacting geographical objects that are characteristic of a specific territory is most clearly manifested. Underestimation of local characteristics makes it impossible to ensure the targeting of management decisions and impacts, as well as the full protection of unique objects in a particular area [3]. To ensure the adequacy of management impacts on nature, taking into account local conditions, it is necessary to apply the principles of territorial management.

## 2. Territory with a special status

When it comes to the use of unique natural objects that are most attractive from the point of view of recreation, recreational use of natural resources and investments, all existing environmental protection mechanisms in the Russian Federation [4] are included in full force to ensure the safety of the facilities and are not sufficient. Environmental and economic activities for the use of unique resources are strictly regulated by special legislation at various levels, establishing a special protection regime and imposing significant restrictions on use. It is appropriate to use the notion of "territory with a special status" [5] - part of the territory of Russia, in relation to which a special system of state regulation for the formation and maintenance of certain parameters peculiar only to a given territory is normatively established, or unique to achieve state interest, the system of power and delimitation of powers.

The most striking example of a "territory with a special status" is Lake Baikal - a UNESCO World Natural Heritage site that attracts not only the close attention of tourists, but also potential investors. Baikal is the deepest lake on the planet with ultra-fresh water and unique ecosystems located along its banks.

The main document that provides legal regulation in the field of its protection and basic rules of economic activity is the Federal Law "On Lake Baikal Protection" (of May 1, 1999 No. 94-FZ, only one federal law adopted for a particular natural object) and related legislative and legal acts. It should be noted that, despite a significant number of regulatory documents of different levels and departmental affiliations, researchers in the field of environmental law note the weak points of environmental legislation as one of the main conditions for securing Lake Baikal [6].

Any water body in the Russian Federation has legally defined zones: a coastal strip (20 m from the coastline), a coastal protective strip (30-50 m, depending on the slope of the shore of the water body), water protection (50-200 m, depending on the length of the river and 500 m near the lake) (Water Code) and fish protection (RF Governmental regulation, of October 6, 2008 No. 743). Within their borders, a special regime for the implementation of economic and other activities is established to prevent pollution, clogging, siltation of these water bodies and depletion of their waters, as well as to preserve the habitat of aquatic biological resources and other objects of the fauna and flora.

Lake Baikal, as a territory with a special status, having its own system of government regulation, falls into even more limitations. The boundaries of the site of the World Natural Heritage "Lake Baikal" coincide with the boundary of the Central Ecological Zone of the Baikal Natural Territory (CES BNT). The RG Government edict No. 368 (of march 5, 2015) equated the water protection zone of Lake Baikal with the CEP of the BNT in order to minimize possible negative impacts, reduce existing man-made loads on the water body, preserve the biological diversity of Lake Baikal's unique ecological system and optimize the use of potential opportunities for the development of protected areas.

As a result of the imposition of environmental laws, Baikal falls into the system of severe environmental constraints of economic activity, which makes the sustainable economic development of these territories extremely difficult [7]. At the same time, the regime of strict restriction of economic activity allows to maximally ensure the protection and conservation of Baikal, thereby implementing the government's environmental policy in respect of the World Natural Heritage site.

In the most difficult position was the 500-meter coastal zone of the lake Baikal, because in addition to the specially established restrictions on activities in the territory with a special status (CES BNT and water conservation zones), there are general limitations on the fish protection zones, coastal protective bands and coastal stripes of general use. Part of the territory falls under the additional effect of limitations imposed on the conduct of economic activities in the territory of specially protected natural reservation.

## 3. Methodology

In conditions of various environmental restrictions, economic activity on a territory with a special status without damage to nature is possible only on the basis of the territorial principle of environmental management. The main task of territorial management is the simultaneous consideration of natural, economic and social factors of the development of the territory. Management is carried out by a system of territorial bodies of management of various levels that are responsible for the state of the environment in the jurisdictional territories; see to the implementation of state environmental programs and environmental measures. The territorial management takes into account the interconnection of environmental components, ensures control over the use of natural resources by nature users and encourages them to use resource-saving technologies.

For the effectiveness of territorial management, a tool is needed that accurately and visually regulates relations in the use of resources and lands, a mechanism that visually represents the operation of environmental regulations and reflects the contradictions in real and permitted use. The quality of such management largely depends on the usefulness and timeliness of the receipt of characteristics about the territory, on the visibility and availability of their presentation. In the tasks of territorial management, the information, with the help of which decisions are made and their consequences are assessed, acquires a system-forming significance.

To make decisions in the current situation of multi-level environmental restrictions, we have developed a tool for legal environmental zoning - an instrument of information (legal) support for decision-making in the mechanisms of territorial management that ensures the adequacy of managerial decisions through the consideration of objective and stable features of the region [8]. Legal zoning is an expression of the modern system of relations to the land at the state, regional and local levels, enshrined in laws and regulations. It regulates relations in the use of land and resources at the level of the constituent entity of the Federation and the municipality (district), allows to visualize the effect of environmental regulations and implement a long-term forecast of territorial development. With its help, the adequacy and targeting of management decisions and impacts on a particular territory is ensured by unambiguous and accurate interpretation of multi-level environmental restrictions on the conduct of economic activities for each specific site (local location features and targeted management, rather than regulation of the situation as a whole).

## 4. Results and Discussion

At the heart of legal zoning lies the division of the territory into a certain number of zones, where for each land plot within these zones, certain regulations are established. It is clear that it is possible to draw boundaries for the implementation of a normative act only when the structure of the natural basis of the territory is known. For example, water protection zones are defined on both sides of the river in width, depending on the extent of the river (figure 1). Obviously, even in this simplest case it is necessary to know the location of the river and its natural characteristics. The geoinformation systems (GIS) containing the entire volume of basic information (topographic and landscape basis, natural and economic infrastructure, space images, forest taxation database, socio-economic characteristics of populated areas, etc.) are the best tool for obtaining a complete and clear picture of the structure and potential of the territory, what is potentially necessary for solving various tasks of territorial management [9].

Each information object contained in the GIS has a replenishing set of basic and unique characteristics, which allows, taking into account legislative environmental regulations, automatically assign it to the appropriate environmental category with strictly defined regulations for conducting economic activities. Such regulations can be: types of permitted use of environmental management facilities, the maximum loads on the territory, the restoration potential of the territory. Laws are, as it were, refracted on the terrain and act in accordance with how significant and meaningful their application is for a particular object. Thus, a map of legal environmental zoning is obtained, which visualizes the regulatory and legal framework for a particular territory. The system of laws becomes not in itself, but oriented to a particular object of use.

**Figure 1.** Water protection zones of rivers and their sources.

For visualization of normative and legal information, methods of through geographical mapping of the territory are used, the logic of which is based on the unity of the three coordinate principles "nature - economy - population". In this case, GIS technologies not only allow to speed up the processing of information, but also harmoniously supplement the standard operations with databases with the advantages of visualization and the ability to solve spatial analysis problems using electronic maps and involving mathematical modeling and through mapping methods [10].

A multi-layered map of the legal zoning of the Slyudyanka district of the Irkutsk region, which is part of the CEZ BNT, reflecting the features of the legal regulation of nature protection activities was created.

It can be seen (figure 2), as a result of the imposition of environmental laws, the entire territory of the region falls into the system of severe environmental constraints of economic activity, among which it is difficult to find a suitable place for conducting economic activities. However, such territories are found. Simultaneous visualization of all current regulatory and legal acts of all levels in the form of a map of legal environmental zoning makes it possible to identify areas on which it is possible to conduct legally permitted economic activities.

IOP Conf. Series: Earth and Environmental Science **190** (2018) 012056 doi:10.1088/1755-1315/190/1/012056

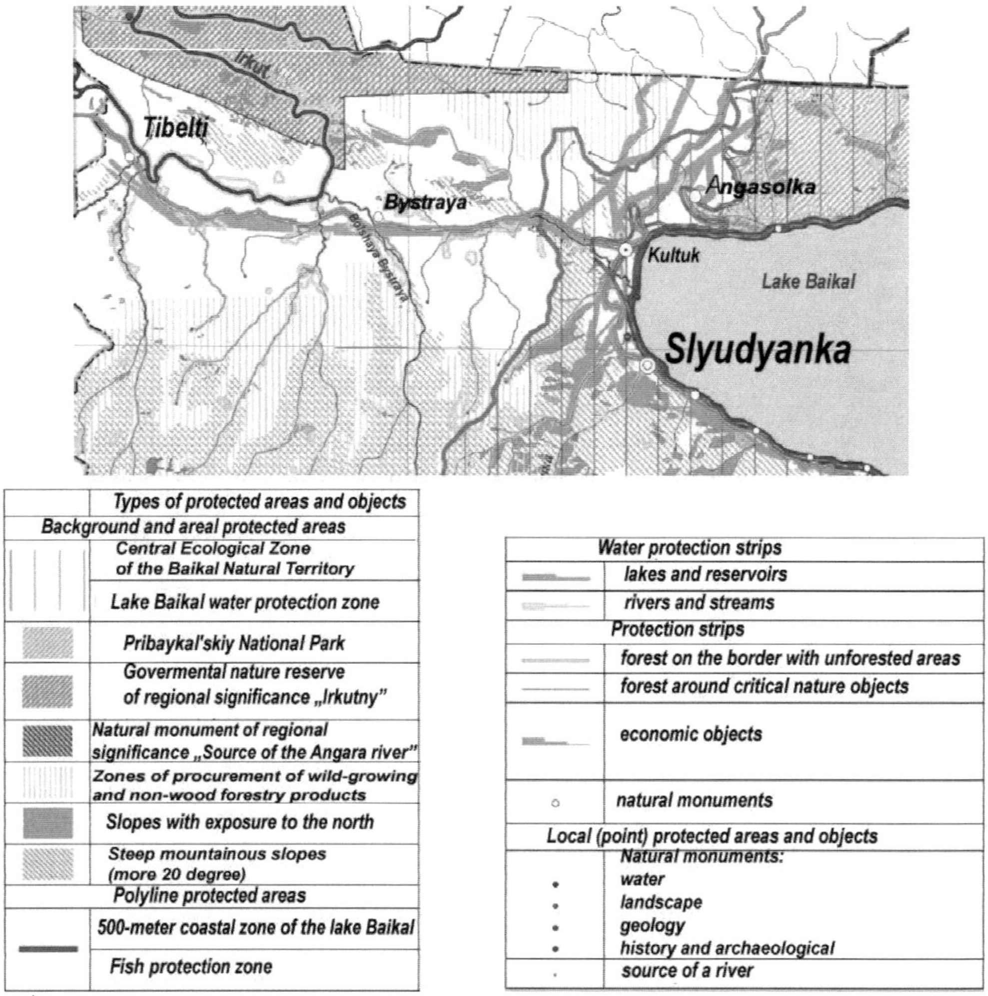


**Figure 2.** Fragment of the legal zoning map of Slyudyanka district of the Irkutsk region.

At the next stage, the management of nature use in the territory should be carried out taking into account the category (purpose) of land (established and regulated by the Land Code of the Russian Federation and other legislative acts and federal laws, including laws established in Russian Federation and established in certain subjects of the Russian Federation) and also the permitted use of land. The legal regime of lands established by the Land Code of the Russian Federation should be adjusted taking into account environmental restrictions identified as a result of the legal environmental zoning procedure. The final determination of the type of permitted use of the land parcel assumes the definition of a specific purpose for the use of this land plot, taking into account all possible restrictions, including the assignment of those types of activities that can be carried out on the land plot by its proprietor, types of objects that can be located on land and types of rights property.

## 5. Conclusion

Since in the territories with special status the same land area falls under the influence of several legal regimes: a specific category of land, a territorial zone and a zone with special conditions for the use of territories, it is the proposed territorial management mechanism that will overcome the difficulties in determining its priority legal regime and organize its targeted and sustainable use.

IOP Conf. Series: Earth and Environmental Science **190** (2018) 012056 doi:10.1088/1755-1315/190/1/012056

**References**

[1] Yaborov V T 2011 *East Agrarian Bulletin* **2** 56-60
Tolstikov A V 2014 *Journal of Scientific Research Publications* **15** 19-31
[2] Komarov S I and Antropov D V 2017 *Bulletin of Ural Federal University. Series Economics and Management* **16**(1) 66–85
[3] Uskova T B, Voroshilov N V, Gutnikova E A and Kozhevnikov S A 2013 *Socio-Economic Problems of Local Territories* (Vologda: Institute of Socio-Economic Development of Territories RAS) p 196
[4] Lapina M A 2008 *Environmental Law* (Moscow: Konsultant-Plus) p 148
[5] Maksimova I I 2010 *Geography and Natural Resources* **4** 152-5
[6] Titosyan E A and Grin E A 2015 *New Science: Experience, Traditions, Innovations* **7**(2) 245-7
[7] Khandazhapova M L, Dagbaeva D S, Ayusheeva N S, Sanzheev D E, Lubsanova B N and Eremko S Z 2016 *Geography and Natural Resources* **5** 137-43
Potapov L V and Dobrovensky Y P 2016 *Bulletin of the Buryat scientific center SB RAS* **3** 151-7
[8] Myasnikova S I 2007 *Geographical Research in Siberia* vol 4, ed A K Cherkashin (Novosibirsk: Academic publishing house "GEO") chapter 8 p 253-7
[9] Vladimirov I N et al 2002 G*eoinformation System of Territorial Management* (Irkutsk: Institute of Geography SB RAS) p 151
[10] Lesnykh S I and Cherkashin A K 2015 *Geodesy and Cartography* **8** 10-6

# Sino-Russian natural gas cooperation: structure and situation

**Z Li[1,2], J Li[1,2], Y Yang[1,2], V A Kolosov[3,4], Y Ren[1,2*], F Chen[1,2]**

[1]Institute of Geographic Sciences and Natural Resources Research, Chinese Academy of Sciences, Beijing, China
[2]University of Chinese Academy of Sciences, Beijin, China
[3]Institute of Geography, Russian Academy of Sciences, Moscow, Russia
[4]Department of Geography, Moscow University, Moscow, Russia

E-mail: *llfsry@sina.cn

**Abstract**. Sino-Russian natural gas cooperation has an important impact on natural gas trade in both Russia and China. Based on the Herfindahl-Hirschman Index (HHI), this paper has established a natural gas import and export security evaluation system to scientifically assess Sino-Russian gas cooperation with respect to the Sino-Russian natural gas structure and situation under the influence of different factors. The research shows: the diversification of natural gas imports in China has gradually increased. China's natural gas imports from Russia are on the rise; When accounting for the export potential of the importing countries, China's HHIC-PE reveals a wavy upward trend after falling to a low in 2008, and then, upon reaching a peak in 2013, it presents a stable, albeit slightly declining, trend; After accounting for price, the overall HHIC-PE-OP declines, rises and declines again; HHIR shows a sharp decline after it increases, and it then rises again; After accounting for the import potential of exporting countries, the Russian natural gas export risk exhibits two M-type declines; The drop in natural gas prices in recent years has had a significant negative impact on Russian natural gas export security; In the next few years, the sharp increase in Sino-Russian natural gas trade volume will increase the import concentration of both Russia and China, which should guard against the rising natural gas import and export risk.

## 1. Introduction

With China's energy consumption and demand continuing to increase, China has become the world's largest energy consumer [1]. China's natural gas dependence increased to 39% in 2017 and its natural gas consumption and foreign dependence will increase rapidly in the future [2]. Russia is a major oil and natural gas producer and exporter at the global level. Its oil and natural gas exports to Europe account for 65% to 70% of its total exports. Accordingly, Russia must strengthen its energy cooperation with the Asia-Pacific region, especially with China, if it wants to eliminate its excessive dependence on the European energy market and enhance its position in the international energy landscape [3]. In recent years, China and Russia have increased their cooperation in the development of natural gas in the Arctic region. On November 1, 2017, during the 22nd regular meeting between the Chinese and Russian prime ministers, PetroChina and Russia Novatec signed a strategic cooperation agreement, and the two sides will continue to work closely together on the Yamal LNG project. By 2019, after the completion of the Yamal LNG project, the three production lines of the project will supply 4 million tons (equivalent to 6 billion cubic meters) of liquefied natural gas to China each year.

Much research has been performed on the progress, situation, significance, problems and countermeasures of the Sino-Russian energy cooperation project. Yang et al. [3], Sun et al. [4], Yang [5], Xu et al [6] reviewed and summarized the progress of Sino-Russian energy cooperation since the collapse of the Soviet Union. Zhang et al. [7], Chen [8], Wang [3], Guo et al. [9] each analyzed the problems and constraints of Sino-Russian energy cooperation. In recent years, increasingly more studies have begun to use quantitative analysis to evaluate the security of the energy trade [2].

In 1974, the International Energy Agency (IEA) was established and defined energy security as the ability to obtain a sufficient energy supply at a reasonable price [10]. Many scholars started to work on energy security from different angles under the leadership of the IEA [11]. In 1975, Willrich first mentioned the term "energy security" in his book [12]. Maull [13], Deese [14] considered energy security to be the unification of energy economic security and energy ecological security and promoted energy economic security as the basic goal of national energy security. The Asia-Pacific Energy Research Center contends that energy security refers to a state in which energy is available and acceptable [15]. Based on the definition of HHI index and energy security, this paper constructs a natural gas import and export security evaluation system for Sino-Russian natural gas cooperation and scientifically evaluates the security pattern and situation of Sino-Russian natural gas trade between 2005 and 2016 under the influence of different factors and providing scientific guidance for promoting natural gas cooperation between the two countries.

## 2. Models and Methods

### *2.1. Models and Methods*

The most commonly used index for measuring energy import security is the HHI, which is based on the measure of diversity. This paper improves the HHI index and builds a natural gas import risk assessment index that comprehensively reflects China's natural gas import security and creates a natural gas export risk assessment index that reflects the security of Russian natural gas exports.

$$HHI_i = \sum_{j=1}^{n} w_{ij}^2 \quad (1)$$

Natural gas import source country distribution (Wcj), natural gas import source country export potential (PEj) and international natural gas price (OP) are key factors that directly affect China's natural gas import security. HHIC, HHIC-PE, and HHIC-PE-OP are China's natural gas import risk indexes, which, respectively, only consider the distribution, the distribution and the export potential, and, comprehensively, the distribution, the export potential and international prices.

$$HHI_C = \sum_{j=1}^{n} w_{Cj}^2 \quad (2)$$

$$HHI_C - PE = \sum_{j=1}^{n} w_{Cj}^2 \times \frac{1}{PE_j} \quad (3)$$

$$HHI_C - PE - OP = \left(\sum_{j}^{n} w_{Cj}^2 \times \frac{1}{PE_j}\right) \times OP \quad (4)$$

The distribution of natural gas export destinations (WRj), the import potential of natural gas exporting countries (PIj) and the international natural gas price (OP) are important factors that affect the security of Russian natural gas exports. HHIR, HHIR-PE, and HHIR-PE-OP are Russia's natural gas export risk indexes which, respectively, only consider the distribution, the distribution and the import potential, and, comprehensively, the distribution, the import potential and international prices.

$$HHI_R = \sum_{j=1}^{n} w_{Rj}^2 \quad (5)$$

$$HHI_R - PI = \sum_{j}^{n} w_{Rj}^2 \times \frac{1}{PI_j} \quad (6)$$

$$HHI_R - PI - OP = \left(\sum_{j}^{n} w_{Rj}^2 \times \frac{1}{PI_j}\right) \times \frac{1}{OP} \quad (7)$$

### *2.2. Data*

The data were obtained from the 2006 to 2017 BP Statistical Yearbook. Since there are different dimensions between the indicators, the original data must be standardized.

## 3. Results and Discussion

### *3.1. China's natural gas import security*

*3.1.1. China's natural gas import risk index that only includes natural gas import sources (HHIC)* China's HHIC experienced a rapid decline after 2005, reaching its lowest point in 2010 before recovering to a relatively high point in 2013, after which it began to decrease again. (figure 1).Prior to 2009, China's natural gas import sources were clustering. In 2005, China's natural gas imports from Indonesia and Malaysia accounted for 51.51% and 42.66%. Driven by China's energy import diversification strategy, China's major natural gas import sources began to expand (figure 2). In 2016, the number of countries from which China imports natural gas accounting for more than 5% of its total natural gas imports was five. The concentration of China's natural gas imports has gradually decreased. During this same period, the amount of natural gas imported from Russia increased from before 2009 to 2.18% in 2016. The increase in the proportion of imported natural gas from Russia and other countries plays an important role in reducing the risk of natural gas imports.

*3.1.2. China's natural gas import risk index that includes natural gas sources and natural gas export potential (HHIC-PE)* China's HHIC-PE shows a wavy upward trend after hitting a low in 2008 and then, after reaching a relatively high value in 2013, it experiences a relatively stable decline (figure 1). This is significantly different from HHIC as a whole, thus indicating that the export potential of natural gas import sources has an important impact on China's natural gas import security. From 2008 to 2013, the export potential of China's major natural gas importing countries exhibited either stability or a wavy decline. After 2014, the export potential of China's major natural gas importing countries showed a steady or slow upward trend. In 2010, the impact of the increase in the import risk index caused by the decline in exporting power of importing countries exceeded the impact of the decline in the import risk index caused by decentralization. It is further noted that with Russia's emergence and its proportion in China's natural gas import market, Russia's increasing natural gas export potential will play an important role in improving China's import security.

*3.1.3. China's natural gas import risk index that includes natural gas import sources, natural gas export potential and natural gas price (HHIC-PE-OP).* After considering the natural gas import source, export potential and price fluctuations, HHIC-PE-OP revealed a downward trend followed by an increase and then a decrease (figure 1). From 2005 to 2016, international natural gas prices fluctuated drastically, resulting in dramatic fluctuations in China's natural gas import security (figure 3). Before 2008, international natural gas prices exhibited an upward trend, but the outbreak of the financial crisis caused a serious oversupply in the international natural gas market, which caused the price to fall. Since then, with the recovery of the world economy, natural gas demand has gradually recovered, and the huge natural gas supply and demand gap has caused international natural gas prices to increase rapidly. However, since 2013, due to the U.S. shale gas revolution, the overcapacity caused by increased natural gas imports, a slowdown in global economic growth, international natural gas prices have fallen rapidly, which has greatly reduced China's natural gas import risk index in the short term.

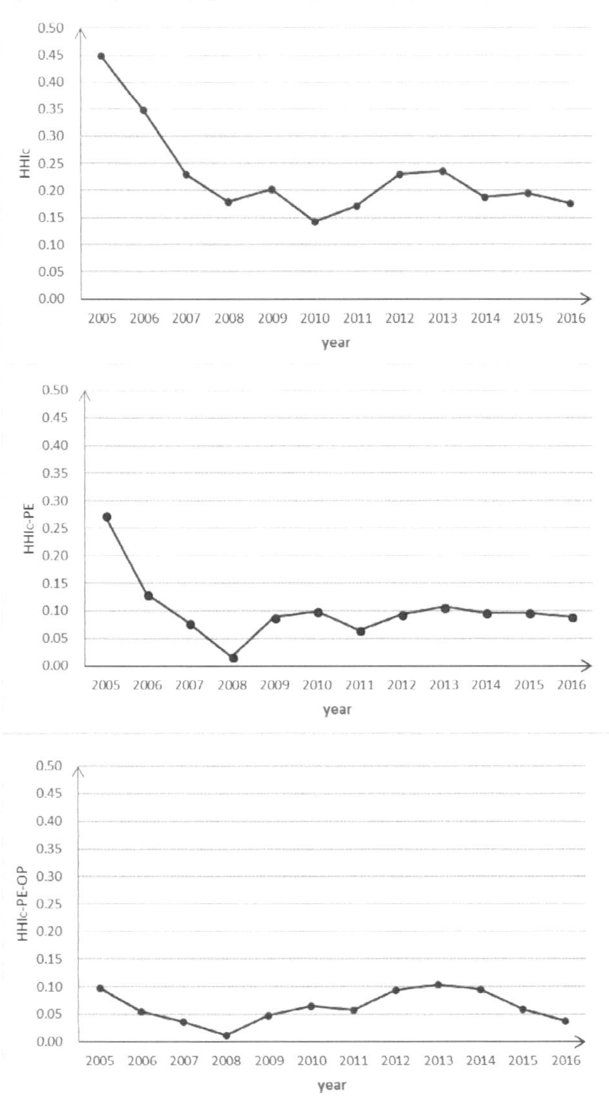


**Figure 1.** HHIC, HHIC-PE and HHIC-PE-OP for the period from 2005 to 2016.

*3.2. Russia's natural gas export security*

*3.2.1. Russia's natural gas export risk index that only includes oil export markets (HHIR)* The HHIR expresses an upward trend prior to 2007 and then a wavy downward trend until 2009, after which it experiences another upward trend (figure 4). Before 2007, Russia's natural gas exports exhibited a trend of concentration. As the target of natural gas exports diversified, the number of Russian natural gas export countries increased from 18 in 2008 to 29 in 2009. As a result, the Russian natural gas export risk index plummeted in 2009. The rapid, high concentration of export targets caused the

Russian natural gas export risk index to rebound from 2010 to 2011. The HHIR continued to rise slowly between 2012 and 2016.

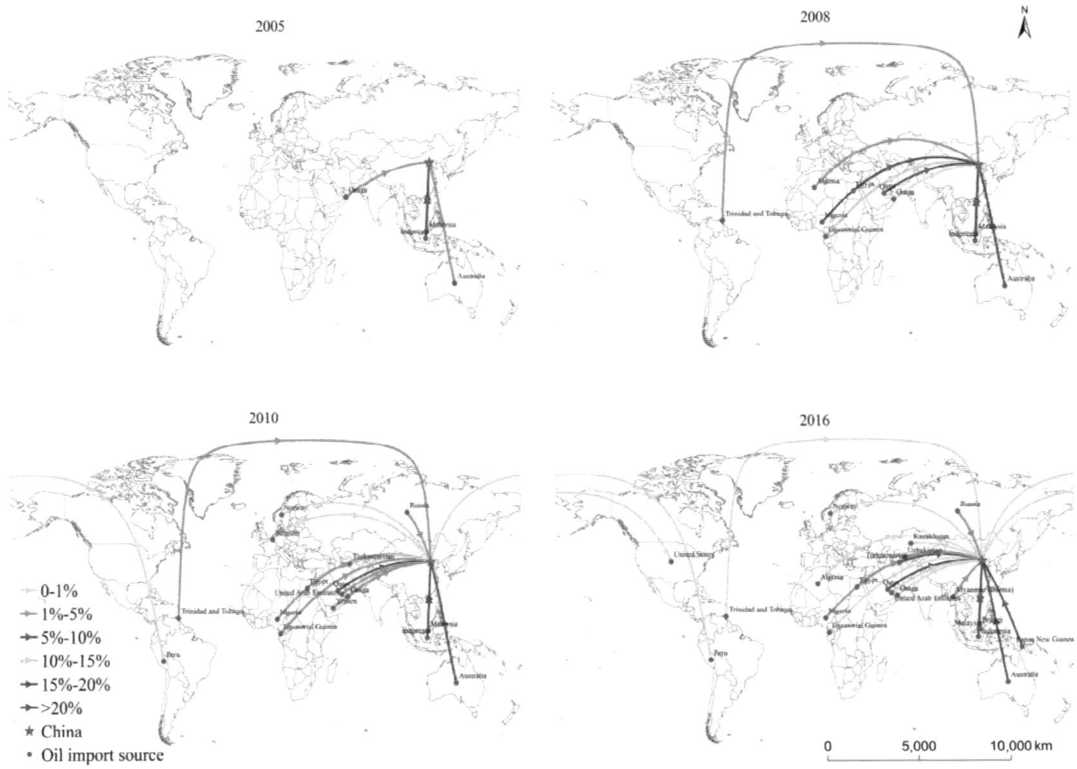


**Figure 2**. The distribution of China's natural gas import sources.

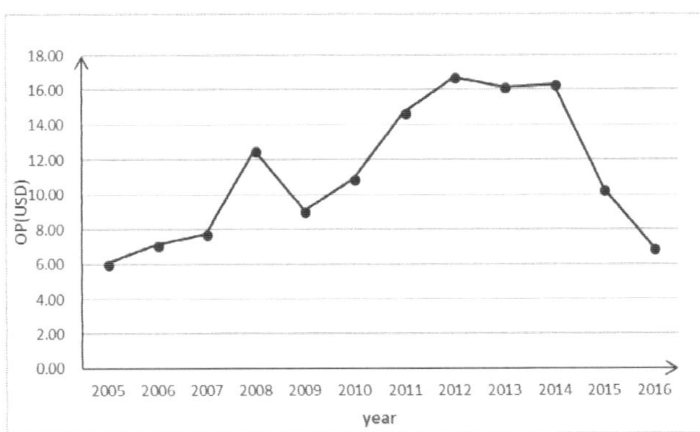


**Figure 3.** OP for the period from 2005 to 2016.

*3.2.2. Russia's natural gas export risk index that includes natural gas export markets and natural gas import potential (HHIR-PI)* From 2005 to 2016, the HHIR-PI showed a wave-shaped downward trend and two M-type evolutions. It peaked in 2007, 2009, 2012, and 2014 and then experienced a downward trend. In 2020, the HHIR-PI was relatively stable (figure 4). Comparing the HHIR and the HHIR-PI, the import potential of Russian natural gas export targets has an important impact on Russia's natural gas export risk. Before 2007, the HHIR-PI experienced steady growth because the demand for natural gas in Russia's main export destinations was gradually decreasing. In 2009, affected by the financial crisis, the global economy was in trouble and energy demand declined substantially. due to factors such as the shale gas revolution from 2010 to 2012, the main export destinations of Russian natural gas were still exhibiting a downward trend. Since 2014, international gas prices have fallen rapidly, thereby stimulating the consumption of natural gas resources. From 2005 to 2016, with China maintaining a strong demand for natural gas resources, natural gas consumption and natural gas imports continued to rise.

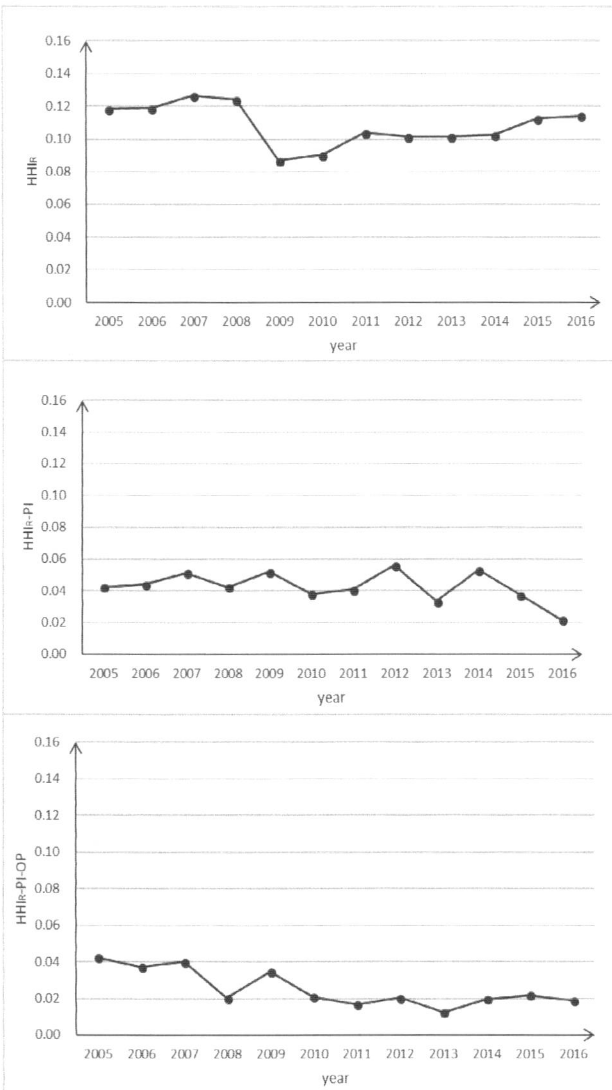


**Figure 4.** HHIR, HHIR-PI and HHIR-PI-OP for the period from 2005 to 2016.

*3.2.3 Russia's natural gas export risk index that includes natural gas export market, natural gas import potential and natural gas price (HHIR-PI-OP)* Before 2008, the HHIR-PI-OP exhibited a wavy downward trend, mainly because of the significant increase in international natural gas prices between 2005 and 2008. The positive effect of the decline in natural gas export risks strengthened the positive effects of the rising import potential of major natural gas exporting countries, while the risk to Russian natural gas exports plummeted. However, in 2009, affected by the financial crisis, international natural gas prices fell, resulting in a sharp increase in the Russian natural gas export risk index. Since then, international natural gas prices have rebounded. Another sharp decline in international natural gas prices greatly increased Russia's natural gas export risk index in 2014. In 2015, the positive effect of the rebound in import potential was completely offset by the negative effects of the collapse in natural gas prices. In 2013, the HHIR-PI-OP began to increase and this continued until 2016.

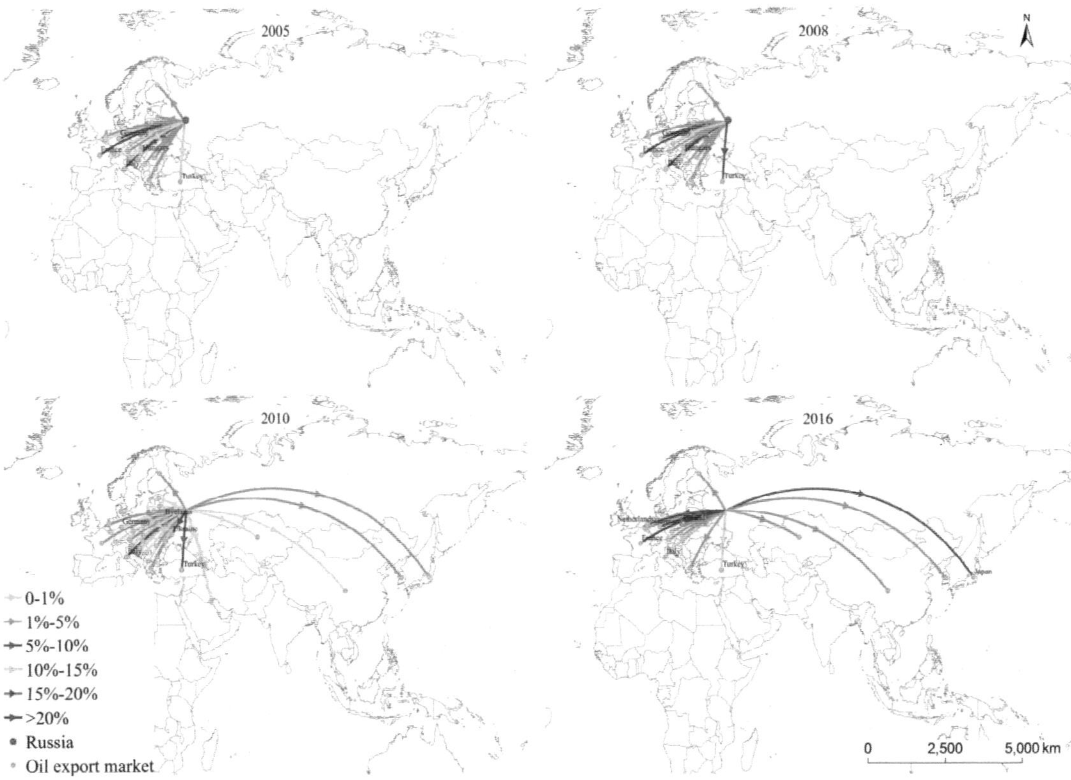


**Figure 5.** The distribution of Russia's natural gas export markets.

*3.3. Prediction of the impact of Sino-Russian natural gas cooperation on the natural gas trade between the two countries under the background of "The Belt and Road"*

In accordance with the natural gas cooperation agreement made by the Chinese and Russian governments and the related plans, progress continues with respect to the China-Russia natural gas pipeline construction and the China-Russia Arctic natural gas cooperation development project. By 2020, China will be importing up to 68 billion cubic meters of natural gas and 4 million tons of liquefied natural gas per year from Russia, totaling approximately 74 billion cubic meters of natural gas. Under the premise that the same amount of natural gas trade continues with other countries, by 2020, China will become Russia's largest natural gas export market, and Russia will become China's largest natural gas import market.

## 4. Conclusion

China's main natural gas import sources are increasingly widespread, the concentration of natural gas imports has gradually decreased and the HHIC is experiencing a downward trend. Among these points, the rising proportion of China's imports of natural gas from Russia has had an important impact on balancing the sources of natural gas imports in China; China's HHIC-PE exhibited a wavy upward trend after hitting a low in 2008. Then, after peaking in 2013, it experienced a slight decline in basic stability; HHIC-PE-OP experienced a decline, then an increase and then another decline, which indicates that international natural gas prices have a huge impact on China's natural gas import risks. At the end of 2017, an insufficient natural gas supply in China caused prices to soar and the natural gas supply to become extremely unstable. In the next few years, China will increase the amount of natural gas it imports from Russia. In accordance with the Sino-Russian gas cooperation agreement and related plans, by 2020, Russia will become the largest exporter of natural gas to China.

Russian natural gas export targets experienced trends of concentration, dispersion and agglomeration. Among Russia's export targets, the natural gas exports to China, when considering the proportion of total Russian natural gas exports, revealed an initial decline and then a continuous increase. To a certain extent, this single trend has an impact on the security of Russian natural gas exports; The HHIR-PI presents two M-type drops; In recent years, international natural gas prices have fallen, and there are obvious side effects on Russia's natural gas export risks; Under The Belt and Road initiative and the construction of the China-Mongolia-Russia Economic Corridor, the Sino-Russian gas pipeline was built and ventilated and the Sino-Russian joint development of natural gas projects in the Arctic region was implemented. While Russia's gas exports will be more concentrated in China, the export risks will not increase too rapidly because China's strong demand for Russian natural gas has an important impact on offsetting the negative effects of rising export risks caused by the falling demand for natural gas in other countries.

**References**

[1] Hao Y B and Tian C S 2014 Sino Russian Energy Cooperation: Progress, Motivation and Influence *N. E. As. For* **23** 71-83

[2] Wang X Y 2015 Research on Evaluation and Countermeasures of Natural Gas Supply Security in China *J. Chin. Univ. Petr (Edit. Soc. Sci)* **31** 6-10

Chen X L 2017 China's Energy Security and Sino-Russian Energy Cooperation *N. E. As. For* **26** 59-73

[3] Wang H Y 2011 Favorable Factors and Restraints in Sino-Russian Energy Cooperation *Acad. J. Rus. Stud* **1** 5-9

Yang Y, Dong S C and Li Z H 2018 Progress, Driving Forces and Challenges of Sino-Russian Energy Cooperation Under the China-Mongolia-Russia Economic Corridor *Res. Sci* **40** 237-49

[4] Sun L D, Qian X K, Jiang X F, Chen R and Su X L 2015 Oil and Gas Cooperation Between China and Russia for 20 Years *Intl. Petr. Econ* **23** 1-9

[5] Yang L 2014 The Course and Prospect of Sino-Russian Natural Gas Cooperation *Rus. Cntrl. As & E. Eur. Mkt* **5** 86-97+128

[6] Xu H F and Shi J M 2014 Sino Russian Rnergy Cooperation *Acad. J. Rus. Stud* **4** 85-92

[7] Zhang H L and Qin P L 2015 The Game in Sino-Russian Energy Cooperation and Its Influence on Geopolitics and Economics *J. Shanghai Univ (Soc. Sci)* **32** 32-44

[8] Chen X Q 2011 The interests' Analysis of Sino Russian Energy Within the Framework of the Shanghai Cooperation Organization *J. Univ. Intl. Rlat* **5** 89-94

[9] Guo R and Hong Y L 2009 Problems and Suggestions of Sino Russian Energy Cooperation *Econ. Rev* **9** 40-3

[10] Agency I E and OECD 1985 *Energy Technology Policy* (Paris: Oecd Publications & Information Center Washington Dc)

[11] Zheng Y 2013 Study on China's Natural Gas Safety Evaluation and Early Warning System *Chin. Univ. Geosci*

[12] Willrich M 1975 *Energy and World Politics* (New York: Free Press)

[13] Maull H 1984 *Raw Materials, Energy and Western Security* (London: The Macmillan Press Ltd)

[14] Deese D A and Nye J S 1981 *Energy and Security* (Cambridge: Ballinger Publishing Co)

[15] Intharak N, Julay J H, Nakanishi S, Matsumoto T, Matsahid E J and Aponte A A 2007 *A Quest*

*for Energy Security in the 21st Century-Resources and Constrains* (Tokyo: Asia Pacific Energy Research Centre)

# Agricultural development status and key cooperation directions between China and countries along "The Belt and Road"

**F Li[1], Q Liu[1,2], S Dong[1], H Cheng[1]**

[1]Institute of Geographic Sciences and Natural Resources Research, Chinese Academy of Sciences, Beijing, China
[2]University of Chinese Academy of Sciences, Beijing, China

E-mail: chenghao0011@163.com

**Abstract**. The Belt and Road initiative will improve the interconnection among Asian, European and Africa, promote international trade and the optimal allocation of factors of production among the counties along the Belt and Road. There are abundant agricultural resources and many agricultural countries along the Belt and Road. Promoting agricultural international cooperation has great important significance in putting forward the Belt and Road initiative construction, ensuring food security and promoting sustainable development of the Belt and Road even all over the world. This study researched on the spatial pattern of agriculture resources, agricultural development advantages and disadvantages of the countries along the Belt and Road and the agricultural trade between these countries and China, and based on this, advanced some strategies for our country in promoting agricultural cooperation to the other countries of the Belt and Road which contains: "Land imports", overseas "high-tech green agriculture park" construction and establish "green agriculture international cooperation and development alliance of the Belt and Road initiative". And then, we put forward some safeguard suggestions. This research took the international perspective and scientific analysis to support the national strategy of "The Belt and Road", and the conclusions will provide a strong theoretical basis and strategic support for the construction of "The Belt and Road".

## 1. Introduction

Along the "Belt and Road" region, there are abundant agricultural resources and numerous agricultural powers [1]. Strengthening agricultural cooperation is of great significance to the implementation of "the Belt and Road" and the maintenance of the "Belt and Road Initiative" area and even the world's food security and development [2]. Therefore, it is necessary to reveal and analysis the agricultural spatial pattern and the international trade situation of these countries of this region and based on this research to put forward the key direction of China's agricultural cooperation with them to promoting the development of all the countries along 'the belt and road' region [3].

IOP Conf. Series: Earth and Environmental Science **190** (2018) 012058 doi:10.1088/1755-1315/190/1/012058

## 2. Models and Methods

This study USES the statistics provided by the World Bank and apply the ARCGIS spatial analysis technology and software to analysis the spatial pattern of agricultural development and international trade along the belt and road region.

## 3. Results and Discussion

### *3.1. Agricultural development situation in countries along "The Belt and Road"*

There are many agricultural countries along "the Belt and Road", accounting for four of the top ten countries in terms of agricultural value added. Among them, China's agricultural added value reached 924.027 billion dollars in 2013, ranking first in the world, India was 337.525 billion dollars, ranking third in the world, Indonesia was 125.402 billion dollars, ranking fifth in the world, Russia is 81.714 billion dollars, ranking the ninth in the world. In addition, the added value of agriculture in Southeast Asian countries such as Pakistan is also more than 50 billion dollars, which is also in the forefront of the world. Agriculture has a relatively high proportion of the national economy of the countries in "the Belt and Road". In particular, Afghanistan, Pakistan, Nepal, Laos, Cambodia and other countries account for more than 20% of GDP. Among the five Central Asian countries, Tajikistan, Kyrgyzstan and Uzbekistan Stan's agriculture also accounts for more than 15% of GDP. It can be seen that agriculture plays an important role in the national economic system of countries in the "Belt and Road Initiative" area in 2013. The "Belt and Road Initiative" area was rich in cereal production. Five of the world's top ten grain producing countries are in the region, and three out of the top five are in the region. The "Belt and Road Initiative" area are the regions where the distribution of agricultural land in the world is relatively concentrated, and the agricultural land accounts for a relatively large proportion of the national area. According to the per capita arable land area, there are a large number of countries with densely populated areas along "the Belt and Road". Almost all countries along the Belt and Road are higher than the per capita arable land area of China. In particular, Russia, Central and Eastern Europe, Central Asia, Laos, Thailand, Cambodia and other countries have high per capita arable land and huge agricultural development potential.

### *3.2. Agricultural development advantages and disadvantages in countries in the "Belt and Road Initiative" area*

#### *3.2.1. Agricultural development advantages and disadvantages of Russia.*

*Advantages*. Russia has abundant agricultural resources, with a total land area of 170.982 million square kilometers, agricultural land accounting for 13% of the country's land area, and the world's largest black belt. Since 2000, agriculture has become the priority of national development policies. The Russian government has increased its support for agriculture. Agriculture, especially grain production, is growing rapidly.

Russia has become the world's third-largest wheat exporter and has an important position in the world food market. In addition, pumpkin feed, potatoes, sugar beets and other production are also among the highest in the world. In 2013, Russian agricultural output increased by 6.2% year-on-year, of which grain and bean crops reached 91.3 million tons, up 28.8% year-on-year, and pig and poultry production reached 12.2 million tons, up 4.8% year-on-year, and grain exports were 19 million tons. (table 1).

*Disadvantages*. Despite the huge production of agricultural products in Russia, the product structure is not complete. A large number of agricultural products such as fruits and vegetables are highly dependent on foreign countries. The countries of Europe, the United States, China and the Commonwealth of Independent States are the main sources of Russian agricultural imports. In 2013, Russian food imports totaled 27,026,900 tons, including 9,503,100 tons of fruits and vegetables, accounting for 35.2%; meat and meat products, 2,241,100 tons, accounting for 9.0%; milk and dairy products, 1,462,800 tons, accounting for 5.4%; fish 893,200 tons, accounting for 3.3%; alcohol and non-alcoholic beverages 1.749 million tons, accounting for 6.5%.

**Table 1**. Production of major crops in Russia.

| | Pumpkin feed | Wheat | Potatoes | Beets | Pasture, silage and corn | Basic vegetables |
|---|---|---|---|---|---|---|
| 2009 | 11820 | 6173.975 | 3113.396 | 2489.2 | 2487.569 | 1482.706 |
| 2010 | 8940 | 4150.758 | 2114.054 | 2225.594 | 1281.361 | 1328.3491 |
| 2011 | 12210 | 5623.999 | 3268.147 | 4764.327 | 3079.612 | 1590.6955 |
| 2012 | 10000 | 3771.964 | 2953.253 | 4505.7 | 2192.3426 | 1568.0274 |
| 2013 | 11060 | 5209.0797 | 3019.9126 | 3932.1161 | 2591.0926 | 1548.5353 |

Every year, Russia imports about 2 billion euros worth of fruits and vegetables from the EU, and imports about 1 billion euros worth of food and agricultural products from the United States. However, it has often changed due to factors such as economic sanctions and anti-sanctions in Western countries, which has affected the stability of agricultural imports.

According to statistics from the Russian Academy of Agricultural Sciences, due to the lag in the development of logistics, warehousing and transportation systems and the lack of technology and equipment, the annual loss of grain is 15 million to 20 million tons, 1 million tons of meat and 7 million tons of milk. Only about 280,000 tractors and 90,000 harvesters are needed in Russia to harvest all the grain.

*3.2.2. Agricultural development advantages and disadvantages of Mongolia*

*Advantages*. Animal husbandry is a traditional advantageous industry in Mongolia. It is the foundation of Mongolia's national economy and accounts for 80% of Mongolia's agricultural added value. It is the main raw material source for Mongolia's processing industry and daily necessities. There are 171,000 former herder families and 327,100 herdsmen. The output value of animal husbandry accounts for 10% of export earnings. In 2014, there were 519.7 million head of livestock in the country, a year-on-year increase of 14.2%. The main varieties are sheep, horses, cattle, camels, etc. (table 2).

**Table 2.** Main varieties of Mongolian animal husbandry.

| Species | Livestock volume | | 2014/2013 | | Account for the total proportion | |
|---|---|---|---|---|---|---|
| | 2013 | 2014 | Increment | Percentage | 2013 | 2014 |
| **Total** | 45144.3 | 51970.4 | 6826.1 | 115.1 | 100 | 100 |
| **Horse** | 2619.4 | 2995.2 | 375.9 | 114.4 | 5.8 | 5.8 |
| **Cow** | 2909.5 | 3412.8 | 503.3 | 117.3 | 6.4 | 6.6 |
| **Camel** | 321.5 | 349.3 | 27.8 | 108.6 | 0.7 | 0.7 |
| **Sheep** | 20066.4 | 23209.5 | 3143.0 | 115.7 | 44.4 | 44.7 |
| **Goat** | 19227.6 | 22003.6 | 2776.1 | 114.4 | 42.6 | 42.3 |

*Disadvantages*. Mongolia is sparsely populated, with poor natural conditions and a bad climate. Although the animal husbandry industry is developed, it still needs to import a large amount of meat and milk every year to meet domestic demand. Agricultural output value only accounts for less than 20% of the total output value of agriculture and animal husbandry. The main crops in Mongolia are wheat, barley, potatoes, cabbage, radishes, onions, garlic, and rapeseed. Wheat and potato production can basically meet domestic demand, and vegetable production can meet about 60% of domestic demand.

Mongolia imports a large amount of grain, meat, vegetables and fruits from abroad every year. Imported agricultural products mainly come from China, Russia, the United States and other countries. China is the largest importer of agricultural products in Mongolia (table 3).

**Table 3.** Imports of major agricultural products in Mongolia in 2013.

| Cereals | | Meat and internal organs | | Edible vegetables and tubers | | Melons and nuts | |
|---|---|---|---|---|---|---|---|
| Country | Volume | Country | Volume | Country | Volume | Country | Volume |
| Global | $11,856,182 | Global | $24,712,179 | Global | $12,988,415 | Global | $12,600,84 7 |
| China | $6,451,404 | United States | $12,425,427 | China | $12,370,745 | China | $5,101,687 |
| Russia | $4,687,566 | Poland | $7,615,986 | Russia | $159,038 | Philippines | $1,626,081 |
| United States | $281,589 | China | $3,353,318 | Korea | $144,439 | Poland | $1,529,577 |
| Korea | $174,849 | Canada | $646,785 | Poland | $84,082 | Belgium | $1,435,528 |
| Japan | $148,754 | Australia | $266,390 | Netherlands | $79,662 | Korea | $513,323 |
| Thailand | $58,920 | Hungary | $230,350 | India | $66,285 | Russia | $380,572 |
| Vietnam | $17,901 | Germany | $53,684 | Kazakhstan | $43,532 | United States | $377,967 |
| Poland | $12,258 | Russia | $53,121 | Germany | $16,269 | Germany | $340,947 |
| Kazakhstan | $11,255 | Japan | $51,774 | United States | $11,489 | Italy | $139,584 |

*3.2.3. Agricultural development advantages and disadvantages of Central Asian countries*

*Advantages*. The five countries of Central Asia are sparsely populated, rich in light, heat, water and soil resources, and have unique resource advantages for developing large-scale agriculture. Among them, Tajikistan and Kyrgyzstan have the most abundant water resources, and Kazakhstan has the most abundant land resources. Therefore, the five countries in Central Asia are well-developed. The main agricultural products are wheat, corn and other food crops and cotton. It is one of the world's major food, cotton producing areas and exporting places. Among them, the planting area of wheat in Kazakhstan and other countries is as high as 90% of the total area.

Cotton planting is also widely distributed in Uzbekistan, Turkmenistan, Tajikistan and other countries, especially Uzbekistan. Cotton planting is the pillar industry of Uzbekistan. Uzbekistan is the fifth largest cotton producer and the second largest cotton exporter in the world. The average annual output of seed cotton is 3.6 million tons, lint is 100-1.2 million tons, and exports are 7-10 million tons. Cotton in Uzbekistan and Turkmenistan and wheat production in Kazakhstan have a greater impact on world exports.

*Disadvantages*. Despite the development of crop production in the Central Asian countries, the output and export volume are relatively high. But due to factors such as agricultural financing difficulties, backward agricultural science and technology, and low degree of agricultural mechanization, the output of agricultural products per unit area of farming is much worse than that of developed countries, and the level of land intensive use is not high (table 4).

**Table 4.** Agricultural production per unit area (kg/ha).

| Country\Year | 2005 | 2006 | 2007 | 2008 | 2009 | 2010 | 2011 | 2012 | 2013 |
|---|---|---|---|---|---|---|---|---|---|
| **Russia** | 1860 | 1887 | 1990 | 2387 | 2281 | 1843 | 2259 | 1858 | 2240 |
| **Mongolia** | 481 | 1098 | 943 | 1383 | 1552 | 1370 | 1485 | 1564 | 1337 |
| **Kazakhstan** | 997 | 1170 | 1327 | 1009 | 1249 | 804 | 1689 | 762 | 1164 |
| **Kyrgyzstan** | 2678 | 2551 | 2513 | 2380 | 3034 | 2610 | 2603 | 2367 | 2904 |
| **Tajikistan** | 2164 | 2159 | 2440 | 2141 | 3119 | 3117 | 2519 | 2891 | 2798 |
| **Uzbekistan** | 4042 | 4112 | 4399 | 4352 | 4634 | 4529 | 4734 | 4645 | 4766 |
| **Turkmenistan** | 2927 | 3288 | 2974 | 2673 | 2429 | 2575 | 2025 | 1778 | 1988 |
| **China** | 5,225 | 5,424 | 5,320 | 5,548 | 5,447 | 5,527 | 5,701 | 5,851 | 5,891 |

### *3.3. Bilateral trade between China and countries in the "Belt and Road Initiative" area*

Countries in the "Belt and Road Initiative" area have close bilateral trade with China in the agricultural sector. Russia, the five countries of Central Asia, India and other countries have brought the country of major agricultural products to China for the "Belt and Road" economy. At the same time, it is also the main export destination of China's agricultural products (figure 1) Among them, the trade volume between agricultural products between Russia and China is the largest, and China's exports to Russia are mainly labor-intensive products such as vegetables, fruits and horticultural products. Russia is the fifth largest fruit consumer market in the world. Due to its full range of fruit exports, nutrient-rich, convenient transportation and low prices, China has become one of Russia's largest fruit importers.

Russia's exports to China are mainly resource-intensive products such as aquatic products. After 2010, feed products and oil crops have been added. In the future, China will have stable demand for Russian dairy products. At present, Russia is the largest importer of Chinese aquatic products, and imports from Russian aquatic products account for about 30% of China's total imports of aquatic products.

## 4. Conclusion

Based on the above analysis, we provide the agricultural cooperation strategy between China and countries in the "Belt and Road Initiative" area.

### *4.1. Jointly building a "high-tech green agricultural park" to comprehensively promote the "Belt and Road Initiative" area national agricultural resources development cooperation*

Focus on capital, talent, industry, equipment, information and other advantages, in Russia, Mongolia, Central Asia countries, Pakistan, Thailand, Myanmar, Malaysia and other countries, focus on the layout, build a group of high-tech-supported, green and organic features Agricultural park. Actively promote the formation of a bilateral cooperation mechanism, the Chinese side provides investment and scientific and technological support, and the partner countries provide policy support and guarantees such as fiscal and taxation, land, development and management rights. Relying on high-tech green agricultural parks, we will carry out agricultural R&D cooperation, transformation of agricultural scientific and technological achievements, promotion of green agriculture, and large-scale operation of agricultural industry.

### *4.2. Strengthening the deep processing cooperation of agricultural products*

Central Asia countries, Mongolia and other countries export a large number of agricultural primary products, but they are limited by the weakness of capital, technology and industrial base. They have not yet formed a sound industrial system in the fields of deep processing of agricultural products, high-end health products and pharmaceutical products. China's agricultural product processing, biomedicine and other aspects have relatively mature technology and industrial systems. The prospect of strengthening cooperation in the field of agricultural product processing in the future is broad.

a.Import

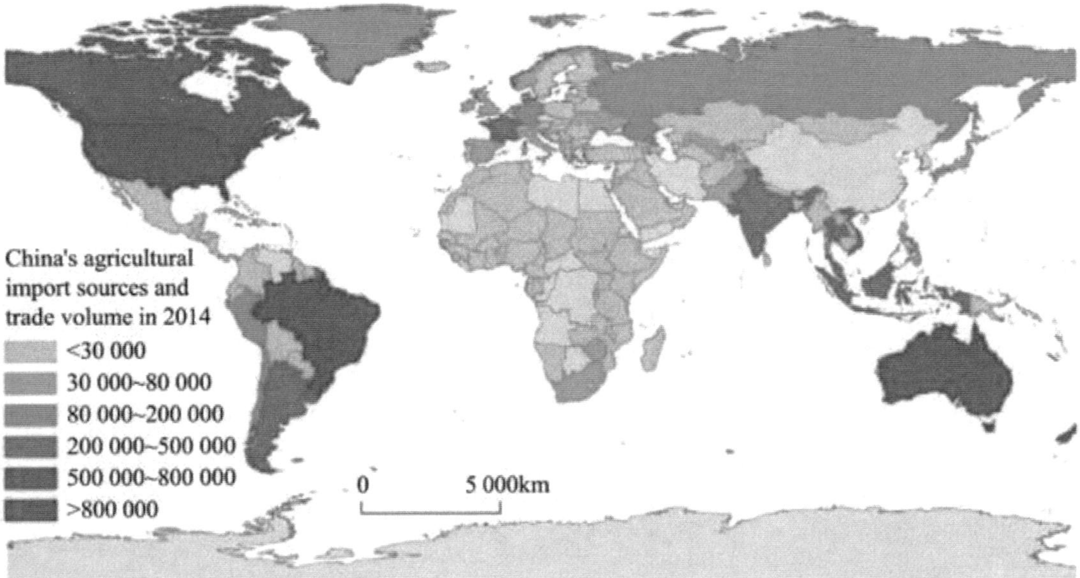


b.Export

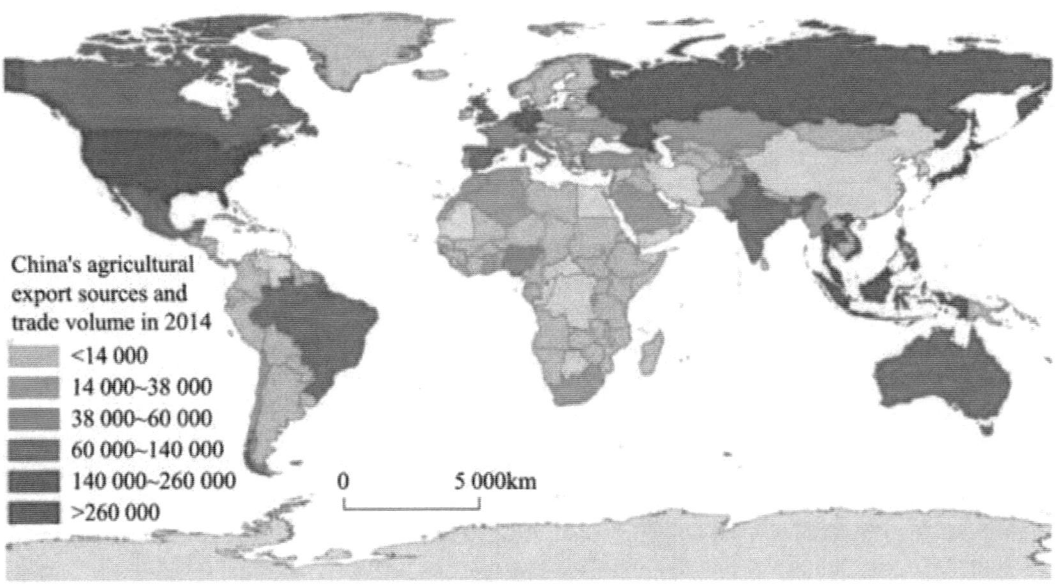


**Figure 1.** China's agricultural import and export in 2014.

### *4.3. Actively promote agricultural financial cooperation*

There are a large number of developing countries along the "Belt and Road", and there are great difficulties in agricultural financing and investment. This leads to inefficient development and utilization despite the abundance of agricultural resources. All countries have given higher policy support and fiscal and tax incentives for agricultural investment, and hope to actively attract foreign investment in agricultural development. Through financial institutions such as the AIIB and domestic policy banks, China can guide funds into the agricultural development areas of countries along the route, and participate in the agricultural development of countries along the route through loans, joint investment, and equity participation.

### *4.4. Strengthen cooperation in agricultural service industry and promote the development of agricultural modernization*

Comprehensively strengthen the extensive and in-depth cooperation between China and countries in the "Belt and Road Initiative" area in the fields of agricultural science and technology research and development, agro-ecological protection and governance, organic agricultural product cultivation, agricultural machinery equipment purchase and leasing, agricultural product intensive processing, biomedicine and other agricultural service industries. Thereby promoting green, modern and sustainable development of the nation.

**Acknowledgements**

Foundation item: National Science and Technology Basic Resources Survey Specialized "Multi-disciplinary Joint Investigation of China-Mongolia-Russia International Economic Corridor" (No.2017FY101300) "Investigation on Economic and Social Development and Investment Environment of China-Mongolia-Russia International Economic Corridor" Project (No.2017FY101304), and National Natural Science Foundation of China (No.41701639). First author: LI Fujia, Associate Professor, E-mail：lifujia2003@163.com

**References**

[1] Zou J, Liu C, Yin G and Tang Z 2015 Spatial patterns and economic effects of China's trade with countries along the Belt and Road *Progress in Geography* **34**(05) 598-605 (in Chinese)
Zhang Y, Yang G and Yang Y 2015 The Belt and Road: an opportunity to strengthen agricultural cooperation between China and central Asia *Journal of International Economic Cooperation* **1** 31-4 (in Chinese)
Song S 2014 Expand agricultural cooperation with other countries under the "the Belt and Road" *Journal of International Economic Cooperation* **9** 63-6 (in Chinese)
[2] Uang Y 2015 Understanding China's "Belt and Road Initiative" as part of its new economic diplomacy *International Economic Review* **1** 48-53 (in Chinese)
Liu H, Yeerken W and Wang C 2015 Impacts of the Belt and Road Initiative on the spatial pattern of territory development in China *Progress in Geography* **34**(5) 545-53 (in Chinese)
Zheng L and Liu Z 2015 Spatial pattern of Chinese outward direct investment in the Belt and Road Initiative area *Progress in Geography* **34**(5) 563-70 (in Chinese)
Gong P, Song Z and Liu W 2015 Commodity structure of trade between China and countries in the Belt and Road Initiative area *Progress in Geography* **34**(5) 571-80 (in Chinese)
Du D and Ma Y One Belt and One Road: The grand geo-strategy of China's rise *Geographical Research* **34**(6) 1005-14 (in Chinese)
[3] Liu W 2015 Scientific understanding of the Belt and Road Initiative of China and related research themes *Progress in Geography* **34**(5) 538-44 (in Chinese)
Dong S, Huang Y, Li Z, Shi G, Mao Q, Li J and Yu H 2014 Economic Development Patterns and Regional Economic Integration Modes for the Silk Road Economic Zone *Resources Science* **36**(12) 2451-58 (in Chinese)

IOP Conf. Series: Earth and Environmental Science **190** (2018) 012059 doi:10.1088/1755-1315/190/1/012059

# Transcontinental bridges VS continentality: the case of Russia and Canada

**A I Lomakina**

Institute of Geography Russian Academy of Sciences, Moscow, Russia

E-mail: lomakina.lomakina@ya.ru

**Abstract**. The article focuses on the role of land bridges as a factor of reduction of continentality, the cases of Russia and Canada are considered. The example of the TransCanada railway, as well as the Transsiberian railway and the Central route of the OBOR, shows that due to relatively cheap tariffs, they will improve the transport and geographical position of inland regions. However, they cannot completely overcome the continentality of these regions and compete on an equal basis with the regions facing the sea or inner waterways accessible for Maritime navigation (the Saint Lawrence and the Great lakes Seaway, Yangtze, the Xi River). The economic effect, similar to that which the coastal regions receive from the benefits of cheap transport routes, is possible in the border Siberian regions because of subsidizing land transportation by the Chinese authorities. At the same time, the development of cross-border transport infrastructure linking adjacent inland regions contributes to the intensification of trade and economic ties between them. Thus, Canada has developed communication between the provinces and the states, and in the future such an intra-continental belt of trade and economic activity may arise in the framework of the Northern course of the OROB.

## 1. Introduction

The access to sea was always counted as benefit trait of economic and geographical location of countries and regions. In the conditions of globalization this trait EGL didn't lose its role, and moreover became more actual. Unequal access to the cheapest marine transport routes predetermines unequal compete advantages of the different countries on the world market, and likewise unequal finance economic indicators of producers and exporters in the framework of the one country.

The significance of the seaside position was evaluated in the classical works on geopolitics (K. Ritter,H. J. Mackinder, Haushofer, Friedrich Ratzel, A. Mahan, P. N. Savitsky, A. G. Dugin, etc.), economic geography and economics (W. Petty, A. Smith, L. I. Mechnikov, N. N. Baranskii, I. M. Maergoiz, V. V. Parsisiusti, etc.). Among recent works, it should be noted the study of L. A. Bezrukov, who developed the first economic and geographical concept of continental-oceanic dichotomy. He also proposed a methodology for assessing the impact of the Maritime situation on international and regional development, most detailed considering the problems, which are associated with the continental Russia [1].

The similar investigations are actual for the other countries, includes giant-countries, which are so familiar with the problems of differentiation and inevitable effect of the "friction of space". This phenomenon is connected with necessity to overcome big distances with relevant time and cost expenditures [2]. The periphery is forms like that, for which is characterized by isolation from the

IOP Conf. Series: Earth and Environmental Science **190** (2018) 012059 doi:10.1088/1755-1315/190/1/012059

most developed and well-developed areas of their own country. This creates a common task for the giant countries of regional policy-the rise of the deep and peripheral areas. Therefore, after China took the initiative of the One Belt and One Road Initiative (OBOR), in the Russian political and scientific discourse, this topic has become extremely popular, although the project itself has caused quite controversial assessments. Some researchers linked the transcontinental bridge with the ways and prospects of economic development of Russia and, above all, the rise of the regions of Siberia and the Far East [3, 4]. Others suggested that it was the OBOR that would be able to offset the costs of continentality [5], and some even saw it as an alternative to sea transportation [6]. Third, without denying the possible benefits, assumed that transport projects within the framework of the OBOR will become an infrastructure for China's economic and political dominance in the Eurasian space [7]. Without trying to touch upon all the problems raised by other researchers, in this article we would like to focus on the following issues. Will land bridges be an alternative to marine transportations? What is their capacity in mitigation of continentality deep regions?

## 2. Results and Discussion

The One Belt and One Road Initiative is a multi-scale project, which works on internal and external strategic interests of the PRC. Domestic interests are connected with the task of development of the Western provinces. The economic backwardness of the internal regions of the country was the result of the economic reforms of the 1980-1990-ies, when all the forces were thrown at the development of coastal areas. "Proximity to the sea" provided the necessary density of foreign trade relations; cheap labor- low production cost, and a special economic regime- a large amounts of foreign investment. As a result, the share of the Western (inland) regions in the total GDP of China in 2000 decreased significantly compared to 1978. To correct this situation, the PRC leadership has prepared a set of measures aimed at the industrial development of the deep provinces, and the development of new transport infrastructure in them, which will contribute to the growth of transport in the Western direction. Initiated by the Chinese authorities, the "flow" of finance from the coastal developed provinces to Xinjiang contributes to the dynamic development of its economy: the region becomes an important energy and resource base of China, development zones and zones of high technologies of state importance are created and successfully functioning in the capital of Xinjiang. A good example of the patronage of the Eastern provinces of the West is "Shanghai automobile company (Xinjiang) "Volkswagen". In 2014, the production of cars reached 10 thousand units, providing the needs of the regions of Central and Western Asia. An interesting example is the "Xinjiang science and technology company Jinfeng", which was founded in Xinjiang, but has gradually become an advanced enterprise on a global scale for the study, development and manufacture of equipment for wind power. It is listed as the 50 largest innovative enterprises in the world. The company's products are sold in Europe, America, Africa and Australia. This economic and technological breakthrough of the province is supported by the development of the transport infrastructure, which connects the Autonomous region with the neighboring States and contributes to the growth of external openness [8, 9].

China's external interests are in some extent a continuation of internal tasks for the development of the Western provinces. In the new paradigm of the country's regional policy, the former periphery becomes a "window to the West", and therefore the main task of the network of transport corridors created outside China is to ensure the widest possible geographical coverage and at the same time reliable sale of goods produced in the West of PRC. It is clear that such a projection of internal tasks on neighboring countries promises certain geopolitical bonuses, but this is not the task of this article.

Another issue actively covered in the domestic publications is the competition between different variants of the silk road and the Transsiberian highway, as well as their ability to compete and create an alternative to marine transport. First of all, we should start with the transport and logistics potential of the Transeurasian corridors. Our interest is attracted by two continental bridges-firstly, the Transsiberian highway as part of the Northern silk road, and secondly, the Central passage of the OBOR-connecting the Central regions of China with Kazakhstan, and further through Russia with Western Europe.

IOP Conf. Series: Earth and Environmental Science **190** (2018) 012059 doi:10.1088/1755-1315/190/1/012059

Contrary to the existing view about competition between Transsiberian highway and the Central course of the silk road, these transport corridors still have different niches and tasks. The first represents a transcontinental corridor linking the ports of the Pacific coast with the Baltic sea, passes through the territory of one country, offers the highest speeds of traffic, but with a limited amount of transit in conditions of approaching the exhaustion of capacity. The Transsiberian railway acts as the framework of the Northern route of the silk road. Unlike the Transsiberian highway, the Central route of the silk road, though associated with a port Lyanyungan in the Pacific, yet is designed to link the inland areas of China with Central Asia through Kazakhstan's Dostyk and Khorgos, and then Russia and further to Western Europe. Of course, the coastal regions of China, associated with the consumer regions with cheap sea bridges, will not send their goods along the land corridor. This route, though shorter than the Transsiberian highway by one thousand km, is still inferior to it in speed. The length of the Russian section, and hence the Russian gain from transit, depends on the chosen transcontinental route. However, in any case, transit revenues will not play any significant role for the budget of transit territories. And the traffic along the Transeurasian corridors has a limited volume.

Despite the increased interest that transcontinental land bridges attract due to the rapid growth of traffic volumes (the rail flow of containers between China and the EU for 2011-2017 increased from 14 to 150 thousand TEU [10]), the water route through the Suez canal is still the main one for cargo flows between the EU and China. The volume of sea container turnover between the EU and China for the period 2009-2016 was generally stable at the level of 10-13 million TEU per year, in the presence of a growth trend – by 31% for the same period. This means that 98-99% of all relevant cargo transportations are carried by sea container ships. Given the average load capacity of one container ship Post-Suezmax carrying more than 12 thousand TEU of containers, the Central route of the OBOR corresponds to two dozens of such containers. The transit flow of containers across the Transsiberian highway not exceeding 150 TEU, against this background, also looks even more modest. Comparison of economic parameters of sea and railway transport clearly shows the advantage of the first. The cost of shipping one container from Shanghai to Hamburg by sea is 2.5 thousand dollars, by rail 9 thousand dollars, by air-30 thousand dollars [10]. The three-fold gap in the economic efficiency of transport, as well as the limited capacity of rail corridors, does not allow them to be considered as a real alternative to marine transport, and the observed increase in transport is due to the subsidization by the Chinese authorities. These subsidies, aimed at zeroing the tariff on the Chinese section of the route for the delivery of goods from the Western and middle provinces to Western Europe (subsidies do not apply to the movement of goods to the ports of export), have reduced the cost of delivery of a 40-foot container by almost half – to 5.5 thousand dollars. This allowed to attract part of the transit container cargo requiring high-speed transportation, as well as containerized cargo to avoid empty container run. However, studies show that the growth of land container traffic is possible only if the volume of subsidies is maintained and increased.

An important task of land corridors is to reducing transport costs, linking neighboring regions with each other, as well as with sea ports and as a result – with the world markets. This is confirmed by the example of such a giant country as Canada. The using of Transcanadian Railways with relatively cheap tariffs allows to reduce the shortcomings of the internal situation of the provinces of Alberta and Manitoba. However, even after receiving transport links to the Center – Ontario and Quebec, Prairie provinces remain closely linked to the United States, and the infrastructure of the Transcanadian trade and transport corridor is used for the supply of export goods to the ports of export on the Pacific coast [11]. This only confirms that the potential to reduce transport costs, and as a consequence of transport and economic continentality is very limited. This is confirmed by the experience of Russia, in which, despite 100 years of operation of the Transsiberian railway, the "friction of space" and the transport and geographical continentality of the regions of Siberia are extremely high. During the USSR period, the shift of Russia's demographic and economic potential to the innercontinental areas was associated with the solution of the problem of complexity and self-sufficiency of the USSR economy. At the post-USSR stage, with the active integration of the national economy into the world economic system, a high degree of continentality caused a significant reduction in the efficiency of the country's economy, which was especially pronounced in the innerland Siberian regions. However, despite the need to overcome the giant land distances, it ultracontinental zone of Russia are the main producers and exporters of fuel and raw materials, being able to adapt to the effects of continentality and weaken it.

In the world practice of Intercontinental countries and regions, a whole range of measures has already been developed, mainly aimed at reducing transport costs: reducing the average distance of transport through the development of interregional and cross-border relations; subsidizing transport and subsidizing intra-continental producers; the mainlining of transport. However, only inland waterways suitable for marine navigation may be a radical solution in spreading the benefits of the seaside position inland. Successful examples include Canada: the Saint Lawrence and the Great lakes Seaway. Lawrence and the Great lakes, accessible to ships and offering year-round (or short-term interrupted) navigation, allows ships to penetrate 3 thousand kilometers inland, directly to the industrial center of the North American continent. The total turnover of canadian ports on inland waterways exceeded 135 million tons (30% of the common Canadian turnover), and the volume of US-Canada «lake trade» exceeded 45 million tons (50% of "marine" trade between US and Canada) [12]. China also has an extensive system of navigable rivers and canals (Yangtze, Xinjiang, Huanghe, Grand canal, etc.), which weakens the negative impact of continentality. In Russia, with its cold freezing shores of the Arctic, not available for shipping, in terms of the placement of the main economic and demographic centres in the innerlandareas at higher latitudes, as removed from sea lanes and borders. This means that in Russia the negative consequences of both Northern and continental situation on economic activity are more tangible.

An important issue is the future of the Siberian regions in view of the OBOR. In a number of publications we can see the expectation of growth of the transit function of the Siberian regions. Even now we can say that this expectation is not fully realized. On the one hand, there is an increase in the absolute flow of transit containers both across the Transsiberian railway (see above) and across border crossings in Zabaikalsk and in Nauska (the absolute flow of containers increased by an average of 3 times). On the other hand, the share ratio shows a decrease in the role of Siberian border crossings: if before 2014 almost all transit rail container traffic between the EU and China passed through Zabaikalsk (95-99%), in 2016 its share decreased to 20%. The share of Kazakhstan's Dostyk border crossing, which grew from 1 to 67%, changed the geography of transit rail transport of container cargo, bypassing the Siberian border terminals and bypassing the Transsiberian railway. Thus, Siberia is no longer a major transit route of OBOR, which in turn limits its influence as a driver of economic development of the Siberian regions.

In addition, the modernization impact of the OBOR does not seem to be any significant, because the planned reconstruction of the Transsiberian railway also does not imply any innovative transformation. From high-speed cargo-and-passenger railways Beijing-Moscow, most likely, they have already refused.

Therefore, we can expect an increase in the load of existing railways and at the same time the growth of goods flow from China to the adjacent Siberian regions. This, in turn, can have both a positive and an adverse effect. First of all, the growth of goods flow will increase the link to the Chinese market due to the fact that the goods will be almost with zero transport costs due to subsidies. Thus, the Siberian regions will benefit from a transport and geographical location similar to the coastal regions: goods will be imported to them with low transport costs, only the costs of export will continue to be high. Orientation of Siberian and far Eastern exports to the adjacent continental regions of China and Mongolia would reduce transport costs by reducing the transport shoulder.

Another effect is indirect, connected not with the transport infrastructure in the Russian regions, but with the rapid development of the neighboring territories of China (catching up with the development relative to the coastal regions).

The growth of the domestic market in Xinjiang, Inner Mongolia and Heilongjiang will lead to the expansion of the Siberian regions's capacity to export not only raw materials, but also industrial goods. Among other things, this will affect food products, which are considered to be high-quality and environmentally friendly. But successful promotion of Russian products requires not only a competent marketing strategy, sufficient volume and appropriate quality, but also the laws protecting licensed products and identify counterfeit ones.

The intensification of trade flows and the development of cooperation in the adjacent territories will lead to the development of North-South relations. A similar effect is observed in Canada, where despite the presence of Transcanadian Railways, the most dense trade relations of innerland regions are observed through the province-state connections [13].

## 3. Conclusion

In response to a question of the study on whether the Eurasian land bridges will be able to compete with Maritime routes, it is safe to say that they will not. This is due to both their technical and economic characteristics. Even if the current pass-through tariff is maintained and the flow of TEU containers continues to increase to 0.5 million by 2020, this volume will not exceed 5% of the number of containers transported by sea.

As for the competition between the TRANS-Siberian railway and the Central route of the silk road passing through Kazakhstan, these trade and transport corridors occupy different niches and perform different strategic tasks. The Transsiberian railway in conjunction with the Northern corridor of the OBOR gives access to the provinces of North-East China to Western Europe, and the Central route connects the Western inland provinces with Europe.

Another issue of the study concerned the possibility of land bridges to mitigate the disadvantages of the Intercontinental situation. The example of the TransCanada railway, as well as the Transsiberian railway and the Central route of the OBOR, shows that due to relatively cheap tariffs, they will improve the transport and geographical position of inland regions. However, they cannot completely overcome the continentality of these regions and compete on an equal basis with the regions facing the sea or inner waterways accessible for Maritime navigation (the Saint Lawrence and the Great lakes Seaway, Yangtze, the Xi River). The economic effect, similar to that which the coastal regions receive from the benefits of cheap transport routes, is possible in the border Siberian regions because of subsidizing land transportation by the Chinese authorities. At the same time, the development of cross-border transport infrastructure linking adjacent inland regions contributes to the intensification of trade and economic ties between them. Thus, Canada has developed communication between the provinces and the states, and in the future such an intra-continental belt of trade and economic activity may arise in the framework of the Northern course of the OROB.

## References

[1] Bezrukov L A 2008 *Continental-Oceanic Dichotomy in the International and Regional Development* (Novosibirsk: Geo) p 369
[2] Trejvish A I 2010 *"Shrinking" of Space: Interpretations and Models* ed S S Artobolevskii and L M Sincerov (Moscow: EHslan)
[3] Stepanov I 2018 The brotherhood of the road ring *EHkspert Sibir'*. Income accessed online on 28th July 2018 via http://expert.ru/siberia/2018/22/bratstvo-dorozhnogo-koltsa
[4] Blagireva E N 2015 New Silk Road. Improving the competitivenesse of Russian regions *Vestnik Akademii* **4** 196-200
[5] Bezrukov L A 2016 The Transsiberian railway and the silk road: global infrastructure and regional development *Ehco* **7** 21-36
[6] Ponomareva N N 2014 The Transsiberian railway as a leading Euro-Asian international transport corridor on territory of Russia *Vestnik of Novosibirsk State Pedagogical University* **20**(4) 57-65
[7] Dergachev V 2017 Chinese strategic initiative "One Belt-One Road". Income accessed online on 30th August 2018 via http://dergachev.ru/geop_events/220517-01.html#.W2GXrS1ePVo
[8] Bazhenova E S 2017 Socio-economic development of Xinjiang – the Outpost of the Western regions of the PRC *Society and State in China* **22** 457-70

[9] Frolova Y Y 2014 China's economic policies in the Xinjiang Uygur Autonomous Region in the context of the interests of Central Asian Countries *Problems of the National Strategy* **4** 65-85
[10] Transport Corridors of the Silk road: the potential growth of traffic through the EEU 2018 (SPb.: CII EABR) p 74
[11] Lomakina A I 2015 Production and export potential of Canadian mining industries in the condition of inner land functioning *Geography and Natural Resources* **4** 171-7
[12] Lomakina A I 2013 Socio-Economic Aspects of Maritime Location of Quebec, Canada *Izvestiya Akademy of Science (Geographical Series)* **3** 14-25
[13] Canadian oil industry in conditions of low prices and landlocked functioning 2017 *Canadian Yearbook. Proceedings of the Russian Society and the Study of the Canadians* ed A N Komarov Vol 21 (Moscow: ROIK) p 88-98

# Role of territorial remoteness, economic and social factors in development of Siberian and northern regions of Russia

**A V Myadzelets**

Sochava Institute of Geography, Siberian Branch of the Russian Academy of Sciences, Irkutsk, Russia

E-mail: anastasia@irigs.irk.ru

**Abstract.** The issue of territorial peripherality of particular regions is considered for the case of Siberian and the northern part of Russia. The peripherality is one of the basic spatio-temporal factors that negatively impacts on economic development, its innovativeness, territorial organization and, through it, on a socio-demographic situation of remote areas. This factor intensifies regional disparity which closely connected with some other geographical factors and may essentially reduce the investment efficiency in the state economy. The impact of the factor is assessed through modeling of connection of different socio-economic and demographic parameters. For the model calculation an indication function is applied. The function connects a generalized integral evaluation parameter of study object state (regional territory) with its individual integral parameters of state (socio-economic characteristics) through coefficients calculated by regression methods. Regional distances and transport accessibility, population life quality index and investment efficiency are chosen as generalized integral evaluation parameters. They reflect complex spatial, economic and social peripheral properties of the investigated area. For the analysis the geoinformation database for every region is created with necessary content of socio-economic statistic information for the period 2000-2015. Results of the research are mapped using GIS-methods. The calculation shows that in spite of the evident peripherality because of the geographical location and low level of transport-logistic infrastructure development there is an opportunity to reduce the factor effect for the regions. It might be reached by increasing of investment efficiency and improvement of population well-being and the life quality.

## 1. Introduction

Active modernization of certain sectors of the economy and lagging of remote territories in the economic development are important contemporary processes in the national economic system of Russia. They cause both rapid unexpected and often opposite natural and socio-economic changes. Such transformations of geographical space, deterioration or improvement of development conditions of regional socio-economic system are considered in the context of territorial remoteness from the center and territorial organization which are real phenomena forming under the influence of various socio-economic factors.

The issue of territorial organization, territorial remoteness and their correlation with social and economic situation is considered for the case of Siberian and northern regions of Russia. The study territory is distant from the central part of the country and differed by difficult geographical conditions. It is reflected on their development and organization of economics, establishment of regional relations and general social attractiveness. Geopolitical situation of last years raises the interest to research and adaptation those areas. Regional geographical environment and socio-economic situation are considered depending on distance between central and remote areas, and interaction between integral characteristics of regional life quality conditions and economic efficiency. Qualitative analysis of spatial differences of regional socio-economic peculiarities and development conditions is fulfilled using mathematic, regression and geoinformation modeling methods. The analyzed period is fifteen years from 2000 to 2015. Used data are published in the official national statistic reports "Regions of Russia".

## 2. Theory and methods

The territorial organization as a geographical process and phenomenon has been studied for a long time [1, 2, 3, 4]. It exists in space and time, connects various factors of the surrounding natural and socio-economic environment and is formed in a natural historical manner [5]. It includes natural, economic and social components, united in interacting territorial geographical systems of different hierarchical level from local to global. The connectivity is manifested as the existence of certain territorial proportions of the production forces, the correlation of production specializations, the interrelationship between the productive and non-productive spheres of the economy and society, taking into account all the prerequisites and limitations of the territorial development [4, 5, 6, 7]. From the other hand, the territorial organization is disclosed through a system of geographical characteristics, information about which is presented as statistical data. They fully and accurately describe the nature, economy and population of the territory through some evaluation function of the combined effect of all factors [8, 9].

The peripherality, or territorial remoteness, is one of the basic spatial-temporal factors that negatively impacts on economic development, its innovativeness, territorial organization and on a socio-demographic situation of remote areas [10, 11]. This factor intensifies regional disparity which closely connected with some other geographical factors and may essentially reduce the investment efficiency in the state economy and level of population life quality.

Some indices are chosen for evaluation and studying the role of territorial remoteness and social and economic factors on regional development. It is spatial peripherality, an integral index of population life quality, and investment efficiency. Spatial peripherality, investment efficiency are indicators of economic development; the integral index of the life quality reflects social well-being and attractiveness of a region. The factors are analyzed through study of complex associations of various socio-economic statistical characteristics. Then comparative assessment is given for every investigated region. These associations are described by both qualitative and quantitative methods such as mathematical modeling and comparative score assessment.

The associations between the socio-economic factors are defined by the mathematical model of an analytic indication function $U=U(x)$, $x=(x_1,x_2,\ldots,x_i,\ldots,x_n)$. The function is reduced to the form $U(x)=a_1x_1+a_2x_2+\ldots+a_ix_i+\ldots+a_nx_n+W(a)$ (1) by some mathematical transformations [12], where $W(a)$ is a transformation function of a number of variables coefficients $a=(a_1,a_2,\ldots,a_i,\ldots,a_n)$. Indicators $x=(x_i)$ are socio-economic factors of territorial development, $U(x)$ is an integral index which reveals the state of a geographical system, for example regional social or economic situation. The model coefficients $a=(a_i)$ characterize association between the factors $x=(x_i)$ and the integral index $U(x)$. Also they show intensity of factor influence on the integral index. The model coefficients are certain properties of local socio-economic conditions [13]. The model structure and coefficients are changed "from region to region" and depend on the geographical situation of the concrete territory. Moreover, using of the indication function model allows to calculate and explain different regional variations of factor association regularities and give some prognosis of economic situation changes.

Three components of model evaluation are chosen for the analysis.

1. Studying the influence of spatial peripherality on the territorial organization is considered through the transport accessibility of regions relative to the central part of Russia. The accessibility index is calculated as the sum of the normalized parameters, such as the distance from Moscow to the

regional centers of the investigated territory, the density of the network of roads and railways, the number of direct flights per day.

2. The integral index of the life quality is analyzed on the basis of a combination of factors of social and economic well-being that comprehensively characterizes the socio-economic and medico-demographic situation in the regions [12, 14]. Four statistic parameters are chosen for assessing the index. It is the average life expectancy of population, the natural increase of population, the total population income, and the crime level. The integral index of life quality is estimated on the basis of the calculated coefficients of the indication model (1). These coefficients reflect the basic connection between socio-economic well-being factors and the integral indicator and reveal their impact on the population life quality of the region.

3. Investment efficiency is estimated based on associations between investment growth or declining and industry and agricultural production in the regions [13]. This connection reflects peculiarities of economic growth or fall in the economy.

All statistic information and calculated parameters are collected as a socio-economic geoinformation database according regional administration division of Russia. Analysis of statistical data is performed using GIS tools or other statistic programs for correlation analysis. The calculated coefficients and estimated parameters are entered into the database structure and then presented in cartographic form.

## 3. Results and discussion

When studying the influence of spatial factors on the periphery, it was assumed that the entire territory of Siberia and the north is peripheral relative to the central part of Russia. According to the analysis three type of peripheraluty are distinguished. It is the nearest, middle and distant spatial periphery (figure 1).

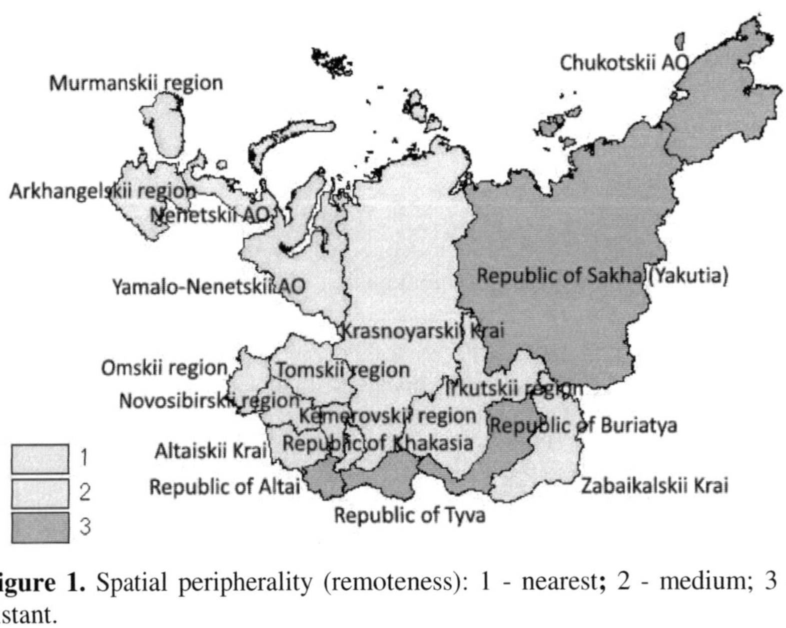


**Figure 1.** Spatial peripherality (remoteness): 1 - nearest**;** 2 - medium; 3 - distant.

The regions of the nearest periphery include the Novosibirskii, Kemerovskii, Omskii, Murmanskii regions, Altaiskii Krai, and Ymalo-Nenetskii Autonomous Okrug (AO). These regions are closest to the central part of Russia. They have good air communication with the capital and are characterized by medium or high density of railroad and paved road net. Arkhangelskii, Irkutskii, and Tomskii region,

Krasnoyarskii and Zabaikalskii Krai, Republic of Khakasia, and Nenetskii AO are referred to the medium spatial periphery. Republic of Sakha (Yakutia), Tyva, Altai, Buryatia, and Chukotskii AO belong to the distant periphery. These regions are far from the center. They have a rare air connection with Moscow, and there is practically no developed railway network.

Estimation of the integral index of the population life quality is made on the basis of the calculated coefficients of the indication model (1) for each region. Then all regions are ranged according the meaning of the coefficients, with range taking into account changes of the life quality conditions during the fifteen period of time from 2000 to 2015. The regions are divided into four groups (figure 2). The first two groups include regions with deterioration of the life quality index that intensifies their peripherality conditions. These are Arkhangelskii region, Republic of Altai, and Nenetskii AO. Regions without obvious changes of the life quality index over the studied period of time belongs to the third medium category. It includes the Krasnoyarskii, Altaiskii, and Zabaikalskii Krai, Republic of Sakha (Yakutia), Novosibirskii, Tomskii, Irkutskii, Omskii, and Kemerovskii regions. Unexpected results are revealed for the fourth group of regions where two of three factors of social and economic well-being have positively changed and affected the quality of life condition. These are Murmanskii region, Republic of Buryat, Khakasia, and Tyva, Ymalo-Nenetskii and Chukotskii AO. Usually they have outsider position in the national socio-economic rating. It is supposed that improvement of the quality of life in these regions is connected with traditional high level of birth rate, but it is weakened by the influence of the spatial remoteness factor.

**Figure 2.** Integral index of the life quality conditions (see region titles at the figure 1): -2 - very low; -1 - low; 0 - medium; 1 - high.

Analysis and diagnosis of economic conditions is given as an example of investment activity in the regions. Results are based on the modeling the indication function (1) and calculating its coefficients. The lowest investment efficiency (group 4 and 5) is defined for the Kemerovskii Region, Republic of Altai and Khakasia, and Chukotskii AO. The highest level of economic efficiency is determined for group 1 and 2 which include Republic of Buryat and Sakha (Yakutia), Krasnoyarskii and Zabaikalskii Krai, Murmanskii Region, and Nenetskii AO (figure 3). The remaining Irkutskii, Tomskii, Novosibirskii, and Omskii regions, Altaiskii Krai, Republic of Tyva, and Yamalo-Nenetskii AO have an average value of the investment efficiency. Their regional conditions also favorably affect the local economic situation and reduce influence of the factor of territorial remoteness.

**Figure 3.** Regional economic efficiency (see region titles at the figure 1): 1 - very high; 2 - high; 3 - medium; 4 - low; 5 - very low.

Thus, despite the pronounced peripheral factor due to the geographical location of the Siberian and northern regions of Russia and the low level of transport and logistics infrastructure, it is possible to reduce the influence of this factor on territorial development. It can be achieved by increasing the effectiveness of investment and improving the conditions of population quality of life, in particular increasing incomes, stimulating measures to increase birth rate, reducing crime level. At the same time, the efficiency of investments is often associated not with proximity to the center, but with the rate of growth of investment and gross industrial and agricultural output.

## 4. Conclusion

Modern territorial usage planning involves improving old and working out new approaches for increasing of the economic development level and favorable social conditions. Management of the process of forming of the territorial organization of society is one of the innovative methods of organizing space. The features of territorial organization and regional development are analyzed by the use of methods of geoinformation and mathematical modeling on the example of remoteness, the integral index of the population life quality, and the investment efficiency. These indicators have local and temporal specificity, i.e. are varied from region to region within certain period of time and determine the direction of economic development of the territory.

In the result of analysis the joint influence of the geographical and socio-economic conditions on economic situation of Siberian and northern Russian regions is shown. The comparative assessment is given depending on the local socio-economic indices for every investigated area. The results of the research allow to carry out not only the assessment of the current situation, but to propose ways of its developing and improving.

## Acknowledgements

The research is supported by the project "Innovative development, territorial organization and increase of population life quality in Siberian and Northern regions of Russia" #16-02-00570(a) funded by the Russian Foundation for Basic Research.

IOP Conf. Series: Earth and Environmental Science **190** (2018) 012060 doi:10.1088/1755-1315/190/1/012060

**References**

[1] Kolossobskii N N 1947 Industrial-Territorial Combination (Complex) in Soviet Economic Geography *Questions of Geography* **6** 133-68
[2] Probst A E 1965 *Efficiency of Territorial Organization of Pruduction* (Moscow: Mysl) p 208
[3] Khrushyov A T 1986 *Geography of Industry in USSR* (Moscow: Mysl) p 416
[4] Khrushyov A T 2009 *Economic and Social Geography of Russia* (Moscow: Dropha) p 607
[5] Annenkov V V 1984 Territorial Organization of Public Reproducation and Direction of its Research *Baltic Region: Geographical Problems of Development* **681** 10-24
[6] Khorev B S 1981 *Territotal Organization of Society* (Moscow: Mysl) p 320
[7] Sharygin M D 2003 *Territorial Society Systems (Regional and Local Levels of Organization and Management)* (Perm: Publishing House of Perm State University) p 260
[8] Baklanov P Y 2007 *Territorial Structures of Economy in Regional Management* (Moscow: Nauka) p 239
[9] Tkachenko A A 2008 Approaches to Creation of General Theory of Territorial Organization of Society *Vestnic Moscovskogo Universitets Seriya 5 Geografiya* **1** 21-6.
[10] Huskey L 2006 Limits to Growth: remote regions, remote institutions *The Annals of Regional Science* **40** 147-55
[11] Huskey L, Maenpaa I and Pelyasov A 2014 Economic systems *Arctic Human Development Report: Regional Processes and Global Linkages* 151-84
[12] Cherkashin A K and Myadzelets A V 2014 Reconstructing a nonlinear dependence of the quality of life on the socioeconomic potential of Siberia's regions *Geography and Natural Resources* **4** 149-60
[13] Cherkashin A K and Myadzelets A V 2015 Comparative Analysis of Investment Changing in Regional Economy *Information and Mathematical Technologies in Science and Management* 209-15
[14] Myadzelets A V 2015 Analysis of socio-economic Well-being Factors of Russian Northern Regions *Arctic Dialog in the Global World* 160-4

IOP Conf. Series: Earth and Environmental Science **190** (2018) 012061 doi:10.1088/1755-1315/190/1/012061

# Social factors of territorial variability of life expectancy in the Russian Federation and features of the Baikal region

**P L Popov, A A Tcherenev**

Sochava Institute of Geography SB RAS, Russia, Irkutsk

E-mail: plp@irigs.irk.ru, tcherenev@irigs.irk.ru

**Abstract.** The article deals with territorial links of the expectancy of the forthcoming life on the aggregate of regions of the Russian Federation with 57 socio-economic phenomena. Phenomena belong to the 8 thematic groups. To identify the social characteristics of the regions of the Russian Federation, the variability of which is most closely related to the variability of life expectancy, the package of programs "Stochastic modeling" was used. It has been established that the life expectancy depends to the greatest extent on the prevalence of crimes against the person, certain infectious and parasitic diseases, abortions, on the level of knowledge of schoolchildren in mathematics and the Russian language, on the level of influence of traditional confessions, on the prevalence of worldview phenomena that are a deviation from traditional religiosity. Factors determining the lowered life expectancy in the Baikal region are identified. The Baikal region is characterized by an increased level of violent crime, an increased number of abortions, a lower level of knowledge of schoolchildren in mathematics and the Russian language, and a lower influence of traditional confessions. These characteristics are reliably combined on the aggregate of regions of the Russian Federation with a reduced life expectancy.

## 1. Introduction

The expected life expectancy at birth is one of the main indicators characterizing the level of development of a country or a group of countries. This parameter is widely used in modern science and is for it the main statistical indicator reflecting the health and living conditions of people [1]. Theoretical and methodological base on this problem are the works of such scientists as J. Vaupel [2], S. Olshansky [3], V. Milholland [4], J. Wilmot [5], B.T. Velichkovsky [6]. Recognition of the value of this parameter is reflected, in particular, in that it is taken into account along with indicators of per capita income and education in calculating the human development index, regularly conducted by the United Nations to assess the level of development of countries and regions of the world.

Life expectancy is related to the notion of "public health". It is believed that the first place among the risk factors for public health is lifestyle; less importance for hereditary factors and pollution of the environment and even less - the work of health authorities and institutions. In this case, the concept of "lifestyle" is difficult to determine. Yu. P. Lisitsyn [7] includes in the content of this concept such components as level, way of life, style and quality of life. A considerable amount of data on the problem of life expectancy has been accumulated. Modern research is carried out mainly in two scientific areas - biology-medical and socio-ecological [8]. The second direction, which includes our work, sometimes also has a link to the territorial natural and climatic factors of the population's living. In our study, only social factors are considered.

IOP Conf. Series: Earth and Environmental Science **190** (2018) 012061 doi:10.1088/1755-1315/190/1/012061

It is common knowledge that life expectancy varies in different countries. It differs by regions of Russia. It is important to know what causes such variability. To try to advance in understanding this problem, we will use various methods to investigate the multifactor statistical relationships between parameters that reflect the territorial variability of life expectancy and the parameters of regions on which this duration may depend. The main task of this paper is to identify the most significant factors affecting the territorial variability of life expectancy in Russia and assess their contribution to the formation of such variability. As quantitative indicators reflecting the influence of various factors on life expectancy, we will consider any of the parameters of regions considered by us in this study separately or certain combinations of these parameters.

## 2. Materials and methods

We conducted a study of the territorial variability of life expectancy in the regions of the Russian Federation, using the "Stochastic Modeling"[9] software package, in which social characteristics of the regions of the Russian Federation were identified, the variability of which is most closely related to the variability of life expectancy. This software package is a convenient tool for investigating statistical dependencies. It includes many different ways to solve this class of problems. The simplest and freely distributed version of it is available through the Internet to any interested person. The use of this program allows in manual or automated mode to formulate mathematically and test various hypotheses about the dependence of variables on each other. The program provides the user with the opportunity to build models of multifactor stochastic relationships between quantitative characteristics possessing different mathematical structure on the basis of real data on their values in joint implementations. It also provides an opportunity for the researcher, on the basis of his expert views on the modeled interrelationships, to set additional constraints on the structure of regression operators. In addition, the package implemented various techniques to more accurately assess the reliability of simulation results, taking into account the nature of the source data and the chosen method of solving the problem.

The initial information for the task was obtained from various sources. Part of the necessary information was taken from the directory "Regions of Russia" for 2015. Based on the above information, a table of joint implementations of a group of 57 quantitative parameters characterizing the life expectancy of people and potentially factors affecting it in 82 regions of Russia in the first half of the second decade of the 21st century was formed. According to 3 regions of the Russian Federation - the Nenets Autonomous District, the Republic of Crimea and Sevastopol, the data are not complete and were not considered in our study. Some of the characteristics used make sense of the average value or share in the total population of the region. Parameters that directly depend on this number are divided by its size.

All the characteristics of the regions present in the initial data were divided into thematic groups. When forming such groups, a certain commonality of the possible influence of the corresponding indicators on the life expectancy was taken into account. Each formed group of indicators in this case can be considered as a parametric description of some particular complex factor of this influence. In accordance with the described principles, 8 such factors were formed. The following are their names (underlined) and the names describing these factors of the initial characteristics of the regions.

1. Crimes. Intentional murder and attempted murder. Intentional serious harm to health. Rape and attempted rape. Theft. Robbery. Dead in an accident. Robbery, Crimes committed by minors. Crimes in the sphere of economy. Crimes related to drugs.

2. The economy. Gross regional product. Per capita income. Products of manufacturing industries. Power generation. Coefficient of funds. Registered unemployment.

3. Education and science. Students of universities. Candidates of Science. Patents for inventions. EGE - the average score in mathematics. EGE - the average score in the Russian language. The success of the best secondary schools. EGE mathematics - the proportion of participants who have overcome the threshold score. EGE mathematics - the ratio of excellent students. EGE Russian

language - the proportion of participants who have overcome the threshold score. EGE Russian language - the ratio of excellent students. Graduates with a certificate of basic general education.

4. Religion and worldview. Orthodox (ROC). Atheists. Believers in signs, fortune-telling, destiny. Loving Russia. Respect the law. Muslims (Islam is indefinite). Muslims (Islam of the Sunni direction). Agnostics. Christians who do not belong to any church. People who believe in an undetermined higher power.

5. Nutrition. Consumption of vegetables and food melons. Consumption of meat and meat products. Consumption of milk and dairy products.

6. Consumption of alcoholic beverages. Consumption of vodka. Consumption of wine. Consumption of beer.

7. Diseases and abortions. Some infectious and parasitic diseases. Neoplasms. Diseases of the blood. Diseases of the endocrine system, eating disorders and metabolism. Nervous diseases. Diseases of the circulatory system. Abortions for 100 deliveries.

8. Demographics. Share of urban population. The proportion of young people who have not reached working age. Fertility. Population. The coefficient of migration increase.

## 3. Results and discussion

As the main factors influencing the territorial variability of life expectancy, statistically more significant are external influences leading to death or a significant deterioration in the health of people, especially at a young age. Such factors are crimes against the person, infectious and parasitic diseases and abortions. This fact can be related both to the intensity of these factors and to the formula for calculating the expected life expectancy.

In regions characterized by an increased level of religiosity and the quality of school education (especially in mathematics), there is also a statistically significant deviation in the life expectancy of people towards increasing it. The economic characteristics of the regions, the characteristics of nutrition, alcohol consumption and demography are not statistically significant or very weakly identified as possible causes that form the territorial variability of life expectancy. Nevertheless, a certain tendency of the positive influence of the level of the economic state of the regions on the life expectancy can be noted. Cause-effect interpretation of established territorial links is a difficult task, given the complexity of the complex of factors affecting life expectancy.

Taking these results into account, one can look at life expectancy indicators in the Baikal region (Irkutsk region, Republic of Buryatia and Trans-Baikal region) and identify factors that determine its differences from the rest of the Russian Federation. The life expectancy of both men and women in the Baikal region is substantially lower than the all-Russian indicators. Table 1 presents data for the Russian Federation and for regions in the Baikal region.

**Table 1.** Socio-economic phenomena that distinguish the Baikal region from the rest of the Russian Federation

| Phenomena | Russian Federation | Irkutsk area | Buryat Republic | Trans-Baikal region |
|---|---|---|---|---|
| **Life expectancy of men (years)** | 64,5 | 60,5 | 62,7 | 61,7 |
| **Life expectancy of women (years)** | 75,8 | 73,4 | 74,5 | 73,4 |
| **Intentional homicide and attempted murder** | 9,8 | 17,1 | 20,6 | 26,6 |
| **Intentional serious harm to health.** | 26,9 | 49,7 | 46,1 | 61,3 |
| **Abortions, cases per 100 births.** | 53,7 | 64,0 | 57,0 | 56,0 |
| **EGE mathematics, average score.** | 47,8 | 44,8 | 46,4 | 39,8 |
| **EGE Russian language, average score.** | 62,6 | 60,2 | 59,8 | 57,9 |

The Trans-Baikal Territory and the Republic of Buryatia, to a lesser extent Irkutsk Oblast, lag behind in the economic development from the average level of the regions of the Russian Federation, but in accordance with our data on the weakness of the influence of economic factors on the territorial variability of life expectancy in Russia, it should be assumed that this gap The Baikal region. According to the results of our studies, the reduced life expectancy has territorial links with the prevalence of infectious and parasitic diseases. But, according to the statistical data on which we rely, the Baikal region as a whole does not differ for the worse from the rest of the Russian Federation in this respect.

The Baikal region is characterized by a high level of crime. In this respect, the constituent entities of the Russian Federation, which are part of the Baikal region, join into the all-Russian pattern found in our study - a high level of violent crime is associated with low life expectancy.

In the Baikal region, the number of abortions is increased, which is typical for regions of the Russian Federation with a reduced life expectancy. Perhaps this factor is significant in the Baikal region. The quality of school education is reduced, which is reflected in the results of the Unified State Exam (USE). This decrease is also typical for regions of the Russian Federation with a low life expectancy.

The regions of the Russian Federation, which are part of the Baikal region, are quite different from the average Russian indicators for a number of phenomena in the worldview sphere (table 2).

**Table 2.** Outlook phenomena that distinguish the Baikal region from the rest of the Russian Federation

| Phenomena | Russian Federation | Irkutsk area | Buryat Republic | Trans-Baikal region |
|---|---|---|---|---|
| **Orthodoxy, ROC (% of survey participants)** | 39,0 | 28,0 | 27,0 | 25,0 |
| **Atheism (% of survey participants)** | 13,0 | 17,0 | 13,0 | 17,0 |
| **Agnosticism (% of survey participants)** | 5,9 | 0,0 | 6,0 | 16,0 |
| **Occult direction (% of survey participants)** | 13,8 | 22,0 | 11,0 | 15,0 |
| **Uncertain religiosity (% of survey participants)** | 25,7 | 37,0 | 25,0 | 28,0 |

Here, the influence of the traditional confessions of Russia, including Orthodoxy (ROC), has been reduced, and the prevalence of forms of outlook that can be regarded as a departure from traditional religiosity has been increased. This is atheism, agnosticism, occultism (belief in signs, fortune telling, destiny), indefinite religiosity ("I believe in God, supreme power, but I do not confess a specific religion").

Earlier, we have already identified with the help of a correlation analysis [11] that the level of influence of traditional confessions (Orthodoxy, Islam) in the region of the Russian Federation is positively related to the level with life expectancy. Islam is not widespread in the Baikal region, so we are not considering it. Buddhism spreads in the Baikal region, but it is difficult to reveal the correlation of levels of its influence on life expectancy on the aggregate of regions of the Russian Federation, due to the scarcity of regions where it is quite common.

Thus, the regions of the Russian Federation that are part of the Baikal region fit into the system of territorial links of life expectancy, identified through the "Stochastic Modeling" software package. Reduced life expectancy, an increased prevalence of crimes against the person and abortion, a reduced quality of school education, and a reduced influence of traditional denominations form one of the characteristic combinations that are found on the aggregate of regions of the Russian Federation.

## 4. Conclusion

The main factors influencing the territorial variability of life expectancy at the level of the aggregate of regions of the Russian Federation are crimes against the individual, infectious and parasitic diseases and abortions.

The increased level of religiosity and the quality of school education (especially in mathematics) have a positive effect on life expectancy.

IOP Conf. Series: Earth and Environmental Science **190** (2018) 012061 doi:10.1088/1755-1315/190/1/012061

The Baikal region, with its characteristic low life expectancy, high prevalence of crimes and abortions, low school performance and low religiosity, joins into the system of territorial links of social phenomena characteristic of the aggregate of regions of the Russian Federation.

## References


[1] Oeppen J and Vaupel J W 2002 Demography. Broken limits to life expectancy *Science* **296** 1029–31

[2] Vaupel J W 2010 Biodemography of human ageing *Nature* **464** 536–42

[3] Olshansky S J, Carnes B A and Cassel C 1990 In search of Methuselah: estimating the upper limits to human longevity *Science* **250** 634–40

[4] Dong X, Milholland B and Vijg J 2016 Evidence for a limit to human lifespan *Nature* **538** 257–9

[5] Wilmoth J R, Deegan L J, Lundström H and Horiuchi S 2000 Increase of maximum life-span in Sweden *Science* **289** 1861–999

[6] Velichkovsky B T 2009 *The Vitality of the Nation* (Moscow: RAMS Press) p 176

[7] Lisitsyn Y P 2002 *Public Health and Health* (Moscow: GEOTAR-media) p 520

[8] Pastukhova E Y and Kochneva O P 2015 Life expectancy in the regions of Siberia: the impact of socio-economic factors *International Journal of Applied and Fundamental Research* **12**(2) 298-301

[9] Chekmarev A A and Ignatov A V 2015 Approbation of software "stochastic modeling" on the example of development of models for predicting the annual number of wolf *Bulletin of Voronezh State University. Series "Geography. Geoecology"* **1** 5–13

[10] Popov P L 2013 Religiosity and socio-economic phenomena: territorial ties in contemporary Russia *Izvestiya IGU, series "Philosophical Sciences"* **1** 71-7

# In limbo between state and corporate responsibility: Transport infrastructure in the oil village Verkhnemarkovo, Irkutskaya Oblast in Russia

**G Saxinger[1], N E Krasnoshtanova[2], G Illmeier[3]**

[1,3]Department of Social and Cultural Anthropology and Austrian Polar Research Institute, University of Vienna, Vienna, Austria
[2]Sochava Institute of Geography SB RAS, Irkutsk, Russia

Email: [1]gertrude.saxinger@univie.ac.at, [2]knesun@mail.ru, [3]gertraud.illmeier@univie.ac.at

**Abstract**. The village of Verkhnemarkovo is located in a remote oil region in the Irkutskaya Oblast in the Siberian part of the Russian Federation. Active socio-economic development of the village began with the discovery of an oil field near settlement in the middle of the 20th century. In the aftermath of the dissolution of the Soviet Union the once thriving oil village faced massive out migration and also decay of the once good transportation infrastructure. This article shows the changes of the provision of transport infrastructure and mobility practices in this village of around 2500 people. It raises the issue of corporate social responsibility (CSR) and that of the state for care and maintenance of roads in particular. The results highlight the state of the village and its inhabitants being in a limbo between the two main bodies who are expected to be responsible for facilitation of good transport infrastructure: the state and the companies operating in the region.

## 1. Introduction

Verkhnemarkovo is a remote village of around 2500 inhabitants in the Irkutskaya Oblast in the Siberian part of the Russian Federation located approximately 50 km north of the regional center Ust-Kut and around 600 km (bee line) north of the district center Irkutsk. It is directly on an oil field, the Markovskoe Mestorozhdenie. This makes the village distinct from the many other small villages in Siberia which very often have poor economic conditions, declining living standards and substantial out migration. This article focuses on transport infrastructure relevant to inhabitants of Verkhnemarkovo. While it introduces the everyday mobilities of the population, reasons for travel and transport conditions, the article raises in particular the important topic of transport for the oil related workforce going back and forth to the oil fields on a rotational shift work basis.

The article tackles the following research question: In what ways are the people of Verkhnemarkovo impacted in their everyday lives and level of mobility by the conditions of transport infrastructure and how does this interplay with questions over responsibility for the provision of good transport conditions. Furthermore, we ask how the local perception of responsibility links to (post-) Soviet conditions in the sense of state paternalism and to new structures under neoliberalism that shape the oil industry.

## 2. Theoretical embedding

Travelling and being mobile is tightly connected to transport infrastructure and thus also a matter of politics. Furthermore, all embodied mobile practices have complex histories and geographies [1]. Larkin's notion of "poetics of infrastructure" [2] highlights that roads are a matter of public discourse as well as quality of life and individual yearnings: "roads and railways are not just technical objects but also operate on the level of fantasy and desire" [2]. This relates to the notion of "enchantment" by infrastructure as described by Harvey and Knox [3] who explored the enthusiasm for roads in terms of their capacity to enchant. People from Verkhnemarkovo are not enchanted, they are rather disappointed or just resigned to the lack of new roads and poor road maintenance. In this sense "infrastructural violence" is a key issue for people dependent on the roads with only little say in such developments. Rodgers and O'Neill [4] researched city infrastructure, but much applies also to a village like Verkhnemarkovo. Materiality, morality and inequality are at the center of their theory. They emphasize not only the aspect of people interacting on an everyday basis with such infrastructure but scale it up to the political economy level to those people and stakeholders who have responsibility for thinking about a just setting as well as those who undermine such thinking [4].

## 3. Methods

This paper is based on geographical and anthropological fieldwork in 2014, 2016 and 2018 in Verkhnemarkovo consisting of participant observation, focus group meetings, qualitative interviews and questionnaire surveys. The semi-structured interviews included youth, mid-aged population and pensioners as well as politicians.

## 4. People and transport infrastructure in the village

### *4.1. Demographic profile of Verkhnemarkovo*

The history of Verkhnemarkovo is closely connected with the oil industry. Until the mid-twentieth century the village Markovo was a big agricultural settlement on the right bank of the river Lena. The way of life in the village changed significantly in 1962 when oil was found on Lena's left bank. From this time onwards the Lenskaya Oil and Gas Exploratory Expedition (Lenskaya expedition) attracted for the next few decades a workforce from all over the Soviet Union. It was not just individual specialists who arrived, but whole families. A new settlement expanded near the oil fields on the left bank of the river which was named Verkhnemarkovo and the older village Markovo was downsized to a micro district. During the years of the Lenskaya expedition, the population increased significantly from 3637 in 1979 to 4259 people in 1989. The crisis of the 1990s changed the situation in the village dramatically. More than ten years later, the population had substantially decreased to 2464 people in 2002. In the following decade, the 2000s, migration outflow slowed and by 2010 the local population counted 2385 people (according to official Census data for these years). In recent years, the population has stabilized and even marginally increased. Today, there are 2467 people living in Verkhnemarkovo [5].The unemployment rate is about three percent. More than 70% of unemployed people are women [6].

### *4.2. Jobs in Verkhnemarkovo*

Verkhnemarkovo has a rather diversified economic base in terms of the labor market, albeit sharing many features of a single-industry oil town. The local school, kindergarten and the village administration employ public workers. The many shops in the village provide substantial jobs to shopkeepers. To some extent small businesses such as a taxi company or small scale transport companies also employ locals. The majority however, especially men, work for the two major oil companies Irkutskaya Neftenaya Kompaniya (INK) and Dulisma as well as their subcontracting companies. People are employed in the oil sector primarily as drivers, heavy vehicle operators and machine operators. A small number of inhabitants are specialists in higher ranks and have rather high paid jobs. The men in the oil fields earn decent money compared to public sector jobs receiving around 40 000 Rubles per month (i.e. 80 000 Rubles per shift) while the latter earn around 30 000 Rubles per month. However, these are average numbers that can vary based on qualifications and level of position.

### *4.3. Transport infrastructure within the village*

The road network within the village is widespread due to the substantial spatial size of the settlement. However, the roads are primarily dirt roads and where they are paved the concrete is crumbling and poorly maintained. The roads are characterized by potholes which allows only very slow driving and bypassing the potholes is dangerous. On the roads are pedestrians, private cars, bicycles, motorbikes, lorries and other oil extraction related heavy vehicles. Responsibility for road maintenance is threefold: federal, regional and municipal government for the various types of roads. At the same time the company INK steps in for certain maintenance, especially for public roads that are of crucial interest to the industry. In some cases INK also organizes snow plowing in winter, but not everywhere and not regularly. The best road conditions occur in winter when snow covers the surface. In summer the roads are dusty while in spring and autumn the roads are muddy. The heavy vehicles used by the oil companies are seen as one of the main reasons for the bad state of the public roads. People would like to see the oil companies paying compensation for their use of the roads.

### *4.4. People's mobility*

In general people are highly mobile within the village and in relation to the regional center Ust-Kut. Everyday duties like going to work or doing shopping can be exhausting on foot due to the distances involved and pedestrians end up in clouds of dust when cars and lorries pass by. The key means of transport is therefore by private car. Given the rather good economic position of people compared to other small and remote villages in Siberia, nearly every household owns a car. Especially for inhabitants of Zayarnovo, a subdivision located a few kilometers outside Verkhnemarkovo, transport infrastructure is essential. A school bus runs on an everyday basis, but people remember when there was bus transport for everyone until the collapse of local state businesses in the 1990s. Since there is no bridge over the river Lena, private boats – no ferry is operating – are the most important means of transport for inhabitants of the Markovo subdivision located on the opposite side of the Lena. In Markovo primarily pensioners live and the few families who have to move into their relatives' houses during the time in spring and autumn when the ice breaks up or is not yet sufficiently frozen (*rasputica*) in order to be able to get to work and to school. There is thus a high dependency on various means of transport in Verkhnemarkovo.

Traveling to the regional center Ust-Kut is essential for health and administrative issues or banking etc. People do grocery shopping and buy other items in Ust-Kut which are more expensive in Verkhnemarkovo or not available at all. The minibus leaves at 7 am for Ust-Kut and returns at 4 pm. This provides only a tiny window for doing all the duties necessary in town. The costs are also substantial especially for pensioners and other people with low income. The trip back and forth by minibus costs 1 500 Rubles, while the more flexible taxi to Ust-Kut is 6 000 Rubles one way. Considering the above-mentioned income level, the latter is a substantial pressure on the household budget. Therefore, many people stay overnight with friends or relatives when they have early appointments at the administration or in the hospital. Such social networks are essential, as is also car sharing.

### *4.5. Movement of Vakhtoviki workers*

Besides a few jobs at the local oil field (Markovskoe Mestorozhdenie) for INK, the majority of the oil related workforce long-distance commutes to oil fields further away from the village such as Dulisminskoe, Yaraktinskoe, Verchnechonskoe or Danilovskoe. A common saying is that "all our men are Vakhtoviki." Vakhtoviki are also known as fly-in/fly-out (FIFO) or drive-in/drive-out (DIDO) workers [7-8]. Primarily men, but also some women – the latter work in the service sector or in administration on the oil field – commute on a roster of 28 days on shift followed by a 30 days' rest period at home. Such a pattern is widespread in the oil and gas industry across Russia [7].

The movement to and from work is different in the four seasons. In summer the Viluy Highway brings inter-regional (coming from all over Russia) and intra-regional (coming from the Irkutskaya Oblast) workers from the regional center Ust-Kut on so-called vakhtovka (a lorry-like truck with a passenger cabin) to Verkhnemarkovo. There they are met and travel further by road to the oil fields when the road conditions are good. When the road connection is unreliable workers from Verkhnemarkovo need to go first to Ust-Kut and are flown from there by helicopter or airplane to the oil fields. In winter the Viluy Highway, for its most part a gravel road, that is turned into a winter road (zimnik) – Mirnynskaya Trassa – that leads to some of the oil fields. Especially in spring and autumn, when roads become muddy (rasputica), helicopter is the favorite means of transport. The Verkhnemarkovo helicopter-airport becomes then a busy hub for workers from Ust-Kut who are transported by vakhtovka to Verkhnemarkovo and get there on the helicopter. This is also the most convenient way to go to work for Verkhnemarkovo people. The travel is facilitated by the oil company and free of charge for workers.

### *4.6. Quality of long-distance transport infrastructure*
For these long-distance commuters and the oil companies a well-functioning transport infrastructure is essential. A substantial problem is road safety. The winter road is described as decent since it has a solid snow cover. However, accidents happen from time to time due to icy conditions and driving at high speeds or risky overtaking. The latter two are also crucial dangers in other seasons when dust is heavily occurring. Very often, as also witnessed by the authors, dangerous overtaking in conditions of no visibility due to the dust puts people at risk and stops the traffic when incidences happen. People do not feel comfortable on the road due to such dangers. Not infrequently blown out tires can also interrupt the ride for quite some time. Furthermore, comfort is low when it comes to transport in vakhtovkas. These heavy vehicles for passenger transport are badly heated and shaky on bumpy roads which impacts on the health conditions of workers who are exhausted after such a trip, but have to go to work the next day. The roads are also used by the timber companies which, as it was frequently reported, sometimes block the roads, delaying transport for a couple of hours.

### *4.7. Maintaining the roads and caring for road safety*
The Viluy Highway is a federal road and is thus maintained through federal government funds. Constant work on the road is visible, but due to the heavy use of the road to transport cargo, passengers, timber and oil operation equipment as well as the high volume of private cars, the road is constantly in a bad condition and potholes make driving exhausting and ruin the vehicles. Furthermore, permafrost and climatic changes contribute to the poor condition of the Viluy Highway. Currently around 35 kilometers of the road outside of Ust-Kut is paved with asphalt. The rest of the highway is gravel. While the government has announced it will extend the asphalt parts, people are waiting for the fulfillment of such promises and are disappointed. The oil companies maintain themselves their own company roads close to the oil fields. In certain incidences, such as is the case with INK in Verkhnemarkovo, companies also maintain public roads that lead to oil production facilities or to the helicopter-airport.

## 5. Conclusion
Everyday mobility is of utmost importance to people in Verkhnemarkovo since the village is spread out over seven kilometers. People use various means of transport such as going on foot or by bicycle, motorbike, car or taxi. The taxi transport and minibus connection to the regional center Ust-Kut is crucial for managing administrative issues or going to the hospital, doing shopping and the like. Even fuel needs to be hauled in from Ust-Kut since there is no gas station in the village. Especially for pensioners public transport which is actually provided at reasonable prices by a private company is

expensive. Therefore, social inequality is visible when it comes to the level of mobility which is severe especially for old people and those without jobs. Stasis is often not a wanted choice but a condition.

The state only weakly takes care of the transport infrastructure, while the INK oil company – as the biggest one operating in the region – takes over certain public responsibilities in the framework of its Corporate Social Responsibility programs (CSR). However, INK maintains primarily only infrastructure that is crucial for its own operations. This has a positive side effect for inhabitants but does not solve the problem. INK does not necessarily intend to step in for actual state responsibilities such as maintaining the roads.

Since many people still remember the former flourishing Soviet state companies, which had the task of providing infrastructure, they still see such a model as the ideal. Thus, they are disappointed by new structures of responsibility which do not meet their mobility demands in terms of quality and convenience. We can speak here of post-Soviet structures that under neoliberal conditions leave the inhabitants of small villages alone with their needs and demands. People feel the state is not interested in their problems and does not allocate sufficient funds for their everyday mobility – which is necessary under conditions of living remotely from urban centers.

At the same time people expect from oil companies support – such as was the case in Soviet times when the state companies looked after towns. However, besides a lot of other CSR activities such as sponsoring social and cultural infrastructure, INK provides primarily maintenance for transport infrastructure where it is crucial for its own operations. Since the roads in town are public, INK does not necessarily see an obligation to care for these too. The absence of good roads and transportation infrastructure in general can be coined as infrastructural violence. People are stuck in limbo between the state and the company, neither taking on the actual responsibility for care of the transport infrastructure which is so much needed in a village such as Verkhnemarkovo. State paternalism – which prevails in the context of the overall economic development of Russia's natural resources and thus investment into such infrastructure is strong – fails in the context of everyday mobilities and related demands of the population.

**Acknowledgements**

This research was carried out in the framework of the Austrian-Russian project "Configurations of 'Remoteness' (CoRe) – Entanglements of Humans and Transportation Infrastructure in the Baykal-Amur Mainline (BAM) Region" financed by the Austrian Science Fund (FWF) [P 27625]. Furthermore, it is related to the project "Innovative development, territorial organization, and quality of life improvements for the population of the Siberian and Arctic regions of Russia" funded by the Russian Foundation for Basic Research #16-02-00570(a). This project is a cooperation between the Department for Social and Cultural Anthropology at the University of Vienna/Austria, the Austrian Polar Research Institute (APRI) and the V. B. Sochava Institute of Geography Siberian Branch of the Russian Academy of Sciences in Irkutsk/Russia. Info: core.univie.ac.at & lifeofbam.com.

**References**

[1] Cresswell T and Merriman P 2013 *Geographies of Mobilities: Practices, Spaces, Subjects* (Surrey: Ashgate)
Cresswell T 2010 Towards a Politics of Mobility *Environment and Planning D: Society and Space* **28**(1) 17-31
[2] Larkin B 2013 The Politics and Poetics of Infrastructure *Annual Review of Anthropology* **42** 327-43
[3] Harvey P and Knox H 2012 The Enchantments of Infrastructure *Mobilities* **7**(4) 521-36
[4] Rodgers D and O'Neill B 2012 Infrastructural Violence: Introduction to the Special Issue *Ethnography* **13**(4) 401-12
[5] The Administration of Verkhnemarkovo settlement of the Ust-Kut municipal district of Irkutsk oblast 2018 *The report of the head of the administration Vlasov K V for the reporting year 2017* (in Russian) Income accessed online on 22th May 2018 via

http://arhiv.verhnemarkovo.ru
[6] The Administration of Verkhnemarkovo settlement of the Ust-Kut municipal district of Irkutsk oblast 2014 *The Passport of Verkhnemarkovo municipal formation for 2013* (in Russian) Income accessed online on 22th May 2018 via http://arhiv.verhnemarkovo.ru
[7] Saxinger G 2017 Infinite travel: the impact of labor conditions on mobility potential in the Northern Russian petroleum industry *New Mobilities and Social Changes in Russia's Arctic Regions* ed M Laruelle (New York: Routledge) 85-103
Saxinger G 2016 Lured by Oil and Gas: Labour Mobility, Multi-Locality and Negotiating Normality and Extreme in the Russian Far North *Extrtractive Industries and Soc*iety **3**(1) 50-9
[8] Storey K 2001 Fly-in/Fly-out and Fly-over: Mining and Regional Development in Western Australia *Australian Geographer* **32**(2) 133-48
Storey K 2010 The Evolution of Commute Work in Canada and Australia *Proc. Int. Conf. Biography, Shift-labour and Socialisation in a Northern Industrial City – The Far North: Particularities of Labour and Human Socialisation* ed Eilmsteiner-Saxinger G and Stammler F (Rovaniemi: Arctic Centre University of Lapland) 23-32
Storey K and Shrimpton M 1989 *Impacts on Labour of Long-distance Commuting Employment in the Canadian Mining Industry* (Institute of Social and Economic Research, Memorial University of Newfoundland)

# Expansion features of Irkutsk suburban areas' development

**E N Serebrennikov**

Sochava Institute of Geography SB RAS, Irkutsk, Russia
Irkutsk Scientific Center of the Siberian Branch of the Russian Academy of Sciences, Irkutsk, Russia

E-mail: togr2010@yandex.ru

**Abstract.** This article is devoted to the problems of spatial development of the suburban zone of the city of Irkutsk in post-Soviet realities. The paper presents some features of suburbanization in Russia, as well as geographic factors that influence the formation and development of the Irkutsk agglomeration, such as high dependence on the poly-highway; a special model of industrial development; Baikal factor. Particular attention is paid to the peculiarities of development in various directions and spatial differentiation of various types of suburbs. The reasons and consequences of such development are considered. A typology of the Irkutsk suburban area is presented based on the nature of the settlements with a brief description.

## 1. Introduction

Urbanization is becoming the leading process in the establishment of the chain and the system of migration. Cities are becoming the centers of population's gravity, points of regional development, and its primary foundation for human migration. During the development of urbanization processes in any large city there arise a moment, when the vector of migration movement alters from urban to suburban areas, in other words, a new urbanization stage takes place, it is called suburbanization [1].

Suburbanization is a process of the growth of suburban areas, as a result of people's migration from a central city or a countryside. Building "satellite" towns has become a classic form of suburbanization. "Satellite" towns are divided into two groups: industrial "satellite" towns and "bedroom" towns. In particular, industrial satellites are located mainly in the periphery of urbanized areas (i.e. away from central cities), although they are bigger and older. And vice versa, despite the fact that "bedroom' towns are younger (they started to develop only after the mass growth of car vehicles), almost all of them are generally located closer to the central city [2].

Russian suburbanization has certain features. When moving to the suburban areas, city dwellers of European countries sell their urban properties and regard their country house as the main residence. On the contrary, in Russia, such property is viewed as the second home, because when moving to a cottage, people prefer to pursue owning their urban flats. Russian urbanization is seasonal. Villages with gardens or dachas as well as country cottages form an alternative system of settlements: their quantity twice exceeds the number of rural settlements. The amount of country house settlements of different sizes is increasing [3]. In this case, it is appropriate to say that suburbanization did not indicate the spread of city lifestyle in rural areas but, on the contrary, it meant the establishment of rural lifestyle in cities [4].

IOP Conf. Series: Earth and Environmental Science **190** (2018) 012063 doi:10.1088/1755-1315/190/1/012063

In Siberian conditions, with its vast territories and far distances between major cities, agglomerations have become in a special demand. Due to satellite towns' close interaction and location agglomerations efficiently accumulate different objects of industry, science, social sphere, culture, etc [5].

## 2. The problem formulation

Irkutsk agglomeration includes both typical and specific features of the system organization, caused by peculiarities of city's location and development. Irkutsk is one of the oldest Siberian centers, developed mostly due to the construction of the Trans-Siberian Railway in the end of the $19^{th}$ century. It used to perform a transit function instead of being the center of the surrounding area's development for a long time, and its potential for the region development was embodied in several highways within eastern and north-eastern directions. For a long time Kachugskiy Trakt (a highway) used to be the most essential one, leading to the Lena's upstreams, it lost its strategic role as a connector of inner districts of the region after the Tayshet-Lena railway had been built in the middle of the previous century. Expansion to the west of Irkutsk was restrained by western Sayan Mountains.

Thus, all Irkutsk "satellite" towns were situated along the Trans-Siberian Railway on the left bank of the Angara River. Besides, they were connected by the federal highway M55, therefore many industrial enterprises had been concentrating there: and towns had been growing. Rural settlements had been developing between the towns: Smolenshchina, Mamoni, Meget, Maksimovshchina. There appeared to be small rural villages outside the city, along the railway and away from industrial manufactures.

However, the situation on the opposite bank was different because of the lack of large highways, northern areas maintained ultimately agricultural relations with a myriad of villages and rural communities. After the collapse of the USSR and without the governmental support distant villages' inhabitants started to concentrate in settlements nearer to Irkutsk.

The features of Irkutsk geographic position have formed several peculiar factors for the suburban areas' development, distinguishing it from other Siberian centers, also being the centers of the regional territories' development.

The first factor among them is a higher dependency from poly-highways including the Moskovskiy Trakt and the Trans-Siberian Railway. The second factor is focused on the fact that the postwar Irkutsk has become more than a center of just development but industrial one. The progressive model for Soviet expansion planning of territorial production complexes was implemented there. Agricultural manufacturing was not on the list of specialization industries of territorial production complexes, as a consequence, the agrarian development of the surrounding territory was being made without any control. In some cases, it remained in down-to-dated forms on collectivization and state manufacturing basis, in other cases, it became a victim of industrial plans and rural population were moved out into other places, as it happened after filling in the storage reservoirs of the large water-power plants.

The third factor is the lake Baikal's closeness, which influences on the system of resettlement around Irkutsk.

## 3. Research method and materials

The cartographic method was used for examination of Irkutsk rural areas, it allowed to detail the rural setting up to each horticulture or a separate homestead. The suburbs of Irkutsk City covers the areas of Irkutskiy, Shelekhovskiy and Angarskiy districts, meanwhile its essential part belongs to Irkutsk district.

Irkutsk rural population had increased in 30 thousand people for 20 post-Soviet years, enlarged almost by one and a half. According to the population Census of 2010, the population contained 84,3 thousand people. The fast growth of district population continued after the Census and by the beginning of 2016, the district population reached the number of 112, 1 thousand people [6]. The majority of Irkutsk district population live in the suburb area (25-30 min away from regional center), and its quantity keeps growing: since 2007 to 2010 the rate of district dwellers of the nearest rural

IOP Conf. Series: Earth and Environmental Science **190** (2018) 012063 doi:10.1088/1755-1315/190/1/012063

areas had grown from 72,1 % to 89%. The positive dynamics is noticed in the majority of municipalities of Irkutsk agglomeration suburbs- Markovskiy, Molodezhniy,Ushakovskiy, Smolenskiy, Maksimovskiy, etc.

Radically parting highways - Moskovskiy, Aleksandrovskiy, Kachugskiy, Goloustenskiy, Baikalskiy, Melnechniy, and Kultukskiy tend to be the core vectors of development. The most developed highways used to be Baikalskiy Trakt, Melnechniy Trakt and, Kultukskiy Trakt. There are several reasons for that: first, they are relatively eco-friendly and situated closer to the Angara's warm bays. Second, they have close communication and location to the city, this is common for Baikalskiy and Melnechniy Trakt. Kachugskiy and Aleksandrovskiy Trakt developed due to such factors as the reconstruction of already existed villages, and lower prices on land and property. The crucial factor is transport purchasing. For example, in Irkutsk the car rate had grown by 38,5% in comparison to 2009. In 2009 this indicator was 223,4 units per a thousand people, and in 2015 it reached 309,1 units per a thousand people. Thus, for example, Khomutovo village had an increase more than by 25% since 2012, Karluk - by 20%, Urik - by 14%.

It should be mentioned that the active migration of citizens and their city lifestyle to settled rural environment contributed to a number of innovations. This fact proves that these villages definitely can not be related to the rural setting. Some kinds of economic activities and small enterprises unfamiliar to rural area started to appear. Among them are cosmetology clinics, car services and car washes, fitness centers and large wholesale retail chains, grooming, etc.

Direction to Baikal - Listvyanka village is considered to be the most prestigious because of its closeness to warm bays of the Angara, developed infrastructure and ecology. This destination has a restricting factor though - Pribaikalskiy National park, it is still going to develop more actively. Even at this moment suburban roads are being reconstructed and lanes on the roads are being broadened. A myriad of high class cottage villages have been built on this direction.

Another direction to Baikal is Kultukskiy Trakt. Despite the fact that this highway is federal, there is no such intensive development of rural areas because of the close presence of Aluminum plant "IrkAZ SAUL" and the transit character of the highway. Two large settlements, Baklashi village and Vvedeshchina are situated away from the highway, near the Irkut river. And quite recently small cottages have appeared there within the programme of multi format building in Shelekhov town and neighboring territories of Shelekhovskiy district, such as Yasnaya Polyana, Gagarin, Familiya and others.

Direction to the north is Kachugskiy Trakt. Kachugskiy Trakt has always had agricultural importance, and after the collapse of USSR the population of distant rural villages started to locate nearer to vilages Karluk, Kuda, Khomutovo, Zapadnoe, Granovshina, etc. Oiyok is considered to be the latest village, about 40 km away from Irkutsk.

One can observe the growth of population there (table 1). Citizens of this district saved quite a strong connection with the land and the household. This direction leads to the Baikal but because objects are too far away from each other, it can only serve as a transit for tourists. And the majority of population will pursue to concentrate in the limits of 25 km from the city.

Suburban areas can be viewed as melting pots for population, changing its structure. If previously they used to be rural citizens, who migrated closer to the city, now they are citizens, who are aimed to live in the suburbs. And, in this case, directions that are not connected with large highways are in high priority.

People seek for the life in the suburbs and do it according to their financial abilities and needs. For some people it is cheap accommodations with affordable municipal transport, for others it is a country house with a small patch of land, situated 20 minutes away from the city and someone values his personal townhouse in a prestigious village with the diverse infrastructure [7].

According to the character of the settlements, Irkutsk suburbs are divided into three zones (figure 1). The first zone is the nearest to the city settlements forming with the city a single system of migration. There the municipal transport is developed in greater extent, also objects of social sphere and all types of infrastructure. In other words, they are bedroom communities, located behind the official edge of the city but creating one unity with it. The second zone is the suburbs in its common definition, means housing owning in the outskirts (a house, a cottage), and a personal car. The third zone are satellites forming boundaries of the suburbs (Oiyok, Listvyanka, Bolshoy Lug).

**Table 1.** Dynamics of the population in Irkutsk suburban areas

| Settlement | 1985 yr | 1996 yr | 2010 yr | 2012 yr | 2016 yr | Increase in % 2010-2016 yr |
|---|---|---|---|---|---|---|
| | | | Vector on the Baikal / Listvyanka | | | |
| **Molodezhnyj** | 246 | 329 | 6640 | 7170 | 8337 | 25,5 |
| **Novaya Razvodnaya** | no data | no data | 1313 | 1395 | 1687 | 28,5 |
| **Dzerzhinsk** | 1896 | 1385 | 2277 | 2477 | 2665 | 17 |
| **Novolisiha** | 282 | 218 | 528 | 562 | 740 | 40,1 |
| **Pivovariha** | 1581 | 2206 | 3506 | 3729 | 4111 | 17,2 |
| **Bol'shaya Rechka** | no data | 2300 | 2612 | 2667 | 2952 | 13 |
| | | | Vector to the North | | | |
| **Karluk** | 1573 | 1879 | 2572 | 2793 | 3356 | 30,3 |
| **Homutovo** | 3705 | 4289 | 12872 | 13793 | 16445 | 27,7 |
| **Oyok** | 2953 | 2802 | 3678 | 3797 | 4013 | 9,1 |
| | | | Vector to the East | | | |
| **Bol'shoj Lug** | 5900 | 5500 | 5015 | 5085 | 5322 | 6,1 |
| **Olha** | no data | no data | 1744 | 1855 | 2064 | 18,3 |
| **Smolenshchina** | 1789 | 2011 | 3214 | 3470 | 4162 | 29,5 |
| **Markova** | 2300 | 3900 | 6930 | 11365 | 20627 | 197 |
| | | | Vector to West | | | |
| **Meget** | 8700 | 8000 | 8828 | 8873 | 8863 | 0,4 |
| **Mamony** | 1167 | 1311 | 3154 | 3178 | 4345 | 37,7 |
| **Malaya Elanka** | 395 | 452 | 774 | 776 | 917 | 18,4 |

One can see in the picture that only the first one forms more or less coherent picture of the suburbs, when further there is an abrupt differentiation between highways. Meanwhile, the land between highways is weakly developed or even unoccupied.

## 4. Results and Discussion

Currently, several types of the development of suburban areas exist. The first type is the decline of villages' growth due to land purchasing or inheriting properties by relatives-citizens. The second type is the development of low-rise buildings, new districts, and bedroom communities. The third one is typical suburban villages with poly-capital buildings, seasonal for living in them. All three types of suburban development are appropriate for Irkutsk, but the character of expansion migration and its optimization is highly individual, as the expansion strongly depends on linear-nodal migration structure.

Of the core factors of suburbanization is its car purchasing. In 2009 this indicator was 223,4 unit per a thousand people, and in 2015 it reached 309,1 unit per a thousand people, i.e. had grown by 38,5%. Mass character of citizens' migration to suburb areas revealed the lack of space quite quickly, it led to the massive cultivation of the lands situated outside the boundaries of localities. Intensive construction of homesteads on the agrarian lands promoted the enlargement of the suburban areas by several times [8].

As a consequence we notice rapid development of unoccupied areas along the major highways. Meanwhile new directions of suburban development are restrained by the lack of funding in infrastructure that would allow compensating linear character of the development; to decrease the overload of current routes; and to allocate in even proportions the populations, objects of social sphere, entertaining and shopping centers.

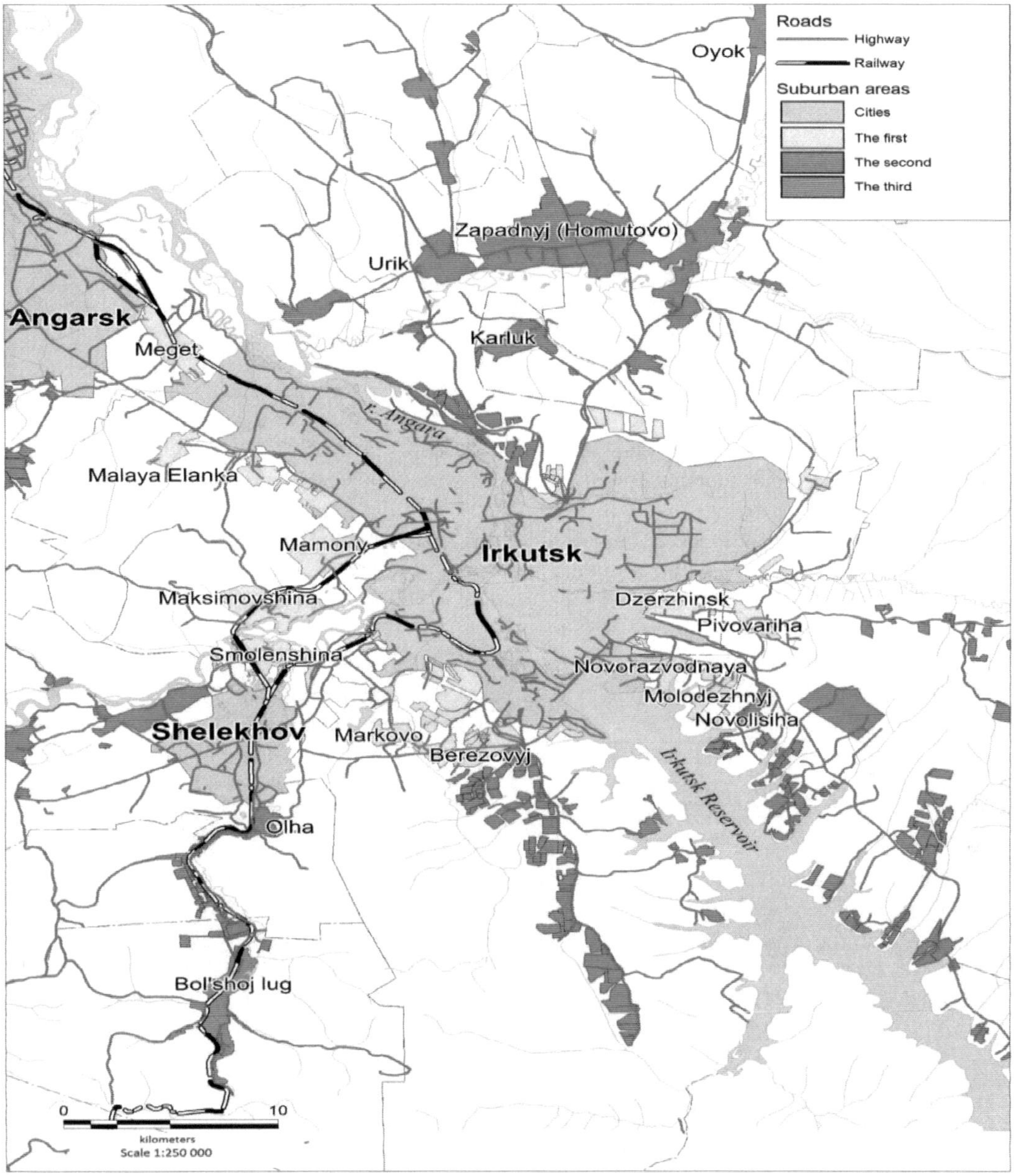


**Figure 1.** Irkutsk suburban area

Methods that can be provided in this case are the restriction on traffic of personal transport in the city center; it is of utmost importance to develop the system of public transport in technological as well as in logistic way. other methods include the creating new bypass routes of public transport on less loaded road sectors, searching for faster and more comfortable routes to key city's places, eliminating of duplicating routes, creating car stations for intercity transport beyond overloaded transport highways, increasing of quantity of public transport vehicles. Another urgent goal is raising the attractiveness of pedestrian zones with the help of the development of greenery, creating comfortable sidewalks, pedestrian crossings.

IOP Conf. Series: Earth and Environmental Science **190** (2018) 012063 doi:10.1088/1755-1315/190/1/012063

These methods are being embodied within the agglomeration program "Safe and qualitative roads". In accordance with the Resolution of the Government of the Russian Federation of 10.10.2016 No. 1050 "On the Organization of Program Activities in the Government of the Russian Federation".

In the process of Irkutsk agglomeration's formation and development, one of the most crucial moment is creating unified real estate market in Irkutsk, Angarsk, Shelekhov, and corresponding districts. Developed market of real estate in the region is characterized by presence of large property developers, and capacity to implement big projects of complex building. A high priority task of agglomeration is urbanizing territories between existing towns, in the long-term perspective it might lead to the establishment of the de-facto unified locality with uninterrupted town building.

If forecasting further direction of the building complex's development within the frames of Irkutsk agglomeration, when choosing the patches for small-scale residential building ( from separate buildings up to several neighborhoods) investors and property builders will prone to patches of land, located beyond city boundaries or directly near to its boundaries- in the suburban zones. Because such building assumes smaller expenses on infrastructure as well as smaller expenses on marketing promotion of the projects.

## 5. Conclusion

Nowadays suburban areas are being developed actively as a zone of population migration and as a zone of additional functions and detachment of particular functions away from the city-center.

One can notice the increasing number of localities, being parts of Irkutsk agglomeration. Positive dynamics is caused by their location near large cities of the agglomeration. The dynamic way of suburban transformation proves the existence of complicated system of massive agglomeration's cooperation. The character of «balance wheel» migrations changes when suburban citizens seek for a job in the central city, and town citizens gradually purchase real estate outside the city having saved their workplaces. Suburbanization forms a new quality of the territory and a new quality of life. The crucial effect of Irkutsk agglomeration's development is its interaction with mutual complementary agglomerations of peripheral zones (agricultural, touristic-recreational, specially protected territories). Periphery of agglomeration's areas near the Baikal is a part of the Baikal national park. The territories of agro-food specialization are developing in Usolskiy and Cheremkhovskiy districts, also, in neighboring to Irkutsk vicinities of the Ust-Ordyn Buryat District. While agrarian territories are significant within the boundaries of Irkutsk region, the Baikal touristic-recreational zones have an important meaning not only in local and regional ways, but also in international one.

Agglomeration processes are controversial. On the one hand, a gradual population's concentration takes place within the whole Irkutsk region. As one of the most economically developed zone, Irkutsk city agglomeration (ICA) attracts population from less developed regional districts. From the other hand, an active de-concentration is going on the territory of the agglomeration: Irkutsk citizens move from overloaded places to less populated localities of the suburb, which offers more attractive life conditions. For the past decades, suburban area of ICA has increased significantly. A greater number of people migrating to the suburb causes new difficulties, wherein earlier existing problems of rural areas are still unsolved.

Two contrary directions for the development can be found in ICA building. In Irkutsk city and its neighborhood there is a highly intensive concentration of residential zone with high-rise buildings. Its consequences are the overload of the existing systems and the infrastructure, the lack of greenery and recreational zones, the discomfort of the environment in general. At the same time, there is a process of extensive enlargement of suburban area with private low-rise houses on patches of land. The area and population is quickly growing, the infrastructure and communications become insufficient for such number of people. Besides, valuable agricultural and recreational territories are being destroyed. Thus, the irrational usage of ICA's regional resources leads to the environment, uncomfortable for people living.

Taking into consideration all the factors above, the following localities with high population growth rate have the greatest potential for the development: Khomutovo direction (Khomutovo

IOP Conf. Series: Earth and Environmental Science **190** (2018) 012063 doi:10.1088/1755-1315/190/1/012063

village, Zapadniy, Urik, Kuda, Granovshina, Pozdnyakova) can be a center of agrarian manufacturing; Markovo village and its outskirts; Meget village; Molodezhniy village and its outskirts. These territories were allotted for creation of zones with the high rate of residential building. Besides, social infrastructure and new workplaces have to be created in these territories.

**References**

[1] Kenneth T J 1985 *Crabgrass Frontier: The Suburbanization of the United States* (NewYork: Oxford University Press) p 432

Lappo G 1978 *Development of City Agglomerations in the USSR* (Moscow: Nauka) p 25

[2] Hayden D 2004 *Building Suburbia: Green Fields and Urban Growth, 1820–2000* (New York: Vintage Books) pp 201-29

Egorov D, Dorofeeva L and Shadrin A 2016 Methodological base of urban agglomerations and their suburban areas *Journal of Economics and Economic Education Research* **17** pp 381-8

[3] Nefedova T and Treivish A 2002 *The Theory of "Differential Urbanization" and Russian Cities' Hierarchy on the Edge of XXI Century* (Smolensk: Oyikumena) pp 71-86

[4] Lesyuta N 2010 The pecularities of the urbanized territories' development in Siberia *Geography and Natural Resources* **1** 119-23

[5] Lesyuta N 2010 *Compression of Socio-Economic Space: New in the Theory of Regional Development and Practice of its State Regulation* (Mockow: Eslan) pp 263-81

[6] Grigorichev K 2012 From Settlement to Suburb: Outskirts of Irkutsk in the Last Third of the XXth – the Early XXIth Century *Izvestiya ISU* **2** 44-51

[7] Zedgenizov A 2016 Suburbanization of the city of Irkutsk: levels and results *ISSN Izvestiya vuzov, Investitsii. Stroitelstvo, Nedvizhimost.* 11-20

[8] Grigorichev K 2017 Suburbanization in Siberian regions as a mechanism of post-soviet transformation *What do We Know about Contemporary Russian Suburbs?* ed A Breslavsky (Ulan-Ude: Publishing House of BSC SB RAS) p 220

# Forest investment in the Russian part of the economic corridor China-Mongolia-Russia

**N M Sysoeva**

The Irkutsk Scientific Center of the Siberian Branch, Russian Academy of Sciences, Sochava Institute of Geography SB RAS, Irkutsk, Russia

E-mail: syssoeva@oresp.irk.ru

**Abstract**. The activity of foreign and joint ventures in the Russian part of the economic corridor is analysed. The enterprises with foreign capital play an insignificant role in the economy of the border regions now in comparison with the developed Siberian territories. Most of the foreign enterprises are registered by individuals, which reduces the innovation potential of investment and contributes to the import of labor. The main donor countries are China and Cyprus. The prevalence of Chinese investors is noted especially in the Zabaikalsky kraj where they dominate in most types of activities. Mongolia is the closest neighbor of this zone but its investors are represented in Buryatia only. Investments in wholesale trade and construction prevail with high concentration of production in coal and metal ore extraction. Such an industry structure hampers the flow of new technologies and the growth of human capital. Prospects for the growth of foreign investment are connected with the expansion of resource extraction including forest resources and their primary processing for export as well as the development of transition activities in the area of the expected economic corridor. It requires a purposeful state and regional policy of attracting investment into technological areas of communications and transport.

## 1. Introduction

The Russian zone of the presumed economic corridor includes several frontier subjects of Eastern Siberia (Zabaikalsky kraj, Republic of Buryatia, Republic of Tyva and the Republic of Altai) as well as the adjacent territories - Khakassia, parts of the Irkutskaja oblast and the Krasnoyarsky kraj, which may also be affected by this project. Specificity of interaction of these territories with the outside world is caused to a great extent by geographical factors, expressed in its comparative remoteness from the main world markets and proximity to a limited number of Asian countries with domination of China.

The idea of developing the economic corridor offers new opportunities for the prospects of the Russian territory adjacent to the border. The access to foreign markets and integration into global economic chains are mainly conditioned by attracting new investment. In the current institutional system of distributing the results of economic activity the peripheral regions of the country have low investment opportunities. Therefore, the problem of attracting funds from abroad is principal especially for territories with a resource orientation of the economy, which is typical for most subjects of the Siberian Federal District (SFD). According to the data of the Central Bank of Russian Federation on foreign direct investment (FDI), all the border regions of Eastern Siberia lag behind other subjects of the SFD. Krasnoyarsky kraj and Irkutskaja oblast are among the leaders in attracting investment from abroad.

To understand the role and possibilities of investing from abroad it is necessary to answer some questions - where the investment has come from till now, who are investors, in which sectors of economic activity do they concentrate, what is the effect of them for the regional economy, and what needs to be done in order to increase their impact on the area development in the future.

## 2. Data and Methods

In most cases foreign investment in Russia is analyzed in macroeconomic terms – its volume and structure by countries and regions [1]. The official statistics of foreign investment in the regional context is incomplete and unstable in recent years. Since 2014 the Federal Service of State Statistics of Russia has stopped publishing data on the volumes and structure of inbound foreign investment, presenting indicators on foreign direct investment (FDI) only, i.e. giving at least 10% of the shares or authorized capital of an organization registered in the country. The Central Bank of RF adheres to this approach in open statistics. Meanwhile, the share of foreign direct investment is relatively small. In 2013 it was only 15.3 % for Russia and 13.5 % for the Siberian Federal District, since the structure of such investments is always dominated by "other foreign investment" (commercial and other loans). However, FDI refers to a relatively stable and efficient part of the transboundary financial flow for the local economy, therefore, their dynamics and factors affecting the level of fundraising are important for the socioeconomic situation and prospects for the development of regions.

A significant part of the investment coming to the regions is not allocated by country of origin in the open statistics of the Central Bank. The volume of unallocated investment, including confidential data, in 2017 was 87% for the Zabaikalsky krai, 97% for Buryatia, 80% for Tyva, and 100% for the Altai Republic. Under the circumstances a more detailed analysis of foreign investment in the region for explaining their motivation and possible prospects can be carried out at the microeconomic level, based on the activities of enterprises with foreign capital.

Data on the enterprises with foreign capital were provided by the national agency of economic information "Prime". These enterprises included two categories - foreign enterprises, i.e. fully owned by foreign investors, and joint ventures where foreign companies and citizens own a part of assets. Foreign enterprises are formed by legal entities or individuals, and this is important for assessing development prospects, since a significant part of the enterprises formed by individuals implements the import of labor more than the import of capital. Despite the fact that amount of enterprises owned by legal entities is significantly smaller throughout the country, the effect for the economy and its export potential is expected just from the foreign companies and transnational corporations.

The indicators of the enterprises include the volume of revenue and income (2016), the type of activity and the country from which the capital has arrived. Based on the volume of revenue, the role of foreign enterprises in the regional economy was analyzed, as well as the concentration of production in large enterprises with annual revenue of more than 100 million rubles.

## 3. Results

In the analyzed four border regions of the Russian Federation, 512 enterprises with foreign capital were registered in late 2016, of which 341 were foreign and 171 joint ventures. Only 63 foreign enterprises are formed by legal entities and 269 by individuals (table 1).

Most economic indicators show lagging of border regions from northern industrial neighbors in the zone of the economic corridor, among them - the effectiveness of foreign investment in the regional economy and per capita, the importance and return on sales of joint ventures. All regions are characterized by a high degree of concentration of production at large enterprises. Some features are noticeable in Tyva with the least number of foreign and joint enterprises, which are represented by several large units in the extractive sector.

The largest number of enterprises with foreign capital is registered in wholesale trade (60 foreign entities and 30 joint ventures) followed by construction, and then by retail trade (table 2). So the priority activity in the zone of the planned corridor is rather intercountry mediation in the supply of goods to local markets than the trade itself. But the share of wholesalers in the sectoral structure of enterprises in the border regions is half as much as in the Siberian Federal District as a whole (15% vs. 30%). The share of builders is higher (13% vs. 9%), as well as of retail dealers. Besides, more significant in the border zone is such activity as the provision of food and beverages - 6% versus 3.5%.

**Table 1.** Enterprises with foreign capital in the zone of economic corridor in comparison with the industrial regions of Eastern Siberia

| | Zabaikal. kraj | Rep. of Buryatia | Rep. of Tyva | Rep. of Altai | Irkutsk oblast | Krasnoyar. kraj |
|---|---|---|---|---|---|---|
| **Total revenue of enterprises with foreign capital in 2016, mln rubl.** | 2215.9 | 9072.0 | 9736.0 | 1863.6 | 69334.2 | 680396.7 |
| **Total revenue per 1 thousand people, mln rubl.** | 8.4 | 2.3 | 30.7 | 8.3 | 28.8 | 236.7 |
| **Revenue to GRP ratio (%)** | 3.5 | 1.1 | 18.7 | 4.0 | 6.5 | 38.5 |
| **Number of foreign enterprises** | 154 | 154 | 7 | 26 | 256 | 404 |
| **The share of foreign enterprises registered by legal entities (%)** | 23 | 8 | 57 | 42 | 30 | 22 |
| **Return on sales of foreign enterprises (%)** | 24.6 | 16.1 | 75.6 | –15.3 | 17.0 | 50.0 |
| **The share of large unities in total revenue of foreign enterprises (%)** | 81 | 58 | 100 | 92.6 | 94 | 97 |
| **Number of joint ventures** | 53 | 69 | 6 | 25 | 130 | 153 |
| **The share of joint ventures in total revenue** | 36.9 | 33.5 | 0.4 | 18.4 | 47.7 | 82.9 |
| **Return on sales of joint ventures (%)** | 19.9 | 25.2 | -30.0 | 9.5 | 53.1 | 50.9 |
| **The share of large unities in total revenue of joint ventures** | 86.7 | 84.6 | 0 | 79.4 | 97.4 | 99.7 |

**Table 2.** Priority activities for foreign capital (by the number of enterprises)

| | Zabaikalsky Kraj | | Republic of Buryatia | | Republic of Tyva | | Republic of Altai | | Total |
|---|---|---|---|---|---|---|---|---|---|
| | foreign | joint | foreign | joint | foreign | joint | foreign | joint | |
| **Wholesale** | 18 | 6 | 30 | 10 | 1 | 2 | 11 | 12 | 90 |
| **Construction** | 28 | 9 | 19 | 6 | 1 | 1 | 4 | 4 | 72 |
| **Retail** | 20 | 6 | 12 | 3 | 0 | 0 | 1 | 0 | 42 |
| **Extraction of mineral resources** | 19 | 6 | 3 | 2 | 3 | 1 | 1 | 1 | 36 |
| **Provision of food and beverages** | 10 | 1 | 15 | 8 | 0 | 0 | 1 | 0 | 35 |
| **Logging** | 5 | 2 | 9 | 8 | 1 | 0 | 1 | 0 | 26 |

So in this zone the diversification of foreign companies' activities takes place due to infrastructure and intermediary branches, which are most accessible for individual business with minimal investment. Therefore, the share of enterprises created by foreign citizens in this zone was 79%, i.e. more than three quarters, while in Siberia this share was 61.5%.

As to industrial production, a significant presence of enterprises with foreign capital is observed in extraction of mineral resources and logging. The processing industry in this zone is relatively developed in Buryatia only, but there are practically no foreign or joint ventures with complex and composite production. The mining industry is represented by extraction of metal ores in Zabaikalsky kraj (19 foreign enterprises and 3 joint ones) and coal mining in the Republic of Tyva at three large enterprises. The largest is the foreign enterprise "Baikalrud" in Zabaikalsky kraj, owned by the Chinese company "Central Asian Silver Polymetal Group Ltd." and developing the Noion-Tologoy deposit of polymetallic ores.

Two types of activity are prevailing in the service sector: consulting on management issues and engineering and technical design. The first one is the phenomenon of Buryatia: there are 28 enterprises of this kind formed by citizens of Mongolia. Perhaps in this case, a more significant role is played by additional activities of these ventures, which include trade, processing of telephone calls, warehousing services, public relations and others. The second type - engineering and technical design - is distributed primarily in Zabaikalsky kraj. Modern business services, important for innovative development, are represented by two joint ventures in telecommunications only.

The largest enterprises of the zone total 19 foreign units with an annual revenue more than 100 mln rub. and 9 of them belong to foreign companies, incl. two from China, four from Cyprus, and by one from Australia, the United Kingdom and Panama. Remaining units belong to citizens of China (8), of Tajikistan and Lithuania. Joint large enterprises include 2 units in the telecommunications service - with an American company in Zabaikalsky Kraj and with a company from Cyprus in Buryatia, 2 - in trade with Belarus and China, one in construction with the Chinese side and one in the financial sector - with Cyprus. In Tyva, almost all the foreign capital is invested in three large coal mining enterprises with founders registered in offshore.

## 4. Discussion

In general, the area of the economic corridor is dominated by representation of Chinese companies and citizens - founders of business, as well as by investors from offshore both for large and small companies. The wide presence of offshore capital is typical for the whole of Siberia, but in the zone of economic corridor it is less than in other regions. The predominance of China is typical for Zabaikalsky kraj and gradually decreases from east to west. In Buryatia the presence of Mongolian founders, mostly citizens rather than companies, is noticeable, investors from Kazakhstan appear in the western part of the corridor zone.

A recipient region usually expects the inflow of capital and new technologies and business practices that accompany it, as well as the growth of human capital [2]. In this zone the activity of enterprises with foreign capital does not give grounds for expecting technological development of production, since foreign investment goes mainly to infrastructure activities or to redistribution of basic industries earnings. This type of relationship with the global markets which is characterized by a small volume of foreign investment and a low proportion of products supplied for export can be called peripheral. A structure of activity of enterprises with foreign capital is aimed mostly at the survival rather than at the expansion and the term "survival" refers also to foreign citizens.

J. Dunning singled out 4 goals for investing abroad, which include the search for raw materials (natural resources), for new markets, for increasing efficiency and for acquiring assets [3]. The first incentive connected with the need to obtain raw materials is prevailing for most Siberian regions and its role is gradually increasing in the border regions including new plans to supply timber to China, which cause disaffected reactions among the local population. The acquisition of assets in the regions analyzed, including real estate, is more relevant for citizens of Mongolia and Central Asia. The sales market is also important for investors, what is reflected by the abundance of investments in trade and intermediary activities.

Prospects for the expansion of foreign investment are largely determined by China. A. Kuznetsov [4] identifies the two most likely trends in relations with China: increasing exports of raw materials and semi-finished products and strengthening transit functions of the territory, including output at enterprises with Chinese capital for further supplies to the west. The second trend is currently being discussed in the framework of the plans for the development of the China-Mongolia-Russia economic corridor, and the expected growth in FDI during implementation of this trend provides a certain risk for this area since it can facilitate the export of profits generated here.

## 5. Conclusion

The development of transit infrastructure probably will not change the specialization of this territory, and foreign corporations will continue to invest primarily in resource projects. The developed regions of the Irkutskaya oblast and the Krasnoyarsky kraj, mentioned in the beginning of the article, will be more attractive for investments into processing industry then the zone of corridor. But attracting foreign investment in infrastructure will allow developing on-site service for maintaining future flows where new forms and scales of intermediation, the sophistication of technologies in communications, transport and other spheres will be required. Regional economies should focus on increasing exports of infrastructure services, which can become an alternative to the export of natural resources of the territory.

And the target for successful development in the corridor zone is to increase the degree of involvement of residents of this zone in international investment projects through stimulation of joint venture creation. Local residents need the help of the state for such a joint participation so that the considerable part of profits from new industries formed with the foreign capital could remain in the zone of corridor and contribute to its development.

## References

[1] Bukalova S V 2017 *Srednerussky Vestnik Obchestv. Nauk* **12**(1) 273-81 (in Russian) doi:10.12737/24803
Davletshin E A, Kotenkova S N 2014 *Fundamental Research. Econ. Scienc.* **9** 1786-9 (in Russian)
Dementjev N P 2017 *Mir Econ. i Upravlenija* **17**(2) 26-36 (in Russian)
Starodubtseva E B 2014 *Financ. Analitika: Probl. i Reshenija* **47** 15-23 (in Russian)
Abramov V L 2017 *Vopr. Region. Econ.* **33**(4) 3-11 (in Russian)
[2] Idrisov G I, Taganov B V 2015 *Ross. Vneshneecon. Vestnik* **1** 3-24 (in Russian)
Falchenko O D 2015 *Ekonomika, obchestvo, chelovek: teoria, metodologia, realnost* part 1 ed Karh D A (Ekaterinburg: Ural. St. Univ.) pp 35-40 (in Russian)
Shebeko K K, Shebeko D K 2016 *Econ. I Banki* **1** 75-83 (in Russian)
Fokina O M, Aleshchenko O M 2017 *Vestn. Voronezh. Agrarn. Univ. Econ. Scienc.* **52**(1) 222-31 (in Russian) doi: 10.17238/issn2071-2243.2017.1.222
[3] Dunning J H 1994 *Transnational Corporations* **3**(1) (NY: United Nations Publ.) pp 23-52
[4] Kuznetsov A V 2017 *Kontury Globalnyh Transformatsyj: Politika, Economika, Pravo* **10**(5) 15-9 (in Russian) doi: 10.23932/2542-0240-2017-10-5-15-29

IOP Conf. Series: Earth and Environmental Science **190** (2018) 012065 doi:10.1088/1755-1315/190/1/012065

# Development of the Transboundary Economic Cooperation within the China-Mongolia-Russia Economic Corridor

**S I Violin**

Department of Regional Economic and Social Problems, Irkutsk Scientific Center of the Siberian Branch of Russian Academy of Sciences, Irkutsk, Russian Federation

E-mail: sviolin78@gmail.com

**Abstract**. The paper deals with the issue of advancing economic cooperation of Russia, China and Mongolia given the increasing number of political and business contacts among the parties in question in the last few years. A recent formal agreement on developing the China-Mongolia-Russia economic corridor provides the framework for implementing potential joint projects in infrastructure, manufacturing, technology, and service sectors. However, there are also a number of challenges that the Russian authorities face, the most important of which is the relative underdevelopment of the eastern regions of Russia compared to north-eastern regions of China. Current economic relations between Russia and China cannot be considered mutually beneficial: Russia is mostly exporting raw and semi-final products and importing final goods including hi-tech products. The paper points out a few ways in which transboundary cooperation can be established while, at the same time, ensuring economic security of border regions.

## 1. Introduction

Economic globalization is the phenomenon that has drawn much of the scholars' attention in the last few decades. Its advocates claim that globalization is beneficial for everyone [1]; others point to its flaws and assert that globalization brought more harm than good to the poor, both in the developing and developed countries [2]. Russia has joined the world economy in the 1990s and according to Stiglitz and some other experts has been on the losing side. Indeed, opening Russian economy to foreign competition proved to be catastrophic for whole industries and put a large share of the population below subsistence level. The early 2000s was the period of stabilization and restoration. The GDP of Russia experienced steady growth, for the large part due to high oil prices and rising exports. After 2014, there have been tensions in the relations between Russia and Western countries that have led to significant distortions in trade. As a result, certain actions of the Russian government show that it is attempting to make a 'Turn to the East' – to shift accents in economic relations from the European Union and the United States to Asian economies, the largest and the most promising of which is China [3].

However, this is a path with many obstacles. On the one hand, Russia is interested in increasing trade and attracting investments from China to make its external relations more diversified. On the other hand, China is perceived as a possible threat to the economic security of the Russian Far East and Siberia. Therefore, it is challenging to formulate a strategy that will be in the best interests of Russian citizens and help develop the eastern part of Russia rather than make this a 'core-periphery' relation where Russia will only play a role of a natural resources exporter [4]. This paper identifies main challenges that the Russian public policy faces in its relations to China and suggest certain ways in which Russian-Chinese economic relations can be beneficial to both parties.

## 2. Challenges in Russia's 'Turn to the East' policy

Given rapid development of Chinese regions that border with Russia, less dynamic development of the corresponding Russian regions poses significant challenges for the governments of those regions, as well as for the federal government. Russia borders with China both in Siberia and in the Far East. However, most of the trade between Russia and China either originates or is carried out through Zabaikalsky krai, the Republic of Buryatia, Irkutsk oblast and Altai krai via transport corridors, of which the railway being the most important. Irkutsk oblast is a large industrial center, together with the Republic of Buryatia and Zabaikalsky krai they are rich in mineral resources. Altai krai and the Republic of Buryatia have strong agricultural complexes. All the aforementioned regions also have a good recreational potential, most well-known being: balneological resorts of Belokurikha in Altai, Tunkinskay valley in Buryatia, Darasun in Zabaikalsky krai, and, of course, Lake Baikal as well as many other prospective tourist destinations. Comparison of dynamics of economic growth between corresponding Russian and Chinese regions is given in the following table.

**Table 1**. Comparison of Gross Regional Products in Statics and Dynamics of Certain Russian and Chinese Regions[a].

| Region | Gross Regional Product (GRP), mln. US dollars[b] | Annual Growth of GRP for the Period of 2011-2016, % |
|---|---|---|
| **Russian Federation:** | | |
| Zabaikalsky Krai | 3932,2 | 0,2 |
| Irkutsk Oblast | 15990,9 | 4,0 |
| Republic of Buryatia | 2980,9 | -0,6 |
| Altai Krai | 7463,2 | 1,7 |
| **People's Republic of China:** | | |
| Inner Mongolia | 272827,6 | 8,3 |
| Heilongjiang | 231560,4 | 6,9 |
| Xinjiang | 145227,8 | 8,9 |

[a] Calculated based on the official data of Russian State Statistics and Chinese State Statistics.
[b] Converted by the author from national currencies using official annual exchange rates.

As we can see from the table, both in absolute terms and in terms of rates of growth Chinese regions are far ahead of corresponding Russian regions. Moreover, the main contributors to economic growth of Chinese regions are secondary industry (manufacturing) and service sectors, whereas in Siberian regions these are natural resources extraction and primary processing. Thus, the quality of growth is also different.

In the last years political, business, scientific, and other ties between Russia and China have received a new impetus to development given rising tension in relations between Russia and Western countries, as well as in relations between China and the USA. Whereas Russia can be confident in scientific and cultural respects in relations with China, they need to offer something more substantial than export of mineral resources in the economic sphere to stand on equal grounds with its neighbor.

Unfortunately, the export of natural resources is precisely the economic agenda formed during the 1990s. Whereas export from Siberian regions to China has changed little from that period, the structure of imports is becoming more complex: during first years after Perestroika Russia mostly imported consumer goods, now, on the other hand, there is a growing share of machinery, equipment, computers and electronics, and other hi-tech products.

In 2009, the joint Program of Cooperation between Russian Far East and Eastern Siberia and North-Eastern Regions of China for the Years of 2009-2018 was signed. Soon after the Program's publication, experts in Chinese studies started to express concerns about its content. The experts noted that the Program activities are much more beneficial for the Chinese party than Russia: the projects that were planned for implementation on the Russian territory were related to extracting and initial processing of natural resources, expanding the handling capacity of border crossings; Chinese projects, however, dealt more with industrial development – producing durable goods [5, 6 p 248-50]. Agricultural projects of the Program also appeared controversial considering the history of rapacious land use by Chinese agricultural producers [7]. Besides that, the Program has set certain preferences for Chinese working migrants [6 p 242-43]. Considering apparent drawbacks in the Program, it has been revised on several occasions. Starting from 2015, it has not been mentioned in official reports of Russian authorities. In any case, 2018 is the last year of the Program implementation.

Another official document worth mentioning here is the Program of Establishing the Economic Corridor China – Mongolia – Russia signed in 2016. The Program's goals are "increasing trade turnover, ensuring produce competitiveness, supporting cross-border transportation, infrastructure development" [8]. It is difficult to consider this program to be a significant step forward in trilateral relations. The Program provides only a framework of future cooperation. Although it contains some projects, in reality these are only references to a possibility of projects' development in the future. For example, the subsection "Transport Infrastructure" of the "Projects List" section of the Program mostly includes activities starting with the phrase: "To research and if economically feasible to start implementing the project…" In other words, the actual implementation of the project may never start. The Program is heavily skewed toward transport infrastructure development (40% of all projects); industrial projects comprise less than 1% of all activities. Thus, the Program mostly aims at increasing trade turnover and transit of goods between Asia and Europe. The goal of ensuring produce competitiveness is not supported by actual activities.

## 3. Prospects of Russia-China economic relations

In the last years, mostly due to deteriorating economic and trade relations with the United States and the European Union, Russian authorities have given closer attention to developing the external economic relations with Asian countries. In these circumstances development of a new program of cooperation between Russian and Chinese regions is desirable, a program that will contain specific activities and projects akin to the Program of 2009. This program should also include Mongolian regions as part of the economic corridor. Apparently, the program should be developed on a new basis though, in which Russia would play a role of an equal partner not an exporter of natural resources for the rapidly developing regions of North-Eastern China. The regions of the Russian Far East and Eastern Siberia should actively participate in program development.

Development of a new program is important for Russia to try to break the undesirable trend of economic exchange of natural resources for finished products. The most worrisome situation has developed in timber industry where Siberian regions export large amounts of timber and China managed to build a number of facilities near Russian borders to process it into construction materials, furniture, and the like. A loss of value-added, potential jobs, profits and taxes for the Russian economy is obvious.

On the other hand, the urgent need for the development of the processing industry in Siberia and the Far East largely remains in the form of declaration. Most large projects that are being implemented or are planned to be implemented in the Siberian regions that we mentioned in the beginning of the paper are dealing with that same export of natural resources. These projects include, for example, 'The Power of Siberia' gas pipeline that will transport gas from Siberia to China, export of oil to China and Japan, development of the Udokansk copper mine in Zabaykalsky krai, metal ores in Buryatia, and others. All of this notwithstanding, Russia has now a unique opportunity to take the negotiations with its Chinese partners concerning future economic cooperation to a new level. A major factor in understanding the future of Russia-China relations is the decision of the Trump administration in the

US to raise import tariffs for Chinese goods [9]. The United States is one of the primary markets for Chinese products so China will have to seek new ways of selling its goods and Russia stands as one of the most prospective candidates for filling at least part of the niche.

Rich natural resources base, which Chinese economy requires for sustained growth, as well as potential markets of Russian regions are the key elements in negotiations on the development of future cooperation of the two countries. However, negotiations on the Russian side will encounter certain difficulties. History of interactions with potential Chinese partners shows that China's officials and businesspeople work in close collaboration and firmly pursue the interests of the country's economy. Particularly, traditionally Chinese investors express the following requirements when asked to participate in joint projects.

Firstly, China is interested in keeping importing natural resources from Russia with consequent processing near the borders to minimize the transportation costs. Besides that, China is interested in importing electric power from Siberian hydropower stations. Correspondingly, these are the priorities for Chinese investors. Secondly, even when Chinese investors are willing to invest in manufacturing, they prefer to hire Chinese workers and demand preferential treatment for the land purchase or lease. There are also certain peculiarities in attracting tourists from China that prefer to stay at Chinese hotels, use the services of Chinese cooks, tour guides, and so on. For this aim, Chinese investors buy pieces of land near popular tourist destinations like Lake Baikal. It is evident that mass tourism of this type is not in the long-term economic interests of Siberian and Far-Eastern regions and the country as a whole and creates significant ecological risks. Of a growing concern are the shadow land-leasing practices for agriculture, both in Siberia and in the Far East. This happens when Chinese nationals do not rent land for the purposes of agriculture openly but make arrangements with Russian farmers and other landowners when the latter do not disclose the details of the deals and do not include these profits in their tax returns. This leads to a growing number of Chinese migrant agricultural workers moving to Russian villages. The concern is that they may stay there permanently and require certain preferences from the Russian government [10].

As we could infer from Table 1, there is a strong imbalance in regional development of the two countries in question. Based on that, the main priority of the federal Russian government in its 'Turn to the East' policy should be enhanced development of its eastern territories. This is an important prerequisite to open trade and other economic relations with stronger counterparts. Certain decisions have already been made on the federal level to ensure that. An important step to attract investment, domestic as well as foreign, was the creation of Territories of Accelerated Social and Economic Development. These territories are administrative locations that enjoy tax and tariffs preferences. The federal government has also increased investments in infrastructure that helped improve the quality and speed of transportation. Certain measures have been implemented to attract workforce to the eastern regions, such as free-of-charge land acquisition in the Far East and one-time money allowance for young specialists willing to work in the countryside. Unfortunately, the measures to increase or at least keep human capital in the eastern regions have not proved to be effective. There is still an outflow of working population from Siberia and the Far East. One of the main reasons for that is the lack of well-paid, high-skilled jobs when universities' graduates just cannot find ways to apply their knowledge and skills in their native regions and are forced to move to the central part of Russia or abroad.

Considering all this, potential cooperation between the two countries can be reached in the following directions.

1. Infrastructure development within the economic corridor.

Infrastructure can be developed in two main ways. First, within the framework of existing projects, such as "The Power of Siberia", it is important to put effort not only into developing the infrastructure of the projects themselves but also the infrastructure of the surrounding territories for the purpose of their further exploration. This will help enhance economic and trade ties, industry, tourism, quality of life, and other spheres. Among the specific joint infrastructural projects, we can suggest enhancing capacity of the Transsiberian and Baikal-Amur railroads, constructing railways and roads to most

IOP Conf. Series: Earth and Environmental Science **190** (2018) 012065 doi:10.1088/1755-1315/190/1/012065

promising mineral deposits, gasification of eastern regions and developing transport infrastructure in local hubs along the economic corridor China – Mongolia – Russia including the upgrade of the airports and river ports infrastructure and roads construction.

Second, special attention should be given to the development of infrastructure of border towns assigning them a hub status. Here large multimodal logistic centers can be created. As for now, most of the imported goods, even those purchased through Chinese Internet stores, are first shipped to Moscow, sometimes to Novosibirsk, and then are distributed around the country. This is a costly and time-consuming process. Creating multifunctional logistic centers near the border will help reduce costs and increase the trade turnover. Federal and regional investments should also be put in development of urban environment. Together, these measures will help attract human capital to the bordering regions – one of the positive effects of the mutually beneficial trade relations.

2. Creating joint ventures in the manufacturing industry.

It is desirable to create joint Russian-Chinese and Russian-Mongolian ventures to process natural resources with the aim of producing high value-added products that could further be exported to the third countries, as well as consumed locally. Given the limitations that the United States and other countries impose on the import of raw materials and primary processed produce, creation of joint ventures of the type mentioned will help our products be more competitive on the world markets. This will be achieved through the joint use of technologies, skilled workforce, natural, financial, and other resources. In this case, the priority should be given to timber processing, metalworking, machinery and equipment, construction materials industry, and food industry.

The Chinese economy no longer enjoys lower level of wages as a competitive advantage. Therefore, utilizing human resources of the eastern part of Russia can be beneficial for all parties. Jobs creation will eventually lead to rising standards of living, which, in turn, will help increase the purchasing power of the local population that will expand markets for consumer products.

3. Advancement of the "green economy."

The third important way of cooperation between Russia and China is the field of "green economy." A common definition of the "green economy" is that it is "the economy aimed at preservation of well-being of society, through effective use of natural resources as well as ensuring the return of end user's products to the production cycle" [11]. Both countries face serious ecological challenges posed by rapid industrialization. For China, it is a relatively new issue and Russia has been dealing with these problems for a few decades now. Addressing ecological challenges assumes search for new technological, organizational and other solutions aimed at sustainable environment, restoration of disturbed ecosystems and striking a balance between economic growth and environmental well-being. A big part of this work is investing in recycling in order to bring back the waste into a production process. This can definitely be the area of mutual interest of the two countries.

## 4. Conclusion

The world's economy is not unipolar anymore. Up to date, Russian external economic relations have mostly been oriented at the European Union and the United States. However, rapid economic growth of Asian economies poses serious challenges as well as presents unique opportunities for the development of the Russian economy. The future of the Russian economy will be greatly affected by its ability to establish mutually beneficial relations with its Asian neighbors, of which China is the most important one. It is vital for the Russian authorities to review the policy of exporting raw materials and semi-final products to China, which stifles local growth and makes economic prospects for the eastern regions bleak. The greatest challenge for Russia is to be able to pour resources into development of eastern regions in order to create a good manufacturing base, significantly raise the standards of living and attract human capital. Without that, the 'Turn to the East' strategy does not seem feasible.

The potential areas of mutual economic interest for Russia, China and Mongolia are: development of transport infrastructure and urban environment within the economic corridor, creation of joint ventures in natural resources processing, and development of the "green economy."

IOP Conf. Series: Earth and Environmental Science **190** (2018) 012065 doi:10.1088/1755-1315/190/1/012065

**References**

[1] Bhagwati J 2005 I*n Defense of Globalization* (New York: Oxford University Press)
Wolf M 2004 *Why Globalization Works* (New Haven: Yale Univeristy Press)

[2] Rodrik D 1997 *Has Globalization Gone Too Far?* (Washington, DC: Institute for International Economics)
Stiglitz J 2017 *Globalization and Its Discontents Revisited: Anti-Globalization in the Era of Trump* (New York: W W Norton & Company)

[3] Blakkisrud et al 2018 *Russia's Turn to the East: Domestic Policymaking and Regional Cooperation* ed Blakkisrud H and Rowe E W (Cham: Palgrave Pivot) p 3

[4] Wallerstein I 2004 *World-Systems Analysis: An Introduction* (Durham: Duke University Press) p 11

[5] Alexandrova M V 2010 The program of cooperation of adjacent territories of Russia and China: history, facts, ways of implementation *Kitay v Mirovoy i Regionalnoy Politike. Istoria i Sovremennost'* [China in the World and Regional Politics. History and the Present Time] vol XVI, ed Safronova E I (Moscow: IDV RAN) p 200-1

[6] Gelbras V G 2011 China and APEC in the context of the global crisis *Vladivostok 2012: ATES i Novye vozmozhnosti Rossii* [Valdivostok 2012: APEC and New Opportunities for Russia] (Moscow: Universitetskaya Kniga)

[7] Gel'bras V G 2016 Chinese labor migration to Russia: what it promises to our country *Asia and Africa Today* **5** 50-4

[8] Ministry of Economic Development of Russian Federation 2016 *The program of establishing the economic corridor China – Mongolia – Russia* (in Russian) Income accessed online on 8th August 2018 via http://minpromtorg.govrb.ru/rus-ch-mn.pdf

[9] Nevelsky A Overchenko M 2018 Trump threatening China with import tariffs worth of $450 billion *Vedomosti* (in Russian) Income accessed online on 9th August 2018 via https://www.vedomosti.ru/economics/articles/2018/06/20/773250-tramp-kitayu-poshlinami

[10] Lee R 2013 The Russian Far East and China: thoughts on cross-border integration *Foreing Policy Research E-Notes* Available at: https://www.fpri.org/docs/Lee_-_Russia_and_China.pdf

[11] Zabortseva T I, Kuznetsova A N and Violin S I 2017 The Potential of a 'green' economy in the socioeconomic development of Irkutsk oblast *Geography and Natural Resources* **38** 380

# Analysis of economic efficiency and eco-efficiency of Chinese star hotels based on SBM model

**B Xia[1,2,3], S Dong[2*], M Zhao[2], Z Li[2] , F Li[2], Y Li[2], H Cheng[2]**

[1]Tourism College, Northwest Normal University, Lanzhou, China
[2]Institute of Geographic Sciences and Natural Resources Research, Chinese Academy of Sciences, Beijing, China
[3]College of Resources and Environment, University of Chinese Academy of Sciences, Beijing, China

E-mail: dongsc@igsnrr.ac.cn

**Abstract**. Star hotels, as the core sector of China's tourism development, the improvement of their economic efficiency is an important way for the tourism industry to operate well. On the other hand, improving eco-efficiency can promote star hotels to take more responsibility for the ecological environment and promote the sustainable development of tourism industry. In this paper, the A Slacks-Based Measure model (SBM), as one kind of data envelopment analysis, is used to construct the economic efficiency and eco efficiency input-output system of Chinese star hotels according to the development characteristics of star hotels in China. By collecting the data in 2014 and using the energy balance sheet to calculate the carbon emission, we further calculate the economic efficiency and eco-efficiency level of the star hotels in 30 provinces in China's main land, and further analysis the input-output spatial pattern and development trend of the eco-economic system in China's star hotels through the Getis-Ord Gi index and coupling analysis. It is found that the economic efficiency and ecoefficiency of Chinese star hotels are consistent with their spatial distribution, and they have a significant linear positive correlation. The spatial distribution of the economic efficiency and eco-efficiency of Chinese star hotels is extremely uneven, and there is no spatial agglomeration. Moreover, the high efficiency area has no obvious positive impact on the periphery.

## 1. Introduction

In the trade of globalization, the competition between Chinese star hotels and the international hotels group is constantly fierce. Star hotel is the main body of the development of China's tourism industry. Its economic efficiency can be more objective and true to reflect the overall development and development potential of the tourism market. In the past ten years, with the promotion of international organizations such as OECD and WBCSD and many scholars, the concept of eco-efficiency has attracted the attention of policymakers [1], researchers and enterprise managers [2], and has been popularized and applied in many fields [3].Huang (2014) pointed out that eco-efficiency is the input output efficiency of a region under the dual constraints of resources and environment [4].

Data envelopment analysis (DEA) has certain advantages in the application of efficiency problems. [5]. The model does not need the unified index unit [6], does not need to clear the function relation, does not need to estimate the parameter [7], also can contain the undesired output index [8], increases the flexibility of the input and output index selection [9], simplifies the measurement process and ensures the integrity of the original information [10]. DEA has many extended models [11]. The main analysis models used in economic efficiency and eco-efficiency analysis are the traditional CCR and BCC models [12]. Non radial SBM model with relaxed variables in objective function [13], DDF model of directional distance function [14], Super efficiency model [15] and so on. This study uses the SBM model, which has the attributes of free dimension and variable unit [16], which can be weighted into input-output variables [17], and can also increase undesired output [18]. It is very suitable for evaluating the eco-efficiency of star hotels.

With the increasing global climate change, the research on carbon emission has become a hot topic in recent years. Taking carbon emissions as the undesired ecological environment index in the input-output eco economic system of star hotels, the eco-efficiency level of star hotels can be further analyzed. It can help to find out the pressure of the star hotels on the ecological environment while achieving economic growth.

Therefore, this paper introduces economic efficiency and eco-efficiency as a tool to measure the economic growth and ecological environment protection of star hotels in China. It analyzes the economic efficiency and eco-efficiency level and spatial pattern of the star hotels in 30 provinces in China's main land in 2014, aiming at expanding the research and application fields of economic efficiency and eco-efficiency, for promoting the green and sustainable development of star hotels.

## 2. Models and Methods

### 2.1. *Analysis of economic efficiency and eco-efficiency: SBM model*

This paper takes 30 provinces in China's main land as the decision units for the evaluation of economic efficiency and ecoefficiency. Each decision unit in the economic efficiency analysis includes input and output vectors, respectively. It is expressed as $x \in R^m$, $y \in R^s$. We definite matrix $X$, $Y$ as $[X] = [x_1, \cdots, x_n]^T \in R^{m\times n}$, $[Y] = [y_1, \cdots, y_n]^T \in R^{s\times n}$, $X>0$, $Y>0$ , and definite production possibility set P as $P = \{(x, y)|x \geq \lambda x, y \leq \lambda Y, \lambda \geq 0\}$.

The basic expression of SBM is [19]:

$$\min P^* = \frac{1 - \frac{1}{m}\sum_{i=1}^{m}\frac{s_{i0}^-}{x_{i0}}}{1 + \frac{1}{s}\sum_{r=1}^{s}\frac{s_r^+}{y_{r0}}} \tag{1}$$

$$St. x_0 = X\lambda + s^- \tag{2}$$

$$y_0 = Y\lambda - s^+ \tag{3}$$

$$\lambda \geq 0, s^- \geq 0, s^+ \geq 0 \tag{4}$$

This study calculates the eco-efficiency of China's star hotels by SBM model constructing a Undesirable outputs. Each decision unit includes three vectors: input, expected output and undesirable output. It is expressed as $x \in R^m, y^g \in R^{s_1}, y^b \in R^{s_2}$. We definite matrix $X$, $Y^g$, $Y^b$ as : $[X] = [x_1, \cdots, x_n]^T \in R^{m\times n}$, $[Y^g] = [y_1^g, \cdots, y_n^g]^T \in R^{s_1\times n}$ *and* $[Y^b] = [y_1^b, \cdots, y_n^b]^T \in R^{s_2\times n}$, $X>0$, $Y^g>0$, $Y^b>0$, and definite production possibility set P as $P = \{(x, y^g, y^b)|x \geq \lambda x, y^g \leq \lambda Y^g, y^b \geq \lambda Y^b, \lambda \geq 0\}$.

The SBM-Undesirable model based on variable returns to scale expressed as [19]：

$$P^* = \min \frac{1 - \frac{1}{m}\sum_{i=1}^{m}\frac{s_i^-}{x_{i0}}}{1 + \frac{1}{s_1 + s_2}\left[\sum_{i=1}^{s_1}\frac{s_r^g}{y_{r0}^g} + \sum_{i=1}^{s_2}\frac{s_r^b}{y_{r0}^b}\right]} \tag{5}$$

Where, $s$ represents the slack variable of input and output, $\lambda$ is a weight vector. The objective function about $s^-$, $s^g$, $s^b$ is strictly diminishing, and $0 \leq P^* \leq 1$. For specific decision making units, if and only if $P^* = 1$, and $s^-$, $s^g$, $s^b$ are all 0, the ecoefficiency is effective. If $P^* < 1$, or $s^-$, $s^g$, $s^b$ are not all 0, it shows that the eco-efficiency of decision making units is invalid, and there is the necessity of improving input output variables.

### 2.2. *Spatial pattern analysis: Getis-Ord Gi* Index*

Getis-Ord Gi$^*$ was proposed by Getis and Ord. It can find sub regions with different attribute values in the study area. It expressed as:

$$G_i^* = \frac{\sum_{j=1}^{n} \varphi_{i,j} x_j - \left(\frac{1}{n}\sum_{j=1}^{n} x_j\right) \times \left(\sum_{j=1}^{n} \varphi_{i,j}\right)}{\sqrt{\frac{\sum_{j=1}^{n} x_j}{n} - \left(\frac{1}{n}\sum_{j=1}^{n} x_j\right)^2} \times \sqrt{\frac{n\sum_{j=1}^{n} \varphi_{i,j}^2 - \left(\sum_{j=1}^{n} \varphi_{i,j}\right)^2}{n-1}}} \tag{6}$$

Where, $x_j$ is the attribute value of factor $j$, $\varphi_{i,j}$ is the spatial weight between elements $i$ and $j$, and $n$ is the total number of elements. According to the Gi* index, it can be showed the location of spatial high value (hot spot) or low value (cold spot) elements in space. The hot spots and cold points of the regional economic efficiency and eco-efficiency of the star hotels in China can represent the regions with high concentration of economic efficiency and eco-efficiency and the regions with low concentration.

### 2.3. *Indicators and data*

The economic input index of traditional production function mainly includes labor force and capital, output index output value. The input and output indicators of the economic efficiency of the star hotels in China mainly include the traditional labor input and capital input, as well as the number of star hotels which can reflect the overall scale of investment, and the output is income. The eco-efficiency of star hotels is selected from the input-output index of traditional economic efficiency, and the carbon emission level which can reflect the energy consumption and the environmental impact is added as the waste output (Table 1).

**Table 1.** Input-output indexes of eco-efficiency.

| | Name of index | Data sources | Unit |
|---|---|---|---|
| | labor | statistical yearbook | persons |
| **Input** | capital investment | statistical yearbook | bln yuan |
| | number of star-hotels | statistical yearbook | - |
| **Output** | income | statistical yearbook | bln yuan |
| **Undesirable output** | carbon emissions | calculate | million ton |

Through the energy balance of the provinces, the consumption level of all kinds of energy in wholesale and retail and accommodation is converted into standard coal, and the conversion coefficient comes from the energy statistics yearbook of China. Combined with the total value of wholesale and retail sales and the added value of accommodation and catering in 2014, the unit added value standard coal consumption was calculated. According to the carbon emission coefficient of standard coal in IPCC, the added value carbon emission coefficient of wholesale and retail accommodation unit was calculated. Finally, it multiplied with the income of star hotels in all provinces, and calculated the total carbon emission of the star hotels in 2014.

## 3. Results and Discussion

### *3.1. Spatial pattern of economic efficiency and eco-efficiency of Chinese star hotels*

This study divides the economic efficiency and eco-efficiency (both expressed E) of Chinese star hotels into five levels: inefficiency area (0.00<E<0.40), low efficiency area(0.41<E<0.60), areas to be improved(0.61<E<0.80), relative efficiency area(0.81<E<0.99), high efficiency area(E=1). The results of the analysis are shown in Figure 1. The high efficiency areas of economic efficiency include Beijing, Tianjin, Shanghai, Zhejiang, Fujian, Guangdong, Qinghai, Ningxia, The high efficiency areas of eco-efficiency include Beijing, Tianjin, Shanghai, Jiangsu , Zhejiang, Fujian, Guangdong, Hainan, Qinghai, Ningxia. The spatial distribution pattern of economic efficiency and eco-efficiency is the same.

In a word, the high efficiency area of economic efficiency and eco-efficiency are mainly concentrated in the eastern coastal areas where China's economic development is relatively developed. It further shows that the economic efficiency and eco-efficiency of China's star hotels are positively related to the level of China's economic development. The inefficiency areas of economic efficiency are Hebei, Yunnan, The inefficiency areas of eco-efficiency are Shanxi, Inner Mongolia, Sichuan, Yunnan, Gansu, Xinjiang, Ecological inefficiency is mainly located in the less developed western provinces of China. Good location and high level of economic development are the main constraints on the development of China's star hotels' economic efficiency. The eco-efficiency level of the star hotels in China is low due to the influence of various factors such as the restriction of the eco-technology level and the rationality of the structure of the resource elements in the western region.

### *3.2. Getis-Ord Gi* index analysis of economic efficiency and eco-efficiency of China's star hotels*

According to the above research, the economic efficiency and eco-efficiency of the star hotels in China may have a certain spatial correlation. Therefore, this study introduces the Getis-Ord Gi* index analysis method to further analyze the spatial pattern characteristics of the economic efficiency and eco-efficiency level of star hotels in China.

According to the calculation of Getis-Ord Gi* index, the Getis-Ord Gi* index is divided into 7 categories according to the significant level of spatial positive and negative correlation (figure 2). The economic efficiency is a transitional region with no significant correlation, while Yunnan and Chongqing are the cold spots with 10% eco-efficiency agglomeration significance. The economic growth and environmental impact of China's star hotels have no obvious spatial agglomeration characteristics, showing a certain spatial dispersion. It shows that the imbalance between economic efficiency and eco-efficiency is obvious. And the economic efficiency and eco-efficiency of the better development areas did not achieve the proliferation and spillover effects on the surrounding areas. This is contrary to the trend of the development of hotel industry in other developed countries.

### *3.3. Spatial coupling analysis of economic efficiency and eco-efficiency of Chinese star hotels*

The coupling of economic efficiency (EE) and eco-efficiency (CE) is analysed in this research (figure 3). It is further proved that the economic efficiency and eco-efficiency of the star hotels in 30 provinces in mainland China are basically the same. The study found that the economic efficiency and eco-efficiency were all obvious concave points in Yunnan. The most convexity appeared in 6 provinces in Guangdong, Beijing, Tianjin, Fujian, Zhejiang, Shanghai, and the eco-efficiency and economic efficiency of four provinces of four points were both 1. Through the analysis of economic efficiency and eco-efficiency, the scatter point analysis found that the eco-efficiency and economic efficiency have obvious linear positive correlation, and the contribution rate of unit eco-efficiency to economic efficiency is 0.7852. This study further divides the coupling economic efficiency and eco-efficiency of the star hotels of 30 provinces into 4 types with the relative effective level of 0.8 as the dividing line. Type I is low carbon development type (EE>0.8, CE>0.8), mainly including 6 provinces: Guangdong, Beijing, Tianjin, Fujian, Zhejiang and Shanghai. Type II is high carbon development type (EE>0.8, CE<0.8), there is null.

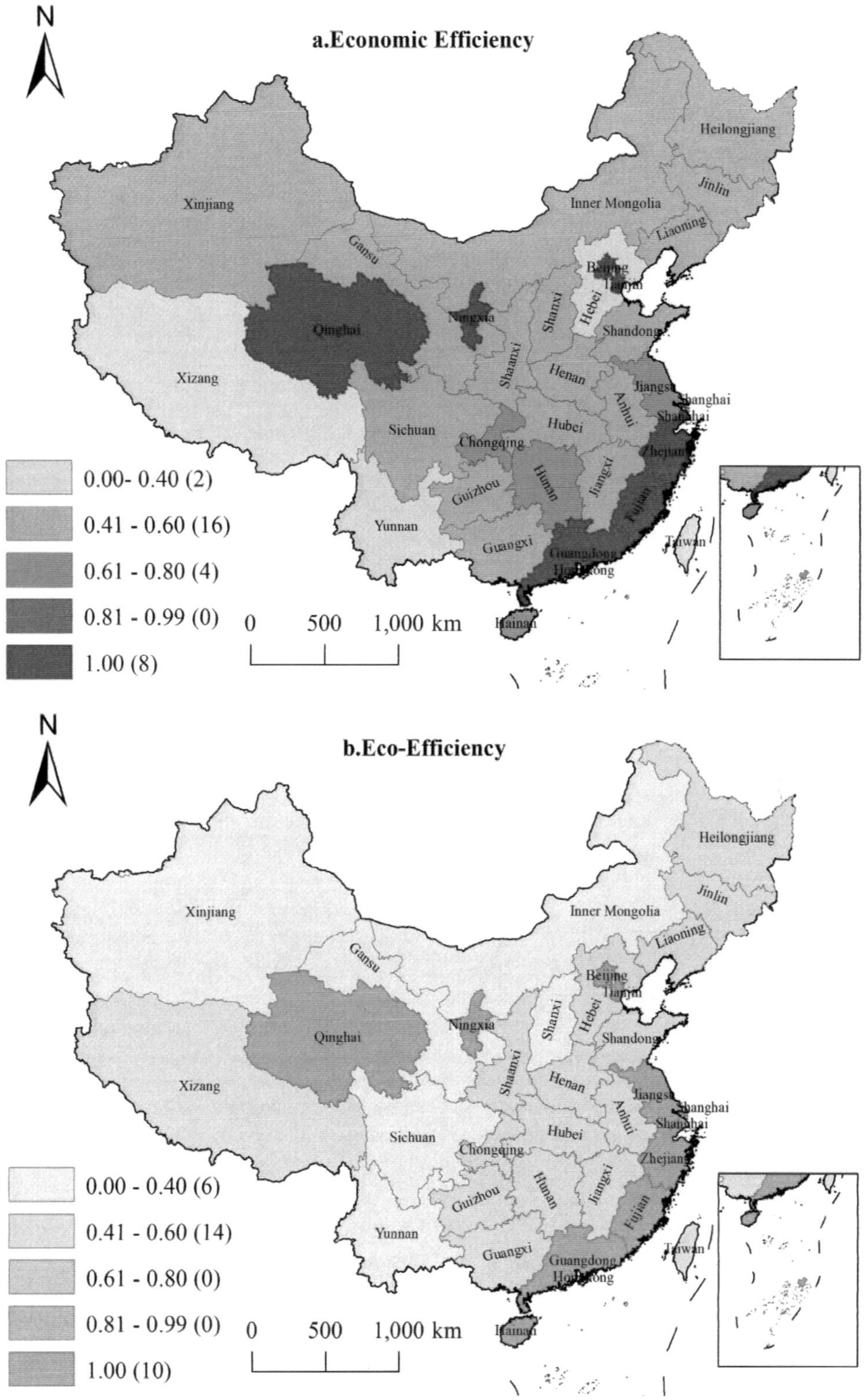


**Figure 1.** The spatial pattern of economic efficiency and eco-efficiency of star hotels in China

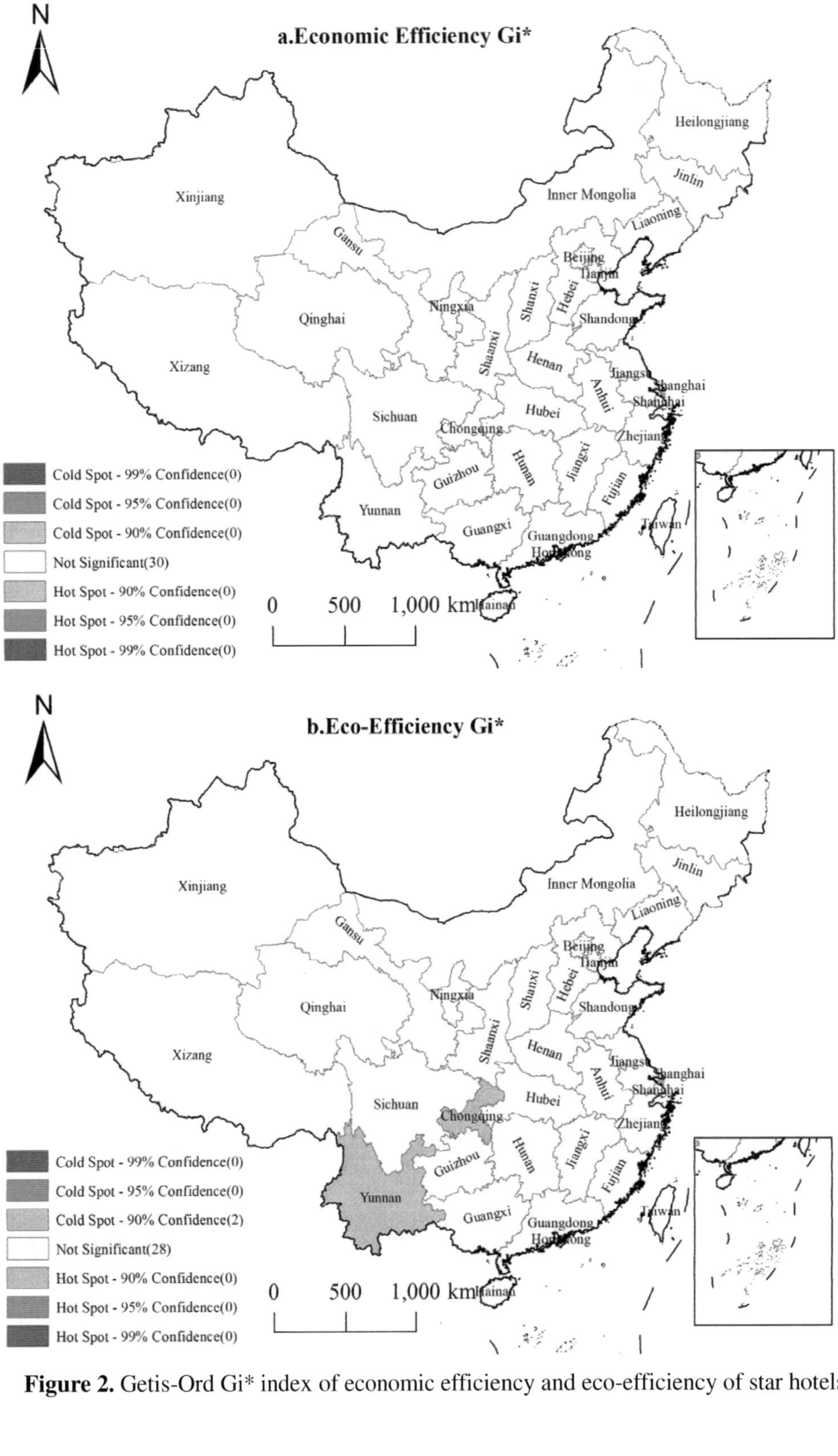


**Figure 2.** Getis-Ord Gi* index of economic efficiency and eco-efficiency of star hotels in China

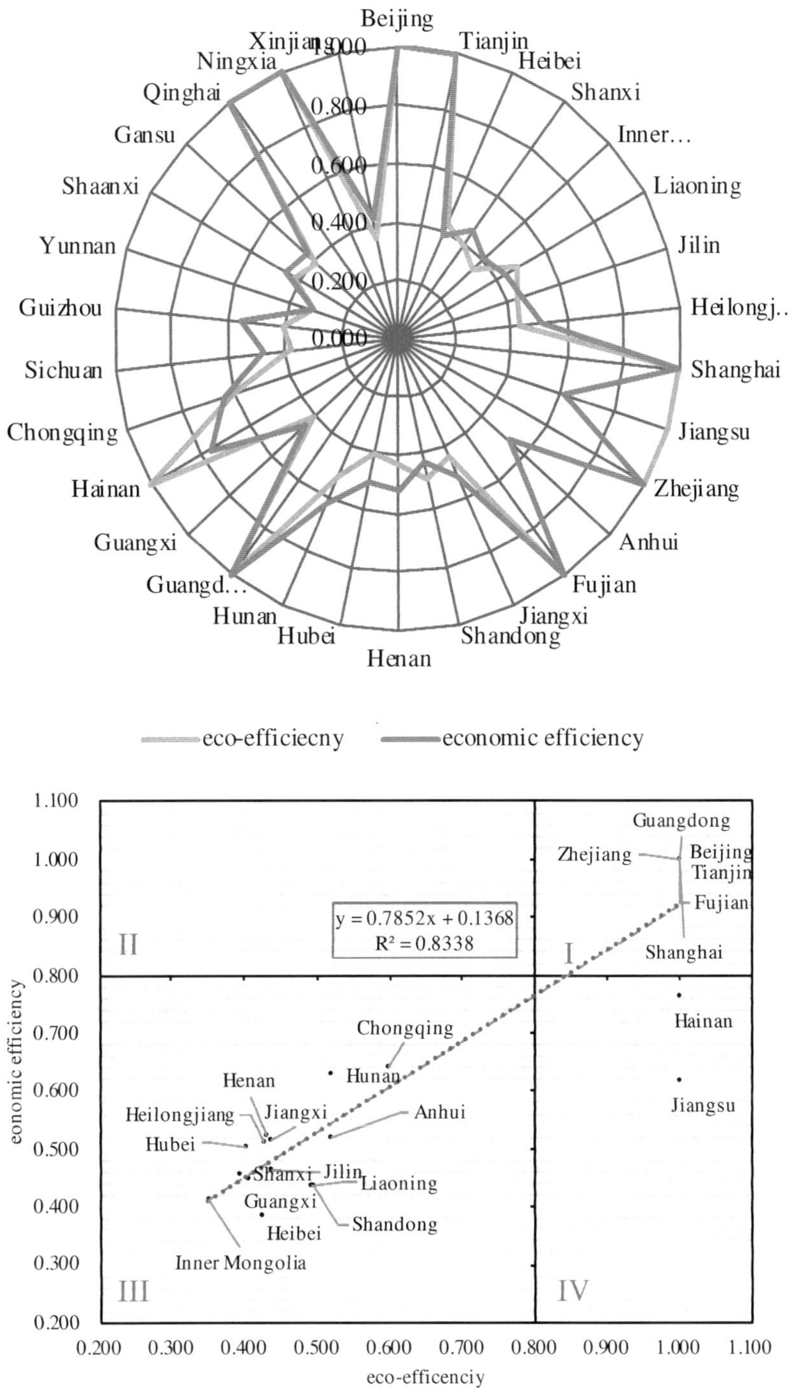


**Figure 3.** Spatial coupling pattern of eco-efficiency and economic efficiency of star hotels in China

Type III is high carbon and low efficiency (EE<0.8, CE<0.8), most of the development of star hotels in China is this type, and it includes 22 provinces, such as Chongqing, Hunan, Anhui, Jiangxi and so on. Type IV is low carbon and low efficiency (EE<0.8, CE>0.8), mainly including 2 provinces of Hainan and Jiangsu. Type I area should be the demonstration area of green sustainable development for star hotels in China. It should actively butt with the world famous hotel group, introduce advanced talent and high-end technology of economic innovation and development, adjust the management mode and internal resource allocation structure of star hotels, and improve the additional value of hotel service and products, sustainable development of green and low carbon in star hotels. Type III areas should optimize the input and output redundant variables, optimizing the allocation of resources, improve the capital efficiency, improving the quality of human capital, optimizing the ecological and economic efficiency, and to strive to directly transit to type I area. In order to optimize the area for economic efficiency, type IV areas should optimize the hotel management model, optimize the organization structure, improve the additional value of production and service, and make a steady transition to the type I area.

## 4. Conclusion

The overall development level of economic efficiency and eco-efficiency of Chinese star hotels is basically the same as that of spatial pattern. Beijing, Tianjin, Shanghai, Zhejiang, Fujian, Guangdong, Qinghai and Ningxia are at the same time at the forefront of the best level of economic efficiency and eco-efficiency. Regions with inefficient eco-efficiency and inefficient economic efficiency mainly concentrated in less developed western China.

According to the coupling analysis of the eco economic efficiency of star hotels in various provinces and regions, the development of star hotels in China is further divided into four types: low carbon development type, high carbon development type, high carbon low efficiency type and low carbon low efficiency type. Low carbon development is mainly concentrated in the developed coastal provinces of eastern China. However, the development of star hotels in most provinces in China is low-carbon and low-efficiency of economic efficiency and eco-efficiency.

The spatial agglomeration characteristics of Chinese star hotels are not significant, and there is no obvious spatial agglomeration characteristics of economic growth and environmental impact. This shows that the imbalance of economic efficiency and eco-efficiency between 30 provinces is obvious. Economic efficiency and eco-efficiency in better developed areas do not achieve the diffusion and spillover effects on the surrounding areas. This is contrary to the trend of the development of hotel industry in other developed countries. The scattered spatial development pattern is not conducive to the agglomeration of industrial elements, and is not conducive to the balanced regional development of star hotels.

## References

[1] Mardani A, Zavadskasv K, Streimikiene D, Jusoh A and Khoshnoudi M 2017 A comprehensive review of data envelopment analysis (DEA) approach in energy efficiency. *Renew. Sust. Energ. Rev.* **70** 1298-322
[2] Caiado G, Dias D, Mattos V, Quelhas G and Leal W 2017 Towards sustainable development through the perspective of eco-efficiency - A systematic literature review J*ournal of Cleaner Production* **165** 890-904
[3] Angelakoglou K and Gaidajis G 2015 A review of methods contributing to the assessment of the environmental sustainability of industrial systems J*ournal of Cleaner Production* **108** 725-47
[4] Huang B, Yang G, Cheng G and Wang Y 2014 A comprehensive eco-efficiency model and dynamics of regional eco-efficiency in China *Journal of Cleaner Production* **67** 228-38
[5] Zhou H, Yang Y, ChenY and Zhu J 2018 Data envelopment analysis application in sustainability: The origins, development and future directions *European Journal of Operational Research.* **264**(1) 1-16
[6] Sueyoshi T, Yuan Y and Goto M 2017 A literature study for DEA applied to energy and environment *Energy Economics* **62** 104-24
[7] Färe R, Crosskopf S and Lovell C 1994 *Production Frontiers* (Cambridge: Cambridge University Press)

[8] Jiang Z, Zhu H and Cao Y 2017 Efficiency pattern and spatial strategy of ports in Yangtze River Delta Region *Chinese Geographical Science* **27**(2) 298-310
[9] Richa K, Babbitt W and Gaustad G 2017 Eco-Efficiency Analysis of a Lithium-Ion Battery Waste Hierarchy Inspired by Circular Economy *Journal of Industrial Ecology* **21**(3) 715-30
[10] Pina A and Martinez P 2016 Development and Urban Sustainability: An Analysis of Efficiency Using Data Envelopment Analysis *Sustainability* **8**(2) 15
[11] Liu W, Tian P, Chen J, Lu W and Gao Y 2015 Environmental Performance Analysis of Eco-Industrial Parks in China: A Data Envelopment Analysis Approach *Journal of Industrial Ecology* **19**(6) 1070-81
[12] Rossi S, Colicchia C, Cozzolino A and Christopher M 2013 The logistics service providers in eco- efficiency innovation: an empirical study *Supply Chain Manag.* **18**(6) 583-603
[13] Long X, Zhao X and Cheng F 2015 The comparison analysis of total factor productivity and eco-efficiency in China's cement manufactures *Energy Policy* **81** 61-6
Li L and Liu Y 2017 Industrial Green Spatial Pattern Evolution of Yangtze River Economic Belt in China *Chinese Geographical Science* **27**(4) 660-72
[14] Yang L, Tang K, Wang Z, An H and Fang W 2017 Regional eco-efficiency and pollutants' marginal abatement costs in China: A parametric approach *Journal of Cleaner Production* **167** 619-29
Ramli A, Munisamy S and Arabi B 2013 Scale directional distance function and its application to the measurement of eco-efficiency in the manufacturing sector *Ann. Oper. Res.* **211**(1) 381-98
Molinos-Senante M, Gemar G, Gomez T, Caballero R and Sala-Garrido R 2016 Eco-efficiency assessment of wastewater treatment plants using a weighted Russell directional distance model *Journal of Cleaner Production* **137** 1066-175
[15] Dai Z, Guo L and Jiang Z 2016 Study on the industrial Eco-Efficiency in East China based on the Super Efficiency DEA Model: an example of the 2003-2013 panel data *Applied Economics* **48**(59) 5779-85
Chen N, Xu L and Chen Z 2017 Environmental efficiency analysis of the Yangtze River Economic Zone using super efficiency data envelopment analysis (SEDEA) and tobit models. *Energy* **134** 659-71
[16] Bocken P, Short W, Rana P and Evans S 2014 A literature and practice review to develop sustainable business model archetypes *Journal of Cleaner Production* **65** 42-56
[17] Lozano S, Adenso-Diaz B and Barba-Gutierrez Y 2011 Russell non-radial eco-efficiency measure and scale elasticity of a sample of electric/electronic products *J. Frankl. Inst.-Eng. Appl. Math.* **348**(7) 1605-14
[18] Song T, Cai J, Yang Z, Chen M and Lin J 2016 Urban Metabolic Efficiencies and Elasticities of Chinese Cities *Chinese Geographical Science* **26**(6) 715-30
[19] Cooper W, Tone K and Seiford M 1999 *Data Envelopment Analysis: A Comprehensive Text with Models, Applications References and DEA-Solver Software with Cd-rom* (Kluwer Academic Publishers) p 145–56

IOP Conf. Series: Earth and Environmental Science **190** (2018) 012067 doi:10.1088/1755-1315/190/1/012067

# Development patterns and cooperation paths of tourism industry within the China–Mongolia–Russia Economic Corridor

**M Zhao[1,2], S Dong[1,3], B Xia[1,3,4], H Cheng[1], Y Li[1,3], Z Li[1,3], J Zheng[1,3]**

[1]Institute of Geographic Sciences and Natural Resources Research, Chinese Academy of Sciences CAS, Beijing, China
[2]Faculty of Ecotourism, Southwest Forestry University, Kunming, China
[3]University of Chinese Academy of Sciences, Beijing, China
[4]Tourism College, Northwest Normal University, Lanzhou, China

E-mail: dongsc@igsnrr.ac.cn

**Abstract.** Strengthening tourism cooperation in border areas along the China–Mongolia–Russia Economic Corridor (CMREC) are important and challenging, under "Belt and Road" (B&R) initiative. Research has constructed a tourism competitiveness evaluation model including three dimensions and based on six factors: natural resources, cultural resources, tourism market, social economy, traffic facilities, and policy support. The results show the spatial characteristics of tourism competitiveness ordered as Beijing–Tianjin–Hebei of China, Northeast China, Northwest China, Russian Siberia Federal District, Mongolia, Russian Far East Federal District, and Russian Ural Federal District. Data show that the average tourism competitiveness of the three areas in Northern China are 2.69 times those of the three Russian areas, and 1.74 times those of Mongolia; at the same time, the average tourism competitiveness of Mongolia is 1.54 times that of the three areas of Russia. Based on the evaluation result, this study proposes a theoretical framework for a competition–cooperation model for tourism within the CMREC. Cluster analysis shows that the seven areas are comprise of absolutely dominant regions, comparatively dominant regions and comparatively weak regions. Tourism strategies are proposed in this study, based on the theoretical framework for a competition–cooperation model.

## 1. Introduction

The China–Mongolia–Russia Economic Corridor (CMREC) is one of the core areas within six economic zones those identified as part of the Chinese "Belt and Road" (B&R) initiative. The cross-border areas within the B&R encompass complex and diversified geographic environments, and contain abundant natural and human resources for tourism. These areas also comprise a series of important international channels that link the Eurasian continent with Chinese regions that will enable the countries to open up to the north [1]. Channels within the B&R include cultural highlands in Northeast Asia are a key for the conservation of global biodiversity, and important to the goal of the World Tourism Organization (WTO) to strengthen international cooperation along the Silk Road. Thus, in 2016, China, Mongolia and the Russian Federation signed the Outline of Planning for the Construction of China-Mongolia-Russia Economic Corridor, symbolizing the formal implementation

IOP Conf. Series: Earth and Environmental Science **190** (2018) 012067 doi:10.1088/1755-1315/190/1/012067

of the first multilateral economic cooperation corridor along the B&R [2]. This corridor was mainly constructed to connect the Chinese Silk Road Economic Belt, the Russian Eurasian Economic Union, and Mongolia's proposed "Prairie Road" [3]. The CMREC emphasizes areas such as expanding human communications and deepening tourism cooperation, and aims to encourage the three parties to reach their tourism resource potentials, take international cooperative advantages, expand economic development space, and raise the joint competitiveness of the nations involved in the international tourism market [4]. According to regional economic and ecological theory, competition and cooperation usually co-exist in regional tourism development, it has an important theoretical significance and practical value to evaluate the tourism competitiveness within the border areas between China, Mongolia, and Russia. This paper explores a competition and cooperation development model for the CMREC, with the aim of promoting the initial joint development of tourism along the B&R.

## 2. Models and Methods

### *2.1. Study Area*

The scope of this research gives priority to the border areas of the CMREC, encompassing a total area of around 2 307 000 km$^2$ (including 4 322 000 km$^2$ within China, 1 314 000 km$^2$ within Russia, and 1 564 000 km$^2$ within Mongolia), see figure 1. According to the principle of comparing similarities between natural geographical environments and social economic conditions, the study areas are divided into seven large tourism zones, the Beijing–Tianjin–Hebei Region (Beijing City, Tianjin City, and Hebei Province of China), the Northeast Region (Heilongjiang Province, Jilin Province, and Liaoning Province of China), and the Northwest Region (Inner Mongolia Autonomous Region, Gansu Province, and the Xinjiang Autonomous Region of China), the entirety of Mongolia (Ulan Bator City and 21 provinces), and the Federal District of Ural (Tyumen Oblast, Sverdlovsk Oblast, and Perm Krai of Russia), the Federal District of Siberia (Zabaykalskii Krai, Republic of Buryatia, and Irkutskaya Oblast of Russia), and the Federal District of the Far East (Primorskii Krai, Khabarovskii Kray, and Amurskaya Oblast of Russia).

**Figure 1.** Study Area

### *2.2. Data collection*

The year 2015 is used as a benchmark in this study, and resource data was collected by a global level, including the Directory of World Natural and Cultural Heritages publicized by United Nations Educational Scientific and Culture Organization (http://unesco.org), the Directory of World Biosphere

IOP Conf. Series: Earth and Environmental Science **190** (2018) 012067 doi:10.1088/1755-1315/190/1/012067

Reserves, Non-material Cultural Heritage, World Geoparks, World Museums under the "Man and the Biosphere Programme", Globally Important Agricultural Cultural Heritages Systems as published by the United Nations Food and Agricultural Organization (FAO) (http://www.fao.org), and Internationally Important Wetlands as published by the Internationally Important Wetland Convention Organization (https://rsis.ramsar.org). In addition, data from the tourism market, the social economy, and traffic facilities were sourced from the WTO, the World Travel & Tourism Council (WTTC), the Statistical Yearbook 2016 of China, the Statistical Yearbook 2016 of Mongolia, and the Statistical Yearbook 2016 of Russia. All data for supporting policies were evaluated using the Google Index method.

## 3. Results and Discussion

The tourism competitiveness evaluation of seven tourism regions within the CMREC is divided into a comprehensive evaluation and three dimension evaluation (resources dimension, market dimension, and environment dimension). Thus, clarifying the status occupied by each tourism region as well as the roles and functions in the whole CMREC, it provides the scientific basis for promoting a high-efficiency international tourism industry and provides a future direction for a competition–cooperation model.

### *3.1. Comprehensive evaluation of tourism competitiveness*

*3.1.1. Evaluation of comprehensive tourism competitiveness.* The tourism competitiveness of China is significantly higher than that of either Russia or Mongolia. Indeed, the tourism competitiveness of the Beijing–Tianjin–Hebei Region is the strongest, while that of the Russian Federal District of Ural is the weakest The seven tourism regions placed in the following sequence from strongest to weakest: Beijing–Tianjin–Hebei Region (China); Northeast Region (China); Northwest Region (China); Federal District of Siberia (Russia); Mongolia; Federal District of the Far East (Russia); Federal District of Ural (Russia) (figure 2).

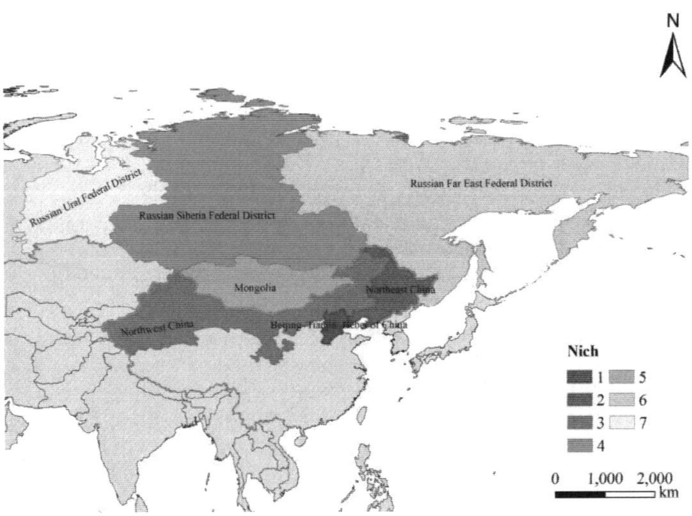


**Figure 2.** Comprehensive evaluation of tourism competitiveness.

*3.1.2. Evaluation of average tourism competitiveness.* The average tourism competitiveness of the Chinese regions considered is 2.69 times that of the Russian regions, and 1.74 times that of the Mongolian regions. At the same time, the average tourism competitiveness of Mongolian regions is 1.54 times that of the Russian regions. The average tourism competitiveness of the Beijing–Tianjin–Hebei Region, the Northeast Region, and the Northwest Region within China is 0.485, while the corresponding value for Mongolia is 0.278, and 0.180 for Russian Federal Districts of Ural, Siberia, and the Far East. Therefore, the general trend of average tourism competitiveness from strongest to weakest is China, Mongolia and Russia.

### *3.2. Dimensional evaluation of tourism competitiveness*
As discussed above, the concept of a "multi-dimensional hyper-volume" of tourism competitiveness originates from the combined action of numerous factors, including resources, markets, and environmental dimensions. The characteristics of each dimension reveal the relative development advantages and disadvantages within seven tourism regions.

*3.2.1. Evaluation of market dimensional factors.* The tourism market within China has huge advantages over other two countries [5]. The total number of tourists and income of the China tourism industry account for 98.12% and 97.86%. The number of inbound tourists within the Chinese sector of this industry is 42.1 times that of the Mongolian sector and 82.8 times that seen in Russia. The inbound tourism income of the Chinese sector of this industry is 50.4 times that of the Mongolian sector and 140.0 times that seen in Russia. Mongolia is the most open of the three countries in terms of tourism services, while Russia has a chronic trade deficit in this industry. There has been a decrease year-on-year in the development of the tourism industry in Mongolia, even though this country is the most open to the outside world in terms of tourism services.

*3.2.2. Evaluation of environment dimensional factors.* This dimension includes social economic, location traffic, and support policy environments. These factors have a total contribution rate of 0.242 to the overall tourism competitiveness of the CMREC, and individual rates of 0.062, 0.066, and 0.114, respectively. In the first place, the overall tourism development environment of the Chinese sector is superior to that seen in Mongolia and Russia. Tourism development has been actively promoted in Mongolia, albeit limited by the national social economic level, while Russia has lagged behind in the construction of traffic infrastructure and lacks supporting policies for tourism development. Secondly, it is clear that restrictive factors promoting the development of tourism within the three CMREC regions are different. The three regions of Northwest China exhibit a minimum social economic level and lack relevant policy support, while severely insufficient for traffic facilities are concentrated obviously within the Russian Federal District of Ural [6].

## 4. Conclusion
Based on the evaluation results of tourism competitiveness, each tourism region within the CMREC occupies a certain space that cause the competition of tourism regions, but the international tourism cooperation will lead to realize the sustainable development of the CMREC [7]. Based on the evaluation comprehensive of tourism competitiveness and cluster analysis, the competition–cooperation tourism model for the CMREC are proposed tends to exhibit a "spindle-shaped" trend of development and exhibits a competitiveness distribution that conforms to a "1-4-2" structure. The Class-I absolutely dominant areas identified are led by the Beijing–Tianjin–Hebei Region and act as growth poles for regional tourism development, driving and radiating benefits into surrounding areas, and generally promoting tourism development within the CMREC. In contrast, the Class-II comparatively dominant areas identified, the three provinces of Northeast China, the three provinces of Northwest China, as well as Siberia and Mongolia, also perform the same driving and radiating functions as their Class-I area counterparts .Due to their own inherent advantages, these regions also link to the Russian Far East and Ural areas that are classified as Class-III regions in this study and complete the competition–cooperation within the CMREC.

### *4.1. Promoting the cooperation mechanism as Triangle Center*
The Chinese National Tourism Administration, the Russian Economic Development Department, and the Mongolian Tourism Bureau have promoted the establishment of a "China–Mongolia–Russia

Economic Corridor Tourism Cooperative" as the main mechanism for the implementation of their cooperation in "developing cultural communication" as laid out in the *Outline of Planning for the Construction of China–Mongolia–Russia Economic Corridor*. This collaboration also aims to establish secretariat divisions in Beijing (China), Irkutsk–Ulan–Ude (Russia), and Ulan Bator (Mongolia) to form a "Triangle Center" that will radiate and drive benefits into regions which are currently relatively weak in tourism [8]. These organizations have the responsibility for planning tourism development within the CMREC, in particular jointly negotiating and discussing policies for tourism cooperation, organizing and promoting market research, developing routes, recommending regions and communicating with these areas as well as with cities, enterprises, and social organizations along the pipeline, undertaking joint marketing and promotional activities, and implementing tourism cooperation projects and investment plans.

### *4.2. Optimization of tourism resources as resource-space-dislocation strategy*

The aspiration of tourism development within the CMREC is to create a series of tourism regions that comprise "One Heart, Three Belts, Four Kernels, and Seven Blocks". Thus, within this structure "One Heart" will comprise the core Beijing–Tianjin–Hebei Region while the "Three Belts" consists of the China–Russia Ecological Tourism Belt (i.e., three provinces in Northeast China, Russian Far East, Siberia, and Ural), the China–Mongolia Cultural Customs Tourism Belt (i.e., Beijing–Tianjin–Hebei, Inner Mongolia, Mongolia, and Lake Baikal), and the China–Mongolia–Russia Ancient Silk Culture Tourism Belt (i.e., Beijing–Tianjin–Hebei, Gansu, Xinjiang, Mongolia, Ural) [9].The "Four Kernels" within the model therefore comprise the city of Vladivostok in the Russian Far East, Lake Baikal in Siberia, and the city of Ulan Bator in Mongolia, as well as the internal border areas of China, Mongolia, and Russia [10]. Finally, the "Seven Blocks" within this model are the Beijing–Tianjin–Hebei Chinese National Culture Block, the Northeast Region Leisure and Holiday Block, the Northwest Region Ancient Silk Road Block, the Mongolia Humanistic Tourism Block, the Siberia Lake Baikal Ecological Tourism Block, the Far East Coastal Natural Experience Block, and the Ural Near Polar Circle Natural Tourism Block.

### *4.3. Enhancing international market as market-stimulation-generalization strategy*

One of development within the CMREC goal should be to jointly enhance the competitiveness of the international tourism market, to cultivate the international common brands of "Silk Road and Tea Road", as well as to develop, design, and integrate a series of international cross-border tourism routes that could include mass sightseeing, the preservation of health, vacations, business surveys, family self-driving, and outdoor adventure tourism [11]. It will also be necessary to simplify visa and customs clearance processes for inbound tourists to relevant cities within China, Mongolia, and Russia, as well as to add mutual visa exemptions and on-arrival application procedures for international tourists, strengthen the connectivity and availability of routes, enable efforts to improve the tourism, leisure, and holiday functions of border cities within the three countries, and enhance the diversity and characteristics of tourism products.

The clear goal of strategy in this area is to develop a CMREC tourism industry belt via a "big tourism model" that will replace traditional industries with this endeavor at the economic forefront, linking primary, secondary, and tertiary industries. This will be achieved by optimizing industrial structures and constructing a large-scale tourism industry system that encompasses the various components as nodes linked together within a chain. Node cities are taken as central points with their corresponding trunk traffic lines as axles that connect points together to enable high-speed railway routes, airway express deliveries, and maritime transportation. It is therefore important to emphasize the construction of infrastructure and tourism service facilities in this context, to interconnect node cities and large scenic spots to enable natural-ecological as well historical-human tourism along the Ancient Silk Road and the Siberia Big Railway. High-speed railway routes connecting Asia-Pacific and European international tourism cities should also be developed to mitigate the highway traffic barriers to international development within the CMREC. Finally, it is also critical to promote the construction of comprehensive mechanisms to guarantee safety in order to protect cross-border and cross-regional tourism between China, Mongolia, and Russia. These measures should include mechanisms to ensure the personal and property safety of tourists, while at the same time emphasizing the protection of local environments within the three countries as well as the necessary means to rescue tourists in emergencies.

**Acknowledgements**
This work was supported by the National Key R&D Program of China [grant number 2017FY101304], [grant number 2017FY101303], [grant number 2017FY101300]; China Postdoctoral Science Foundation [grant number 2018M631559].

**References**

[1] Dong S and Zhao M 2016 Development Mode and Countermeasures for International Ecotourism Zone along the Belt and Road *China Academic Journal* **6** 647-55
[2] Shi Y, Zhong L, Chen T and Yu H 2016 Tourism competitiveness evaluation and spatio-temporal characteristics of Chinese border counties *Chinese Geographical Science* **26** 817-28
[3] Zhang X 2017 Chinese capitalism and the maritime silk road: a world-systems perspective *Geopolitics* **3** 1-22
[4] Soni S K 2018 *China–Mongolia–Russia Economic Corridor: Opportunities and Challenges China's Global Rebalancing and the New Silk Road* (Singapore: Springer) pp 101-17
[5] Ge Q and Zhong L 2014 *China Border Tourism Development Report* ed Peng S C (Beijing: Science Press) chapter 3 pp 38-40
[6] Lin W and Zou Y 2016 A Study on Space Layout and Cooperation Mechanism of Tourism Cooperation among Countries along the "the Belt and Road" *Around Southeast Asia* **4** 48-54
[7] Li S and Chen J 2014 The Cooperation-Competition Model of Regional Tourism Cities of Shandong Based Niche *Theory Economic Geography* **34** 179-85
[8] Zhang X 2011 Study on China and Mongolian tourism cooperation and its development strategy. Russian *Central Asian & East European Market* **6** 32-6
[9] Enkhmend U and Wu W L 2013 Study on the Tourism Industry Between China and Mongolia *Economic Research Guide* **3** 56-62
[10] Dorjsuren A 2017 Tourism resources, challenges, opportunities along the economic corridor between Mongolia, Russia and China *China-Mongolia-Russia Think Tank* pp 23-35
[11] Campi A 2014 Transforming Mongolia-Russia-China Relations: The Dushanbe Trilateral Summit *The Asia-Pacific Journal* **12** 3-11